T0135336

Smart Innovation, Systems and Technologies

Volume 81

Series editors

Robert James Howlett, Bournemouth University and KES International,
Shoreham-by-sea, UK
e-mail: rjhowlett@kesinternational.org

Lakhmi C. Jain, University of Canberra, Canberra, Australia;
Bournemouth University, UK;
KES International, UK
e-mails: jainlc2002@yahoo.co.uk; Lakhmi.Jain@canberra.edu.au

About this Series

The Smart Innovation, Systems and Technologies book series encompasses the topics of knowledge, intelligence, innovation and sustainability. The aim of the series is to make available a platform for the publication of books on all aspects of single and multi-disciplinary research on these themes in order to make the latest results available in a readily-accessible form. Volumes on interdisciplinary research combining two or more of these areas is particularly sought.

The series covers systems and paradigms that employ knowledge and intelligence in a broad sense. Its scope is systems having embedded knowledge and intelligence, which may be applied to the solution of world problems in industry, the environment and the community. It also focusses on the knowledge-transfer methodologies and innovation strategies employed to make this happen effectively. The combination of intelligent systems tools and a broad range of applications introduces a need for a synergy of disciplines from science, technology, business and the humanities. The series will include conference proceedings, edited collections, monographs, handbooks, reference books, and other relevant types of book in areas of science and technology where smart systems and technologies can offer innovative solutions.

High quality content is an essential feature for all book proposals accepted for the series. It is expected that editors of all accepted volumes will ensure that contributions are subjected to an appropriate level of reviewing process and adhere to KES quality principles.

More information about this series at http://www.springer.com/series/8767

Jeng-Shyang Pan · Pei-Wei Tsai
Junzo Watada · Lakhmi C. Jain
Editors

Advances in Intelligent Information Hiding and Multimedia Signal Processing

Proceedings of the Thirteenth International Conference on Intelligent Information Hiding and Multimedia Signal Processing, August, 12–15, 2017, Matsue, Shimane, Japan, Part I

Editors
Jeng-Shyang Pan
Fujian Provincial Key Lab of Big Data Mining and Applications
Fujian University of Technology
Fuzhou, Fujian
China

Pei-Wei Tsai
Swinburne University of Technology
Hawthorn, VIC
Australia

Junzo Watada
Universiti Teknologi Petronas
Teronoh
Malaysia

Lakhmi C. Jain
University of Canberra
Bruce, ACT
Australia

ISSN 2190-3018 ISSN 2190-3026 (electronic)
Smart Innovation, Systems and Technologies
ISBN 978-3-319-87655-9 ISBN 978-3-319-63856-0 (eBook)
DOI 10.1007/978-3-319-63856-0

Printed on acid-free paper

This Springer imprint is published by Springer Nature
The registered company is Springer International Publishing AG
The registered company address is: Gewerbestrasse 11, 6330 Cham, Switzerland

Preface

Welcome to the 13th International Conference on Intelligent Information Hiding and Multimedia Signal Processing (IIH-MSP 2017), which will be held in Matsue, Shimane, Japan, on August 12–15, 2017. IIH-MSP 2017 is hosted by Universiti Teknologi PETRONAS in Malaysia and technically co-sponsored by Fujian University of Technology in China, Taiwan Association for Web Intelligence Consortium in Taiwan, Swinburne University of Technology in Australia, Fujian Provincial Key Laboratory of Big Data Mining and Applications (Fujian University of Technology) in China, and Harbin Institute of Technology Shenzhen Graduate School in China. It aims to bring together researchers, engineers, and policymakers to discuss the related techniques, to exchange research ideas, and to make friends.

We received a total of 321 submissions from Europe, Asia, and Oceania over places including Taiwan, Thailand, Turkey, Korea, Japan, India, China, and Australia. Finally, 103 papers are accepted after the review process. Keynote speeches were kindly provided by Professor Zhiyong Liu (The Institute of Computing Technology, Chinese Academy of Sciences, Beijing, China) on "Cryo-ET Data Processing and Bio-Macromolecule 3-D Reconstruction" and Professor Takashi Nose (Tohoku University, Japan) on "Flexible, Personalized, and Expressive Speech Synthesis Based on Statistical Approaches." All the above speakers are leading experts in related research fields.

We would like to thank the authors for their tremendous contributions. We would also express our sincere appreciation to the reviewers, Program Committee members, and the Local Committee members for making this conference successful. Finally, we would like to express special thanks to the Universiti Teknologi PETRONAS in Malaysia, Fujian University of Technology in China, Swinburne University of Technology in Australia, Taiwan Association for Web Intelligence Consortium in Taiwan, and Harbin Institute of Technology Shenzhen

Graduate School in China for their generous support in making IIH-MSP 2017 possible.

August 2017

Jeng-Shyang Pan
Pei-Wei Tsai
Junzo Watada
Lakhmi C. Jain

Conference Organization

Conference Founders

Jeng-Shyang Pan	Fujian University of Technology, China
Lakhmi C. Jain	University of Canberra, Australia and Bournemouth University, UK

Honorary Chairs

Lakhmi C. Jain	University of Canberra, Australia and Bournemouth University, UK
Chin-Chen Chang	Feng Chia University, Taiwan

Advisory Committee

Yôiti Suzuki	Tohoku University, Japan
Bin-Yih Liao	National Kaohsiung Univ. of Applied Sciences, Taiwan
Kebin Jia	Beijing University of Technology, China
Yao Zhao	Beijing Jiaotong University, China
Ioannis Pitas	Aristotle University of Thessaloniki, Greece

General Chairs

Junzo Watada	Universiti Teknologi PETRONAS, Malaysia
Jeng-Shyang Pan	Fujian University of Technology, China

Program Chairs

Akinori Ito	Tohoku University, Japan
Pei-Wei Tsai	Swinburne University of Technology, Australia

Invited Session Chairs

Isao Echizen	National Institute of Informatics, Japan
Ching-Yu Yang	National Penghu University of Science and Technology, Taiwan
Hsiang-Cheh Huang	National University of Kaohsiung, Taiwan
Xingsi Xue	University of Birmingham, UK

Publication Chairs

Chin-Feng Lee	Chaoyang University of Technology, Taiwan
Tsu-Yang Wu	Fujian University of Technology, China
Chien-Ming Chen	Harbin Institute of Technology Shenzhen Graduate School, China

Electronic Media Chairs

Tien-Wen Sung	Fujian University of Technology, China
Jerry Chun-Wei Lin	Harbin Institute of Technology Shenzhen Graduate School, China

Finance Chair

Jui-Fang Chang	National Kaohsiung University of Applied Sciences, Taiwan

Program Committee Members

Toshiyuki Amano	Nagoya Institute of Technology, Japan
Supavadee Aramvith	Chulalongkorn University, Thailand
Christoph Busch	Gjøvik University College, Norway
Canhui Cai	Hua-Qiao University, China
Patrizio Campisi	University of Roma TRE, Italy
Turgay Celik	National University of Singapore, Singapore

Thanarat Chalidabhongse	King Mongkut Institute of Technology Larbkrabang, Thailand
Chi-Shiang Chan	Asia University, Taiwan
Kap-Luk Chan	Nanyang Technological University, Singapore
Bao-Rong Chang	National University of Kaohsiung, Taiwan
Feng-Cheng Chang	Tamkang University, Taiwan
Chien-Ming Chen	Harbin Institute of Technology Shenzhen Graduate School, China
Shi-Huang Chen	Shu-Te University, Taiwan
Yueh-Hong Chen	Far East University, Taiwan
L.L. Cheng	City Univ. of Hong Kong, Hong Kong
Shu-Chen Cheng	Southern Taiwan University of Science and Technology, Taiwan
Hung-Yu Chien	Chi Nan University, Taiwan
Jian Cheng	Chinese Academy of Science, China
Hyunseung Choo	Sungkyunkwan University, Korea
Shu-Chuan Chu	Flinders University, Australia
Kuo-Liang Chung	National Taiwan University of Science and Technology, Taiwan
Hui-Fang Deng	South China University of technology, China
Isao Echizen	National Institute of Informatics, Japan
Masaaki Fujiyoshi	Tokyo Metropolitan University, Japan
Pengwei Hao	Queen Mary, University of London, UK
Yutao He	California Institute of Technology, USA
Hirohisa Hioki	Kyoto University, Japan
Anthony T.S. Ho	University of Surrey, UK
Jiun-Huei Ho	Cheng Shiu University, Taiwan
Tzung-Pei Hong	National University of Kaohsiung, Taiwan
Jun-Wei Hsieh	National Taiwan Ocean University, Taiwan
Raymond Hsieh	California University of Pennsylvania, USA
Bo Hu	Fudan University, China
Wu-Chih Hu	National Penghu University, Taiwan
Yongjian Hu	South China University of Technology, China
Hsiang-Cheh Huang	National Kaohsiung University, Taiwan
Du Huynh	University of Western Australia, Australia
Ren-Junn Hwang	Tamkang University, Taiwan
Masatsugu Ichino	University of Electro-Communications, Japan
Akinori Ito	Tohoku University, Japan
Motoi Iwata	Osaka Prefecture University, Japan
Jyh-Horng Jeng	I-Shou University, Taiwan
Kebin Jia	Beijing University of Technology, China
Hyunho Kang	Tokyo University of Science, Japan
Muhammad Khurram Khan	King Saud University, Kingdom of Saudi Arabia
Lei-Da Li	China University of Mining and Technology, China

Li Li	Hangzhou Dianzi University, China
Ming-Chu Li	Dalian University of Technology, China
Shu-Tao Li	Hunan University, China
Xuejun Li	Anhui University, China
Xue-Ming Li	Beijing University of Posts and Telecommunications, China
Zhi-Qun Li	Southeast University, China
Guan-Hsiung Liaw	I-Shou University, Taiwan
Cheng-Chang Lien	Chung Hua University, Taiwan
Chia-Chen Lin	Providence University, Taiwan
Chih-Hung Lin	National Chiayi University, Taiwan
Jerry Chun-Wei Lin	Harbin Institute of Technology Shenzhen Graduate School, China
Shin-Feng Lin	National Dong Hwa University, Taiwan
Yih-Chaun Lin	National Formosa University, Taiwan
Yuh-Chung Lin	Tajen University, Taiwan
Gui-Zhong Liu	Xi'an Jiaotong University, China
Haowei Liu	Intel Corporation, California
Ju Liu	Shandong University, China
Yanjun Liu	Feng Chia University, Taiwan
Der-Chyuan Lou	Chang Gung University, Taiwan
Guang-Ming Lu	Harbin Institute of Technology, China
Yuh-Yih Lu	Minghsin University of Science and Technology, Taiwan
Kai-Kuang Ma	Nanyang Technological University, Singapore
Shoji Makino	University of Tsukuba, Japan
Hiroshi Mo	National Institute of Informatics (NII), Japan
Vishal Monga	Xerox Labs, USA
Nikos Nikolaidis	Aristotle University of Thessaloniki, Greece
Alexander Nouak	Fraunhofer Institute for Computer Graphics Research IGD, Germany
Tien-Szu Pan	Kaohsiung University of Applied Sciences, Taiwan, Taiwan
Ioannis Pitas	Aristotle University of Thessaloniki, Greece
Qiang Peng	Southwest Jiaotong University, China
Danyang Qin	Heilongjiang University, China
Kouichi Sakurai	Kyushu University, Japan
Jau-Ji Shen	Chung Hsing University, Taiwan
Guang-Ming Shi	Xi'dian University, China
Yun-Qing Shi	New Jersey Institute of Technology (NJIT), USA
Nobutaka Shimada	Ritsumeikan University, Japan
Jong-Jy Shyu	University of Kaohsiung, Taiwan
Kotaro Sonoda	National Institute of Information and Communications Technology, Japan
Yi Sun	Dalian University of Technology, China

Contents

Multimedia Security and Its Applications

A Survey of Reversible Data Hiding Schemes Based on Two-Dimensional Histogram Modification

Chin-Feng Lee[1], Jau-Ji Shen[2(✉)], and Yu-Hua Lai[2]

[1] Department of Information Management, Chaoyang University of Technology, Taichung, Taiwan
lcf@cyut.edu.tw

[2] Department of Management Information Systems, National Chung Hsing University, Taichung, Taiwan
jjshen@nchu.edu.tw

Abstract. In recent years, in addition to one-dimensional reversible data hiding many two-dimensional histogram modification methods have been proposed. In this paper, various two-dimensional histogram modification reversible data hiding method are studied, analyzed, and compared, providing useful information for researchers in this field. The analysis and comparison of these include two-dimensional histogram modification reversible data hiding methods based on: difference-pairs, pairwise prediction error expansion, and PVO and pairwise prediction error expansion.

Keywords: Histogram modification · Reversible Data Hiding (RDH) · Prediction-Error Expansion (PEE) · Two-Dimensional Histogram Modification (2DHM)

1 Introduction

Reversible data hiding technology involves secret information embedded into images, which cannot be perceived during the process of transmission. This is to ensure information safety and avoid any interception, destruction, and tampering. The embedded secret-message image is called a stego-image. This can be extracted and recovered from the original image.

In 2006, a histogram shifting and modification (HSM) method was proposed by Ni et al. [1]. Using statistics, it obtains the number of times each pixel value occurs and then defines the highest frequency occurrence as the peak point and the lowest as the zero point. Shifting the pixel value between the peak and zero-value points results in a vacated position in which to embed the secret message. In the same year, Lee et al. proposed a difference histogram shifting and modification (dHSM) method [2] to improve upon Ni's HSM method by making use of the correlation between adjacent pixels to enhance the embedding capacity (EC).

In 2013, Li et al. proposed a pixel value ordering method based on prediction error expansion [9], which produces high visual quality reversible data hiding. In 2014, Peng

J.-S. Pan et al. (eds.), *Advances in Intelligent Information Hiding and Multimedia Signal Processing*, Smart Innovation, Systems and Technologies 81,
DOI 10.1007/978-3-319-63856-0_1

[7] proposed to improve Li's embedded method, wherein the maximum and minimum prediction errors were calculated similarly to that of Li's method but a little different from Li's, numbering each pixel in the block. Subtracting a larger numbered pixel value from a smaller one generates the prediction error value, and bins 0 and 1 are used for embedding. Generally, the frequency at which the peak point is zero is the highest. Thus, compared with the method of [9], the EC for the Lena image increased by 0.14 bpp, and the image quality improved by 0.61 dB, with a capacity of 10,000 bits.

All the above methods used a one-dimensional histogram for embedding and expansion, a good two-dimensional histogram that can improve image quality and EC while reducing the distortion caused by shifting pixels. Therefore, in this paper we study several methods of two-dimensional histogram modification (2D HM) in embedding technology, including the difference-pairs method based on 2D HM of reversible data hiding proposed by Li et al. [4]. Ou et al. proposed a pairwise prediction-error expansion based on 2D HM reversible data hiding [5] and a PVO and pairwise prediction error expansion based on 2D HM reversible data hiding [6].

2 Related Works

In this section, reversible data hiding based on 2D HM is introduced. The following methods may encounter an overflow/underflow problem. To avoid this during the process of embedding data, the pixel value is modified from 0 or 255 to 1 or 254, and the location map $LM_{(i)} = 1$ recorded as unavailable, where $i = 1, 2, \ldots, n$. Otherwise, $LM_{(i)} = 0$. To calculate the noise level (NL), pixels adjacent to the predicted value are defined as the reference pixel value, and the sum of the absolute differences in the horizontal and vertical reference value is used to obtain NL. If NL is more than a threshold T, this block is deemed a rough block, which can easily lead to distortion making it unsuitable for embedding and shifting.

2.1 A 2D HM Reversible Data Hiding Based on a Difference-Pairs

In 2013, Li et al. proposed a 2D HM reversible data hiding method based on a difference-pairs [4], which improves the difference-histogram method in 2006 [2]. Li et al. modified the coordinate direction and combined the gradient-adjusted-prediction (GAP) method proposed by the Fallahpour in 2008 [9] to further increase the embedded data and reduce the shifted pixel values. In this method, the per pixel-pair is modified by, at most, one-pixel value, and each pixel value by at most 1. The GAP predictor determines if the pixel belongs to the horizontal or vertical edges. The GAP predicted value z is computed according to seven adjacent pixel values as shown in Formulas (1), (2), and (3). For a pixel-pair (x, y), two difference values $d_1 = x - y$ and $d_2 = y - z$ are computed and defined as (d_1, d_2). Figure 1 shows the embedding rules.

$$d_v = |V_1 - V_5| + |V_3 - V_6| + |V_4 - V_7|, d_h = |V_1 - V_2| + |V_3 - V_4| + |V_4 - V_5| \quad (1)$$

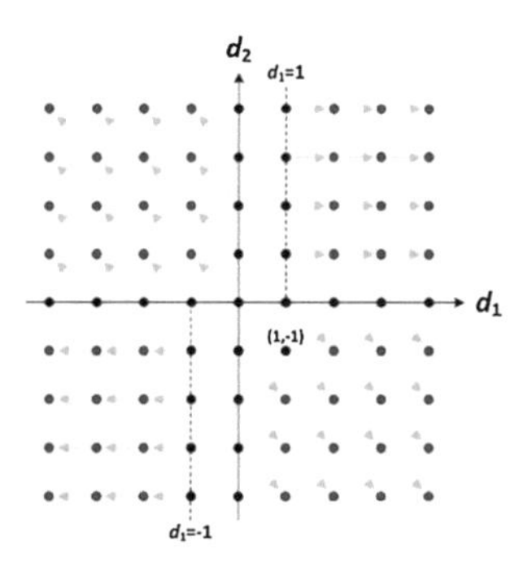

Fig. 1. Two-dimensional histogram modification for data embedding

$$u = \lfloor \frac{(V_1 + V_4)}{2} + \frac{(V_3 - V_5)}{4} \rfloor \tag{2}$$

$$z = \begin{cases} V_1, & \textit{if } d_v - d_h > 80 \\ \frac{(V_1 + u)}{2}, & \textit{if } d_v - d_h \in (32, 80] \\ \frac{(V_1 + 3u)}{4}, & \textit{if } d_v - d_h \in (8, 32] \\ u, & \textit{if } d_v - d_h \in [-8, 8] \\ \frac{V_4 + 3u}{4}, & \textit{if } d_v - d_h \in [-32, 8) \\ \frac{V_4 + u}{2}, & \textit{if } d_v - d_h \in [-80, -32] \\ V_4, & \textit{if } d_v - d_h < -80 \end{cases} \tag{3}$$

Example 1. First, the pixel pair $(x, y) = (160, 161)$ is chosen, and the ten pixel values adjacent to (x, y) are defined as the reference pixels. Compute $NL = 19$, as shown in Fig. 2. When $T = 20$, $NL < T$ is a smooth block and data can be embedded. We then calculate the predicted value using Formulas (1), (2), and (3), and obtain $d_v = 8, d_h = 4$, $u = 161$, and $z = 161$. Then, the difference-pair $(d1, d2) = (-1, 0)$ for a pixel-pair (x, y) is computed. The secret message $s = 1$, the modified difference-pair $(d1, d2) = (-2, 1)$. Finally, we get a stego-image as shown in Fig. 3.

160_x	161_y	163_{V_1}	161_{V_2}
161_{V_3}	159_{V_4}	159_{V_5}	159
160_{V_6}	162_{V_7}	159	159

Fig. 2. Cover image

160	162	163	161
161	159	159	159
160	162	159	159

Fig. 3. Stego-image

2.2 A 2D HM Reversible Data Hiding Based on Pairwise Prediction Error Expansion

In 2013, Ou et al. proposed a 2D HM reversible data hiding method based on pairwise prediction error expansion [5]. Combined with the rhombus prediction method of Sachnev et al. [8] it is used to compute the prediction error value. Ou designed a 2D HM rule for embedding and shifting, and in this method the pixel-pair is modified at most by a two-pixel value, each pixel value at most by 1, and the secret message is represented by a ternary numeral. Scan the cover image *I* from left to right and top to bottom according to Formula (4) to classify the image into black and white blocks, the results show as a checkerboard, then use the rhombus prediction to calculate the predicted value $E^{'}_{(i)}$ of each black block according to Formula (5). The rhombus prediction method takes the average value of the four nearest neighboring pixels (R_1, R_2, R_3, R_4) as shown in Fig. 4 from the prediction value E_1 using Formula (5). The predicted value $E^{'}_1$ of the predicted original value E_1 is obtained.

	R₁		
R₂	E₁	R₄	
	R₃	E₂	R₅
		R₆	

Fig. 4. The scan order for black block pixels

$$P_{(i)} = \begin{cases} black\,block, I(i+j)\,mod\,2 = 0 \\ white\,block, I(i+j)\,mod\,2 = 1 \end{cases}, \tag{4}$$

$$E^{'}_{(i)} = \lfloor \frac{R_1 + R_2 + R_3 + R_4}{4} \rfloor \tag{5}$$

In addition to the first row and column of the cover image, divide the cover image into 2 × 2 sized blocks and order the prediction error values according to black block last scanned, then obtain the prediction error pair sequence $(d_1, d_2, \ldots, d_n)$. The two adjacent prediction error values (d_{2i}, d_{2i-1}) occur as a pair of pairwise prediction errors $A_{(i)}$, where $i = 1, 2\ldots, n$. For 4 × 4 sized block, calculate the sum of the absolute differences of the diagonal pixels between the white block pixels and obtain the noise level (*NL*) of the block. According to Fig. 5 embed the data in $A_{(i)}$.

Example 2. Assuming Fig. 6 is the cover image *I*, we divide the image into black and white blocks using Formula (4) and calculate the prediction value to obtain $E^{'}_1 = 160$ and $E^{'}_2 = 158$ according to Formula (5). Subtract the prediction value from the pixel value of the original image to obtain the prediction error value $d_1 = 1$ and $d_2 = -1$, and compute a pair of pairwise prediction error values as $(1, -1)$. After that, the noise level of the block is obtained by calculating the sum of absolute differences of the

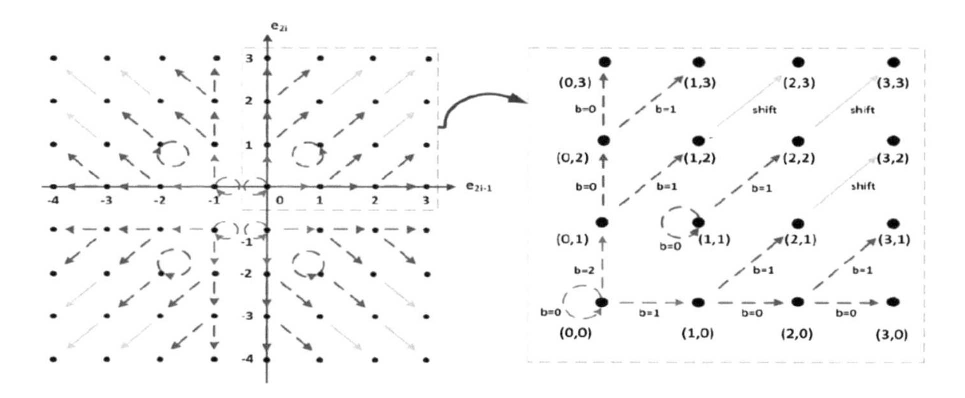

Fig. 5. Two-dimensional histogram modification for illustrating the data

diagonal pixels between the white block pixels: $NL = |161 - 161| + |161 - 162| + |162 - 159| + |159 - 161| + |162 - 156| + |156 - 156| + |156 - 159| + |159 - 161| + |162 - 158| = 21$. When $T = 25$, $NL < T$ is a smooth block, and data can be embedded. Finally, the data is embedded into the cover image. The secret message $s = 1$, the pairwise prediction error is $(d_1, d_2) = (2, -2)$ and the stego-image is as shown in Fig. 7.

160	161 R_1	163	161
161 R_2	159 E_1	159 R_4	160
160	162 R_3	159 E_2	156
158	159	156	157

Fig. 6. Cover image

160	161	163	161
161	158	159	160
160	162	160	156
158	159	156	157

Fig. 7. Stego-image

2.3 A 2D HM Reversible Data Hiding Based on PVO and Pairwise Prediction Error Expansion

In 2016, Ou et al. proposed a 2D HM reversible data hiding method based on pixel value ordering (PVO) and pairwise prediction error expansion [6]. Here it's combined with the IPVO method of Peng et al. [7] and the one-dimensional (1D) histogram modified to a two-dimensional representation. Ou's approach is to use two 2D HMs as the embedding and shifting rules, f1 and f2, for images with different smoothness, and predict the maximum and minimum prediction error pairs within the block to embed data according to a side-matching prediction method. In this method, the per pixel-pair is modified by, at most, a two-pixel value, and each pixel value at most by 1. The secret message is represented by a ternary numeral. The side match prediction sorts the per pixel values within the block in ascending order so the adjacent pixel values move closer to each other and calculates the maximum and minimum prediction error value for each block, the result of the calculation is only produce a positive value. The embedding procedure is described as follows:

Step 1: Divide the cover image I into $n_1 \times n_2$ sized blocks, calculate the noise level (NL) from the block size $(n_1 + 2) \times (n_2 + 2)$, and compute the sum of the absolute differences in the horizontal and verticals in the neighboring pixels except the block.

Step 2: Number each pixel in the block. Scan the $n_1 \times n_2$ sized block from left to right and top to bottom, and sort the pixels $(p_1, p_2, \ldots, p_{(n_1 \times n_2)})$ within the block in ascending order where an ordered sequence $p_{\sigma(1)} \leq p_{\sigma(2)} \leq \cdots \leq p_{\sigma(n_1 \times n_2)}$. $(p_{\sigma(n_1 \times n_2)}, p_{\sigma(n_1 \times n_2 - 1)})$ and $(p_{\sigma(1)}, p_{\sigma(2)})$ denote the maximum and minimum pixel pairs, respectively. Here, $p_{\sigma(n_1 \times n_2 - 2)}$ and $p_{\sigma(3)}$ are taken as the reference pixels of the maximum and minimum pixel pairs, respectively.

Step 3: Calculate the maximum and minimum prediction error pairs, as shown in Formulas (6) and (7).

$$\begin{cases} e^1_{max} = p_u - p_{\sigma(n_1 \times n_2 - 2)} \\ e^2_{max} = p_v - p_{\sigma(n_1 \times n_2 - 2)} \\ d_{max} = \left(e^1_{max}, e^2_{max}\right) \\ u = min(\sigma(n_1 \times n_2 - 1), \sigma(n_1 \times n_2)) \\ v = max(\sigma(n_1 \times n_2 - 1), \sigma(n_1 \times n_2)) \end{cases} \tag{6}$$

$$\begin{cases} e^1_{min} = p_{\sigma(3)} - p_u \\ e^2_{min} = p_{\sigma(3)} - p_v \\ d_{min} = \left(e^1_{min}, e^2_{min}\right) \\ u = min(\sigma(1), \sigma(2)) \\ v = max(\sigma(1), \sigma(2)) \end{cases} \tag{7}$$

Step 4: Define thresholds $T1$ and $T2$. Then, based on the rules of Formula (8), choose a 2D HM to embed and shift the different smoothness images, where $f1$ and $f2$ denote the 2D HM by Ou et al. [6] and Peng et al. [7], shown in Fig. 8.

$$\begin{cases} f1, & if\ LM_{(i)} = 0\ and\ NL \leq T1 \\ f2, & if\ LM_{(i)} = 0\ and\ T1 < NL \leq T2 \end{cases} \tag{8}$$

Step 5: Repeat Steps 2 to 4 until all the secret message is embedded.

Example 3. Assuming Fig. 9 is the cover image I, divide it into 2×3 sized blocks. Number each pixel in the block. Calculate the noise level $NL = 50$, and sort the pixels within the block in ascending order as $(159_{(5)}, 159_{(6)}, 160_{(1)}, 161_{(2)}, 161_{(4)}, 163_{(3)})$, where the maximum pixel pair is $161_{(4)}$ and $163_{(3)}$, and the minimum pair is $159_{(5)}$ and $159_{(6)}$. Then, compute the maximum and minimum prediction error pairs as, respectively,

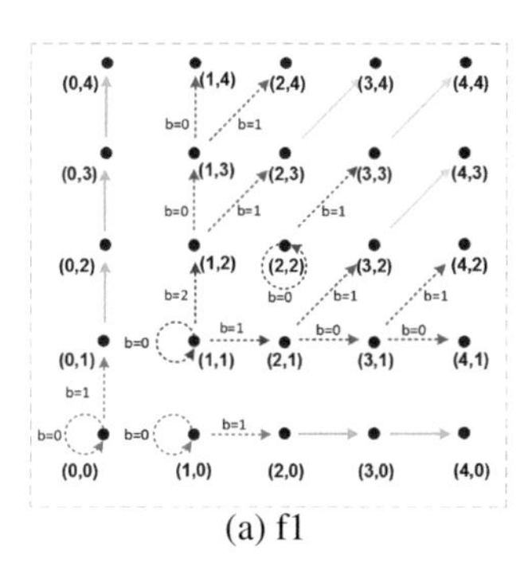

(a) f1

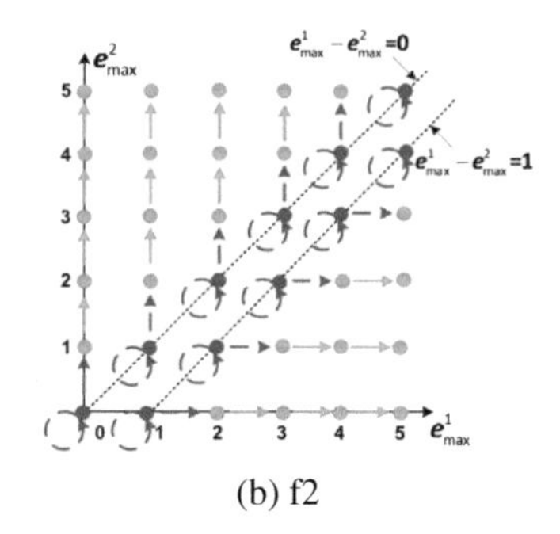

(b) f2

Fig. 8. Two-dimensional histogram modification for data embedding procedure

$d_{max} = (2, 0)$ and $d_{min} = (1, 1)$, where $e^1_{max} = 2$, $e^2_{max} = 0$ and $e^1_{min} = 1$, $e^2_{min} = 1$. When $T_1 = 57$ and $T_2 = 94$, $NL < T_1$ uses the rules of $f1$ to embed the data, as shown in Formula (8). The secret message $s = 2$, the maximum and minimum prediction error pairs are $d_{max} = (3, 0)$ and $d_{min} = (1, 2)$, respectively and the stego-image is as shown in Fig. 10.

160_{P_1}	161_{P_2}	163_{P_3}	161	165
161_{P_4}	159_{P_5}	159_{P_6}	160	160
160	162	159	156	159
158	159	156	157	159

Fig. 9. Cover image

160_{P_1}	161_{P_2}	164_{P_3}	161	165
161_{P_4}	159_{P_5}	158_{P_6}	160	160
160	162	159	156	159
158	159	156	157	159

Fig. 10. Stego-image

3 Comparison of the Experimental Results Based on 2D HM Reversible Data Hiding

In this section, we introduce the experimental results based on the one-dimensional IPVO [7] and three 2D HM methods. Table 1 shows a comparison of the performance of each method, including the noise level determination, prediction method, and number of thresholds, 2D HM is used as the embedding rule, prediction results, pixels can be embedded in the number of message, PSNR and the method is used of 1D or 2D. We observe that Ou's two-dimensional PSNR method [6] and EC is better than Peng's one-dimensional method [7], as the embedding capacity increased by 0.07bpp. For PSNR of 2D HM in the smoother image, Ou's method using PVO and side match prediction [6] is higher than Li's method at1.13 dB, but in the rough image, Ou's method [5] is the best and is higher that Ou's method in [6] i.e., 0.29 dB. For EC of 2D HM, Ou et al.'s method [5] uses pixel correlation and Ou et al.'s method [6] uses pixels in ascending order so that the pixels move closer to each other, leading to more

Table 1. The common and differences points among the methods based on 2D HM

		Li et al. [4]	Qu et al. [5]	Qu et al. [6]	Peng et al. [7]
Noise level		Ten pixels adjacent to the predicted value are defined as the reference pixel value, and the sum of the absolute differences in the horizontal and vertical reference value	The sum of the absolute differences of the diagonal pixels between the 9 pixel-pair	The sum of the absolute differences in the horizontal and verticals in the neighboring right side two rows and the bottom two columns pixels except the block	Subtract the second smallest value from the second largest value $(p_{\sigma(n-1)} - p_{\sigma(2)})$
Prediction method		Gradient-Adjusted-Prediction (GAP)	Rhombus Prediction	Side Match Prediction	Improved Prediction-Error Expansion (IPEE)
Feature		The per pixel-pair is modified by, at most, one-pixel value, and each pixel value by at most 1	The pixel-pair is modified at most by a two-pixel value, each pixel value at most by 1, and the secret message is represented by a ternary numeral	The per pixel-pair is modified by, at most, a two-pixel value, and each pixel value at most by 1. The secret message is represented by a ternary numeral	The per pixel-pair is modified by, at most, two-pixel value, and each pixel value by at most 1
Threshold		1	1	2	1
2D HM		1	1	2	X
Prediction results		Have positive and negative	Have positive and negative	Only positive	Have positive and negative
Bits at most be carried		1	2	2	1
Payload (bpp)		0.14	0.26	0.26	0.19
PSNR (dB) with Payload = 0.038 (bpp)		Lena 59.78	Lena 59.75	Lena 60.91	Lena 60.47
		Airplane 63.18	Airplane 63.76	Airplane 63.68	Airplane 62.96
		Baboon 53.96	Baboon 55.21	Baboon 54.92	Baboon 53.55
		Barbara 59.67	Barbara 59.48	Barbara 60.86	Barbara 60.54
PSNR	Lena	Qu et al. [6] > Peng et al. [7] > Li et al. [4] > Qu et al. [5]			
	Airplane	Qu et al. [5] > Qu et al. [6] > Li et al. [4] > Peng et al. [7]			
	Baboon	Qu et al. [5] > Qu et al. [6] > Li et al. [4] > Peng et al. [7]			
	Barbara	Qu et al. [6] > Peng et al. [7] > Li et al. [4] > Qu et al. [5]			
2 dimensional		Yes	Yes	Yes	No

messages being embedded. Both of these methods can also embed two bits using 2D HM. Therefore, Ou et al.'s [5] and Ou et al.'s [6] methods have high embedding capacity. Overall, the method of Ou et al. [6] is relatively good.

4 Conclusions

Using the characteristics of the histogram compared with one-dimensional histogram-based methods, this paper shows the method based on 2D HM has better performance. The above methods are used for embedding and expanding in smooth images, however, we should also try to embed data in rough images. In areas of smooth data, human vision is sensitive to and easily aware of embedded data but is not so sensitive for rough images. Therefore, it makes sense to embed most data in the edges of the rough image area according to this feature.

Acknowledgements. This research was partially supported by the Ministry of Science and Technology of the Republic of China under the Grants MOST 105-2221-E-324-014.

References

1. Ni, Z., Shi, Y.Q., Ansari, N., Su, W.: Reversible data hiding. IEEE Trans. Circuits Syst. Video Technol. **16**(3), 354–362 (2006)
2. Lee, S.K., Suh, Y.H., Ho, Y.S.: Reversible image authentication based on watermarking. In: Proceedings of the IEEE ICME, pp. 1321–1324 (2006)
3. Li, X.L., Li, J., Li, B., Yang, B.: High-fidelity reversible data hiding scheme based on pixel–value-ordering and prediction-error expansion. Sig. Process. **93**(1), 198–205 (2013)
4. Li, X., Zhang, W., Gui, X., Yang, B.: A novel reversible data hiding scheme based on two-dimensional difference-histogram modification. IEEE Trans. Inf. Forensics Secur. **8**(7), 1091–1100 (2013)
5. Ou, B., Li, X., Zhao, Y.: Pairwise prediction-error expansion for efficient reversible data hiding. IEEE Trans. Image Process. **22**(12), 5010–5021 (2013)
6. Bo, O., Li, X., Wang, J.: High-fidelity reversible data hiding based on pixel-value-ordering and pairwise prediction-error expansion. J. Vis. Commun. Image Represent. **39**, 12–23 (2016)
7. Peng, F., Li, X., Yang, B.: Improved PVO-based reversible data hiding. Digit. Signal Process. **25**, 255–265 (2014)
8. Sachnev, V., Kim, H.J., Nam, J., Suresh, S., Shi, Y.Q.: Reversible watermarking algorithm using sorting and prediction. IEEE Trans. Circuits Syst. Video Technol. **19**(7), 989–999 (2009)
9. Fallahpour, M.: Reversible image data hiding based on gradient adjusted prediction. IEICE Electron. Express **5**(20), 870–876 (2008)

Steganographic Image Hiding Schemes Based on Edge Detection

Chin-Feng Lee[1], Jau-Ji Shen[2(✉)], and Zhao-Ru Chen[2]

[1] Department of Information Management, Chaoyang University of Technology, Taichung, Taiwan
lcf@cyut.edu.tw

[2] Department of Management Information Systems, National Chung Hsing University, Taichung, Taiwan
jjshen@nchu.edu.tw

Abstract. In 2010, Chen et al. proposed a steganography mechanism using a hybrid edge detector. The hybrid edge detector is combining the Canny and the fuzzy edge detectors. However, although this method has a high embedding payload (2.10 bpp) it is an irreversible data hiding scheme. A considerable amount of research has since been conducted to provide an improved method of data hiding that removes LSBs of the image pixel value and performs edge detection. This increases the payload as it is not necessary to store edge information in the stego image. This paper conducts a comprehensive study of irreversible and reversible data hiding based on the image texture data hiding method, and compares embedding capacities and visual image qualities.

Keywords: Data hiding · Irreversible data hiding · Reversible data hiding · Edge detector technique

1 Introduction

The internet has now developed to such an extent that information security needs to be extremely proficient with respect to using social media, internet banking, and sending and receiving highly confidential emails. Data hiding ensures that the confidential information transmitted is not found and destroyed. There are two types of data hiding schemes: irreversible information hiding and reversible information hiding. The former uses a process whereby the stego image cannot be restored to the original image after secret data has been removed from it, and the latter involves restoring the stego image to the original image after secret data has been removed from it. As the human vision system is less sensitive to changes on the edges of images, some researchers have hidden most of the secret data on the edge of an image; this gives the stego image a better quality. For example, Chen et al. [1] proposed an innovative steganography scheme that uses the LSB (Least Significant Bit) replacement method and a hybrid edge detector that combines a fuzzy edge detector and Canny edge detector. Furthermore, in the same year, Lee et al. [2] proposed a method to improve this concept. The edge image is generated

J.-S. Pan et al. (eds.), *Advances in Intelligent Information Hiding and Multimedia Signal Processing*, Smart Innovation, Systems and Technologies 81,
DOI 10.1007/978-3-319-63856-0_2

by four MSBs (Most Significant Bits) of the image pixel value, and it is thus not necessary to record the location of the edge pixels; this greatly improves the embedding capacity and image quality of the stego image. In 2017, Bai et al. [3] also proposed a method to improve the shortcomings inherent in Chen et al. study; by using the 3 MSBs of the original image pixel value they obtained the edge image without needing to hide the edge information, which improved both the stego image quality and embedding rate (the embedding rate can reach 3.11 bpp).

However, some researchers have proposed hiding secret data in the smooth area of the image to obtain a better image quality. For example, in 2010, Hong and Chen [4] proposed a reference pixel distribution mechanism (RPDM) to detect the complex and smooth area of the original image. Here the secret data are embedded in the pixel error of the smooth area of the image using interpolation, so that the image quality can be improved in the reversible case.

This paper is organized as follows. Section 2 introduces irreversible and reversible data hiding techniques based on image texture; a detailed analysis and comparison of the methods are presented in Sect. 3; and concluding remarks are in Sect. 4.

2 Data Hiding Schemes Based on Image Texture

2.1 Irreversible Data Hiding Schemes

High payload steganography mechanism using hybrid edge detector. In 2010, Chen et al. proposed a method that firstly generates an edge image using a hybrid edge detector. Firstly, an edge image, E, is obtained from the cover image, I, using a hybrid edge detector. Then, the edge image, E, is divided into non-overlapping blocks. The block is called a "n-pixel block" with n pixels represented as $(P_1, P_2, \ldots P_n)$, respectively. The first pixel value P_1 of each block uses the LSB replacement method to record the edge information of the other pixels from P_2 to P_n in the block. Each edge pixel embeds 'x' bits of the secret message using the LSB replacement method, and each non-edge pixel embeds 'y' bits of the secret message using LSB replacement method. It is of note that there are more secret message bits in the edge pixel than in the non-edge pixel.

Furthermore, the process used to extract the secret message is as follows (with examples to follow). Step 1: Herein, the stego image, I', is divided into non-overlapping blocks with n pixels in each block; this block is called a "n-pixel block" and the n pixels are represented as $(P'_1, P'_2, \ldots P'_n)$, respectively. Step 2: The status of the pixels is extracted from P'_2 to P'_n, and is then classified into edge pixels and non-edge pixels. The 'x' LSBs of the edge pixels and the 'y' LSBs of the non-edge pixels are then extracted and concatenated as the secret message, S.

Example 1. The pixel A has a pixel value of {[1 0 1 0 0 0 1 1], [0 1 1 1 1 0 0 0], [1 1 1 1 1 1 1 1], [0 0 1 1 1 1 0 0]}, which is P_1, P_2, P_3, and P_4, respectively, and the secret message S = '0 0 1 0 1 0 1'. In this case, n is set to 4, so that the image A is a "4-pixel block."

Firstly, after the cover image, A, is detected using the hybrid edge detection method, the edge image, E, is generated. Assuming that P_2 and P_4 are edge pixels in the image

A, the status of P_2, P_3, and P_4 are '1 0 1', respectively. The 3 LSBs of the pixel value P_1 are then replaced by '1 0 1', so that the stored state of P_1 becomes P'_1 and the pixel value is [1 0 1 0 0 **1 0 1**]. We also assume that the parameters x and y are 3 and 1, respectively. P_2 and P_4 are embedded in 3 bits of the secret message, and P_3 is then embedded in the 1 bit of the secret message. The new pixel values of P_2, P_3, and P_4 are subsequently [0 1 1 1 **0 1 1**], [1 1 1 1 1 1 **0**] and [0 0 1 1 1 **1 0 1**], respectively. Thus, a stego image A' is generated with pixel values of {[1 0 1 0 0 1 0 1], [0 1 1 1 0 1 1], [1 1 1 1 1 1 0], [0 0 1 1 1 0 1]}. The embedding procedure in this example is represented in Fig. 1, and the process of extracting secret data is shown in Fig. 2.

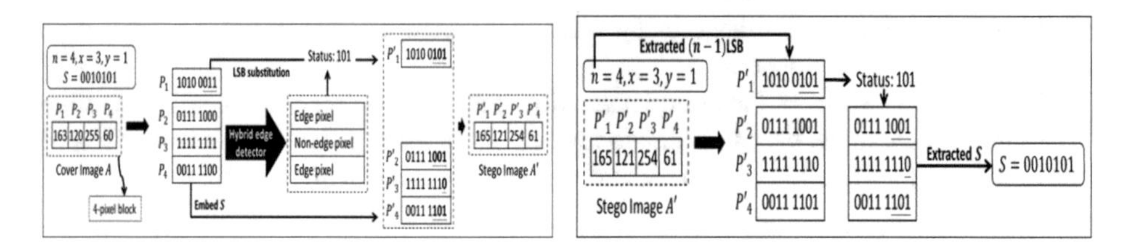

Fig. 1. Example of embedding procedure. **Fig. 2.** Example of extraction procedure.

Data Hiding Scheme with High Embedding Capacity and Good Visual Quality Based on Edge Detection. In 2010, Lee et al. [2] proposed a method that aimed to improve the limitations inherent in the study of Chen et al. [1] Firstly, the 4 MSBs of the pixel values are extracted from the cover image, *I*, using Formula (1) to obtain the reference image I_R. *M* and *N* are the width and height of the cover image, respectively. For pixels $I(i,j)$ satisfying $i = 1, 2, \ldots, M$ and $j = 1, 2, \ldots, N$. Then, the edge image, *E*, is obtained from the edge information of the reference image, I_R, using the ED-MOF method by Kang and Wang's edge detection method [5], where each pixel $E(i,j) \in \{0, 1\}$. If $E(i,j) = 1$. The relative pixels in the cover image, *I*, are then the edge pixel, otherwise they are the non-edge pixels. The secret message, *S*, is embedded in the LSBs of the cover image, *I*. The number of embedded bits in edge pixel is "*x*", and the number of embedded bits in non-edge pixel is "*y*". However, the 4 MSBs of the pixel of the cover image cannot be modified because of the edge detection, and thus, *x* and *y* cannot be greater than four and $x < y$.

$$I_R(i,j) = \left\lfloor \frac{I(i,j)}{16} \right\rfloor \times 16 \tag{1}$$

The process of extracting the secret message begins by applying ED-MOF method on the reference image, I_R to obtain an edge image, *E*, where I_R comes from the 4 MSBs of the stego image, I', using Formula (1). If the position of the pixel is an edge pixel, *x* LSBs are extracted from the pixel, and if the pixel is a non-edge pixel, *y* LSBs are extracted from the pixel. When all the extracted LSBs are combined, they become the secret message, *S*.

Example 2. Firstly, the pixel value of the cover image, *A*, is {[1 0 1 0 0 0 1 1], [0 1 1 1 1 0 0 0], [1 1 1 1 1 1 1 1], [0 0 1 1 1 1 0 0]}, which is denoted as P_1, P_2, P_3, and P_4, respectively, and the secret message S = '0 0 1 0 1 0 1 1 1 0'. The 4 MSBs of each pixel are first extracted by Formula (1) and are recorded to the reference image, A_R. The edge information of the reference image, A_R, is then detected using the ED-MOF method, and the edge image, *E*, is obtained. If we assume that the edge image, *E*, is {0, 1, 0, 1}, then P_2 and P_4 are edge pixels. In this case, the parameters x and y are set to 3 and 2, so P_1 and P_3 are embedded in 2-bit secret messages. P_2 and P_4 are embedded in 3-bit secret messages. The embedded pixel value is {[1 0 1 0 0 0 **0 0**], [0 1 1 1 1 **1 0 1**], [1 1 1 1 1 1 **0 1**], [0 0 1 1 1 **1 1 0**]}, which is the stego image, *A*′. The embedding procedure in this example is shown in Fig. 3, and the process of extracting secret data is shown in Fig. 4.

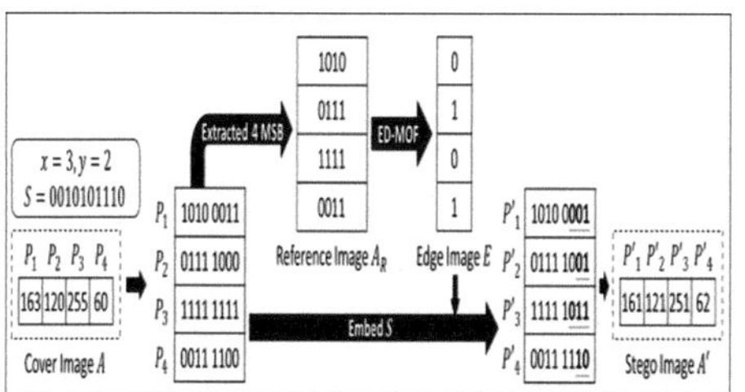

Fig. 3. Example of embedding procedure.

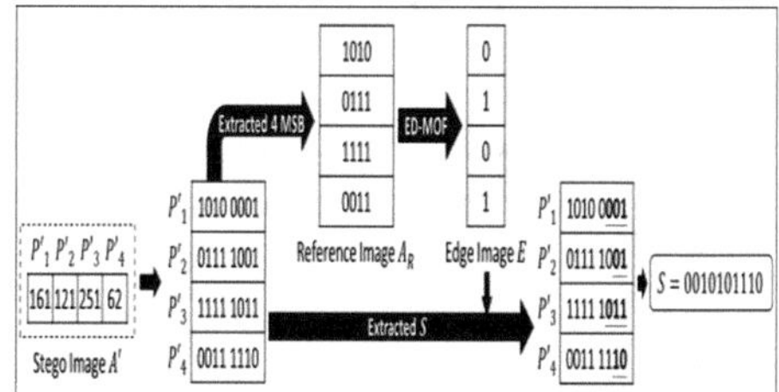

Fig. 4. Example of extraction procedure.

A high payload steganographic algorithm based on edge detection. In 2017, Bai et al. determined a limitation in Chen et al.'s method; that the index pixels occupy the embedding space of the image. Similar to Lee *et al.* [2], they thus designed a mechanism to make full use of the embeddable pixels in the cover image, and experimental results have shown that the embedding mechanism achieves better results under different edge detectors, such as Canny, Sobel, and Fuzzy. The process of embedding and secret extraction is as follows by an example.

Example 3. It is supposed that there are four pixels of the cover image, *A*, which are {[1 0 1 0 0 0 1 1], [0 1 1 1 1 0 0 0], [1 1 1 1 1 1 1 1], [0 0 1 1 1 1 0 0]} are P_1, P_2, P_3, and P_4, respectively, where x = 4 and y = 2 and the secret message *S* = '0 0 1 0 1 0 1 1 1 0 0 1'. Firstly, the pixel value of the cover image, *A*, is extracted and recorded in the new image, A_{MSB}. The pixel value of the new image, A_{MSB}, is {[1 0 1 0 0 0 0 0], [0 1 1 0 0 0 0 0] [1 1 1 0 0 0 0 0], [0 0 1 0 0 0 0 0]}. The edge image, *E*, is generated by the new image, A_{MSB}, using the edge detector. We then assume that after edge detection, P_2 and P_4 are edge pixels, and P_1 and P_3 are non-edge pixels. Therefore, the secret message, *S*, is embedded in the 4 LSBs of the pixel values P_2 and P_4 using the LSB substitution method, and the two bits are embedded in P_1 and P_3. Finally, the embedding of x and y in the last four pixels of the cover image, *A*, is complete. Therefore, the pixel values of the stego image,*A*′, are {[1 0 1 0 0 0 0 0], [0 1 1 1 1 0 1 0], [1 1 1 1 1 1 1 1], [0 0 1 1 1 0 0 1]}. The embedding procedure in this example is shown in Fig. 5, and the process of extracting secret data is shown in Fig. 6.

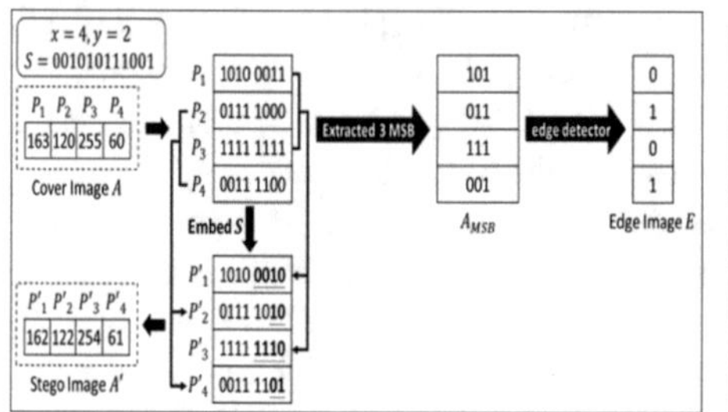

Fig. 5. Example of embedding procedure.

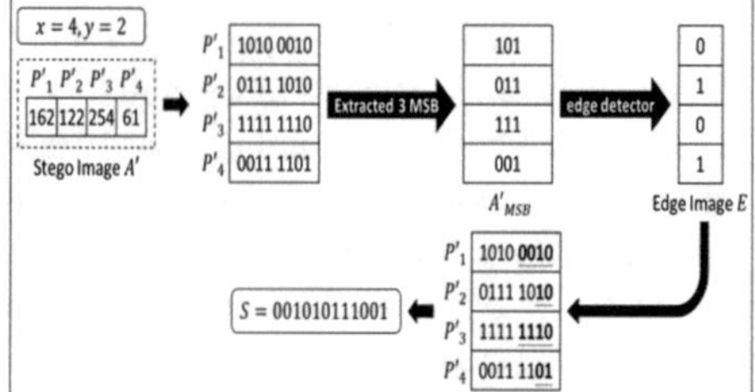

Fig. 6. Example of extraction procedure.

2.2 Reversible Data Hiding Scheme

Reversible data embedding for high quality images using interpolation and RPDM. In 2011, Hong and Chen proposed a reversible data hiding technique based on image interpolation and detection of smooth and complex regions of the cover image. This method enables the most secret message to be hidden in the smooth area of the image rather than in the complex area. To detect the smooth and complex areas in the image, a reference pixel distribution mechanism (RPDM) is proposed, and a range function is used to evaluate the smoothness of the image. The range function Range$(x_1, x_2, \ldots, x_n)$ is defined as the absolute difference between the maximum value and the minimum value of the given value $x_1, x_2, \ldots, x_n$. For given a of pixel P, if the range of the upper, lower, left, and right reference pixels of P is smaller than the threshold, T_0, then the pixel P is within the smooth region; however; if the range is larger than the threshold, T_1, the pixel P is within the complex region. Assuming that the size of the cover image, I, is $M \times M$, the detailed steps involved in the RPDM are described below.

Step 1: According to Formula (2), an $M \times M$ sized image, B, is generated for recording the position of the reference pixel, where Δ is the number of pixels between reference and non-reference pixels. If the pixel is a reference pixel, then the relative position of the image B is recorded as 1; if not it is recorded as 0. In Fig. 7, $\Delta = 3$.

$$B_{i,j} = \begin{cases} 0, & \bmod(i, \Delta) = 0 \text{ and } \bmod(j, \Delta) = 0 \\ 1, & \text{otherwise} \end{cases} \tag{2}$$

Step 2: If the conditions $B_{i,j} = 0, \bmod(i/\Delta + j/\Delta, 2) = 0$, and $\text{Range}\left(I_{i-\Delta,j}, I_{i,j-\Delta}, I_{i+\Delta,j}, I_{i,j+\Delta}\right) < T_0$ hold, then $I_{i,j}$ is within a smooth region. Update $B_{i,j} = 1$.

Step 3: The pixel value, $I_{i,j}$, satisfying $(0 \le i + \Delta, j + \Delta \le M)$ in the cover image, I, shows that the pixel position has a pixel value in the right and lower intervals Δ distance. If $I_{i,j}$ is $\text{Range}(I_{i,j}, I_{i+\Delta,j}, I_{i,j+\Delta}, I_{i+\Delta,j+\Delta}) > T_1$, then $I_{i,j}$ is within a complex region. In this case, the pixel values $B_{i',j'}$ in the range $(B_{i',j'}, i \le i' \le i + \Delta, j \le j' \le j + \Delta)$ are all set to 0, which means that $I_{i,j}$ is a reference pixel.

The cover pixels of 0 or 255 are modified to 1 or 254. The location of the pixel is then recorded in a location map, and the location map is compressed by run-length encoding; the result is expressed as L_M. And L_s is the bit stream of concatenate L_M and the secret message, S. The modified original image, I'', is used to generate a bitmap, B, using RPDM. And classify the bits of the image, B, into B_0 and B_1 classes. The pixel position, $(i,j) \in B_0$, since $B_{i,j} = 0$; $(i,j) \in B_1$, because $B_{i,j} = 1$. An interpolation method is apply to generate the interpolation image, P and the interpolation error $E_{i,j} = I''_{i,j} - P_{i,j}$ is calculated using when satisfying $(i,j) \in B_1$.

The embedding rules are as follows. If $E_{i,j} = p_e^+$ or $E_{i,j} = p_e^-$, secret data bit b L_s is embedded through Formula (3); if $E_{i,j} \neq p_e^+$ or $E_{i,j} \neq p_e^-$, the interpolation error is shifted by Formula (4) and the result is recorded in E'. In Formula (3), b is extracted one bit from L_s and p_e^+ and p_e^- is the interpolation error value that occurs most frequently and second-most frequently, respectively. Finally, the modified interpolation error, E', and the interpolation image, P, are added as the stego image, I'.

$$E'_{i,j} = \begin{cases} E_{i,j}, b = 0 \\ p_e^+ + 1, b = 1 \text{ and } E_{i,j} = p_e^+ \\ p_e^- - 1, b = 1 \text{ and } E_{i,j} = p_e^- \end{cases}, \tag{3}$$

$$E'_{i,j} = \begin{cases} E_{i,j} + 1, E_{i,j} > p_e^+ \\ E_{i,j} - 1, E_{i,j} < p_e^- \end{cases} \tag{4}$$

The process of extracting the secret message and restoring the image is using RPDM to generate the image, B, from the stego image, I'. Next, the interpolation image, P, is obtained with reference to the image, B. The embedded secret message can be obtained using $E'_{i,j} = I'_{i,j} - P_{i,j}$, and the embedded bits, b, is extracted from the pixel value, $E'_{i,j}$, in the interpolation error, E', using Formula (6). After all the bits, b, have been extracted, L_M and S are obtained from the lengths of L_M and S.

$$b = \begin{cases} 0, \; E'_{i,j} = p_e^+ \text{ or } E'_{i,j} = p_e^- \\ 1, \; E'_{i,j} = p_e^+ + 1 \text{ or } E'_{i,j} = p_e^- - 1 \end{cases} \tag{6}$$

After the secret message, S, has been obtained, the interpolation error, E', with the embedded secret message is restored to the interpolation error, E, without embedding the secret message, according to Formula (7),

$$E_{i,j} = \begin{cases} E'_{i,j} - 1, E'_{i,j} > p_e^+ \\ E'_{i,j} + 1, E'_{i,j} < p_e^- \\ E'_{i,j}, \text{otherwise} \end{cases} \tag{7}$$

By adding the interpolation error, E, and the interpolation image, P, the modified cover image, $I''_{i,j}$, is obtained. Eventually, the cover image, I, is restored by modifying the pixel values, $I''_{i,j}$, of 1 and 254 back to 0 and 255 according to L_M.

Example 4. Assuming that the image to be embedded is A, the thresholds for smoothing and complex regions are $T_0 = 5$ and $T_1 = 50$, respectively, and the secret message, S, is 0110. Firstly, the binary image, B, is generated by RPDM, and the interpolation image, P, is obtained using the binary image, B, as a reference. The cover image, A, and the interpolation image, P, are subtracted to obtain an interpolation error image, E. The secret message, S, is embedded into the interpolation error image, E, by the Formula (4). The modified interpolation error image, E', is then added to the interpolation image, P, as the stego image, A'. The embedding procedure in this example is shown in Fig. 7 and the process used to extract the secret message is shown in Fig. 8.

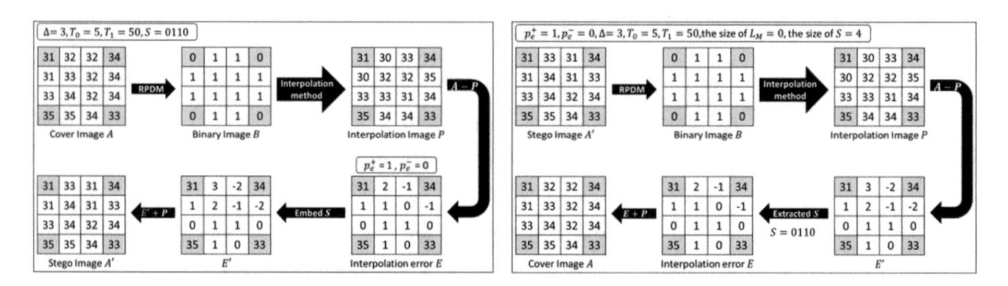

Fig. 7. Example of embedding procedure.

Fig. 8. Example of extraction procedure.

3 Comparison and Analysis

A considerable amount of research has focused on embedding a secret message in the texture features of the image to ensure reversibility and to achieve a better image quality and higher payload. In addition, a complex or smooth region within the embedded secret message has also been proposed. Here we compare the image quality (PSNR) and embedding rate (bpp) of each parameter according to whether the method is reversible, characteristic, or whether it has an image texture detection technique, as shown in Table 1.

In relation to the irreversible data hiding method in Table 1, under the parameter setting, the result of Chen et al. method [1] is PSNR = 37.5 dB and payload = 2.10 bpp. In relation to the limitations of Chen et al. study, Lee et al. proposed an improvement, in which the edge information is not recorded in the image, hybrid edge detection is not used, and only the pixel value of the four MSBs uses ED-MOF edge detection. In addition, the secret message is embedded in the 4 LSBs of the pixel value to increase the amount of embedding. It can be seen that the PSNR of Lena is increased by about 2.89 dB and that the embedding rate is increased by 0.04 bpp in the case of the reduced parameter, y. Bai et al. also improved the limitations inherent in the study of Chen et al., but this method is not limited to the edge detection method. We compare the results using Canny, Sobel, and Fuzzy, respectively, and although the results of the image quality are slightly

Table 1. Comparison of data hiding schemes based on image texture detection.

	Chen et al. [1]	Lee et al.[2]	Bai et al. [3]			Hong and Chen [4]
Parameters	$x = 4, y = 3, n = 2$	$x = 4, y = 2$	$x = 4, y = 3$	$x = 4, y = 3$	$x = 4, y = 3$	$T_0 = 8, T_1 = 60$
Block size	2×2	–	–			4×4
PSNR (dB)	Lena	Lena	Lena	Lena	Lena	Lena
	37.50	40.39	38.34	38.34	40.16	49.98
Embedding rate	2.10	2.14	3.11	3.05	2.79	0.18

lower than those of Lee et al., the embedding rate is improved and the maximum embedding rate is 3.11.

In relation to reversible data hiding, Hong and Chen's method proposed RPDM to detect the image texture. In this method, the secret information is embedded in the smooth region of the image rather than in the complex region of the image edge. Because it is reversible information hiding, so the embedded rate than the irreversible three methods are much lower. The embedding rate is 0.18 bpp, but the image quality is relatively high, with PSNR of 49.98 dB.

4 Conclusions

This paper studies the use of image texture to embed a secret message via data hiding technology. Chen et al. proposed an innovative method that uses hybrid edge detection to obtain edge information and LSB replacement to hide most of the edge information and the secret message in the edge pixels. However, the methods of Lee et al. and Bai et al. use the MSBs of extracted pixels to detect the edges, and embedding edge information is thus not necessary. As a whole, the embedding rate of Bai et al.'s method is superior and it reaches 3.11 bpp when Canny edge detection is used. However, this method is an irreversible data hiding method. As a reversible method, the image quality of Hong and Chen's method is superior, but because of reversibility the amount is lower. Therefore, despite the improvements made in image-based texture data hiding technologies in recent years, the embedding capacity requires a considerable amount of future research and is an important and challenging subject for researchers in this field.

Acknowledgements. This research was partially supported by the Ministry of Science and Technology of the Republic of China under the Grants MOST 105-2221-E-324-014.

References

1. Chen, W.J., Chang, C.C., Le, T.H.N.: High payload steganography mechanism using hybrid edge detector. Expert Syst. Appl. **37**(4), 3292–3301 (2010)
2. Lee, C.F., Chang, C.C., Tsou, P.L.: Data hiding scheme with high embedding capacity and good visual quality based on edge detection. In: 2010 Fourth International Conference on Genetic and Evolutionary Computing (ICGEC), pp. 654–657. IEEE (2010)
3. Bai, J., Chang, C.C., Nguyen, T.S., Zhu, C., Liu, Y.: A high payload steganographic algorithm based on edge detection. Displays **46**, 42–51 (2016)
4. Hong, W., Chen, T.S.: Reversible data embedding for high quality images using interpolation and reference pixel distribution mechanism. J. Vis. Commun. Image Represent. **22**(2), 131–140 (2011)
5. Kang, C.C., Wang, W.J.: A novel edge detection method based on the maximizing objective function. Pattern Recogn. **40**(2), 609–618 (2007)

A New Data Hiding Strategy Based on Pixel-Value-Differencing Method

Hui-Shih Leng(✉)

National Changhua University of Education, Changhua, Taiwan
lenghs@cc.ncue.edu.tw

Abstract. The Pixel-Value-Differencing (PVD) method is a popular irreversible data-hiding scheme proposed by Wu and Tsai (2003). The design of the interval range plays an important role in this method. It is intuitive to design the interval range using the width of the power of two, and the number of bits that can be embedded in a block of two consecutive pixels is determined based on the width of the interval range to which the difference value belongs. In 2013, Tseng and Leng broke the rules in that they designed the interval range based on a perfect square and achieved a higher payload and lower distortion. Based on Tseng and Leng's idea, we propose a new data-hiding strategy based on the PVD method for an optimal design of the interval range.

Keywords: Pixel-Value-Differencing method · Interval range · Perfect square number

1 Introduction

The Pixel-Value-Differencing method (PVD) is a popular irreversible data-hiding scheme proposed by Wu and Tsai [1] (2003). It conveys a large number of payloads while still maintaining the consistency of the image characteristics after data embedding, and provides high imperceptibility to a stego-image. In the process of embedding a secret message, a cover image is partitioned into non-overlapping blocks of two consecutive pixels. A difference value is calculated based on the values of the two pixels in each block. The new difference value is then replaced by the sum of the lower bound of an interval and the secret message. This method was designed in such a way that the modification is never out of the interval range. The selection of the interval range is based on the sensitivity characteristics of human vision to gray value variations, from smoothness to contrast.

It is intuitive to design the interval range using the width of the power of two and the number of bits that can be embedded in a block of two consecutive pixels, as determined based on the width of the interval range to which the difference value belongs. The PVD method embeds fewer secret messages in the smooth areas than in the edge areas. Because most of the blocks belonging to a smooth area lead to a low payload, Wu et al. [2] (2005) proposed an improved PVD method: increase the payload using the least-significant-bit (LSB) substitution method [2] to embed a secret message in a

J.-S. Pan et al. (eds.), *Advances in Intelligent Information Hiding and Multimedia Signal Processing*, Smart Innovation, Systems and Technologies 81,
DOI 10.1007/978-3-319-63856-0_3

smooth area. Most studies related to this issue [3–8] have focused on increasing the capacity by using the LSB method, which leads to another problem, i.e., detection using RS-Steganalysis [9].

In 2013, Tseng and Leng [10] broke the rules in designing the interval range using a width of the power of two. Their interval range were designed based on a perfect square number, and can achieve a higher payload and lower distortion. Based on Tseng and Leng's idea, we propose a new data-hiding strategy based on the PVD method for an optimal interval range design.

The remainder of this paper is organized as follows. Section 2 briefly describes related works. Section 3 presents the proposed method on how to optimally design the interval range based on the PVD method. Section 4 offers a theoretical analysis of the payload and distortion. Finally, Sect. 5 provides some concluding remarks regarding this paper.

2 Literature Review

In this section, we briefly describe Wu and Tsai's method, as well as the PVD method proposed by Tseng and Leng.

The PVD method first partitions a cover image into non-overlapping blocks of two consecutive pixels, with states p_i and p_{i+1}. Then, for each block, the difference value is calculated as $d_i = |p_{i+1} - p_i|$, where d_i ranges from zero to 255. Different interval range and an embedding algorithm are applied in their method.

2.1 Wu and Tsai's PVD Method

Wu and Tsai first proposed the PVD method in 2003. Their embedding algorithm can be described as follows:

1. Search the interval range for d_i to determine how the bits will be embedded. Let $d_i \in R_i$, in which $R_i = [l_i, u_i]$, where l_i and u_i are the lower and upper bounds of R_i. Two sets of interval range are shown in Fig. 1.
2. Denote the number of embedding bits $k = \lfloor \log_2(u_i - l_i) \rfloor$. Read a secret message of k bits and transfer the bits into decimal values, s.
3. Calculate the new difference value $d_i' = l_i + b$. Note that both d_i and d_i' are located within the same range, R_i.
4. Average d_i' for p_i and p_{i+1}. We then obtain the stego pixels p_i' and p_{i+1}' through the following formula:

$$(p_i, p_{i+1}) = \begin{cases} (p_i + \lceil m/2 \rceil, p_{i+1} - \lfloor m/2 \rfloor, & \text{if } p_i \geq p_{i+1}, d_i' > d_i \\ (p_i - \lceil m/2 \rceil, p_{i+1} + \lfloor m/2 \rfloor, & \text{if } p_i < p_{i+1}, d_i' > d_i \\ (p_i - \lceil m/2 \rceil, p_{i+1} + \lfloor m/2 \rfloor, & \text{if } p_i \geq p_{i+1}, d_i' \leq d_i \\ (p_i + \lceil m/2 \rceil, p_{i+1} - \lfloor m/2 \rfloor, & \text{if } p_i < p_{i+1}, d_i' \leq d_i \end{cases}, \tag{1}$$

where $m = |d'_i - d_i|$.

Steps 1 through 4 are repeated until all secret messages are embedded and the stego image is produced.

During the extraction procedure, the new difference d'_i is calculated from each block, and using the same interval range, we can then determine which range R_i that d'_i belongs to and find the lower bound, l_i. Compute $s = d'_i - l_i$ and transfer s into binary format. Repeat until all secret messages are completely extracted.

R_i	R_1	R_2	R_3	R_4	R_5	R_6
range	[0,7]	[8,15]	[16,31]	[32,63]	[64,127]	[128,255]
width	8	8	16	32	64	128
payload	3	3	4	5	6	7

(a) Focus on high payload

R_i	R_1	R_2	R_3	R_4	R_5	R_6	R_7	R_8	R_9	R_{10}	R_{11}	R_{12}	R_{13}
range	[0,1]	[2,3]	[4,7]	[8,11]	[12,15]	[16,23]	[24,31]	[32,47]	[47,62]	[63,94]	[95,127]	[128,191]	[192,255]
width	2	2	4	4	4	8	8	16	16	32	32	64	64
payload	1	1	2	2	2	3	3	4	4	5	5	6	6

(b) Focus on low distortion

Fig. 1. Two sets of Wu and Tsai's interval range: focusing on (a) a high payload and (b) low distortion.

2.2 Tseng and Leng's PVD Method

It is intuitive to design the interval range using a width of the power of two. In 2013, Tseng and Leng broke the rules and designed the interval range based on a perfect square number (Fig. 2), and achieved a higher payload and lower distortion than We and Tsai' method.

For each difference value d_i, choosing the nearest perfect number n^2, we then have interval range $n^2 - n \leq n^2 < n^2 + n$ for $n \in [1, 16]$. The width of this range interval is $(n^2 + n) - (n^2 - n) = 2n$, and the payload for each interval range is $k = \lfloor \log_2(2n) \rfloor$. For each interval range $[n^2 - n, n^2 + n)$, if the width of the interval is larger than 2^k, then it is partitioned into two sub-ranges: $[n^2 - n, n^2 + n - 2^k]$ and $[n^2 + n - 2^k + 1, n^2 + n - 1]$.

The embedding algorithm is described as follows:

1. If $d_i \in [240, 255]$, then the new difference value $d'_i = d_i + secret(4)$, where $secret(4)$ indicates four bits to-be-embedded secret messages transfer into decimal format. For example, suppose four bits to-be-embedded secret message is '1001', then $secret(4) = 9$ in decimal format. The PVD method then applies the new difference value x using formula (1).
2. Check whether d_i is located in the initial subrange of R_i. If so, a value of x exists in the first subrange, which is $LSB(x, k + 1) = secret(k + 1)$, where $LSB(x, k + 1)$ indicates four least-significant bits of x. For example, denote $d_i = 32 \in R_6, k + 1 = 4$,

and $32 = (00010\underline{0000})_2$, if four bits of a secret message to be embedded is '0000', then $secret(4) = 0$ and $LSB(32,4) = secret(4)$. Otherwise, a value of y must exist in the second subrange such that $LSB(y,k) = secret(k)$. Because of the width of the second subrange is 2^k all combinations of k bits are concluded. For example, for the second subrange [34,41], where $k = 3$, all values of $LSB(y,3)$ are illustrated as follows: $34 = (000100\underline{010})_2$, $35 = (000100\underline{011})_2$, $36 = (000100\underline{100})_2$, $37 = (000100\underline{101})_2$, $38 = (000100\underline{110})_2$, $39 = (000100\underline{111})_2$, $40 = (000101\underline{000})_2$, and $41 = (000101\underline{001})_2$. Next, the PVD method is applied to the new difference value y using formula (1).

In the extraction procedure, calculate the new difference d'_i from each block. If $d'_i \in [240, 255]$, secret messages $s = LSB(d'_i - l_i)$. Otherwise, if $d'_i \in R_i$, check whether d'_i is located in the first subrange or the second subrange, and then extract secret messages of $k + 1$ or k bits.

R_i	range	width	1st subrange	width	payload	2nd subrange	width	payload
R_1	[0,1]	2						1
R_2	[2,5]	4					4	2
R_3	[6,11]	6	[6,7]	2	3	[8,11]	4	2
R_4	[12,19]	8						3
R_5	[20,29]	10	[20,21]	2	4	[22,29]	8	3
R_6	[30,41]	12	[30,33]	4	4	[34,41]	8	3
R_7	[42,55]	14	[42,47]	6	4	[48,55]	8	3
R_8	[56,71]	16						4
R_9	[72,89]	18	[72,73]	2	5	[74,89]	16	4
R_{10}	[90,109]	20	[90,93]	4	5	[94,109]	16	4
R_{11}	[110,131]	22	[110,115]	6	5	[116,131]	16	4
R_{12}	[132,155]	24	[132,139]	8	5	[140,155]	16	4
R_{13}	[156,181]	26	[156,165]	10	5	[166,181]	16	4
R_{14}	[182,209]	28	[182,193]	12	5	[194,209]	16	4
R_{15}	[210,239]	30	[210,223]	14	5	[224,239]	16	4
R_{16}	[240,255]	16						4

Fig. 2. Tseng and Leng's interval range.

3 The Proposed Method

Based on Tseng and Leng's idea, we propose a new data-hiding strategy based on the PVD method for an optimal design of the interval range.

Let us review Tseng and Leng's interval range, as shown in Fig. 2: the width of the second subrange is always the power of two, and the width of the first subrange is relatively small and irregular. The proposed method attempts to achieve an optimal design of the range interval by maximizing the width of the first subrange (Fig. 3).

R_i	range	width	R_n^-	width	payload	R_n^*	width	payload	R_n^+	width	payload
R_1	[0,2]	3	[0]	1	2	[1]	1	1	[2]	1	2
R_2	[3,7]	5	[3]	1	3	[4,6]	3	2	[7]	1	3
R_3	[8,14]	7	[8,10]	3	3	[11]	1	2	[12,14]	3	3
R_4	[15,23]	9	[15]	1	4	[16,22]	7	3	[23]	1	4
R_5	[24,34]	11	[24,26]	3	4	[27,31]	5	3	[32,34]	3	4
R_6	[35,47]	13	[35,39]	5	4	[40,42]	3	3	[43,47]	5	4
R_7	[48,62]	15	[48,54]	7	4	[55]	1	3	[56,62]	7	4
R_8	[63,79]	17	[63]	1	5	[64,78]	15	4	[79]	1	5
R_9	[80,98]	19	[80,82]	3	5	[83,95]	13	4	[96,98]	3	5
R_{10}	[99,119]	21	[99,103]	5	5	[104,114]	11	4	[115,119]	5	5
R_{11}	[120,142]	23	[120,126]	7	5	[127,135]	9	4	[136,142]	7	5
R_{12}	[143,167]	25	[143,151]	9	5	[152,158]	7	4	[159,167]	9	5
R_{13}	[168,194]	27	[168,178]	11	5	[179,183]	5	4	[184,194]	11	5
R_{14}	[195,223]	29	[195,207]	13	5	[208,210]	3	4	[211,223]	13	5
R_{15}	[224,254]	31	[224,238]	15	5	[239]	1	4	[240,254]	15	5

Fig. 3. Proposed interval range.

For each difference value $d_i \in [0, 254]$, a value of n exists such that $n^2 - 1 \leq d_i < n^2 + n$, where $n \in [1, 15]$. The width of this range interval is $(n^2 + n) - (n^2 - 1) = 2n + 1$, and the payload for each interval range is $k = \lfloor \log_2(2n + 1) \rfloor$. In other words, the interval range $R_n = [n^2 - 1, n^2 + 2n) = [n^2 - 1, n^2 + 2n - 1]$. Each R_i is partitioned into three parts: R_n^-, R_n^*, and R_n^+, where $R_n^- = [n^2 - 1, n^2 + 2n - 2^k - 1]$, $R_n^* = [n^2 + 2n - 2^k, n^2 + 2^k - 2]$, and $R_n^+ = [n^2 + 2^k - 1, n^2 + 2n - 1]$.

The embedding algorithm is described as follows:

1. If $d_i \in [0, 254]$, search the interval range R_i and determine which subrange (R_n^-, R_n^* or R_n^+) it belongs to (if $d_i = 255$, it remains unchanged).
2. Embed a secret message by choosing the new difference value x in R_i. For example, $d_i = 40 \in R_6 = [35, 47]$ and $k = \lfloor \log_2(2n + 1) \rfloor = 3$, as shown in Fig. 4.

Note that the secret messages are embedded and extracted in reverse order. Denote the secret messages as $s = s_1 s_2 s_3 \ldots = (110\ldots)_2$. Suppose that three-bit secret messages are to be embedded; $(110)_2$ is then fetched from s and converted into $(011)_2$ for the embedding. In addition, $R_6^- \cup R_6^*$ satisfies all combinations of k bits, and $R_6^- \leftrightarrow R_6^+$ is symmetric.

n	n^2	$2n$+1	Range
6	36	13	[35,36,37,38,39,40,41,42,43,44,45,46,47]

$k = \lfloor \log_2(2n+1) \rfloor = \lfloor \log_2 13 \rfloor = 3$

$x = (b_1 b_2 b_3 b_4 b_5 b_6 b_7 b_8)_2,\ s = s_1 s_2 s_3 s_4 \ldots$

$R_6 = R_6^- \cup R_6^{\bullet} \cup R_6^+$

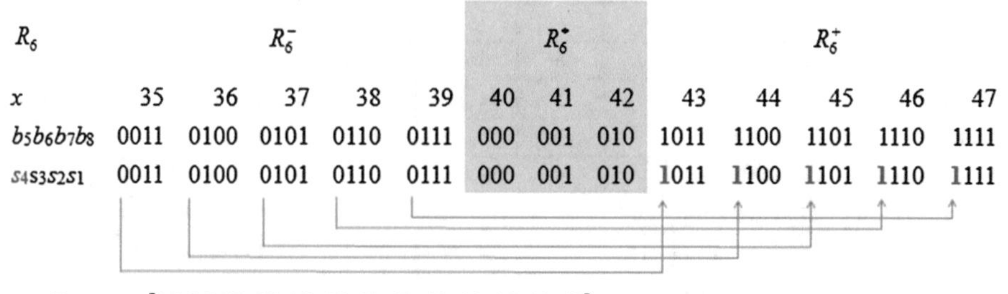

$x \in R_6 \Rightarrow x = [35, 36, 37, 38, 39, 40, 41, 42, 43, 44, 45, 46, 47]$

if $s_4s_3s_2s_1$=0011 ⇒ x'=35
if $s_4s_3s_2s_1$=0100 ⇒ x'=36
if $s_4s_3s_2s_1$=0101 ⇒ x'=37
if $s_4s_3s_2s_1$=0110 ⇒ x'=38
if $s_4s_3s_2s_1$=0111 ⇒ x'=39
if $s_3s_2s_1$=000 ⇒ x'=40
if $s_3s_2s_1$=001 ⇒ x'=41
if $s_3s_2s_1$=010 ⇒ x'=42
if $s_4s_3s_2s_1$=1011 ⇒ x'=43
if $s_4s_3s_2s_1$=1100 ⇒ x'=44
if $s_4s_3s_2s_1$=1101 ⇒ x'=45
if $s_4s_3s_2s_1$=1110 ⇒ x'=46
if $s_4s_3s_2s_1$=1111 ⇒ x'=47

Fig. 4. An example of embedding using the proposed method ($d_i = 40$).

4 Theoretical Analysis

For the sake of fairness, we only provided a theoretical analysis comparing Tseng and Leng's method with the proposed method in terms of the average payload and average error (Fig. 5). The average payload is computed using the following formula:

$$\frac{1}{2n+1} \sum_{j=1}^{3} (C(n_{i,j}) \times n_{i,j}), \tag{2}$$

and the average error for each range is calculated using the following formula:

$$\frac{1}{2n+1}\sum_{i=1}^{3}\frac{\sum_{j=0}^{c(n_{i,j})-1} j}{c(n_{i,j})}, \tag{3}$$

where $C(n_{i,j})$ is the total number within the jth subrange, and n_j is the embedding bits within the jth subrange of R_i.

Ri	width	average payload	average error
R_1	2	1.00	0.50
R_2	4	2.00	1.25
R_3	6	2.33	1.94
R_4	8	3.00	2.63
R_5	10	3.20	3.30
R_6	12	3.33	3.97
R_7	14	3.43	4.64
R_8	16	4.00	5.31
R_9	18	4.11	5.98
R_{10}	20	4.20	6.65
R_{11}	22	4.27	7.32
R_{12}	24	4.33	7.99
R_{13}	26	4.38	8.65
R_{14}	28	4.43	9.32
R_{15}	30	4.47	9.99
R_{16}	16	4.00	-

(a) Tseng and Leng's method

Ri	width	average payload	average error
R_1	3	1.67	0.89
R_2	5	2.40	1.60
R_3	7	2.86	2.29
R_4	9	3.22	2.96
R_5	11	3.55	3.64
R_6	13	3.77	4.31
R_7	15	3.93	4.98
R_8	17	4.12	5.65
R_9	19	4.32	6.32
R_{10}	21	4.48	6.98
R_{11}	23	4.61	7.65
R_{12}	25	4.72	8.32
R_{13}	27	4.81	8.99
R_{14}	29	4.90	9.66
R_{15}	31	4.97	10.32

(b) The proposed method

Fig. 5. Theoretical analysis comparing (a) Tseng and Leng's method with (b) the proposed method in terms of the average payload and average error.

From Fig. 5, it can be seen that the proposed method not only achieves a higher payload but also a lower distortion. For example, with Tseng and Leng's method, $d_i = 32 \in R_6$, and we then obtain an average payload of 3.33 and an average error of 3.97; with the proposed method, $d_i = 32 \in R_5$, and we obtain an average payload of 3.55 and an average error of 3.64.

5 Conclusions

In this paper, based on Tseng and Leng's method, we proposed a new strategy for data hiding based on the Pixel-Value-Differencing method to achieve an optimal design of the interval range. A theoretical analysis proves that the proposed method achieves a higher payload and lower distortion than the previous method.

Acknowledgements. This study was supported by a Research Grant, MOST, from Taiwan's Ministry of Science and Technology (MOST 105-2221-E-018-020-).

References

1. Wu, D.-C., Tsai, W.-H.: A steganographic method for images by pixel-value differencing. Pattern Recogn. Lett. **24**, 1613–1626 (2003). doi:10.1016/s0167-8655(02)00402-6
2. Chan, C.-K., Cheng, L.: Hiding data in images by simple LSB substitution. Pattern Recogn. **37**, 469–474 (2004). doi:10.1016/j.patcog.2003.08.007
3. Wu, H.-C., Wu, N.-I., Tsai, C.-S., Hwang, M.-S.: Image steganographic scheme based on pixel-value differencing and LSB replacement methods. IEE Proc. Vis. Image Signal Process. **152**, 611 (2005). doi:10.1049/ip-vis:20059022
4. Kim, K.-J., Jung, K.-H., Yoo, K.-Y.: A high capacity data Hiding method using PVD and LSB. In: 2008 International Conference on Computer Science and Software Engineering (2008). doi:10.1109/csse.2008.1378
5. Yang, C.-H., Weng, C.-Y., Wang, S.-J., Sun, H.-M.: Varied PVD LSB evading detection programs to spatial domain in data embedding systems. J. Syst. Softw. **83**, 1635–1643 (2010). doi:10.1016/j.jss.2010.03.081
6. Gadiparthi, M., Sagar, K., Sahukari, D., Chowdary, R.: A high capacity steganographic technique based on LSB and PVD modulus methods. Int. J. Comput. Appl. **22**, 8–11 (2011). doi:10.5120/2582-3568
7. Mandal, J.K., Das, D.: A novel invisible watermarking based on cascaded PVD integrated LSB technique. In: Eco-friendly Computing and Communication Systems Communications in Computer and Information Science, pp. 262–268 (2012). doi:10.1007/978-3-642-32112-2_32
8. Sabokdast, M., Mohammadi, M.: A steganographic method for images with modulus function and modified LSB replacement based on PVD. In: The 5th Conference on Information and Knowledge Technology (2013). doi:10.1109/ikt.2013.6620050
9. Fridrich, J., Goljan, M., Du, R.: Reliable detection of LSB steganography in color and grayscale images. In: Proceedings of the 2001 Workshop on Multimedia and Security New Challenges - MM&Sec 2001 (2001). doi:10.1145/1232454.1232466
10. Tseng, H.-W., Leng, H.-S.: A steganographic method based on pixel-value differencing and the perfect square number. J. Appl. Math. **2013**, 1–8 (2013). doi:10.1155/2013/189706

Data Hiding Scheme Based on Regular Octagon-Shaped Shells

Hui-Shih Leng(✉)

National Changhua University of Education, Changhua, Taiwan
lenghs@cc.ncue.edu.tw

Abstract. In 2006, Zhang and Wang proposed a novel data hiding method, namely the exploiting modification direction (EMD) scheme. In their scheme, the secret message in a $(2n + 1)$-ary notational system is carried by n cover pixels, where n is a system parameter, and at most one pixel is increased or decreased by 1. The EMD scheme exhibits two disadvantages: the secret message must be converted into a non-binary system, and the embedding capacity is limited. In this study, a new data hiding scheme, based on a regular octagon-shaped shell, is proposed to overcome the EMD scheme's shortcomings.

Keywords: Data hiding · Exploiting modification direction · Regular octagon-shaped shell

1 Introduction

Data hiding is a technique used to embed a secret message into cover media for secret communication. A data hiding scheme's performance is evaluated by embedding capacity (or payload) and the visual quality of the stego image, between which there is a trade-off, since these two factors are inversely proportional to one another.

In 2006, Zhang and Wang [1] proposed a novel data hiding approach, namely the exploiting modification direction (EMD) scheme. In this scheme, the secret message in a $(2n + 1)$-ary notational system is carried by n cover pixels, where n is a system parameter, and at most one pixel is increased or decreased by 1. In order to increase embedding capacity, Chao et al. [2] proposed an image data hiding scheme using diamond encoding. In this scheme, for each cover pixel a pair is applied to calculate the diamond characteristic value, which is then modified to the secret message, and can be obtained by adjusting pixel values. In addition, Kieu and Chang [3] improved the EMD scheme by exploiting eight modification directions to hide more secret bits in a cover pixel pair at a time. Although the above works succeeded in increasing embedding capacity, the most serious problem still exists – that is, the secret message must be converted into a non-binary notational system in advance. To address this issue, Chang et al. [4] introduced a turtle shell-based data hiding scheme. In this scheme, a reference table is generated; then, based on this table, each cover pixel pair is used to embed the secret message with the guidance of turtle shells. The embedded secret message can be extracted precisely from the stego image by using the same reference table and the turtle shell. Liu et al. [5]

J.-S. Pan et al. (eds.), *Advances in Intelligent Information Hiding and Multimedia Signal Processing*, Smart Innovation, Systems and Technologies 81,
DOI 10.1007/978-3-319-63856-0_4

extend turtle shell matrix structure to a different matrix model to meet different embedding capacity and image quality needs. Liu et al. [6] also conducted work on high-capacity turtle shell-based data hiding. Furthermore, Kurup et al. [7] proposed a data hiding scheme based on octagon-shaped shells to achieve a higher embedding capacity than the turtle-based shell scheme, while maintaining visual quality.

The remainder of this paper is organized as follows. Section 2 provides a brief literature review, while Sect. 3 describes the proposed method. The experimental results are presented in Sect. 4, and finally, Sect. 5 provides concluding remarks.

2 Literature Review

In this section, we briefly describe Zhang and Wang's EMD scheme, Chang et al.'s turtle shell-based scheme, and Kurup et al.'s octagon-shaped shell-based scheme.

2.1 Zhang and Wang's EMD Scheme

The main concept of this scheme is embedding a $(2n + 1)$-ary notational system secret message in a cover pixel pair, where only one pixel is increased or decreased by 1. Before data embedding takes place, the secret message must be converted into a $(2n + 1)$-ary notational system. Then, an extraction function f is defined as a weighted sum modulo $(2n + 1)$:

$$f(g_1, g_2, \dots, g_n) = \left[\sum_{i=1}^{n} (g_i \cdot i)\right] \bmod (2n+1), \tag{1}$$

where $(g_1, g_2, \dots, g_n)$ is an n-tuple cover pixel pair.

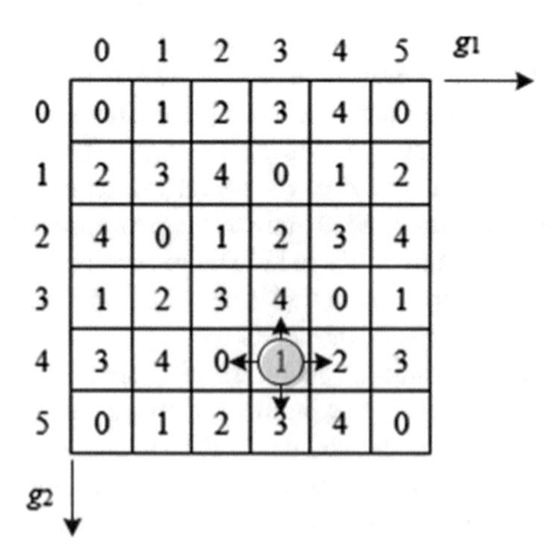

Fig. 1. Reference matrix of Zhang and Wang's EMD scheme ($n = 2$).

Formula (1) generates a reference matrix, as shown in Fig. 1, and maps each secret message to a cover pixel pair.

For example, if a cover pixel pair (3, 4) is calculated using formula (1), we have *f*(3, 4) = 1. If the secret message is 1, no change is required and the stego pixel pair is still $(g_1', g_2') = (3, 4)$. If the secret message is 0, the stego pixel pair is adjusted to $(g_1', g_2') = (2, 4)$. If the secret message is 2, the stego pixel pair is adjusted to $(g_1', g_2') = (4, 4)$. If the secret message is 3, the stego pixel pair is adjusted to $(g_1', g_2') = (3, 5)$. If the secret message is 4, the stego pixel pair is adjusted to $(g_1', g_2') = (3, 3)$.

The secret message can easily be extracted by calculating the stego pixel pair's extraction function.

2.2 Chang et al.'s Turtle Shell-Based Scheme

As opposed to the EMD scheme, in Chang et al.'s scheme, reference matrix is constructed using a special rule instead of the extraction function.

The reference matrix is constructed with several turtle shells, as shown in Fig. 2.

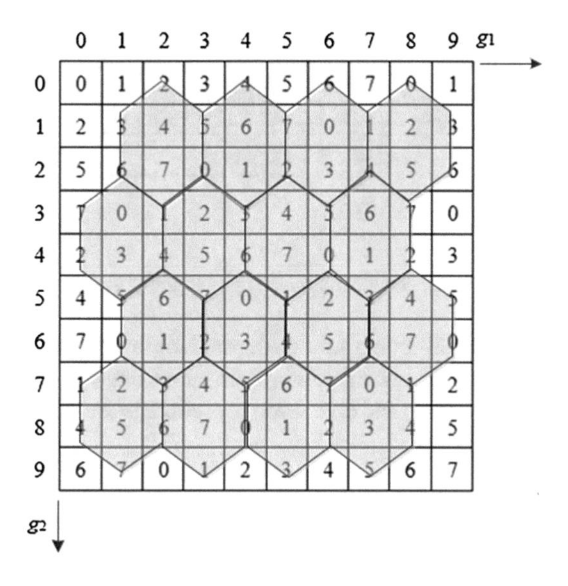

Fig. 2. Reference matrix based on turtle shells.

A turtle shell is a hexagon composed of eight different digits, including six edge and two back digits ranging from $(000)_2$ to $(111)_2$. According to the reference matrix mapping, three secret bits can be embedded in each cover pixel pair. The embedding and extraction procedure is the same as that of the EMD scheme.

Liu et al. [5] extend turtle shell matrix structure to a different matrix model to meet different embedding capacity and image quality needs. But their scheme use non-binary notation systems. Liu et al. [6] also conducted work on high-capacity turtle shell-based data hiding. Different from the EMD scheme, their scheme not only use a reference matrix but also use a location table references to different positions, so the embedding and extraction procedure is different.

2.3 Kurup et al.'s Octagon-Shaped Shell-Based Scheme

Kurup et al.'s scheme attempts to increase embedding capacity by constructing a reference matrix with octagon-shaped shells, as shown in Fig. 3.

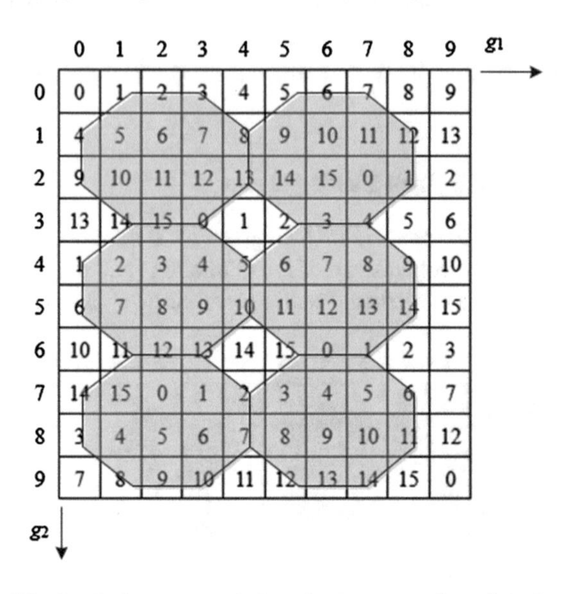

Fig. 3. Reference matrix based on octagon-shaped shells.

An octagon-shaped shell contains 16 digits ranging from $(0000)_2$ to $(1111)_2$. According to the reference matrix mapping, four secret bits can be embedded in each cover pixel pair. The embedding and extraction procedure is the same as that of the EMD scheme.

3 Proposed Method

The advantage of Change et al. and Kurup et al.'s schemes is that the secret message does not need to be converted into a non-binary notational system.

If an octagon-shaped shell is not regular, the worst case will increase distortions; if the octagon shape is regular, the distortions will be decreased. The problem can be described as follows.

$$n^2 - 4 = 2^k, \tag{2}$$

where n is the octagon size and we wish to solve the value k.

If we have $k = 5$ and $n = 6$ to satisfy formula (2), we construct the reference matrix with regular octagon-shaped shells, as shown in Fig. 4.

$g_2 \downarrow$ / $g_1 \rightarrow$	0	1	2	3	4	5	6	7	8	9	10	11
0	0	1	2	3	4	5	6	7	8	9	10	11
1	5	6	7	8	9	10	11	12	13	14	15	16
2	11	12	13	14	15	16	17	18	19	20	21	22
3	17	18	19	20	21	22	23	24	25	26	27	28
4	23	24	25	26	27	28	29	30	31	0	1	2
5	28	29	30	31	0	1	2	3	4	5	6	7
6	1	2	3	4	5	6	7	8	9	10	11	12
7	7	8	9	10	11	12	13	14	15	16	17	18
8	13	14	15	16	17	18	19	20	21	22	23	24
9	19	20	21	22	23	24	25	26	27	28	29	30
10	24	25	26	27	28	29	30	31	0	1	2	3
11	29	30	31	0	1	2	3	4	5	6	7	8

Fig. 4. Reference matrix based on regular octagon-shaped shells ($n = 6$).

A regular octagon-shaped shell ($n = 6$) contains 32 digits ranging from $(00000)_2$ to $(11111)_2$. According to the reference matrix mapping, five secret bits can be embedded in each cover pixel pair. The embedding and extraction procedure is the same as that of the EMD scheme. A regular octagon-shaped shell can therefore embed more secret messages and maintain good stego image visual quality.

4 Experimental Results

In order to evaluate the performance of the proposed method, we tested four images selected from the USC-SIPI image database, namely 'Barbara,' 'Boats,' 'Lena,' and 'Peppers,' as shown in Fig. 5.

(a) Barbara (b) Boats (c) Lena (d) Peppers

Fig. 5. Four test images.

The peak signal-to-noise-ratio (PSNR) was applied to evaluate the stego images' visual quality, and is defined as follows.

$$PNSR = 10\log_{10}\frac{255^2}{MSE}(\text{dB}) \tag{3}$$

and

$$MSE = \frac{1}{h \times w}\sum_{i=1}^{h}\sum_{j=1}^{w}(g_{i,j} - g_{i,j}')^2, \tag{4}$$

where w and h are the width and height of the cover image, respectively, and $g_{i,j}$ and $g_{i,j}'$ represent the pixel value of the cover pixel image and stego image, respectively.

Table 1 summarizes the results of the visual quality and payload of the stego images in the proposed scheme.

Table 1. Experimental results of visual quality and payload.

	MSE	PSNR (dB)	Payload (bpp)
Barbara	3.2436	43.0206	2.5
Boats	3.2538	43.0069	2.5
Lena	3.2577	43.0017	2.5
Peppers	3.2685	42.9873	2.5

It can be seen from the results that the proposed method achieves a payload of 2.5 bpp which is higher than previous scheme and maintains good stego image quality (greater than 42.98 dB).

5 Conclusion

In the EMD scheme, n cover pixels carry one $(2n + 1)$-ary notational system secret message. This scheme exhibits two major shortcomings: the secret message must be converted into a non-binary system, and the embedding capacity is limited. The proposed method not only maintains the binary format of the secret message, but also achieves a higher payload.

Acknowledgement. This study was supported by the MOST research grant, MOST from Taiwan's Ministry of Science and Technology (MOST 105-2221-E-018-020-).

References

1. Zhang, X., Wang, S.: Efficient steganographic embedding by exploiting modification direction. IEEE Commun. Lett. **10**(11), 781–783 (2006). doi:10.1109/lcomm.2006.060863
2. Chao, R., Wu, H., Lee, C., Chu, Y.: A novel image data hiding scheme with diamond encoding. EURASIP J. Inf. Secur. **2009**, 1–9 (2009). doi:10.1155/2009/658047

3. Kieu, T.D., Chang, C.: A steganographic scheme by fully exploiting modification directions. Expert Syst. Appl. **38**(8), 10648–10657 (2011). doi:10.1016/j.eswa.2011.02.122
4. Chang, C.C., Liu, Y., Nguyen, T.S.: A novel turtle shell based scheme for data hiding. In: 2014 Tenth International Conference on Intelligent Information Hiding and Multimedia Signal Processing (2014). doi:10.1109/iih-msp.2014.29
5. Liu, L., Chang, C., Wang, A.: Data hiding based on extended turtle shell matrix construction method. Multimedia Tools Appl. (2016). doi:10.1007/s11042-016-3624-7
6. Liu, Y., Chang, C., Nguyen, T.: High capacity turtle shell-based data hiding. IET Image Proc. **10**(2), 130–137 (2016). doi:10.1049/iet-ipr.2014.1015
7. Kurup, S., Rodrigues, A., Bhise, A.: Data hiding scheme based on octagon shaped shell. In: 2015 International Conference on Advances in Computing, Communications and Informatics (ICACCI) (2015). doi:10.1109/icacci.2015.7275908

A Web Page Watermarking Method Using Hybrid Watermark Hiding Strategy

Chun-Hsiu Yeh[1], Jing-Xun Lai[2], and Yung-Chen Chou[2(✉)]

[1] Department of Management Information Systems,
Chung Chou University of Science and Technology, Changhua 510, Taiwan R.O.C.
chyeh@dragon.ccut.edu.tw

[2] Department of Computer Sciences and Information Engineering,
Asia University, Taichung 41354, R.O.C.
kevin50406418@gmail.com, yungchen@gmail.com

Abstract. On the Internet, various communication channels are readily available for companies, communities, and individuals. As companies have increasingly made use of social websites and instant messaging technologies for marketing and public relations, additional effort has been put in place to create interactive web contents to attract online audience. It is easy for anyone to simply copy and modify existing web contents for their own use. A common approach to protecting the copyright of a companys web content is digital watermarking with copyright information. The watermarking technique is a good way to achieve the goal of copyright protection. This paper presents a web page copyright protection method by integrating Cartesian product combination, CSS and HTML tag capitalization method, HTML attribute combination method, HTML attribute quotation mark method, and CSS attribute value embedding method. The experimental results demonstrate that the proposed method success achieved the goal of copyright protection for web pages.

Keywords: HTML · CSS · Data hiding · Digital watermarking

1 Introduction

With advances in the digital information era, digital communications devices are becoming more important in our lives. Because an increasing amount of information is now readily available on the Internet, users are staying on the Internet longer. To entice more users to visit their websites, companies create attractive and interactive web contents. In addition to designing eye-catching user interface, it is also important to implement the latest web architecture standards (for example, HTML5 and CSS3). By using the latest web architecture and design standards, a website becomes more flexible and is able to host various multimedia files. To protect intellectual content on a website, one can use data hiding techniques to verify content copyright ownership. Currently, you can embed

J.-S. Pan et al. (eds.), *Advances in Intelligent Information Hiding and Multimedia Signal Processing*, Smart Innovation, Systems and Technologies 81,
DOI 10.1007/978-3-319-63856-0_5

secret data in various types of files (for example, image, voice, video, document, and HTML files). This paper discusses how to embed watermarks into HTML files using CSS arguments, HTML tags (including capitalization and CSS), and HTML space characters. In addition to discussing robustness as the most important factor for a watermarking technique, this paper also discusses how to prevent watermark data from being destroyed. The data embedding method to embed watermark data into HTML file is an effective way for maintaining the visual quality of watermarked HTML codebase.

A number of HTML data hiding methods have been proposed to hide data in HTML files; Yang and Yang had proposed using tags and attribute quotation marks in an HTML file to embed secret data (information); Huang et al., had proposed the use of HTML attributes as sequence of tags to embed secret data in an HTML file; Sui and Luo had proposed using different capitalizations of an HTML tag to represent the secret data, and Chou et al., proposed a data hiding method using Cartesian production with blank spaces in an HTML file.

Digital watermarking is a common copyright protection technique to prevent illegal use of web contents. Our data hiding method can hide both program source codes [8] and hypertext mark-up language (HTML) [2–7,9,11,12] for an existing web page. In addition, if a dispute were to arise over the copied web content that is modified and hosted on another website, the proposed method can prove who the rightful owner of the copied content should be. We propose a novel method that embeds watermark data multiple times in the same HTML file using five data embedding methods. To increase the accuracy of recovered watermark data, we can extract the embedded watermark data using the five methods with a voting system. Based on the proposed method, we can achieve the goal of proving the ownership of online contents and protecting the intellectual property of a company.

2 Related Works

2.1 Hamming Code

Richard Hamming invented the Hamming code that is used as a linear error-correction code [10]. By adding a checksum (p_1, p_2 and p_3) in data, one can detect and fix data errors in a single bit. Hamming coding can detect single-bit errors and use parity checksum for data recovery. However, Hamming code is unable to detect and correct double bit errors. The checksum bit p_1 is used to encode p_1, d_1, d_2 and d_4. The checksum bit p_2 is used to encode p_2, d_1, d_3 and d_4. And the checksum bit p_3 is used to encode p_3, d_2, d_3 and d_4. The checksum values can be obtained by using Eq. (1).

$$\begin{aligned} p_1 &= (d_1 + d_2 + d_3)\ mod\ 2, \\ p_2 &= (d_1 + d_3 + d_4)\ mod\ 2, \\ p_3 &= (d_2 + d_3 + d_4)\ mod\ 2. \end{aligned} \tag{1}$$

2.2 HTML Quotation Mark Tags

In 2010, Yang and Yang presented a method using HTML quotation mark tags in an HTML file to embed a secret message [11]. Although HTML tags are crucial to constructing a web page, using different quotation marks within HTML tags does not affect web browsing experience. For example, <font color = blue>, <font color = ‘blue’> and <font color = “blue”> are treated the same way in a web browser. Yang and Yang used this feature to embed secret messages in HTML files.

2.3 HTML Attribute Combinations

Huang et al., presented a method using permutations of HTML attributes to embed a secret message in an HTML file [5]. For example, <font color = “green” size = “3” face = “Arial”>, <font color = “green” face = “Arial” size = “3”>, <font face = “Arial” color = “green” size = “3”>, and <font face = “Arial” size = “3” color = “green”> are render than same output treated the same way in a web browser and they do not affect the web browsing experience. The different combinations of HTML attributes is are used to embed a secret bit. There are six combinations for the font attributes Each font attribute can embed $\lfloor \log_2 6 = 2 \rfloor$ bits.

2.4 Embedding Data Using Different Capitalization in HTML Tags

Sui and Luo presented a method in 2004 [9] to use different capitalization in HTML tags for embedding a secret message. A static web page is composed of many HTML tags (for example, “<HTML>”, “<BODY>”, and “<OL>”). Because HTML tags are case-insensitive, capitalization in HTML tags does not affect the rendering of the content in a web browser. For example, “<HTML>” and “<HTml> render the same output in a web browser. Sui and Luo use this case-insensitive characteristic of HTML tags to embed secret messages. An uppercase letter is used to express the secret bit “1” while a lowercase letter is used to express the secret bit “0”. For example: “<hTml>” represents the secret message “0100”.

2.5 Watermark Data Embedding Using Cartesian Product

Chou et al., presented a data hiding method by using a Cartesian product to code the space between words in an HTML file [1]. Because there are multiple ways to code a space character in an HTML file, a web browser compiles the code to display a space character in HTML. However, users are not aware of the coding difference on a web page. Table 1 lists the possible space character codes. By using a Cartesian product and data segmentation, a secret message can be embedded in the space characters in an HTML file. α is a segment that contains a space character in the HTML file, and NC is the availability of the total number of space characters. We can create a pairing of the space character

Table 1. Space character code

No.	Code	No.	Code	No.	Code	No.	Code
0	type space;	7		14		21	
1		8		15		22	
2		9		16		23	
3		10		17		24	
4		11		18		25	
5		12		19			
6		13		20			

codes using the Cartesian product. Then, a Cartesian product combination is used to obtain a different combination of the space character code to replace the space character code in the HTML file.

Therefore, β is applied to obtain segments of the secret message. Each segment of the secret message must have $\beta = \lfloor \log_2 NC^k \rfloor$ bits, where k is the number for each pair. We can convert the segmented bits of the secret message into decimal format and find the corresponding index in the Cartesian product.

3 The Proposed Method

The proposed method uses five watermark embedding methods to embed the same watermark to achieve the robustness requirement. The browsing qualities of the watermarked HTML file and the original HTML file are the same. Watermark data extraction will result in five sets of watermark data because of the five methods used for data embedding.

In addition, we utilize (7, 4) Hamming code to generate the checksum data for the watermark and then embed the watermark and checksum data using the five data embedding methods Cartesian product combination, CSS and HTML tag capitalization method, HTML attribute combination method, HTML attribute quotation mark method, and CSS attribute value embedding method. Because the watermark data is embedded using five data embedded methods, when we extract the watermark data, we obtain five sets of data. After watermark data have been extracted, the correctness of the watermark data will be checked by adopting Hamming checksum code. After that, the voting system is adopted to check the correctness of watermark data to get the final watermark data.

3.1 Embedding in CSS Values

With the development of the Web 2.0 technology, cascading style sheets (CSS) can be used to improve the look-and-feel of a web page. CSS is a style sheet language used for describing the presentation of web pages. Minor changes in CSS are hardly noticeable to the users. Thus, we can take advantage of this feature in CSS to embed a watermark. For example, embedding watermark data in the last bit of the text color code (red, green, or blue). Changes in the last bit do not significantly change the text color and it is not easy for users to detect watermark data in the color code. Thus, we can effectively prevent tampering of the watermark data. The key steps of the proposed HTML file watermarking are summarized as follows:

Input: HTML code H, Watermark W
Output: Watermarked HTML code H
Step 1: Generate checksum data W for W using (7, 4) Hamming coding
Step 2: Embed W into H to generate H^1 using Cartesian product combination method
Step 3: Embed W into H^1 to generate H^2 using CSS and HTML tag capitalization method
Step 4: Embed W into H^2 to generate H^3 using HTML attribute combination method
Step 5: Embed W into H^3 to generate H^4 using HTML attribute quotation mark method
Step 6: Embed W into H^4 to generate H^5 using CSS attribute value embedding method
Step 6.1: Specify the color code (red, green, or blue) in the SPAN tag.
Step 6.2: Replace the last bit of the color code (red, green, or blue) with the watermark bit.
Step 7: Output H^5 as H

3.2 Extracting Watermark Data

Extracting the watermark is a reversed procedure of the watermark embedding process. The five sets of watermarks data extracted from the watermarked HTML file can be obtained by applying Cartesian product embedding method, CSS and HTML tag capitalization method, HTML attribute combination method, HTML attribute quotation mark method, and CSS attribute value embedding method. For Cartesian product embedding method, parse the article content in the watermarked web page and check the pattern of the blank space characters to extract the watermark. For the CSS and HTML tag capitalization method, check the capitalization of the HTML and CSS tags to extract the watermark data. After that, the fifth copy of the watermark data can be extracted by getting the LSB bits of the color attribute values to form the extracted watermark data.

By embedding the same watermark data in the same HTML file using the five data embedding methods, we can compare the data bits from each method and the bit with the highest frequency is used in the final watermark data. Even if the watermark has been tampered with, it is still possible to recover the original watermark data using the voting system.

4 Experiment Simulation

To evaluate the performance of the proposed method, we implemented the proposed method using Octave 3.6.4. The web page content in our simulation is the 1841 U.S. President William Henry Harrison's inauguration speech, with a total of 8460 word. If each work in the speech is segmented within a <span> tag, the total payload is 16,920 bits. Figures 1(a) and (b) show the watermark data and a part of the original web page, respectively. Figure 1(c) show the results of watermarked HTML rendering in Chrome and IE. As we can see, the watermarked HTML file has the same visual quality in the web browsers tested. Figure 1(d) show a portion of the watermarked HTML file and the extracted watermark HTML source code.

(a) Digital watermark (size: 64x64)

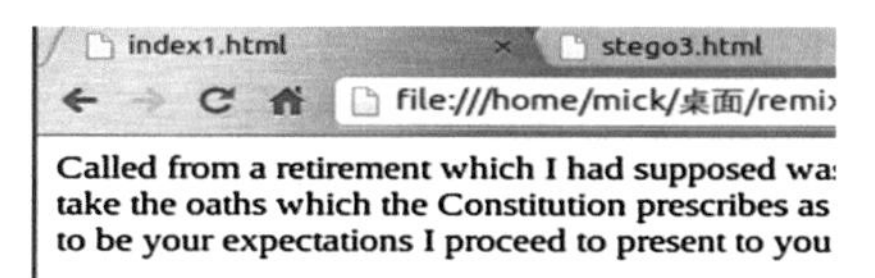

(b) Original HTML

index1.html
stego3.html
file:///home/mick/桌面/remix_
Called from a retirement which I had supposed was
take the oaths which the Constitution prescribes as a
to be your expectations I proceed to present to you a
It was the remark of a Roman consul in an early peri
before and after obtaining them, they seldom carryin
in the lapse of upward of two thousand years since t
governments would develop similar instances of viol

(c) Watermarked HTML (Chrome)

(d) Extracted watermark

Fig. 1. The watermark, original HTML, watermarked HTML, and extracted watermark

5 Robustness Assessment

To test the stability of this method, we designed a series of attack programs to try to alter the watermarked HTML code while not affecting browsing experience. The original watermark data size is 64*64.

Simulation Attack A. The first simulation attack changes all single quotes to double quotes in HTML attributes and all uppercase letters to lowercase letters in HTML tags. Even with these changes, we can still restore the original watermark data with 100% accuracy (refer to Fig. 2(a)).

Simulation Attack B. After simulation attack A, we launched simulation attack B to change the space character "	" to " " in the HTML file. With the two simulation attacks, the extracted watermark data is 99.4% accurate compared to the original watermark data (refer to Fig. 2(b)). In this case, we are not able to recover 24 bits of the original watermark data.

Simulation Attack C. Continuing from simulation attack b, this attack changes the CSS value "101011" to "111010". In this case, the extracted watermark is 97.9% accurate compared to the original watermark data (refer to Fig. 2(c)). Only 86 bits of the original watermark data are not recoverable.

Simulation Attack D. Continuing with simulation attack C, this simulation attack changes the permutation of the font attributes from (face, color, size) to (color, size, face). Compared to simulation attack C, only 140 bits cannot be restored to the original watermark data and the extracted watermark data is 96.5% accurate (refer to Fig. 2(d)).

Simulation Attack E. Continuing with the previous simulation attack, we change the space characters " " to "	" in the HTML file and obtained an extracted watermark that is 95.0% accurate, which is still of high quality (refer to Fig. 2(e)).

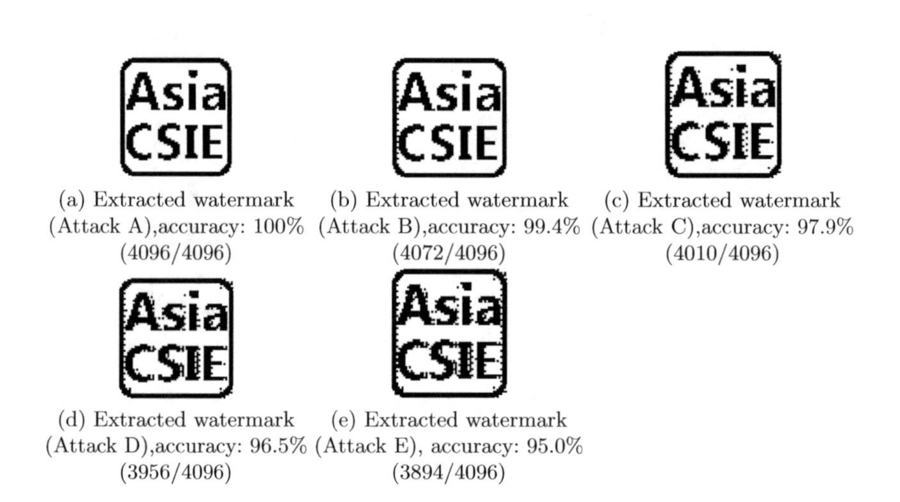

Fig. 2. The accuracy of different attack simulations

6 Conclusion

Since five data embedding methods are used, we must determine the smallest payload among the five methods. In the previous section, we presented five simulation attacks to show the high stability level of our proposed method. The advantage of integrating the five data embedding methods in our proposed method is the ability to restore and recover the original watermark using Hamming code and the voting system, even if the watermark has been tampered with. Although we have tried to modify the watermark data extraction process by first implementing the voting system, recovering the data, and then removing the checksum bits of Hamming code (7, 4), the result is not satisfactory. Thus, our original proposed data extraction process is far more suitable to achieving the goal of copyright protection for HTML files.

References

1. Chou, Y.-C., Huang, C.-Y., Liao, H.-C.: A reversible data hiding scheme using Cartesian product for html file. In: 2012 Sixth International Conference on Genetic and Evolutionary Computing (ICGEC), pp. 153–156. IEEE (2012)
2. Chou, Yung-Chen, Lin, Iuon-Chang, Hsu, Ping-Kun: A watermarking for html files based on multi-channel system. Int. J. Secur. Appl. **7**(3), 163–174 (2013)
3. Dey, S., Al-Qaheri, H., Sanyal, S.: Embedding secret data in html web page. arXiv preprint arXiv:1004.0459, pp. 474–481 (2009)
4. Huang, H., Sun, X., Li, Z., Sun, G.: Detection of hidden information in webpage. In: 2007 Fourth International Conference on Fuzzy Systems and Knowledge Discovery, FSKD 2007, vol. 4, pp. 317–321. IEEE (2007)
5. Huang, H., Zhong, S., Sun, X.: An algorithm of webpage information hiding based on attributes permutation. In: 2008 International Conference on Intelligent Information Hiding and Multimedia Signal Processing, IIHMSP 2008, pp. 257–260. IEEE (2008)
6. Lai, J.-X., Chou, Y.-C., Tseng, C.-C., Liao, H.-C.: A large payload webpage data embedding method using CSS attributes modification. In: Advances in Intelligent Information Hiding and Multimedia Signal Processing: Proceeding of the Twelfth International Conference on Intelligent Information Hiding and Multimedia Signal Processing, Kaohsiung, Taiwan, 21–23 November 2016, vol. 1, pp. 91–98. Springer (2017)
7. Lee, I.-S., Tsai, W.-H.: Secret communication through web pages using special space codes in html files. Int. J. Appl. Sci. Eng. **6**(2), 141–149 (2008)
8. Lee, I.-S., Tsai, W.-H., et al.: Security protection of software programs by information sharing and authentication techniques using invisible ASCII control code. IJ Netw. Secur. **10**(1), 1–10 (2010)
9. Sui, X.-G., Luo, H.: A new steganography method based on hypertext. In: 2004 Proceedings of the Asia-Pacific Radio Science Conference, pp. 181–184. IEEE (2004)
10. Tarillo, J.F., Mavrogiannakis, N., Lisboa, C.A., Argyrides, C., Carro, L.: Multiple bit error detection and correction in memory. In: 2010 Proceedings of the 13th Euromicro Conference on Digital System Design: Architectures, Methods and Tools (DSD), pp. 652–657. IEEE (2010)
11. Yang, Y., Yang, Y.: An efficient webpage information hiding method based on tag attributes. In: 2010 Seventh International Conference on Fuzzy Systems and Knowledge Discovery (FSKD), vol. 3, pp. 1181–1184. IEEE (2010)
12. Zhang, X., Wang, S.: Steganography using multiple-base notational system and human vision sensitivity. IEEE Signal Process. Lett. **12**(1), 67–70 (2005)

Integrated Health Check Report Analysis and Tracking Platform

Tzu-Chuen Lu(✉), Wei-Ying Li, Pin-Fan Chen, Run-Jing Ren, Yit-Ing Shi, HongQi Wang, and Pei-Ci Zhang

Department of Information Management, Chaoyang University of Technology, Taichung 41349, Taiwan
tclu@cyut.edu.tw

Abstract. In order to help individuals effectively manage and record these health check physiological measurement data, this research developed an "Integrated Health Check Report Analysis and Tracking Platform" together with H&B Health Centers. Using this platform, the public can query health check data and analysis charts from health check centers through a website. The information will include suggestions from doctors, nutritionists and representatives of various health check categories. The meanings behind various data can be explained to the general public using illustrations.

The "Integrated Health Check Report Analysis and Tracking Platform" can retain health check data from recent years and the system will provide sharing functions so that friends and family are also able to care for the patient. The system will also target various abnormal test data and provide nutritional recommendations in the specific category. This will prevent the patient from using the wrong medicine or remedies, resulting in more serious consequences. Using the health check data from the complete history of records from the platform, it can provide analysis and trend graphs on various data to allow laypersons to understand their own health conditions and trends in simple ways via graphics. In addition, using H&B Health Center's big data combined with personal lifestyle assessment and health check data over the years will allow automatic generation of routine health check recommendations and reminders for regular checks.

Keywords: Health check · Physiological measurement data · Web platform · Data analysis

1 Introduction

As the level of knowledge and health awareness of the general public rises, the common sense of health checks is becoming increasingly popular. Public awareness of disease is no longer just about diagnosis and treatment and the concept of health check to inform people of the possibility of disease in advance is becoming more and more ingrained in the public psyche. This perspective also conforms with the idea of health awareness and disease prevention advocated by the concept of preventative medicine. It is hoped that this will help in early detection and treatment of diseases. However, in the face of the

J.-S. Pan et al. (eds.), *Advances in Intelligent Information Hiding and Multimedia Signal Processing*, Smart Innovation, Systems and Technologies 81,
DOI 10.1007/978-3-319-63856-0_6

large number of health check items, the public is often confused and may not have a good understanding of the contents of the health check results.

When faced with a long and tedious health report, people usually prefer to have someone capable to explain the details to them and answer questions such as "According to the report, what should I be aware of?" and "Should I go back for further diagnosis or treatment?" Excellent health check centers should have expert staff to explain the results to patients after completion of health checks or arrange for further follow-up actions and reviews. However, the explanations given by expert staff may sometimes be too technical so that the general public may forget some of the technical terms, or it may be the case that there are too many health issues and the patient forgets about them. When patients receive a call, they may find it troublesome or may be afraid that the expert staff may not have enough time for them. Furthermore, the objective of health check is to manage an individual's health. Only by comparing past health records, can people avoid blindly following trends and doing unnecessary and meaningless health checks. Otherwise, even with a thick stack of medical data and clinical suggestions, it is meaningless if the importance of the results are not understood. However, test reports normally only provide a single test result and do not retain data from the past, and reports from health check centers are usually given in paper form which is easily damaged or lost. Making an electronic file requires inputting the data manually, which can be troublesome.

It would be very convenient if health check centers can provide online health check information querying service. However, most health check systems currently can only query the latest health check data for individuals. If the data are to be shared with friends and family or if children want to inquire about the health data of the elders, this cannot be done. Moreover, there is no historical data analysis to show whether the health conditions have improved or not, and there are no detailed explanations on the report. Furthermore, in the face of so many data, laypersons may not understand the meaning of such data. The study aims to develop the "Integrated Health Check Report Analysis and Tracking Platform" together with Bio-check H&B Health Centers. The public can use the platform to query health check data and analysis charts at Bio-check related test units. The system includes recommendations from doctors and nutritionists and representatives of various check item categories. Illustrations are used so that the public can understand the significance of various data. Furthermore, the system also has data sharing functions and can provide health check data to friends and family members authorized by the user. All data will be subject to confidentiality measures to prevent disclosure of confidential information.

2 Related Works

Currently, there are many websites and apps developed for medical uses, such as the "Taiwan Medical Travel" app which provides convenient real-time medical tourism information. It displays the most representative tourist attractions in each area, as well as information on medical institutions. "Medical Wizard" was developed by the government of New Taipei City. It integrates the latest news, medical institutions, registration information, health check information, vaccination, New Taipei City online medical services, disease references, and other services and information. The China Medical University Hospital has a mobile registration system which includes an introduction of the hospital, online registration, registration query, visit progress query, directions, a "which department should I visit" function, and other functions. The "Show Chwan Health Care System Online Registration" can query Show Chwan Health Care System hospitals' for real-time outpatient schedules, outpatient information, hospital contact information, instant registration for returning patients, and health information.

Some apps can be connected wirelessly to medical instruments such as blood pressure and blood glucose monitors, and provide recommendations based on these data. For example, "Instant Heart Rate" can detect the user's heart rate just by placing a finger in front of the camera, "Tactio Health" can measure blood pressure and record the data to be used as reference by doctors, and "Glucose Buddy" can measure and help users control blood glucose levels. There is also an app that can detect melanoma (a type of skin cancer) for users and search for a suitable doctor called "MelApp".

This study assists partner units to establish an "Integrated Health Check Report Analysis and Tracking Platform" to allow users to query their own health check data in recent years. The platform will also provide clinical recommendations for various health check items and recommendations by doctors and nutritionists. The public can check their own health check data and health trends on the platform at any time. The health check data can also be shared with certain family members and friends so that they can care for the patients' health. The platform can also provide analysis and recommendations on abnormal test data. The platform pays attention to individual privacy; hence it is only for subjects who agree to upload their health check data and accept health management.

Main functions of the system include:

1. Importing health check data intermediate file from the HBIS system onto the "Integrated Health Check Report Analysis and Tracking Platform" regularly.
2. Health check data include all physiological measurements. These measurements can help in understanding the user's current health status. In particular, data accumulated over a long period can be used effectively by users or doctors as reference.
3. Data graph creation: instantly displays and analyzes data graphs to let the user understand his/her own health conditions. When the user needs to see the changes in physiological measurements over time, he/she can choose the data for a certain time period and display them in a line graph. This is a fast, convenient, and effective way to analyze the changes in the values and assist in long term tracking.
4. Provide analysis and suggestions for various abnormal test values.
5. Share data with authorized friends and family members.
6. Provide various nutritional suggestions and a follow-up mechanism using analysis of H&B Health Center's big data combined with personal lifestyle assessment and health check data over the years to automatically generate routine health check recommendations and reminders for regular checks.

3 Integrated Health Check Report Analysis and Tracking Platform

H&B Health Centers started off as a professional testing company which utilizes a business management model to cooperate with hospitals to run a testing and health check business. Its mission is to introduce advanced professional knowledge and skills so that people in Taiwan can enjoy world-class professional service. Collaborating hospitals have expanded from a few in the northern regions to the southern regions that now cover the entire country. Collaborating hospitals include Taipei Medical University Hospital, Keelung Hospital, Taichung Hospital, Feng-Yuan Hospital, Ministry of Health and Welfare Hospital, Changhua, Nantou Hospital, Tsaotum Psychiatric Centre, Kaohsiung Municipal United Hospital, Shuang Ho Hospital, Wan Fang Hospital, Puli Christian Hospital, Saint Mary's Hospital Luodong, Tainan Sin Lau hospital, and others.

In order to effectively manage the organizational structures for so many hospitals, we have created six different user levels for the "Integrated Health Check Report Analysis and Tracking Platform": Chief Supervisor, manager, IT staff, nurse, patient and relatives. The organizational structure chart is shown below (Fig. 1).

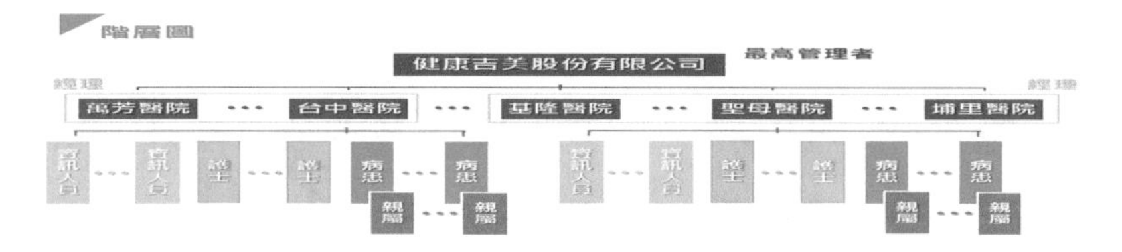

Fig. 1. H&B Health Center organizational structure chart

The main function modules in the platform are shown in the figure below. The platform will import health check data from the HIS systems from various hospitals onto the platform's servers. This is done automatically and regularly or initiated manually. Members can use the platform to view their own data and share them with friends and family. They can also use the online consultation function to talk to nurses (Fig. 2).

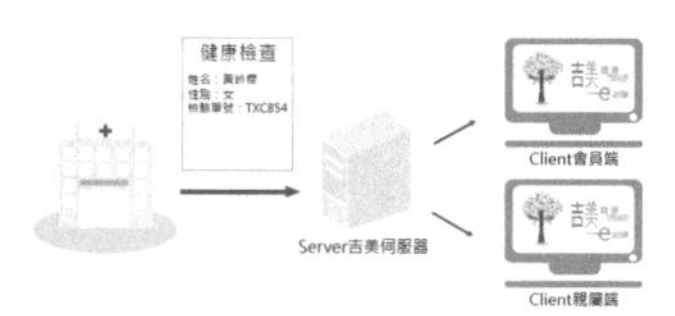

Fig. 2. Platform structure diagram

In order to implement such a structure, the platform is divided into front-end and rear-end sections.

3.1 Front-End Functions

Using the "Health Check Report History" function from the front-end, patients can view the histories of their health check data. If they had a health check in a certain year, the timeline will show the corresponding data. Clicking on it will display the health check report summary table. If there are abnormal values in the report, the system will automatically detect them and display the category of the abnormal values in red letters on the right side of the screen.

When the user clicks on "Sub-Item Report," a detailed report will be displayed. Selecting "Measurement Record" will display the history of measurement data and a trend graph (Figs. 3 and 4).

Fig. 3. Sub-item report

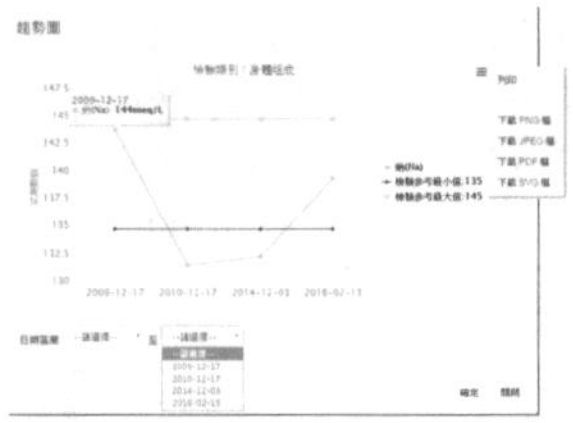

Fig. 4. Measurement record trend graph

Selecting the "View Historical Status" button will list out the test values from the previous three years. Here, the time where the abnormal test value began appearing can be seen (Figs. 5 and 6).

Fig. 5. Health check report history

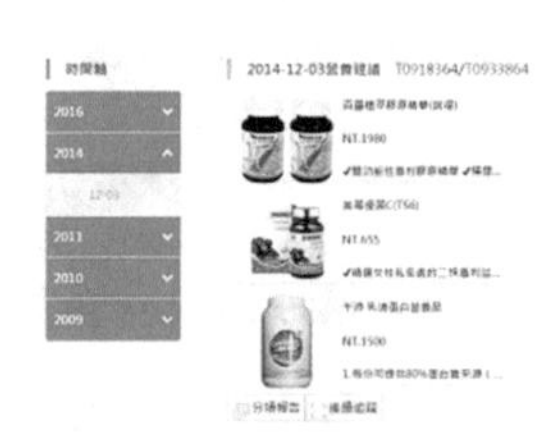

Fig. 6. Nutritional supplement recommendations

Clicking "Follow-Up Tracking" will remind the patient to go to a certain category for follow-up tracking and viewing. "Nutritional Suggestion" will display doctor recommended nutritional supplements.

It will also provide a message counseling service where the user can communicate with a customer service representative online (Figs. 7 and 8).

Fig. 7. Nutritional supplement recommendation online counseling

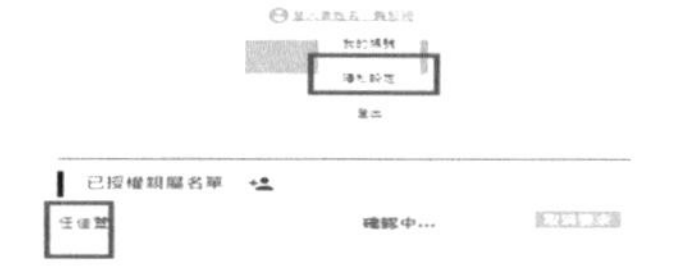

Fig. 8. Privacy settings

If the user wants to share the report with friends and family, he/she can go to the "Privacy" settings and add friends and family member's account numbers.

3.2 Rear-End Functions

"Item Management" is mainly used to set up all health check items for H&B Health Centers. Item types are categorized as numerical, text, special value, and different other statuses. The corresponding maximum value, minimum value, gender, etc., for each type of data are different and this system provides definitions for different types of data. The figure below shows the test item data type settings (Fig. 9).

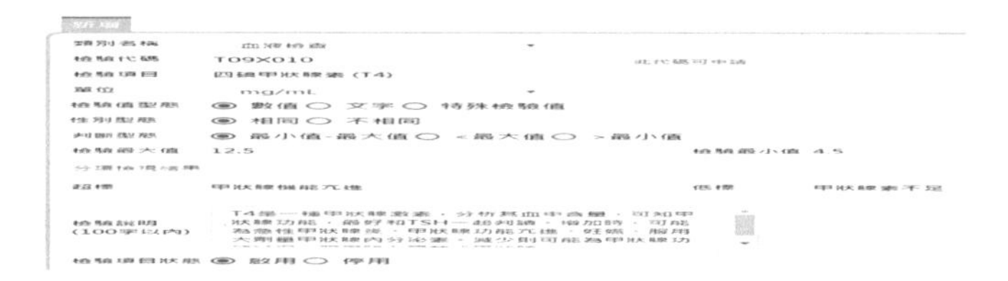

Fig. 9. Item data type settings

When the patient's test value is higher than the maximum value or lower than the minimum value, the system will automatically display this exception and recommend health materials specified for the item and transfer the user to the follow-up tracking category.

The "Medical Records Management" function imports health check data from the HIS system to the platform's servers. There are two methods for importing this data. The first is directly exporting them from the HIS system, then importing the Excel file into the server. The other method is to use the scheduling system. The system can be set to automatically import the exported data (Fig. 10).

medical_list_num	hospital_id	lis_update	inspect_no	medical_id	medical_name	prescription_id	hospitalized_id	prescription_depart	prescription	status	issuingpractitioner
1	台中	HIS批價為0	T0918711	00917144	龔郁琦	09812031683	09812030039	門診	內科	全民健保	林毓宏
2	台中	HIS批價為0	T0918711	00917144	龔郁琦	09812031683	09812030039	門診	內科	全民健保	林毓宏
3	台中	HIS批價為0	T0918711	00917144	龔郁琦	09812031683	09812030039	門診	內科	全民健保	林毓宏
4	台中	HIS批價為0	T0918711	00917144	龔郁琦	09812031683	09812030039	門診	內科	全民健保	林毓宏
5	台中	HIS批價為0	T0918711	00917144	龔郁琦	09812031683	09812030039	門診	內科	全民健保	林毓宏
6	台中	HIS批價為0	T0918711	00917144	龔郁琦	09812031683	09812030039	門診	內科	全民健保	林毓宏

Fig. 10. Imported data

"Abnormal Medical History Management": nurses can manage the patient's abnormal medical history using the abnormal data. The system will list out records with abnormal test data and nurses can edit or add comments to the medical records. The methods for follow-up tracking, and the category and frequency of tracking can be set for the test item (Figs. 11 and 12).

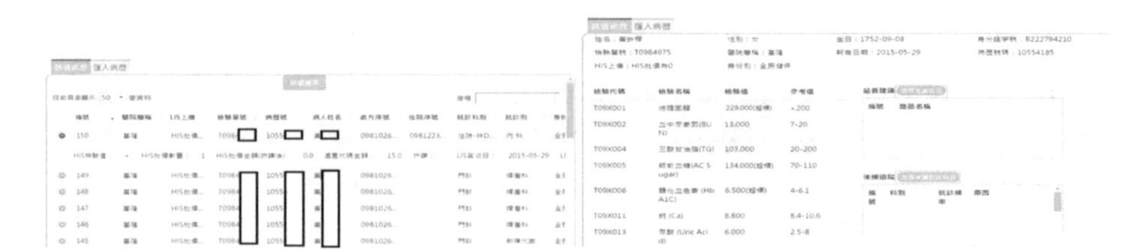

Fig. 11. Abnormal medical history list

Fig. 12. Abnormal medical history management

Alternatively, a particular diet can be recommended when the data of the health check item are abnormal (Figs. 13 and 14).

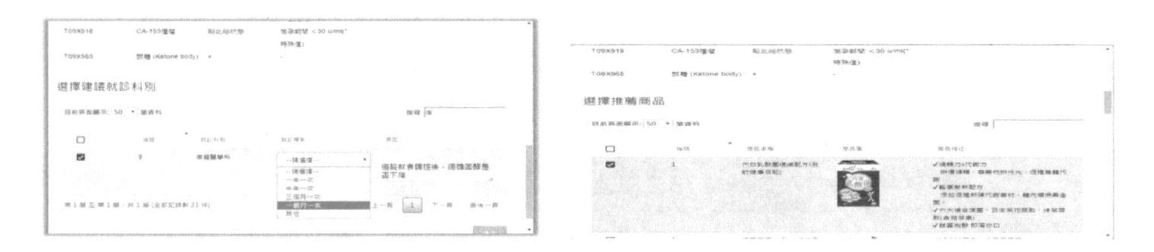

Fig. 13. Category and frequency of follow-up tracking recommendations

Fig. 14. Nutritional product recommendations

Furthermore, general comments can be made regarding the entire health check report and doctors can write evaluations (Fig. 15).

Fig. 15. Doctor's comments

4 Conclusion

The main objective of this study is to help H&B Medical Centers to establish a health check report management system to allow users or managers to use the web platform to view and manage the system via the internet, thus saving paper and time while increasing convenience. For the future, it is hope that the system can be used to integrate health reporting by hospitals all over Taiwan and not just only by hospitals collaborating with H&B Health Centers. This will allow health check reports to be obtained and viewed more conveniently and health reporting digitized.

The main contributions of this study are as follows:

1. Contributions to the national information industry: the web platform to provide various health check data analysis and trend graphs for patients and allows them to learn about their own health conditions at any time. The health check unit can provide the public with a variety of recommendations and health consultations. This study combines information technology with professional medical knowledge, which has made great contributions to interdisciplinary cooperation and Taiwan's information technology industry.
2. Contributions to the national manufacturing industry: The system will inform patients of clinical recommendations for various data. Health check staff can also use this system to perform follow-up tracking or issue warnings to let patients understand their risks in advance so they can control their eating habits and reduce risk factors which in turn reduce the strains on medical resources. The health check analysis results can also be used as very practical references for researchers conducting follow-up studies and clinical testing.
3. Contributions to national academia: The practical application of the system established in this study by the medical industry successfully took information technology from the academia and applied it in the industry, which made positive contributions to the value of academia.
4. Benefits for participants: All of the participating research assistants learned how to devise websites and data collection methods and learned how to take theoretical technology and apply it to a practical system. IT staff also gained medical knowledge via this study. The most important benefit is that different participants can provide their own insights regarding the same problems, and participants appreciated the basic spirit of academic research. In addition, all participants have taken this opportunity to learn the importance of teamwork and cultivated themselves to become information research and development talents with specialized knowledge.

Acknowledgements. This study was supported by a Research Grant, MOST, from Taiwan's Ministry of Science and Technology (MOST 105-2622-E-324-001 -CC3).

References

1. Tseng, C.Y.: Hemodialysis key features mining and patients clustering technologies. Thesis, Chaoyang University of Technology, Taichung, Taiwan, R.O.C. (2010)
2. Science and the Death - You Do Not Know the Health Check. http://140.123.13.96/retired/health_knowledge/health_check.pdf
3. Wu, M.S., Huang, C.Y.: Five conceptions of health. Good Heart J. **60**, 17–20 (2012)
4. Why Do Health Checks? http://www.yumin.com.tw/HMC/ha/index.php
5. The Significance of a Health Check. http://www.tpech.gov.taipei/ct.asp?xItem=1112335&CtNode=15449&mp=109151

6. Wu, C.M.: Kidney disease care program, effectively reducing the incidence of dialysis. Central Health Insurance Board, vol. 178. http://www.nhi.gov.tw/epaper/ItemDetail.aspx?DataID=2438&IsWebData=0&ItemTypeID=5&PapersID=205&PicID=,2011
7. Lin, J.L.: How to detect renal function? Green Health Network (2011). http://www.greencross.org.tw/kidney/symptom_sign/kid_func.html
8. Hu, C.F.: Treatment options peritoneal dialysis is more convenient, why 90% patients choice hemodialysis. Joint News Network Health Medicine (2009) http://mag.udn.com/mag/life/storypage.jsp?f_ART_ID=180611
9. Chen, C.C.: Know kidney disease to win kidney life. Tri-Service General Hospital, vol. 94 (2011). http://tsgh-ejournal.ndmctsgh.edu.tw/PaperData.aspx?ID=947&Category=41
10. You, J.X.: Alternative therapy for chronic kidney disease. Wonderful Broad Bean Information (2006) http://www.capd.org.tw/main1_1_4.htm

A Study of the Multi-Organization Integrated Electronic Attendance System

Xi-Qing Liang, Wei-ying Li, and Tzu-Chuen Lu(✉)

Department of Information Management, Chaoyang University of Science and Technology, No. 168, Jifeng East Road, Wufeng Township, Taichung 41349, Taiwan
tclu@cyut.edu.tw

Abstract. Since 1997, internet-based e-government has been promoted by the Research and Development Review Board of Executive Yuan, R.O.C., and the application of electronic forms is one of the e-government processes. Electronic form system includes integrated document management and files management systems and is integrated with an electronic attendance system as well. In this way, other than merely achieving creating a paperless system, it also reduces the manpower required for paper-form transmission and shortens document delivery time.

This study took Tainan city government and its subordinate organization as the study objects. Owing to the merging of Tainan County and Tainan City a couple of years ago, their old electronic attendance systems were redesigned and reconstructed into a single system. Through this example, this study investigated how to design an electronic attendance system which could integrate various business characteristics.

Keywords: Electronic attendance system · Electronic forms · Tainan municipal government

1 Introduction

Electronic form application refers to the digitalization of document management systems, file management systems and electronic attendance systems in governmental organizations during the process of office automation. Besides a drastic reduction of printed paper forms, it also shortens the delivery time of paper-form signature and for approval; hence, it enhances overall administrative efficiency. The electronic attendance system is used by the personnel units of the government to carry out all kinds of personnel absence, attendance, travel, and leave management. Employees connect to the electronic attendance system via the Internet, and apply for leave using various travel and leave forms online. Through the integration of physical sign-in/sign-out equipment, the attendance status of employees is completely presented. Through the electronic attendance system, unit supervisors can sign and approve employees' leave slips online. The personnel management unit can manage employees' nonattendance and leave through the system, and provide the basis for governmental supervisors' personnel management decisions.

J.-S. Pan et al. (eds.), *Advances in Intelligent Information Hiding and Multimedia Signal Processing*, Smart Innovation, Systems and Technologies 81,
DOI 10.1007/978-3-319-63856-0_7

Tainan County and Tainan City were merged in 2010. After the merger, the number of staff posts were increased, and the workplaces were separated in Yonghua and Minchi administrative centers. These increased the difficulty for staff attendance/absence management for the personnel unit. Originally, Tainan County and Tainan City had their own attendance management requirements and electronic attendance systems. After the merger of county and city, the government was forced to combine the electronic attendance systems originally independently operated. This included the conversion of different organizational personnel and travel/leave data in the original electronic attendance systems of both sides; the formulation of a revised organization attendance management system after the merger (such as the time of nonattendance, reasons for leave, and the leave slip signature/approval in each unit). In addition, after the upgrade of the system, the personnel management authority of primary and secondary governmental organizations were changed to self-manage. The institutional changes, the planning and development of electronic attendance system and other issues, had actually given the Tainan Municipal Government Personnel Department considerable challenges after the merger.

Eventually, the Tainan Municipal Government Personnel Department decided to abandon the original electronic attendance system. To facilitate management, an electronic attendance system which encompasses and satisfies the needs of all the governmental organizations in Tainan city was planned. This study analyzes and classifies the official business characteristics of Tainan city government and its subordinate organizations, and proposes corresponding system functions for various business characteristics to ensure that the development and implementation of the electronic attendance system can be fully applied to Tainan municipal government and its subordinate organizations.

2 Literature Review

Prior to 2005, staff absence and attendance control of Tainan Municipal Government was done via paper-based sign-in/sign-out books. Staff of an organization or a unit would go to a fixed place to sign in and sign out. Management of applications for travel and leaves was done through Leave Cards specially assigned to an employee. When an employee intends to go on vacation or to travel of any kind, they have to fill in proper reasons for the type of travel or leave on their individual Leave Card, which was circulated manually to the heads of all levels for paper-based leave slip signatures and approval. Besides the fact that attendance management could not be carried out instantly, the personnel staff also had to spend a lot of time every day to deal with routine attendance errands and paper-based file management. Furthermore, at the end of the year, when it was desirable to conduct annual performance appraisal for staff, the Leave Cards as well as the paper-based sign-in/sign-out books of all colleagues in the specific year must be retrieved for manual and meticulous review. The relevant nonattendance and leave statistics were eventually provided to unit supervisors for assessment. Not only were they prone to manual statistical errors, but also consumed a high cost of manpower resources.

By the end of 2005, in response to the e-government movement, IT vendors commissioned by Tainan Municipal Government imported a new electronic attendance system to Tainan. The original architecture of the system is shown in Fig. 1. All employees sign in to work and sign out to leave using card readers [5]. Browsers linking to the electronic attendance system was used for employees to apply for all kinds of travel/leave applications and expense reimbursement online. All auditing of the travel/leave statistics and nonattendance/attendance discrepancies were controlled by the Electronic Attendance System. All the nonattendance/attendance travel/leave, and overtime records of individuals could be queried online. All the organization supervisors and personnel managers also conducted online leave-slip sign-offs through the Electronic Attendance System, and managed all nonattendance/attendance, travel/leave and overtime records for employees.

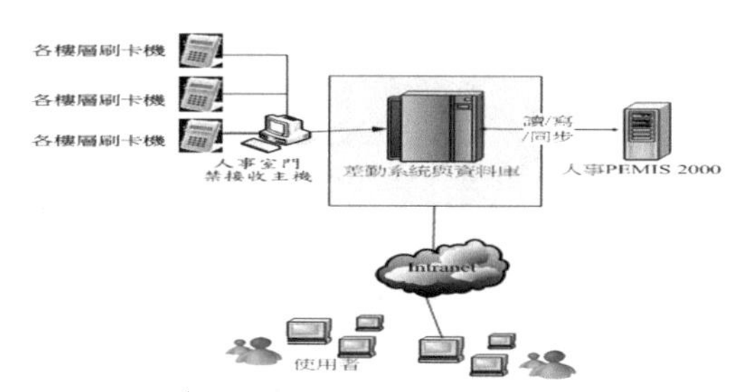

Fig. 1. The system architecture diagram of the original electronic attendance of Tainan city

At the end of 2010, in order to reduce the time and labor cost required to develop the electronic attendance systems for organizations nationwide, the General Office of Personnel Administration of the Executive Yuan developed a nationwide shared version of an attendance electronic form system for internal organizational use called the WebITR system [1], and provided transferal services to the governmental organizations. The WebITR system is divided into six main functions, namely, the attendance operation, expense operation, personal information, signing/approval notice, shift work and on-duty work. The system function content is depicted as follows:

(1) The attendance operation includes general leave, group leave, overtime sheet, group overtime, business trip sheet, group business trips, business leave sheet, group business leave, leave data, overtime data, attendance data and diligence/laziness statistics.
(2) The expense operation includes personal overtime fee printout request, departmental overtime fee printout request, business trip expenditure request, business holiday expenditure request, on-duty fee request and overtime in holiday fees.
(3) Personal information includes basic information and duty agents.
(4) Signing/approval notices cover business trips, pending approval, on behalf approval, completed, approved, approved on behalf, and form inquiries.
(5) The shift work includes shift rostering, shift transfer application, shift overtime, shift scheduling record inquiries and shift attendance.
(6) On-duty work includes on-duty rostering, on-duty compensatory leave maintenance, on-duty fee request, on-duty fee printing, and on-duty attendance.

3 The Electronic Attendance System of Tainan Municipal Government

After the merger of Tainan County and Tainan City, all the organizations of the city are now in accordance with the organizational structure of Direct Municipalities [2], as shown in Table 1.

Table 1. Organizational structure of Tainan city government

Unit classification	Name of organization
Primary unit	Secretariat, Office of Legal Affairs, Office of Information and International Relations, Commission on Ethnic Affairs, Research and Development Assessment Committee, Personnel Office, Office of the Comptroller, and Ethnics Office
Primary organization	Bureau of Civil Affairs, Education Bureau, Agriculture Bureau, Economic Development Bureau, Tourism Bureau, Works Bureau, Water Conservancy Bureau, Social Bureau, Labor Bureau, Lands Bureau, Urban Development Bureau, Bureau of Cultural Affairs, Department of Transportation, Health Bureau, Environmental Protection Bureau, Police Department, Fire Services Department, and Finance & Taxation Bureau
Affiliated organization	District office, health office, household office, and land office

In this study, the attendance characteristics of all organizations are divided into seven categories. To meet the attendance management characteristics of all the organizations, the electronic attendance system shall provide different system corresponding functions during the design phase targeting the seven aforementioned kinds of attendance characteristics. The various characteristics are analyzed as follows:

(1) Characteristics of general public affair departments
   (a) The working hours are calculated as 8 h daily. The core working time is defined by each unit respectively. The actual working dates of all organizations are based on the calendar table for administrative organizations issued by the Personnel Administration General Office.
   (b) 20 h of paid overtime per month. Project overtime can be up to 70 h. There is no ceiling for compensational resting hours. For staff covered by labor law, the overtime limit is 46 h per month.
   (c) The rules for business trips and leaves depend on the status of the staff and the leave rules for civil servants, leave rules for staff of Executive Yuan and its affiliated organizations, and labor laws.

The basic system functions [3] are shown in Table 2.

(2) Characteristics of the Fire Services Department
   (a) The Fire Services Department and its subordinate units divide the staff working hours into two types: office-work and field-work. The office staff are on a two-day off per week basis. As for field staff, to meet service table requirements, there are the “1-day work, 1-day rest system” and the “2-day work, 1-day rest

system" [4]. For the "1-day work, 1-day rest system", service hours are from 8 am until 8 am the next day. The service time is 24 h consecutively, then off for 24 h. A leave of 24 h is counted as one day. For the "2-day work, 1-day rest system", its service hours are from 8 am until 8 am the next day. The service time is 48 h consecutively, then off for 24 h. A leave of 48 h is counted as one day. Firefighters who perform the service according to the service allocation table and actually perform the field service may receive overtime pay. The number of overtime hours are calculated according to the service rotational on-duty table, with a ceiling of 8 h per day and a ceiling of 100 h per month.

(b) Those whose vacation was cancelled owing to natural disasters or special reasons can arrange for compensatory off time (in corresponding hours or days) later.

For staff with these characteristics, the electronic attendance system shall be able to provide service center scheduling function, field staff scheduling function, overtime pay application, overtime pay reimbursement function, and so forth.

(3) Characteristics of the Environmental Protection Bureau

(a) Working days for the cleaning team members shall be in compliance with the "1-day work, 1-day rest system". The working hours are managed and scheduled by the cadres. Hence, the daily commute time may not be exactly the same.

(b) The system shall be able to provide a night snack fee, cleaning bonus, and driving safety bonus for the cleaning team members. The night snack fees are paid to team members working from 21:00 pm until 06:00 the next day according to the specific day in the on-duty table. But those who apply for overtime in this period cannot receive the night snack fee. The cleaning bonus is NT$8,000 per month. If a team member has been absent in the specific month, a portion of the total sum is deducted according to the number of days absent. Driving safety bonus is calculated according to scheduled cleaning car driving days of the team member in the specific month. For staff with these characteristics, the electronic attendance system must be able to provide management and reimbursement of night snack fees, cleaning bonus and driving safety bonus.

(4) Police characteristics

The service system of police stations is divided into two categories, back-office staff and field staff. The back-office staffs are similar to common civil servants as the "five-day workweek system" is applicable to them. The policemen in the service directing center are subject to a "1-day work, 2-day rest system". The policemen in the control center and the field staff are subject to the "1-day work, 1-day rest system". The service hours of the so called "1-day work, 1-day rest system" are from 8:00 am to 8:00 am the next day. The service hours are 24 consecutive hours; afterwards, they are allowed to rest for 24 h. For the staff, a leave of 3 h is converted into a 1-hour leave formally. The service hours of the so called "1-day work, 2-day rest system" are also from 8:00 am to 8:00 am the next day. The service hours are 24 consecutive hours; afterwards, they are allowed to rest for 48 h. Although the service hours for the staff subject to this system is also 24 h, a leave of 24 h must be counted as 48 h formally [6]. Field staff may request for overtime fees of up to 100 overtime hours per month. However, as an alternative, 52 accumulated overtime hours can be converted to apply for rewards.

(5) Characteristics of the Health Bureau
   (a) Staff from the Health Bureau are often required to travel on business trips at night and to work overtime at the night. It is therefore necessary to provide travel expenses for overnight travel, and to allow staff to apply for overtime fees or overtime compensatory rest hours.
   (b) There are numerous project reimbursements available. For overtime or travel expense reimbursement, it is necessary to provide budget control and a summary mechanism for the units to ensure that the overspending of the budget account will not happen. For staff of with characteristics, the electronic attendance system must be able to provide a summary of travel expenses, a summary of overtime payments, and facilitate budget account management.

(6) The characteristics of the Bureau of Cultural Affairs
The staff of the Cultural Center take turns to work on holiday days. On Mondays, besides formal civil servants, the rest of the staff are on holiday. As to daily working hours, they are on for a 4-hour, 8-hour and 12-hour basis. In cooperating with performance units, staff may have to work overtime overnight and work overtime until the next morning. The system should be able to account for overnight overtime work and overnight nonattendance/attendance exceptions.

(7) The characteristics of the Education Bureau
   (a) The civil servants in schools use the calendar year system. Teachers who also work as part time administrators and the general teachers use the school year system. The calculation period of leave days and holidays is from August 1 to July 31 of the following year.
   (b) Teachers applying for leave must be assessed in accordance with the curriculum. When courses have been arranged during the period of absence, prior to the leave application, the teacher is required to properly complete the online class transfer tasks.
   (c) Teachers are divided into two categories, teachers with who work in administration part time and general teachers. Teachers who work in administration part time are required to work under the sign-in/sign-out management; their leave days are calculated according to seniority. The general teachers do self-management in accordance with class time.
   (d) During winter or summer vacations, flexible working measures are implemented in schools. Furthermore, the staff or teachers who work in administration part time have the option to choose vacations in summer or winter time. The number of days are in accordance with school regulations. For staff with these characteristics, the electronic attendance system must be able to provide a platform for applying for summer and winter vacations, class transfer notes and a scheduling & transferring class system for teachers.

Table 2. Functions of electronic attendance system

Function module	System functions
Travel/leave application	Business leave form, business trip form, leave application form, overtime application form, forgot-to-punch-card application form, go-abroad application form
Expenditure application	Application for business trip expenses, application for overtime pay, application for education allowance for children, application for overtime pay without leave, application for marriage, funeral, or maternity allowance
Diligence/laziness query	Attendance record query, overtime record query, travel/leave record query
Diligence/laziness management	Daily attendance, exception record management, attendance card record query, and Taiwan traveler card management
Basic setting	Overtime data setting, holiday data setting, job title data setting, class data setting, job grade data setting, leave data setting, and personnel data management
On-duty management	On-duty classes, on-duty setting, and on-duty expense reimbursement
Scheduling management	Scheduling settings and scheduling data query
Process operation	To be signed/approved file box, to be reviewed file box, signed/approved file box, and return file box

After analyzing the attendance characteristics for different organizations, the planning of the Tainan municipal electronic attendance system is finally divided into 9 modules, namely, leave application form, application of various expenses, inquiry of basic diligence/laziness, management of diligence/laziness in personnel affairs, system basic setting, approval process, scheduling system, scheduling & transferring class system and on-duty system. Furthermore, via the integration of RFID card readers and other heterogeneous systems, the whole system becomes more comprehensive.

(1) Combining with RFID card readers: While introducing the electronic attendance system, the Tainan municipal government also introduced the RFID card readers. When a staff signs in or signs out via the RFID card reader, the reader screen will show the name of the card holder and his sign in/out time. Meanwhile, the name is also displayed at the top of the LED display.

(2) Combining with heterogeneous systems: In the system design phase, the convenience for staff and system integrity were put into consideration. Hence, in the design and planning phase of the electronic attendance system, the approval process system, the business entrance network and the document system were already integrated [7].

  (a) Combining with the portal: When a staff inputs his account number and password and logs into the portal, the system home page will show all the authorized applications, the staff's signed and approved attendance documents and his sign-in/sign-out status for the day. When he selects an item in the electronic attendance system, the system will sign in the electronic attendance system right away. It is not necessary to key in the account number and password once more.

(b) Combining with the approval process system: When a staff completes a leave slip application and sends it out, the system will call the approval process system. According to the organization level, the approval process system will send the leave slip to-be-approved information to the to-be-signed documents of the authorized supervisors. The supervisors will then check and sign the leave slip online. The system completely records the approval time, names, and signing comments of all parties.

(c) Integrated document system: After applying for leave slip on the electronic attendance system, the document system will obtain the duty information for the current staff through the Web Service provided by the electronic attendance system and will automatically transfer the staff's authority in signing documents to the duty agent.

4 Conclusions

Most of the current organizational electronic attendance systems are developed for a single organization's attendance requirements. When the system is imported to organizations with different attendance characteristics, it is often necessary to make significant modifications or even to redesign it for customized use. This will result in high development costs, and the maintenance of subsequent versions is a heavy burden for the company.

The electronic attendance system proposed in this study was analyzed, designed and implemented according to the Tainan Municipal Government and its subordinate organizations. From December 2010, the system was introduced into the Tainan Municipal Government and its subordinate organizations in four stages. At present, all the organizations have been successfully implemented online. This electronic attendance system with multi-organizational characteristics has been proved to possess integrity, practicality and reliability. The major errands of the public affairs personnel include personnel information, attendance, performance appraisal, rewards and punishment, education and training, personnel promotion, salary issuance and retirement pension management. This study only the implemented the personnel data management and electronic attendance management. In the future, the relevant fields of personnel information systems will be studied and we will continue to develop the yet to be implemented personnel business system, so as to attain a comprehensive information or overall personnel management.

Acknowledgements. This study was supported by a Research Grant, MOST, from Taiwan's Ministry of Science and Technology (MOST 105-2221-E-324-020-).

References

1. Xuezhen, L.: Introduction and Popularization of WebITR System. Personnel Monthly 320 (2012)
2. Global Information Network of Tainan City Government. http://www.tainan.gov.tw/tainan/department_list.asp
3. Zhijun, G.: The Study in the Promotion of the Attendance Electronic Form System of the Taipei Municipal Government Fire Department - Viewpoints Integrating Technology Acceptance and Use with Theories, Master's Thesis of the Eleventh Master's Course of Administration, School of Social Science, National Chengchi University, Taipei (2011)
4. Huiru, C.: Using RFID to Build Access Control and Attendance Management System - Taking the Central Area Vocational Training Center as an Example, Master's Thesis, Master's Degree Program in Information Engineering, Asian University, Taichung (2010)
5. Minhua, W.: The Study of Users' Satisfaction in Electronic Attendance System - Taking the Kaohsiung Municipal Government Police Station as an Example, Master's Thesis, Information Management Department, Yoshimori University, Kaohsiung (2005)
6. Jianming, G.: Integration and Analysis in Using Software Development Mode (SCRUM) to Implement Heterogeneous System: Taking Attendance Integration Process as an Example, Master's Thesis, Information Engineering Institute, National Chung Cheng University, Chiayi County (2017)

A Content Analysis of Mobile Learning on Constructivism Theory

Ling-Hsiu Chen[1(✉)], I-Hsueh Chen[1], Po-Hsuan Chiu[1], and Hsueh-Hsun Huang[2(✉)]

[1] Department of Management Information System, Chaoyang University of Technology, 168, Jifeng East Road, Wufeng District, Taichung 41349, Taiwan
ling@cyut.edu.tw

[2] PhD Program in Strategic Development of Taiwan's Industry, Chaoyang University of Technology, 168, Jifeng East Road, Wufeng District, Taichung 41349, Taiwan
kentintw@gmail.com

Abstract. The development of network and mobile technologies contributed to the emergence of mobile and ubiquitous learning. Therefore mobile and ubiquitous learning are attracting both academic and public interest in the recent year. Owing to mobile learning is defined as "the processes of coming to know through conversations across multiple contexts among people and personal interactive technologies", this study conducts a content analysis to view the relative researches that apply this technology to facilitate various activities of learning. Those activations of learning can be designed in formal and informal and personalized learning environments during classroom lectures or outside of the classroom. The advantage of portable technologies can help learners to connect various learning activities both in formal and informal personalized learning environments.

Generally, previous research in mobile learning are focus on the lower or younger learners, the broad application of mobile and ubiquitous learning in higher education settings is limited. Further the mobile learning design can be design and allow learner selects and transforms information, constructs hypotheses, and go beyond the information given. Therefore this research try to analyze the mobile learning research that are focus on higher education setting and based on constructivism theory.

Keywords: Mobile learning · Ubiquitous learning · Content analysis · Constructivism theory

1 Introduction

The development of network and mobile technologies contributed to the emergence of mobile Learning (ML). ML aims to provide technological supports for collaborative learning and can be defined as an interdisciplinary research field that includes a branch of the learning sciences and educational technology research concerned with studying how people can learn together with the help of computer [9]. Mobile learning is more and more popular in education setting for it is convenience, expediency, immediacy

J.-S. Pan et al. (eds.), *Advances in Intelligent Information Hiding and Multimedia Signal Processing*, Smart Innovation, Systems and Technologies 81,
DOI 10.1007/978-3-319-63856-0_8

and supervising teachers' opinions. First, it is convenience, trainees had opportunity to use their time even when they were on the move. Second, from the expediency point of view trainees they had to work at home and there they did not have access to Internet other than via the phone, just like they used phones in shops when they had to check if there was particular foodstuff at school. Third, trainee can get a great idea on the way to university in the very early morning and was happy to be able to share this idea with colleagues straightaway. Mobile devices were a great tool to give feedback while observing other trainees. Finally, during training the mobile device helped supervising teachers a lot. The development of network and mobile technologies contributed to the emergence of mobile Learning (ML). One of the most widely accepted definitions of mobile learning is "the processes of coming to know through conversations across multiple contexts among people and personal interactive technologies" [8]. Mobile and portable technologies are conceived either as tools that allow learners to access information irrespective of their physical context, for example on a bus [1] or, as a way to provide learners with location-based information, for example while they are exploring a butterfly garden [5].

Mobile learning's ubiquitous characteristic has been considered to be one of the most promising applications of modern information and communication technology toward the improvement of teaching and learning. While, the mobile learning allow learner selects and transforms information, constructs hypotheses, and go beyond the information, that is the mobile learning can improve learners' knowledge by designing based on the aspect of specific teaching/learning theory (ex. constructionist learning). Thus, the main purpose of conducting mobile learning is to improve learners' learning performance therefore the significant importance of theoretical aspects in mobile learning, the aim of this study is to conduct a literature review examining the research and conceptual aspects of mobile learning.

2 Search Strategies and Processes

This study only analyzed papers that were identified as "articles" in the SSCI. Other types of papers such as "book reviews," "reviews," and "editorial materials" were all excluded from this study. For paper selection, the search terms combined the concepts of mobile and ubiquitous learning and higher education in the meta field "topic", which included the search in article titles, abstracts, author keywords, keywords and plus fields. More precisely, the terms mobile learning, m-learning and ubiquitous learning were combined (AND) with higher education or university. In addition, the database ERIC was searched in February 2015 because this source contains a broad selection of articles specific to the fields of education and learning sciences. Finally this study scans of Google scholar using the same key terms.

Based on these articles, a content analysis was carried out, using article abstracts and publication information indexed in the SSCI. The selected articles was processed by two doctoral researchers in educational technology and further validated by a professor in the field. For the remaining publications, the full texts were retrieved and reviewed against the following six criteria: (1) sound methodological design; (2) higher education setting; (3) involvement of mobile technology; (4) educational orientation;

(5) primary study designs based on mobile learning activities. This procedure identified 20 articles that were judged to be relevant to the topic of mobile learning. The analysis of 20 articles' abstract and outline, 20 articles were picked out as applied construction theory. In the final phase, after careful complete article analysis, 5 articles were identified as our study objects.

3 Results

Among the articles, they conducted both quantitative and qualitative data, and most of them used questionnaires as their primary data collection techniques or utilized learners' log files or discussions as the data sources for analysis. In addition, 2 of them developed learning application and using multiple platforms. Several educational researchers have argued about the commonalities between quantitative and qualitative research, and have advocated mixed method research as a new research paradigm.

Table 1. The results of research on mobile learning

Author (year)	Aims/Objectives of study	Methods	Results and conclusions
Seppälä et al. [7]	To discover the opinion of using mobile device for teaching and learning	Device: Communicator, digital cameras Participants and research setting: 11 students (ages between 20 and 25), and 5 teachers Data collection: discussions Data analysis: discourse analysis	1. Students have positive responses on this learning mode for it is *convenience*, expediency, immediacy 2. Teachers' opinions of mobility and the use of mobile devices were also mostly positive for the use of mobile device helped their work a lot
Garrett et al. [2]	To design, implement and evaluate a PDA based tool to support reflective learning in nursing and medical students' learning	Device: PDA Participants and research setting: 6 final year Nurse Practitioner Students and 4 final year Medical Students at UBC Data collection: Questionnaire, discussions Data analysis: Statistical, discourse analysis	The wireless PDA can support and improve clinical learning, and at this point the clinical reference and the communication tools seem to provide optimum value to the student's learning

(*continued*)

Table 1. (*continued*)

Author (year)	Aims/Objectives of study	Methods	Results and conclusions
Lan et al. [4]	To develop a mobile learning system, mobile interactive teaching feedback system (MITFS) and then compare the leaning effectiveness between the mobile-learning and the online asynchronous learning	Device: smartphone Participants and research setting: 40 first year university students Data collection: Logs files of online discussion, questionnaires Data analysis: Statistical, discourse analysis	1. The mobile devices can facilitate and assist learners' social construction of the knowledge process 2. Based on the problem-based learning, learners can construct their knowledge through different learning activities 3. The mobile device offers a new discussion strategy compare to the online asynchronous interactions
Schepman et al. [6]	To compare the leaning effectiveness among multi-platform-cloud-based note-taking software	Device: individual networked PC Participants and research setting: 61 undergraduate students Data collection: Questionnaire Data analysis: Statistical, discourse analysis	1. The mobile devices can facilitate and assist learners' social construction of the knowledge process 2. Based on the problem-based learning, learners can construct their knowledge through different learning activities such as raising questions, collecting and sharing information, discussing with peers, and discovering the solution to a problem 3. Apart from the support of the online asynchronous interactions, the mobile device offers a new discussion strategy. For instance, learners can facilitate learning activity in the outside world and further acquire the context-aware learning materials to enhance their learning experience

(*continued*)

Table 1. (*continued*)

Author (year)	Aims/Objectives of study	Methods	Results and conclusions
Wang et al. [10]	To explore whether the mobile device would have time management and learning benefits for doctoral students	Device: Mobile Pocket PC smartphone Participants and research setting: 6 doctoral research students from the School of Nursing participated Data collection: Questionnaires Data analysis: Discourse analysis	These multiple and variable contexts have significant impact on learners' appropriation of new technologies and related uses

In the context of mobile learning, Jonassen (1996) proposed three major research types and identified in the Handbook of Research for Educational Communication and Technology [3]. Those research types are experimental research, descriptive research, and development research. Experimental research is a research it designed an experimental group and a control group to test hypotheses regarding certain treatments. Descriptive research is a study that gathered data from events (ex. log record) or participants' responses (ex. attitude, discussion) to describe, explain, validate or explore a particular issue. Developmental research is a research which systematically studied the design, development, and evaluation process of certain educational interventions. This study based on those research types to displays the methods of these articles. This research also presents the aim (objectives) and the main results of these researches. The analyzed results are presented in Table 1.

Acknowledgments. The authors would like to thank the Ministry of Science and Technology of the Republic of China, Taiwan for financially supporting this research under contract No. MOST 105-2511-S-324-001-.

References

1. Chen, G.D., Chang, C.K., Wang, C.Y.: Ubiquitous learning website: scaffold learners by mobile devices with information-aware techniques. Comput. Educ. **50**(1), 77–90 (2008)
2. Garrett, B.M., Jackson, C.: A mobile clinical e-portfolio for nursing and medical students, using wireless personal digital assistants (PDAs). Nurse Educ. Prac. **26**, 647e654 (2006)
3. Jonassen, D.: Computers as Mindtools for Engaging Critical Thinking and Representing Knowledge. Prentice Hall, Columbus (1996)

4. Lan, Y.-F., Tsai, P.-W., Yang, S.-H., Hung, C.-L.: Comparing the social knowledge construction behavioral patterns of problem-based online asynchronous discussion in e/m-learning environments. Comput. Educ. **59**, 1122e1135 (2012)
5. Liu, Gi-Zen, Hwang, Gwo-Jen: A key step to understanding paradigm shifts in e-learning: towards context-aware ubiquitous learning. Br. J. Educ. Technol. **41**(2), E1–E9 (2010)
6. Schepman, A., Rodway, P., Beattie, C., Lambert, J.: An observational study of undergraduate students' adoption of (mobile) note-taking software. Comput. Hum. Behav. **28**, 308e317 (2012)
7. Seppälä, P., Alamäki, H.: Mobile learning in teacher training. J. Comput. Assist. Learn. **19**, 330e335 (2003)
8. Sharples, M., Taylor, J., Vavoula, G.: A theory of learning for the mobile age. In: Andrews, R., Haythornthwaite, C. (eds.) The Sage Handbook of eLearning Research. Sage, London (2007)
9. Stahl, G., Koschmann, T., Suthers, D.: Computer-supported collaborative learning: an historical perspective. In: Sawyer, R.K. (ed.) Cambridge Handbook of the Learning Sciences, pp. 409–426. Cambridge University Press, Cambridge (2006)
10. Wang, R., Wiesemes, R., Gibbons, C.: Developing digital fluency through ubiquitous mobile devices: findings from a small-scale study. Comput. Educ. **58**, 570e578 (2012)

An Independence Mechanism Design for the Software Defined Device

Ling-Hsiu Chen[1(✉)], I-Hsueh Chen[1], Po-Hsuan Chiu[1], and Hsueh-Hsun Huang[2(✉)]

[1] Department of Management Information System, Chaoyang University of Technology, 168, Jifeng E. Rd., Wufeng District, Taichung 41349, Taiwan
ling@cyut.edu.tw

[2] Development of Taiwan's Industry, Chaoyang University of Technology, 168, Jifeng E. Rd., Wufeng District, Taichung 41349, Taiwan
kentintw@gmail.com

Abstract. Owing to there are still has difficulties of lacking a convenient way to manage various devices and provide customize service either by an appliance or recreational equipment. Therefore, it is necessary to develop a system (or device) to control and management this problem. For economic considerations, it is inevitable to have a new system (or device) for this requirement. Preferably, it can be achieved by using available or easily made hardware with properly designed software architecture for control. In present, it is made by customizing circuits with only one specification for specific equipment. It is still lacking a integrate system or device can provide and integrate and manage those various respond or feedback data. To achieve this goal, this research proposes a device with control programming system also defined data in a device database and an instruction database to solve this problem.

Keywords: Software defined · Device independence · Wireless transmission · Device and instruction database

1 Introduction

The advent of the information age has added many new scientific and technological knowledge, but household equipment has problems of inconvenient management of control devices and difficulties in function customization, no matter it is an appliance or recreational equipment. Therefore, a control and management system is desired. For economic considerations, it is inevitable to have a new system (or device) for this requirement. Preferably, it can be achieved by using available or easily made hardware with properly designed software architecture for control. For a single product in the market, it is made by customizing circuits with only one specification for specific equipment. There is no control device or system to integrate all control functions in one unit. There is no control device or system to feedback data. A feature of the present study is using another device with newly added software defined data in a device database and an instruction database without being affected by the equipment. With the device control

J.-S. Pan et al. (eds.), *Advances in Intelligent Information Hiding and Multimedia Signal Processing*, Smart Innovation, Systems and Technologies 81,
DOI 10.1007/978-3-319-63856-0_9

programming system from the concept of programming system for device control, the goal can be achieved. The device control programming system is based on a model of Device Independence [1] and provided in the present study. Based on the concept of Software-Defined, the device control programming system of programming system for device control doesn't have to concern characteristics and class of the equipment when programming. It is a procedure or mechanism which makes the application procedure of the software can be applied to run on the local equipment. No more consideration is required on its hardware architecture or software environment. As long as the equipment whose application programming interfaces can be obtained, it can be applied by the programming system for device control in the present study, further reducing the threshold of development. The concept of programming system for device control in the study is implemented in robotic arms and robots to test the feasibility and advantages of SDDI. Results come out to show that programming system for device control is able to control equipment which is capable of wireless communication by programming the functions of wireless communication.

2 Relative Works

The previous technology is divided into two kinds, first one is printed circuit board, sometime everybody use printed circuit board also just control TV, Air conditioner or toy which cannot teaching programing language, everybody use the controller to control TV, Air conditioner or toy, there maybe do not have definite time, so we discover many paper want to use bluetooth [8], WIFI [3, 7], RF [6] change the communications protocol, but do not have any paper talk about use programming interface [2] for control TV, Air conditioner or toy. Second one is open hardware, refer to Fig. 1 [4, 9] which subsequent development or applications of these system development boards is to control the hardware on these system development boards through C programming language or a more low-level language, most systems on the system development board only function or control one specific device, and they are not able to setup and manage more devices at the same time.

If a subsequent developed product of the system development board can be applied on more devices with similar functions, or a board of an available device can be further modified and more widely used in other products and applications, it would be great news to the developers and users. It is not only to reduce development cost (labor and materials cost), but squeezes the time for new products launched into the market. Users enjoy the benefits of familiar interface so that to learn the new products fast. Control limitation of the products would not be a problem.

The requirements mentioned above is subject to a well-known problem: it is difficult to learn and apply a low-level programming language used to control the hardware of devices. Therefore, reducing development threshold, based on a low-level programming language and specific development boards, further to construct an easily used architecture and system suitable for every system development boards and make it possible for everyone to use a high-level programming language or interface to finish the management or development of devices, is much desired for the developers.

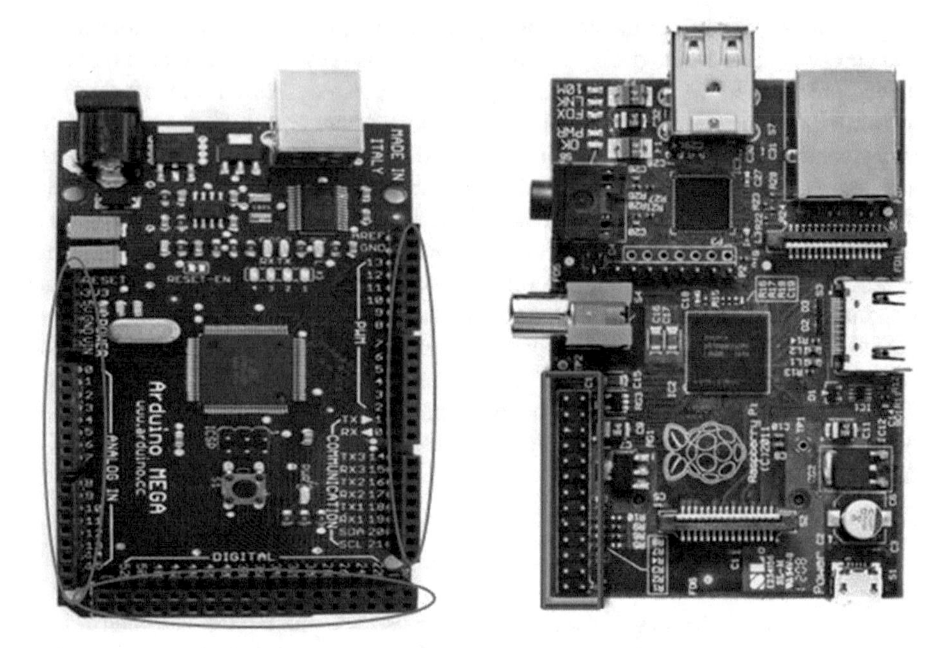

Fig. 1. The boards for illustration of pin forms

3 Research Design

In order to settle the problems mentioned above, a programming system for device control is disclosed as Fig. 2.

The system includes: a development board scanning module, for scanning I/O pins of a development board, acquiring an operating function of the development board, specifications of controllers and hardware, definition of the I/O pins, and functions set for the controllers, recording the acquired data mentioned above in a development board functional database, and transmitting the definition of the I/O pins; a web server, remotely connected with the development board scanning module, including: a device database, for defining and accessing the specifications of controllers and hardware, I/O pins, and functions set for the controllers of a plurality of devices and/or boards; and an instruction database, for accessing instructions for setting the functions for the controllers; and a working host, connected to the development board scanning module and the web server, having the development board functional database for providing a platform for programing codes of a high-level programming language and an application interface used by the high-level programming language to program and control the operating function, and an application programming interface packaging module, for encapsulating the application interface used by the high-level programming language with data in the device database, the instruction database, and/or the development board functional database; wherein the application interface compiles the codes into instructions coded by a low-level programming language used to communicate with the development board after receiving the codes of the high-level programming language.

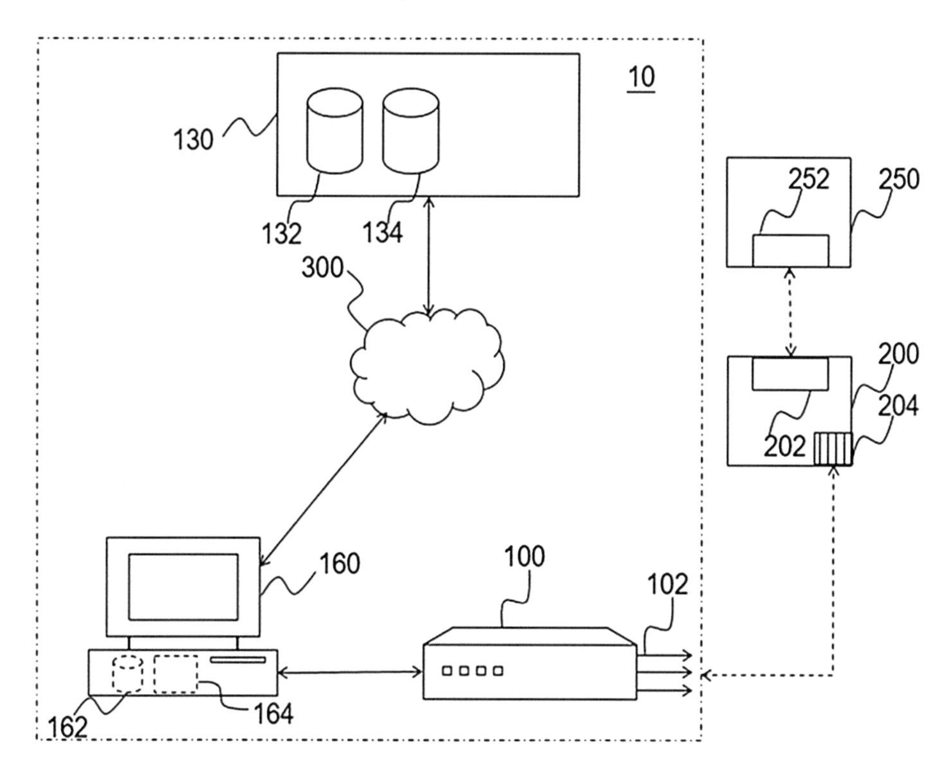

Fig. 2. The schematic diagram of a programming system for device control

According to the present research, the web server judges if the development board has been defined in the device database by the received definition of the I/O pins. If a result of the judgment is yes, record the specifications of controllers and hardware of the defined devices or boards, and functions set for the controllers in the development board functional database. The specifications of controllers and hardware of the defined devices or boards, functions set for the controllers, and instructions for setting the functions for the controllers are provided to the application programming interface packaging module, for encapsulating the application interfaces used by the high-level programming language. If a result of the judgment is no, record the acquired specifications of controllers and hardware of the development board, definition of the I/O pins, and functions set for the controllers in the development board functional database. The specifications of controller and hardware, functions set for the controllers, and instructions with respect to set the controller functions of the development board acquired by scanning are provided to the application programming interface packaging module, for encapsulating application interfaces used by the high-level programming language.

The operating function is sent to the development board from an external board via a connecting channel. The connecting channel may be a wireless communication interface or a wired transmission interface. The wireless communication interface may be an infrared communication module, a Wi-Fi communication module, a Bluetooth communication module, a RF communication module, a LTE communication module, or a WiMAX communication module. The wired transmission goes through a USB (Universal Serial Bus) port, an eSATA (external Serial Advanced Technology Attachment) port, or a Thunderbolt port. The operating instruction is defined by users.

Preferably, the working host is a laptop computer, a desktop computer, or a tablet. The application programming interface packaging module may be software running in the working host, or hardware installed in the working host.

The application programming interface packaging module in the present research is a key element to compile codes from a high-level programming language to a low-level programming language. Due to the compilation, it is possible to use an easy high-level programming language or interface to finish device control or development.

4 Conclusion

Programming system for device control activate information technology, beacause if use it for Internet of Things, it can more popular defined everything, for developers, there is no need for additional customization reduce many costs, another one it can use for programming language education, student can practice computer thinking, we will use this system in the future for Implemented in an educational environment and actually measured optimize student achievement.

Acknowledgments. The authors would like to thank the Ministry of Science and Technology of the Re-public of China, Taiwan for financially supporting this research under contract No. MOST 105-2511-S-324-001-.

References

1. Acín, A., Brunner, N., Gisin, N., Massar, S., Pironio, S., Scarani, V.: Device-independent security of quantum cryptography against collective attacks. Phys. Rev. Lett. **98**(23), 230501 (2007)
2. Browne, S., Dongarra, J., Garner, N., Ho, G., Mucci, P.: A portable programming interface for performance evaluation on modern processors. Int. J. High Perform. Comput. Appl. **14**(3), 189–204 (2000)
3. Bychkovsky, V., Hull, B., Miu, A., Balakrishnan, H., Madden, S.: A measurement study of vehicular internet access using in situ Wi-Fi networks. In: Proceedings of the 12th Annual International Conference on Mobile Computing and Networking, pp. 50–61, September 2006
4. Carpentier, S.C., Witters, E., Laukens, K., Van Onckelen, H., Swennen, R., Panis, B.: Banana (Musa spp.) as a model to study the meristem proteome: acclimation to osmotic stress. Proteomics **7**(1), 92–105 (2007)
5. Lantz, B., Heller, B., McKeown, N.: A network in a laptop: rapid prototyping for software-defined networks. In: Proceedings of the 9th ACM SIGCOMM Workshop on Hot Topics in Networks, p. 19, October 2010
6. Lee, T.H.: The Design of CMOS Radio-Frequency Integrated Circuits. Cambridge University Press, Cambridge (2004)

7. Marzetta, T.L.: Noncooperative cellular wireless with unlimited numbers of base station antennas. IEEE Trans. Wirel. Commun. **9**(11), 3590–3600 (2010)
8. Miller, B.A., Bisdikian, C.: Bluetooth Revealed: the Insider's Guide to an Open Specification for Global Wireless Communication. Prentice Hall PTR, Upper Saddle River (2001)
9. Sheinin, A., Lavi, A., Michaelevski, I.: StimDuino: An Arduino-based electrophysiological stimulus isolator. J. Neurosci. Methods **243**, 8–17 (2015)

An Ontology-Based Herb Therapy Recommendation for Respiration System

Hung-Yu Chien[1(✉)], Jian-Fan Chen[1], Yu-Yu Chen[1], Pei-Syuan Lin[1], Yi-Ting Chang[1], and Rong-Chung Chen[2]

[1] Department of Information Management, National Chi Nan University, Nantou, Taiwan, R.O.C.
hychien@ncnu.edu.tw

[2] Department of Information Management, ChaoYang University of Technologies, Taichung, Taiwan

Abstract. Air pollution has become a major threat today and the related respiration illnesses have deteriorated the wellness of many people. Herb, as one option of the medicine and also as diet, can provide very promising solution to the above threats. However, even though herbs are easily accessible in every cultures and areas, the knowledge of applying herbs on improving health is complicated and it takes quite lots of efforts to acquire the knowledge. In this paper, we design an ontology-based herb therapy recommendation web for respiration system health.

Keywords: Herb · Ontology · Respiration · OWL · Recommendation · Treatment · Java · PHP

1 Introduction

Air pollution has become one major threat to public health globally, especially in some east Asian countries for the past decade. Taiwan, one among these countries has suffered the pollution and the related illnesses. Sore throat, asthma, bronchitis, sinusitis, laryngitis, etc., are some of the common symptoms and illnesses. As the issue of PM 2.5 [1, 2] has been discussed widely, more and more people would like to take some prevention and to seek some alternative medicines that are more holistic and less bad side effects for long-term usage.

Herb has been an indispensable source of nutrition and medicines in many cultures. There are many wisdoms and knowledge around the world that specialize in using herbs as diets and as medicines; to name a few, Indian Ayurvedic medicines, Thai's herb knowledge [18–20], Chinese herb medicines, and so on. Herb, as one of the alternative medicines, has several quite inspiring merits. Herb can be used as medicine and diet; herb can be used to extract many health-enhancing essential oils; it can be used in bathing and as air-improving substances.

However, as the mainstream medication system dominates the markets and it affects how the general public absorb the knowledge, and therefore, affect how people take care

J.-S. Pan et al. (eds.), *Advances in Intelligent Information Hiding and Multimedia Signal Processing*, Smart Innovation, Systems and Technologies 81,
DOI 10.1007/978-3-319-63856-0_10

of their health and wellness. Many people used to take pills off-the-shelf, even if some herbs are quite accessible and health-beneficial.

Even though some cultures used to apply some herbs in their diet and in simple treatments. The comprehensive knowledge of herbs and their actions is quite huge, interrelated and needs carefully discerned. Nowadays it is not difficult to access herbs or many herb-related products like nutrition supplements. However, without ease access to comprehensive knowledge of herbs or access to experts, it is still very difficult and challenging for the general public to apply them in daily life [3, 26].

We, therefore, aim at sorting out the knowledge and building an easy-access web system that provide trusted and comprehensive knowledge of herb for improving respiration system wellness. We study the books from Hoffmann [21, 22] who is one prestigious herbalist. We extract the knowledge and express the knowledge using the Web Ontology Language (OWL) [4]. The knowledge we focus on include biological knowledge of herbs, their actions and related treatments, etc. We use the Protégé tool [5, 8], Jess [9], SWRL [10] to build the knowledge. We then develop programs using PHP and Java to build a recommendation web and to provide a friendly platform for interested users and various products/ services providers to interact.

2 Related Work

An expert system is a knowledge-based computer system that applies domain knowledge and rules to answer users' queries on domain questions. It greatly improves the dissemination and accessibility of the knowledge, answers people's questions or provides recommendations, as computers can unceasingly provide its services. Therefore, various expert systems like [5–7, 23, 27] have been designed and implemented to solve various domain challenges.

MYCIN [29], developed in the 1970s, is a well-known infectious disease distinguishing diagnosis system. Through a series of interface conversations, MYCIN provides infectious disease diagnosis as well as antibiotic dosages and treatment. Developed out of INTERNIST-I, Quick Medical Reference (QMR) [17] is an in-depth information resource that helps physicians to diagnose adult diseases. It provides electronic access to more than 750 diseases representing the vast majority of the disorders seen by internists in daily practice as well a compendium of less common diseases. Roventa and Rosu [24] used Prolog to develop an expert system for kidney disease diagnosis providing probable evaluation models to assist medical specialists making diagnosis decision.

The above systems focused on diagnosis. Some other systems were designed for diet recommendation for specific kind of patients. [11, 14, 15, 23] respectively constructed a general dietary recommender system for chronic patients. Chen, Huang, Bau, and Chen [12], based on ontology and SWRL, implemented an anti-diabetic drugs recommendation system for diabetes patients. Lee et al. [15] developed a dietary assessment system using fuzzy techniques and a domain ontology with fuzzy set layer extension to evaluate diet healthiness.

Ontology [4, 13] is the study of representing/applying/inferring knowledge in specific domains. It has been widely applied in solving several domain challenges. Some examples are like tourism recommendation [5, 6], ontology-based tour guidance [13], Herb-related systems [18–20], or medicine-related systems [12, 17], and diet-related systems [11, 14, 15].

Some works apply ontology to extract the knowledge of herbs to improve human health. Kato et al. [18], based on ontology and Semantic Web Technologies, designed a Thai herb recommendation system that consider a patient's symptoms, chronic diseases, Thai herb actions, and the user's taste preference. Their similar publication [20] focuses on representation of Thai herbs and their actions using ontology.

3 The Process and System Design

3.1 The Process of Knowledge Acquisition

The knowledge extraction and the representation are two of the important factors for the effectiveness of a recommendation system. The main source of the knowledge is Hoffmann's prestigious book titled "Easy breathing" [22]. Hoffmann was a lecturer for ecology in the Universities of Wales, an herbalist, ex-Director of the California School of Herbal Studies, and a fellow of Britain's prestigious National Institute of Medical Herbalists. We study the books, compare and analyze the information. Then we sort out the knowledge, according to our goals/functions of our system. Finally, we create the ontology (knowledge) using a friendly and convenient tool protégé [8] for building ontology in the OWL language [25]. The process is depicted in Fig. 1.

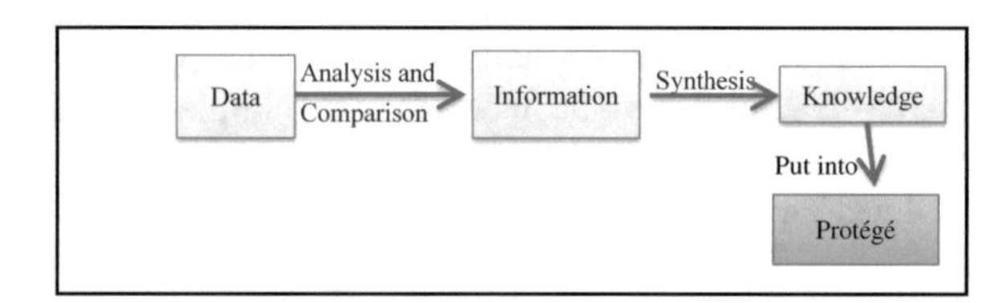

Fig. 1. The knowledge acquisition process

3.2 The System Architecture

In addition to strong domain knowledge expertise, availability and friendliness are key factors in designing an expert/recommendation system. Choosing open source development platform/API like Protégé [8], OWL API [25] and Java/PHP to a web helps the portability of our system and increases the availability of the service. The potential users of our web for respiration system health include people with respiration ailment and the general public who wants to enhance their wellness. When designing the system, we consider the usage scenarios, the practicability and the friendliness.

The system architecture is depicted in Fig. 2. An PHP web is built which accepts users' queries/clicks to trigger java programs; these japa programs interacts with an inference engine [16] and an OWL database, via OWL api. The engine accepts the queries, reads the OWL database, outputs the results to the PHP webs.

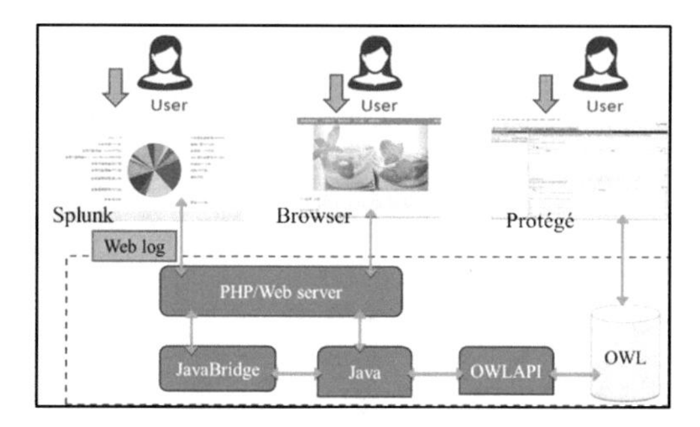

Fig. 2. The system architecture

3.3 The Domain Ontology

The main source of the knowledge is Hoffmann's prestigious book titled "Easy breathing" [22]. To extend the ontology with some add-on features like flavor, energy, popular name and pictures, we consulted other sources like [22] (American Botanical Council).

The relation of main classes of our ontology is depicted in Fig. 3. The listings of classes, object properties and data properties in protégé are respectively shown in Fig. 4 (a)(b)(c). Some key classes are described as follows.

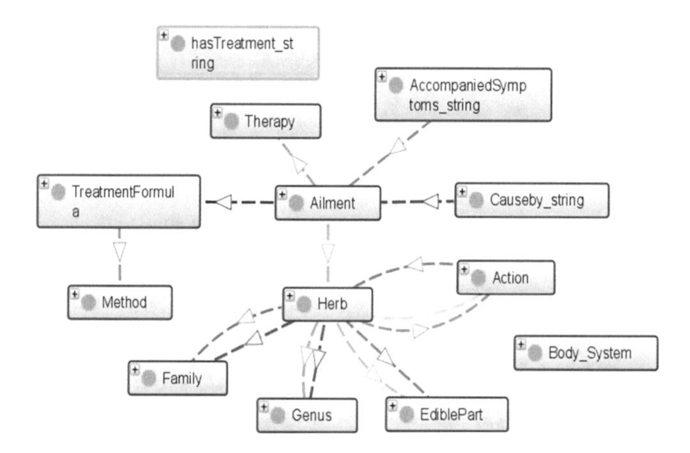

Fig. 3. The relation of classes of our ontology

The Class Action is designed for describing the changes that occur in the body or in a bodily organ as a result of herb's functioning. Some instances of the class are astringent, anti-inflammatories, antimicrobials, antispasmodics, etc. The properties of this class include hasDescription (which provides the explanation of an Action), hasNotice (which describes what one should notice about this Action), hasHerb (what herbs have such an Action), hasBodyPart (which organ this Action has effect on) and hasIntensity (the intensity of this Action).

The Class Ailment consists of two sub-classes: Symptom and Disease. This class describes an ailment and its related properties. These properties include hasDescription, accompanied Symptoms (which provides potential symptoms of this ailment), hasCaution (which reminds users what they should notice about this ailment), treatedBy (which lists potential therapies), and effectiveHerb (which lists popular herbs for treating this ailment).

4 The Functions and User Interface

Our system is designed for herbalist, knowledge engineer and also the general public; therefore, a well-designed interfaces and usage scenarios (functions) could boost the performance of the web.

Figure 4 depicts the main functions of the web. They include the biological knowledge of herbs, the actions, the therapies, and symptoms, the causes, the recipes, the Blogs, the discussion room, the membership management, etc.

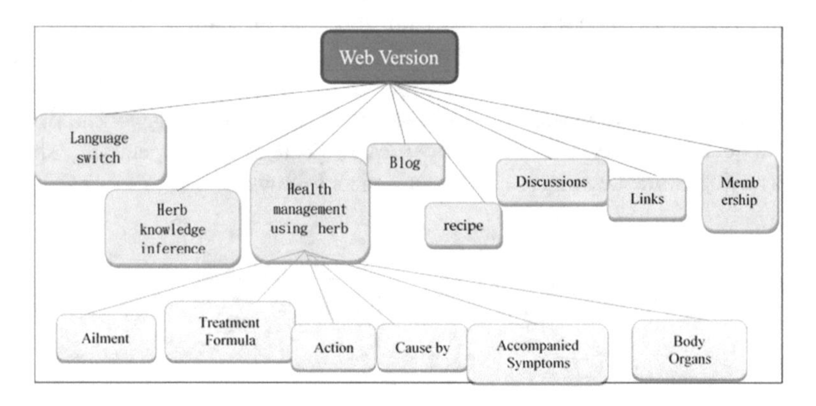

Fig. 4. The main functions of the web

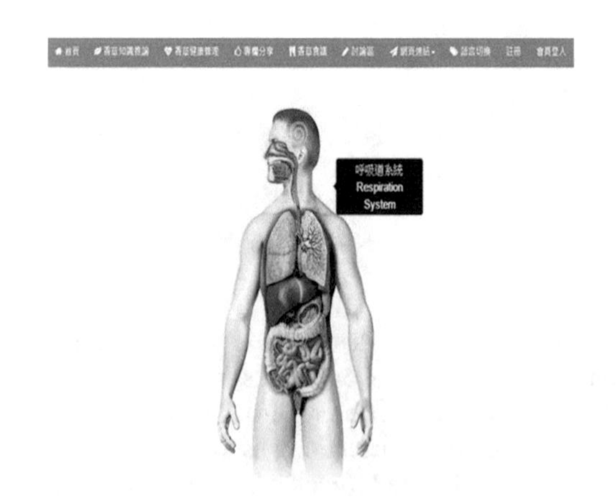

Fig. 5. Respiration system

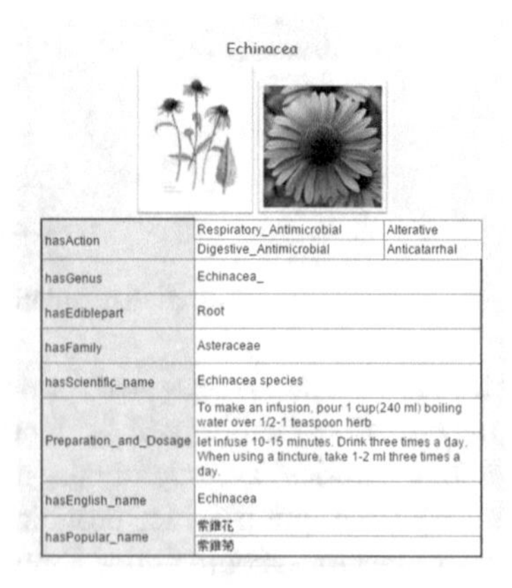

Fig. 6. Herb–Echinacea

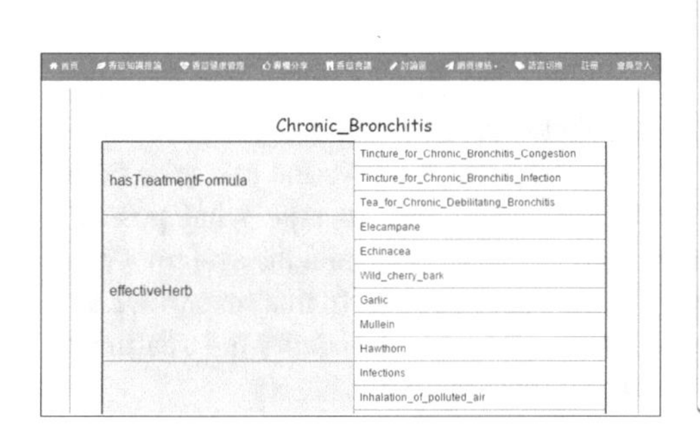

Fig. 7. Chronic Bronchitis

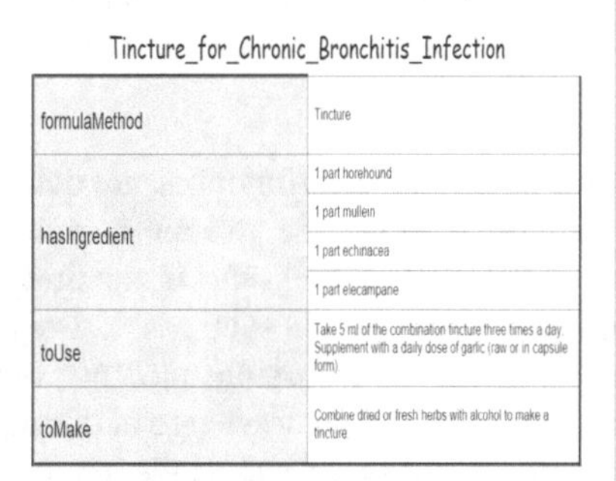

Fig. 8. The tincture treatment

These functions are briefly summarized as follows.

1. Query an herb and its properties: Users can click on a specific herb instance, and the system answers the specified properties like herb's actions, edible parts, local names, etc.
2. Query an action and its properties: Sometimes uses are not familiar with herbs, but they need to acquire the knowledge what herbs own a specific action. This function is designed for this scenario. For example, if a user is interested in carminatives, then he/she can select carminatives and the system will respond the herbs that own this action.
3. Query an ailment and its properties: From this function, users select an ailment instance, and the system answers the properties for it. The properties include (1) the possible causes, (2) the life-style treatment, (3) the cautions, (4) the possible causes, (5) the potential symptoms, (6) a series of therapies, (7) effective herbs, and so on. Each therapy provides (1) the conditions to apply this treatment, (2) the cautions for this therapy, and (3) the effects.
4. Query a treatment formula and its properties like (1) its ingredients, (2) how to make it, and (3) how to use it.
5. Use symptoms to query possible ailments. Sometimes users have no idea what herbs they are interested and what ailments they might have, but they observe some symptoms they have. So this function is quite useful. A user provides the symptoms he has, and the system lists all potential ailments for those specified symptoms. And, from the listed aliments, the user can further select the ailment and check the corresponding therapies and herbs.
6. Use causes to query potential ailments. Sometimes people do not feel well but are not sure what ailments they have. In addition to providing symptoms, they can also provide the causes (the life styles or habits), and the system will answer possible ailments and the corresponding symptoms. For example, a user might feel high pressure and feel fatigue, and he would like to know what ailments might result in.
7. The recipes: Users, in addition to the knowledge of herbs, may need the knowledge of how to cook/apply these herbs in their diet.
8. Blog/Discussion Room: Experts and interested users can share their information and knowledge via the blogs/discussion room.

9. Links: this part provides a platform for various service providers (book authors and product/service providers) to interact with users.

The user interfaces

Some user interfaces/functions are described as follows.

Figure 5 shows the various systems of human. Figure 6 shows the effective herbs and treatments for chronic bronchitis. Sometimes users are not sure what possible ailments he/she has; in such cases, users can input the symptoms, and the system would respond with some possible ailments; Fig. 7 shows one such case. In this scenario, users are interested in the knowledge of chronic bronchitis, and they can query the treatments for this illness; Fig. 8 shows the tincture treatment for this illness.

5 Conclusions

In this paper, we have designed an ontology-based recommend web for improving respiration system health. The feedbacks from users confirm its practicality and usefulness. Applying herbs in daily diet and as alternative treatments even though is not new, but it still needs great promotion and education as many people have used to go to the hospitals to get quick and easy pills to kill the symptoms. It is still a long and great challenge to promote the knowledge. With this wen, we do have a great and better opportunity.

Acknowledgements. This project is partially supported by the National Science Council, Taiwan, R.O.C., under grant no. MOST 103-2221-E-260-022 and 105-2221-E-260-014.

References

1. Wardoyo, A.Y.P., Juswono, U.P., Noor, J.A. E.: Measurements of PM2.5 motor emission concentrations and the lung damages from the exposure mice. In: 2016 International Seminar on Sensors, Instrumentation, Measurement and Metrology (ISSIMM), pp. 99–103 (2016)
2. Zhang, X., Shen, H., Li, T.: Effect characteristics of Chinese New Year fireworks/firecrackers on PM2.5 concentration at large space and time scales. In: 2016 4th International Workshop on Earth Observation and Remote Sensing Applications (EORSA), pp. 179–182 (2016)
3. Chien, H.-Y., Cho, Y.-C., Yeh, J.-Y., Coa, Y.-T., Tsai, P.-C., Chen, Y.-Y.: An ontology-based herb expert system for digestion system wellness. In: International Conference on Advanced Information Technologies, Wufun, Taiwan, 23 April 2016
4. Web Ontology Language. http://en.wikipedia.org/wiki/Web_Ontology_Language. OWL Web Ontology Language Overview. http://www.w3.org/TR/owl-features/
5. An example ontology for tutorial purpose-travel ontology. http://protege.cim3.net/file/pub/ontologies/travel/travel.owl
6. Li, X.-T., Chen, M., Wang, X.-F.: Construction method of travel knowledge database based on ontology. In: Proceedings of the 3rd International Conference on Intelligent Human-Machine Systems and Cybernetics 2011, Hangzhou, China, pp. 182–185 (2011)
7. Choi, C., Cho, M.-Y., Kang, E.-Y., Kim, P.-K.: Travel ontology for recommendation system based on semantic web. In: Proceedings of the 8th International Conference of Advanced Communication Technology 2006 ICA0T 2006, Phoenix Park, Korea, pp. 623–627 (2006)

8. Protégé. http://protege.stanford.edu/
9. Jess- the Rule Engine for the Java Platform. http://herzberg.ca.sandia.gov/
10. The Semantic Web Rule Language (SWRL). http://protegewiki.stanford.edu/wiki/SWRLTab
11. Chi, Y.-L., Chen, T.-Y. Chen, Tsai, W.-T.: A chronic disease dietary consultation system using OWL-based ontologies and semantic rules. J. Biomed. Inf. (2014). doi:10.1016/j.jbi.2014.11.001
12. Chen, R.-C., Huang, Y.-H., Bau, C.-T., Chen, S.-M.: A recommendation system based on domain ontology and SWRL for anti-diabetic drugs selection. Expert Syst. Appl. **39**, 3995–4006 (2012)
13. Chien, H.-Y., Chen, S.-K., Lin, C.-Y., Yan, J.-L., Liao, W.-C., Chu, H.-Y., Chen, K.-J., Lai, B.-F., Chen, Y.-T.: Design and implementation of zigbee-ontology-based exhibit guidance and recommendation system. In: Int. J. Distrib. Sens. Netw. (2013). doi:10.1155/2013/248535
14. Chen, Y., Hsu, C.-Y., Liu, L., Yang, S.: Constructing a nutrition diagnosis expert system. Expert Syst. Appl. **39**, 2132–2156 (2012)
15. Lee, C.-S., Wang, M.-H., Acampora, G., Hsu, C.-Y., Hagras, H.: Diet assessment based on 930 type-2 fuzzy ontology and fuzzy markup language. Int. J. Intell. Syst. **25**, 1187–1216 (2010)
16. HermiT OWL Reasoner. http://hermit-reasoner.com/
17. Quick Medical Reference (QMR) – Open Clinical. http://www.openclinical.org/aisp_qmr.html
18. Kato, T., Maneerat, N., Varakulsiripunth, R., Kato, Y., Takahashi, K.: Ontology-based E-health system with thai herb recommendation. In: Proceedings of the 6th International Joint Conference on Computer Science and Software Engineering (JCSSE2009), pp. 172–177 (2009)
19. Nantiruj, T., Maneerat, N., Varakulsiripunth, R., Izumi, S., Shiratori, N., Kato, T., Kato, Y., Takahashi, K.: An e-health advice system with Thai herb and an ontology. In: Proceedings of the 3rd International Symposium on Biomedical Engineering, pp. 315–319 (2008)
20. Kato, T., Kato, Y., Maneerat, N., Varakulsiripunth, R.: Providing Thai herbal recommendation based on ontology and reasoning. In: Proceedings of 3rd Conference on Human System Interactions (HSI), Rzeszow Poland, pp. 217–222, 13–15 May 2010
21. Biography: Hoffmann, D.L., BSc (Hons), MNIMH. http://www.healthy.net/Author_Biography/David_L_Hoffmann_BSc_Hons_MNIMH/61
22. Hoffmann, D.: Easy breathing, Storey Books, US (2000). ISBN:1-58017-250-4
23. Li, W.W., Li, V.W., Hutnik, M., Chiou, A.S.: Tumor angiogenesis as a target for dietary cancer prevention. J. Oncol. (2012). doi:10.1155/2012/879623
24. Roventa, E., Rosu, G.: The diagnosis of some kidney diseases in a small prolog Expert System. In: Proceedings of the 3rd International Workshop on Soft Computing Applications, pp. 219–924 (2009)
25. The OWL API. http://owlapi.sourceforge.net/
26. American Botanical Council. http://abc.herbalgram.org/
27. Buchanan, B.G., Shortliffe, E.H.: Rule Based Expert Systems: The MYCIN Experiments of the Stanford Heuristic Programming Project. Addison-Wesley, Boston (1984). ISBN 978-0-201-10172-0

Robust Optimal Control Technology for Multimedia Signal Processing Applications

En-Chih Chang[1(✉)], Shu-Chuan Chu[2], Vaci Istanda[3], Tien-Wen Sung[4], Yen-Ming Tseng[4], and Rong-Ching Wu[1]

[1] Department of Electrical Engineering, I-Shou University, Kaohsiung City, Taiwan, R.O.C.
{enchihchang,rcwu}@isu.edu.tw

[2] School of Computer Science, Engineering and Mathematics, Flinders University, Adelaide, SA, Australia
jan.chu@flinders.edu.au

[3] Yilan County Indigenous Peoples Affairs Office, Yilan City, Taiwan
biungsu@yahoo.com.tw

[4] College of Information Science and Engineering, Fujian University of Technology, Fuzhou, China
tienwen.sung@gmail.com, swk1200@qq.com

Abstract. This paper develops a robust optimal control technology for multimedia signal processing (MSP) applications. The proposed control technology combines the advantages of nonsingular finite-time convergence sliding mode control (NFTCSMC) and genetic algorithm (GA). The NFTCSMC has finite system state convergence time including nonsingular merit unlike infinite-time exponential convergence of classic sliding mode control, but the chattering still occur under a highly uncertain disturbance. For multimedia signal processing applications, the chattering causes high voltage harmonics and inaccurate tracking control. To enhance the performance of multimedia signal processing, the GA is well adopted to tune the control gains of the NFTCSMC so that the chattering can be removed. Experimental results are given to conform that the proposed control technology can lead to high-quality AC output voltage and fast transient response. Because the proposed control technology is easier to realize than prior technologies and gives high tracking accuracy and low computational complexity algorithm, this paper will be of interest to designer of related MSP applications.

Keywords: Multimedia signal processing (MSP) · Nonsingular finite-time convergence sliding mode control (NFTCSMC) · Genetic algorithm (GA) · Chattering · Voltage harmonics

1 Introduction

Uninterruptible power supply (UPS) is a very significant device to personal computers in achieving the stability of multimedia signal processing (MSP) [1]. A high-performance UPS is dependent upon the static inverter-filter arrangement, which is used to

J.-S. Pan et al. (eds.), *Advances in Intelligent Information Hiding and Multimedia Signal Processing*, Smart Innovation, Systems and Technologies 81,
DOI 10.1007/978-3-319-63856-0_11

convert a DC voltage to a high-quality AC output voltage of low harmonic distortion (THD) and fast dynamic response. To achieve above-mentioned requests, proportional integral (PI) controller is usually used. However, when the system using PI controller under the case of a variable load rather than the nominal ones, cannot obtain fast and stable output voltage response [2]. Several control technologies derived for UPS systems are reported in the literature, such as wavelet transform technique, deadbeat control, repetitive control, and so on [3–5]. But, these technologies are difficult to realize and have complicated algorithms. Sliding mode control (SMC) is a robust trajectory-tracking method due to its insensitivity to internal parameter variations and external disturbances. Sliding mode control of UPS systems can be found in numerous publications, however the asymptotic stability of the sliding mode by linear sliding surface is ensured in infinite time [6, 7]. Recently, a nonsingular finite-time convergence sliding mode control (NFTCSMC), which employs nonlinear sliding surface is developed instead of linear sliding surface. By the use of the nonlinear sliding surface, the NFTCSMC has finite system state convergence time and there is no singular problem [8, 9]. However, the chattering is a serious shortcoming for the practical realization of NFTCSMC. The chattering incites unmodeled high-frequency plant dynamics, and sometimes even makes the controlled system be unstable. Genetic algorithm (GA) is a stochastic search technique that guides the principles of evolution and natural selection in a population towards an optimum using genetics [10, 11]. Hence, the control gains of the NFTCSMC can be properly determined by GA, and then the chattering will be removed. By combining NFTCSMC with GA, the proposed control technology yields a multimedia signal processing-based closed-loop UPS system with low total harmonic distortion and fast transience under different types of loading. Experimental results demonstrate the feasibility and advantages of using the proposed control technology.

2 System Modeling

The output voltage v_o of the UPS inverter, shown in Fig. 1, can be forced to track a sinusoidal reference voltage v_{ref}, by applying the proposed control technology. If the desired output voltage is v_{ref}, and a state variable $x_e = \left[x_{e1} = v_o - v_{ref}\ x_{e2} = \dot{x}_{e1}\right]^T$ related to the tracking error, the error state equation can be obtained as

$$\begin{bmatrix} \dot{x}_{e1} \\ \dot{x}_{e2} \end{bmatrix} = \begin{bmatrix} 0 & 1 \\ -a_1 & -a_2 \end{bmatrix} \begin{bmatrix} x_{e1} \\ x_{e2} \end{bmatrix} + \begin{bmatrix} 0 \\ b \end{bmatrix} u + \underbrace{\begin{bmatrix} 0 \\ -a_1 v_{ref} - a_2 \dot{v}_{ref} - \ddot{v}_{ref} \end{bmatrix}}_{d} \tag{1}$$

where a_1 is $1/LC$, a_2 equals $1/RC$, b stands for $1/LC$, and d represents the disturbance. As can be seen from (1), the control signal u must be designed well so that x_{e1} and x_{e2} can be converged to zero. The design concept of this proposed control technology is to modify the classic SMC by introducing nonsingular finite-time convergence criterion and GA, so as to resolve infinite-time convergence and chattering.

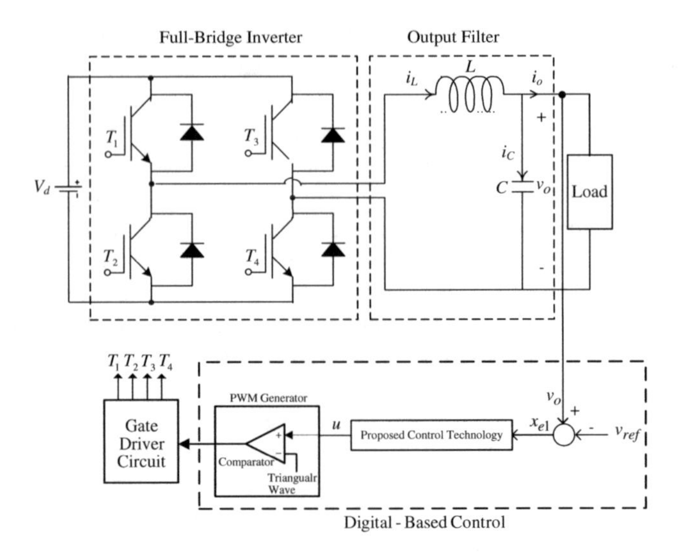

Fig. 1. Structure of UPS inverter.

3 Control Design

3.1 Nonsingular Finite-Time Convergence Sliding Mode Control (NFTCSMC)

For the error dynamics (1), the robust sliding function is defined as

$$\sigma(t) = x_{e1}(t) + \frac{1}{\delta} x_{e2}^{\frac{p}{q}}(t) \tag{2}$$

where $\delta > 0$, $p > q$, p and q are positive odd numbers ($1 < p/q < 2$), and a sliding mode reaching equation $\dot{\sigma} = -k\sigma - \varepsilon \operatorname{sgn}(\sigma)$ is constructed.

Then, the control law u can be expressed as

$$u(t) = u_{eq}(t) + u_s(t) \tag{3}$$

with

$$u_{eq}(t) = b^{-1}[a_1 x_{e1} + a_2 x_{e2} - \delta \cdot \frac{q}{p} x_{e2}^{2-\frac{p}{q}}] \tag{4}$$

$$u_s(t) = -b^{-1}\left[k\sigma + (l_g + \varepsilon)\operatorname{sgn}(\sigma)\right], k > 0, \varepsilon > 0 \tag{5}$$

where u_{eq} denotes the equivalent control with non-singularity, and u_s displays the sliding control for compensating the perturbation influences. Thus, the system will be driven to the sliding mode $\sigma = 0$ and converged within finite time, and the perturbation $d(t)$ is bounded as $|d(t)| < \Psi, \forall t \geq 0$.

The NFTCSMC σ converges to zero within finite time; On the other hand, when the (2) is reached, then the states of the system (1) will converge to zero within finite time. However, NFTCSMC has chattering in the UPS system design. This is because of the changeable load, so once the loading is a severe uncertain condition, the system (1) will not provide accurate tracking performance, even causes the instability of multimedia signal processing. To remove chattering, the control gains of the NFTCSMC can be tuned by GA as follows.

3.2 GA-Based Tuning

3.2.1 Initial Population

An initial population is built as

$$x(n) = x_{\min}(n) + RANDOM \cdot (x_{\max}(n) - x_{\min}(n)), \quad n = 1, 2, \ldots, j \tag{6}$$

where $RANDOM$ is bounded by 0 and 1, and $x_{\min}(n)$ and $x_{\max}(n)$ denote minimum and maximum values, respectively.

3.2.2 Fitness Function

To get the minimum tracking error, the objective function, J can be designed as

$$J = \int_0^\infty (W_1 |x_{e1}| + W_2 u^2) dt \tag{7}$$

where W_1 and W_2 symbol both weight values. The fitness function, F can be expressed as $F = 1/J$.

3.2.3 Selection

Let the generation size be g, and define the adaptive degree of individual k be $F(k)$, the proportion of individual k can be obtained as

$$P(k) = F(k) \Big/ \sum_{k=1}^{g} F(k), \quad k = 1, 2, \cdots, g \tag{8}$$

3.2.4 Crossover

Choose a crossover site, K_{crs} within $0 \leq K_{crs} < M$ randomly, then by keeping the genes of the parent creatures between position 1 and K_{crs} unaltered and interchanging the genes of the parent creatures between position $K_{crs} + 1$ and M, two new creatures yield

$$\begin{cases} x^{crs,z,h} = \rho x^{s,z,h}(n) + (1-\rho)x^{s,z,w}(n) \\ x^{crs,z,w} = (1-\rho)x^{s,z,h}(n) + \rho x^{s,z,w}(n) \end{cases} \tag{9}$$

where n is bounded by $K_{crs} + 1$ and M, and $0 \leq \rho \leq 1$.

3.2.5 Mutation

By the use of the mutation operation, the $(z + 1)$th population can be constructed as

$$x^{z+1,h}(n) = x^{crs,z,h}(n) + \kappa \xi_n \tag{10}$$

where κ is bounded by -1 and 1, and ξ_n represents the mutation amplitude of the nth gene.

4 Experiments

The proposed system parameters are listed as follows: DC-bus Voltage $V_d = 220$ V; Output Voltage $v_o = 110$ V; Output Frequency $f = 60$ Hz; Filter Inductor $L = 1.5$ mH; Filter Capacitor $C = 20$ μF; Switching Frequency $f_{sw} = 18$ kHz; Rated Loading $R = 12\,\Omega$. The parameter setting for GA is below: An initial generation equals 200; Crossover: generally, the value of crossover probability is between 0.6 and 1. Therefore, a probability of crossover equals 0.8. Mutation: generally, the value of mutation probability is between 0.0001 and 0.1. The values of L and C filter parameters are supposed in suffering from 10% ~ 200% of nominal values as the UPS inverter system is under 12 Ω resistive loading; Figs. 2 and 3 show output-voltage waveforms of the UPS inverter controlled by the proposed control technology and the classic SMC. The proposed control technology is more insensitive to parameter variations than the classic SMC. To test the transient behavior of the UPS inverter using the proposed control technology,

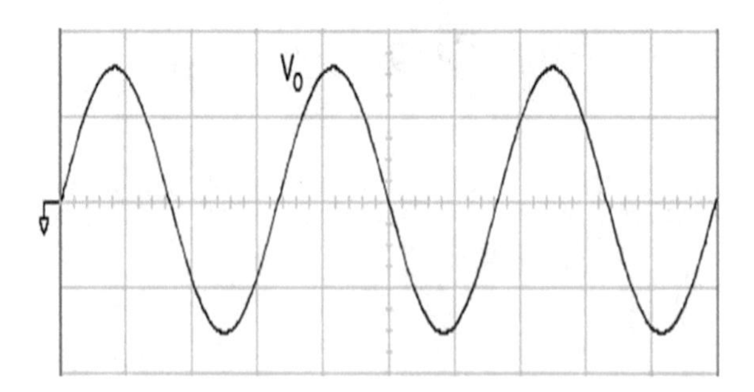

Fig. 2. Output-voltage under *LC* parameter variations obtained using the proposed control technology (100 V/div; 5 ms/div).

Fig. 4 shows the output-voltage and the output-current for step load change. As can be seen, a fast recovery of the steady-state response is obtained. However, the waveforms obtained using the classic SMC under the same loading shown in Fig. 5, display visible voltage sag at 90° firing angle.

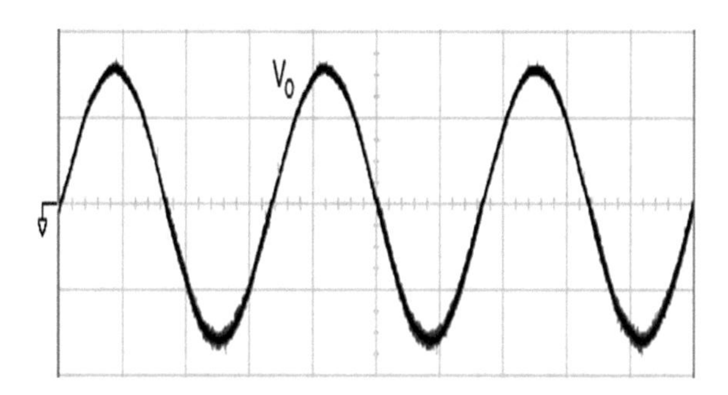

Fig. 3. Output-voltage under *LC* parameters variations obtained using the classic SMC (100 V/div; 5 ms/div).

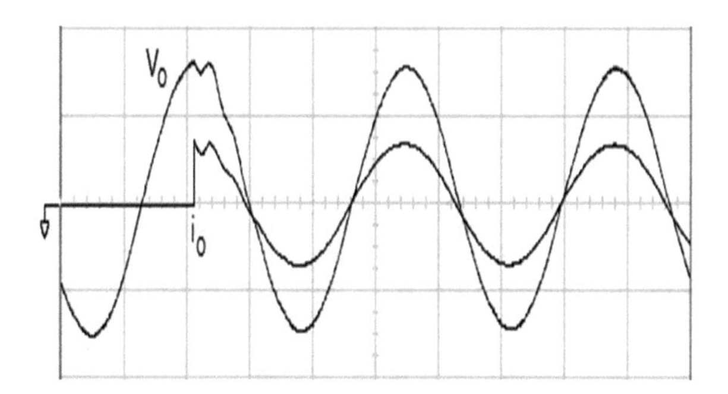

Fig. 4. Output-voltage and output-current under step load change obtained using the proposed control technology (100 V/div; 20 A/div; 5 ms/div).

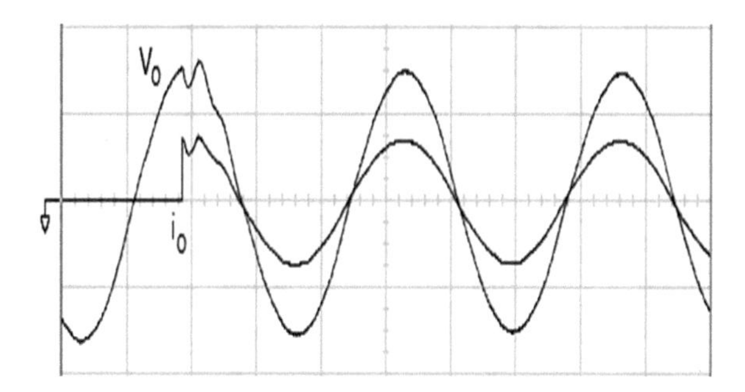

Fig. 5. Output-voltage and output-current under step load change obtained using the classic SMC (100 V/div; 20 A/div; 5 ms/div).

5 Conclusions

In this paper, a high performance UPS inverter by associating NFTCSMC with GA is proposed. Relative to the classic SMC, the NFTCSMC without singular problem can force the system tracking error to reach the original within finite time. The GA is used to determine optimal control gains of the NFTCSMC, thus removing the chattering, which exists in the NFTCSMC while a severe uncertain disturbance is applied. Experimental results show that low total harmonic distortion, fast dynamic response, chattering alleviation, and steady-state error reduction are achieved in the proposed controlled UPS system under transient and steady-state loading.

References

1. Luo, F.L., Ye, H.: Power Electronics: Advanced Conversion Technologies. CRC Press, Boca Raton (2010)
2. Rebeiro, R.S., Uddin, M.N.: Performance analysis of an FLC-based online adaptation of both hysteresis and PI Controllers for IPMSM drive. IEEE Trans. Ind. Appl. **48**(1), 12–19 (2012)
3. Saleh, S.A., Rahman, M.A.: Experimental performances of the single-phase wavelet-modulated inverter. IEEE Trans. Power Electron. **26**(9), 2650–2661 (2011)
4. Hu, J.B., Zhu, Z.Q.: Improved voltage-vector sequences on dead-beat predictive direct power control of reversible three-phase grid-connected voltage-source converters. IEEE Trans. Power Electron. **28**(1), 254–267 (2013)
5. Chen, D., Zhang, J.M., Qian, Z.M.: An improved repetitive control scheme for grid-connected inverter with frequency-adaptive capability. IEEE Trans. Ind. Electron. **60**(2), 814–823 (2013)
6. Tan, S.C., Lai, Y.M., Tse, C.K.: Sliding Mode Control of Switching Power Converters: Techniques and Implementation. CRC Press, Boca Raton (2012)
7. Wai, R.J., Lin, C.Y., Huang, Y.C., Chang, Y.R.: Design of high-performance stand-alone and grid-connected inverter for distributed generation applications. IEEE Trans. Ind. Electron. **60**(4), 1542–1555 (2013)
8. Nekoukar, V., Erfanian, A.: A decentralized modular control framework for robust control of FES-activated walker-assisted paraplegic walking using terminal sliding mode and fuzzy logic control. IEEE Trans. Biomed. Eng. **59**(10), 2818–2827 (2012)
9. Zou, A.M., Kumar, K.D., Hou, Z.G., Liu, X.: Finite-time attitude tracking control for spacecraft using terminal sliding model and chebyshev neural network. IEEE Trans. Syst. Man Cybern. Part B Cybern. **41**(4), 950–963 (2011)
10. Holland, J.H.: Adaptation in Natural and Artificial Systems. The University of Michigan Press, Ann Arbor (1975)
11. Hasanien, H.M., Muyeen, S.M.: Design optimization of controller parameters used in variable speed wind energy conversion system by genetic algorithms. IEEE Trans. Sustain. Energy **3**(2), 200–208 (2012)

Wearable Computing, IOT Privacy and Information Security

Implementation of an eBook Reader System with the Features of Emotion Sensing and Robot Control

Jim-Min Lin[1(✉)], Jan-Hwa Hsu[1], and Zeng-Wei Hong[2]

[1] Department of Information Engineering and Computer Science, Feng Chia University, Taichung City, Taiwan
jimmy@fcu.edu.tw

[2] Department of Computer Science, Faculty of Information and Communication Technology, Universiti Tunku Abdul Rahman, Petaling Jaya, Malaysia
hungcw@utar.edu.my

Abstract. The paper reports the implementation of an eBook reader with emotion sensing feature through Kinect sensor device. Recently, more and more traditional paper books have been transformed into an electronic form which is so called the electronic book (eBook). Currently, eBook contents could be in the form of combining various media through visual and audio communications. However, it still lacks the capability of real interactions between human and eBook to respond to user's emotion as a person is reading eBooks. Therefore, we will apply emotional sensing technology to enhance the human-eBook interactions. So that an eBook reader device can sense a person's reaction that reflects the situation and meaning as described in eBook contents. In this paper, the implementation of a novel eBook reader system integrating a Kinect sensor and the associate system development kit (SDK) will be reported. We will use a widely used face recognition technology, Active Appearance Model (AAM), for detecting user's facial expression and emotions. With such a new feature, users may have funs and more realistic interactions with eBooks as a person is reading eBooks.

Keywords: Human-Computer Interaction (HCI) · Emotion sensing · Facial expression detection · e-book reader · Kinect

1 Introduction

The Human-Computer Interaction (HCI) technology gets more and more attentions [1]. The Emotional communication and application play an important key in the process of HCI. A good emotional application can help users to relieve stress or increase the interests and thus promoting human interaction on the machine [2, 3]. Research [4] reports that, in the humanoid robot performance process, through the rich variety of similar human emotions, the audience can feel the meanings of the robot action. Therefore, researchers tried to apply robots in many applications like education to increase the learner interactions and attention. For example, in reading storybooks or books, a robot can stimulate user's learning motivation through human-robot interactions/games [5, 6].

J.-S. Pan et al. (eds.), *Advances in Intelligent Information Hiding and Multimedia Signal Processing*, Smart Innovation, Systems and Technologies 81,
DOI 10.1007/978-3-319-63856-0_12

Additionally, most of the current researches design the robot performance unidirectional from the robot to the human and lack of sense of audiences' responses. Therefore, a good HCI design indeed requires not only a certain human-robot interaction programming ability for performing robot actions [7] but also needs the sense of user's emotions as the feedback to the robots.

The purpose of early e-books is to reduce the size of the contents of the paper books and saved in a file, and also to provide a fast browsing of electronic content. With the development of electronic technology, e-book contents can no longer be limited to plain text. Interactive e-book development, allowing users to interact with the graphics in the book.

Many e-book research focused on the use of education. For example, study [8] explores the relevance of learning methods through reading e-books and the improvement of reading ability for students. Research [9] explores the future development of e-books through the comparison of the marketing strategies of major e-book reader manufacturers. Through the interaction of people and e-books in study [10], user's reading interest is thus enhanced. But all of these studies lack the sense of physical interaction with a user and neither the reaction to user emotions. So, we will try to add the emotional sensor function and involve the help of the robot into the interactive mode of existing e-book. Using the robot as an interactive performance and sensing the user's emotions in the e-book reader, users can be involved in e-book story situation in-depth more.

The proposed eBook reader system provides an interactive robot motion programming and authoring tool that supports emotion sensing feature using Kinect and responds to user using robot's actions, as shown in Fig. 1. Through of use of the proposed eBook reader system, better HCI experience will be obtained.

Fig. 1. Structure of the proposed system

2 Kinect Facial Expression Detection Technique

Kinect is a somatosensor developed by Microsoft XBOX360 [11]. The skeleton tracking system with color image and depth image sensing feature allows users to manipulate the system without handheld device. Kinect is equipped a color camera, microphone array, and a rotatable base. It has the features of capturing human body movements and real time close-up face recognition. In this study, emotion detection is accomplished by face recognition feature of Kinect. From the official sample code [12–14], a Kinect coordinates (X, Y, Z) is obtained through the calculation of depth image, color image, and skeleton system, as shown in Fig. 2.

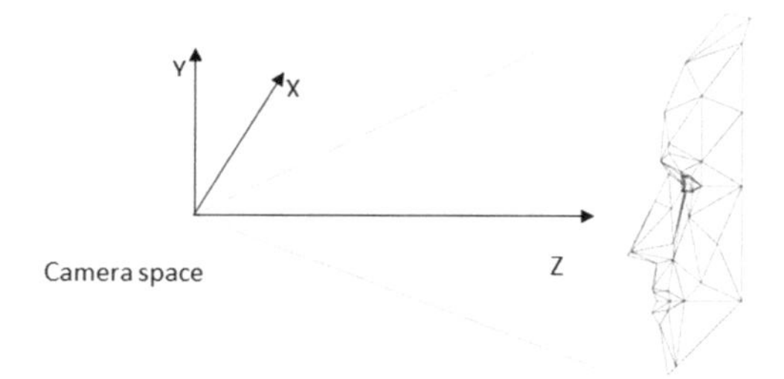

Fig. 2. Facial feature marking using AAM in Kinect

The active appearance models (AAM) [15, 16] is used to mark facial feature values in Kinect. As shown in Fig. 3, these characteristic points represent the shape. The study will define the facial emotions represented by the expression.

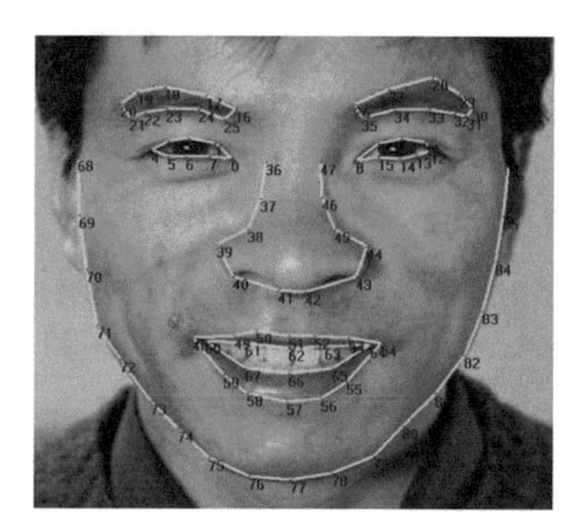

Fig. 3. Active appearance models (AAM)

3 Emotion Detection

3.1 Experiment Environment

The system is composed of PC, Robot, and Kinect. The system configuration is listed as Table 1.

Table 1. Configuration of the proposed system

Item	System configuration
Kinect	Kinect
	SDK for windows 1.8
	Visual studio 2012
Robot	Innovati Robotinno 1
	Touch keys
	Bluetooth 100 M
	innoBASICworkshop2
PC	Windows 7 OS
	CPU i3, RAM 8 GB

3.2 Facial Expression Feature Extraction

Kinect facial expression detection flow chart used in this study is as follows: after the program is started, all the data, including staging data will be initialized. Then Kinect starts the image capturing functions: color image, the depth of the image, and the skeleton subsystem. The data stream will be copied to the staging data area and calculated using the Kinect face tracking function. If the face feature values are obtained, the output will be displayed on the color image part. The detection process is continued until the face feature value is determined.

Based on AAM, Kinect face tracking could output users 87 2D face characteristic points.

Then, as shown Fig. 4, 16 key characteristic points will be adopted out from the 87 2D face characteristic points for identifying the emotions of a user. The user's emotion is identified by instantaneous face feature point matching. The result will be returned if the threshold is reached. Here's how to define a user's emotions:

1. The user-defined emotional name;
2. The corresponding facial feature data defined by the matching user's mood, that is, the 16 key feature values recorded, are shown in Fig. 4.

The user emotion data is obtained through the facial characteristic data associated with a corresponding user emotion type, say smile, anger, sadness, and happy.

After the user face characteristic data is obtained, the system will then match these data stream captured from Kinect with emotion data in the database. The matching algorithm is as follows:

Step 1: Building a predefined facial emotion data file through a learning process for each user.

Step 2: Calculating the Euclidean distance and differences for the 16 facial characteristic points and the saving these data into an emotion data file.

Step 3: The calculated data is compared with Kinect data stream to determine whether the minimum Euclidean distance and the difference ($\pm$A) are equal, then the matching emotion name is output as the result.

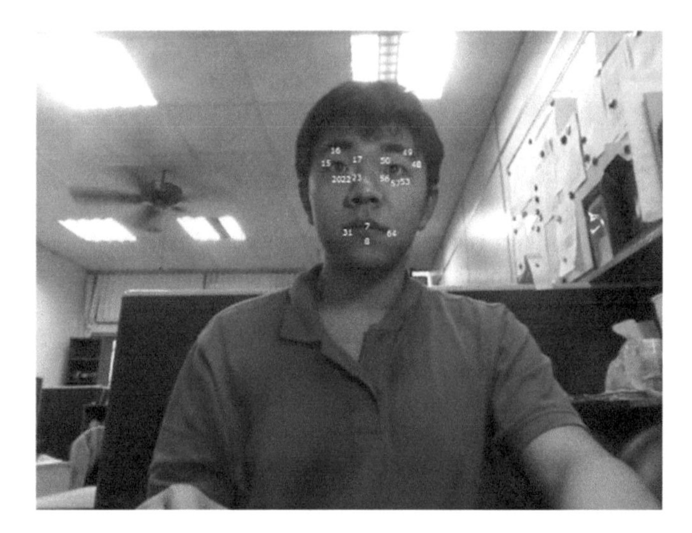

Fig. 4. Key feature points

3.3 Emotion Detection

Emotional sensing is derived from the results of the user's facial expression recognition. It should be aware that, before one start to do this experiment, facial expressions and emotional sensation results may be different. Facial expression is based on the user's own facial information, including facial movement and eigenvalue changes, recorded into several facial expressions. However, human emotions can change for a variety of reasons. So to get user's emotional information, the system will generate a corresponding emotional file as the base for future emotional recognition. In addition to immediate emotional recognition, the system can do continuous recognition for every 5 s. the user's emotional distinction, through the interval to sensitize the user's emotions and infer the results of the test. For example, as shown in Fig. 5, X axis is for the time seconds, Y axis is for the number of detection of specific emotions. In the case in Fig. 5, the system

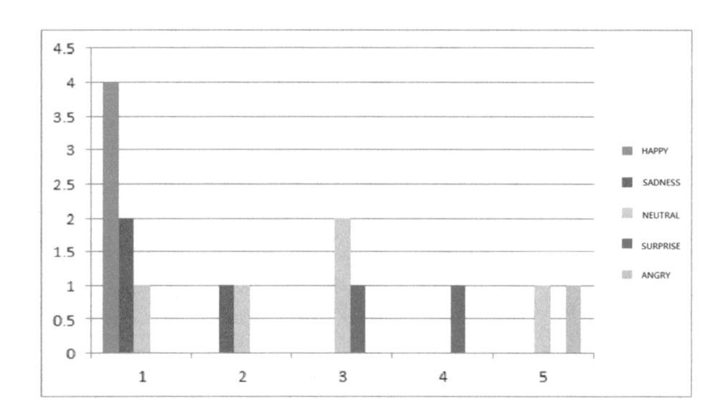

Fig. 5. Emotion detection results

will identify the user's mood to be happy because the "HAPPY" is detected for 4 time which is larger than the detected times of other moods.

3.4 Auxiliary Emotion Detection Using Mouth Shapes

In this study, we found that the effect of reflecting the emotion of the user is not satisfied by just using AAM. So it will help the system to enhance the recognition rate by the other emotion detection methods which are mainly based on the mouth type detection and the voice detection.

According to the famous psychologist Paul Ekman's research [26], facial emotion is mainly expressed by the lower half of facial muscles. So the system could instantly estimate the user mood according to not only the data through the Emotional file, but also to the AAM for lower half of mouth type as the auxiliary detection. The AAM based feature point calculation of the mouth of the lower half is mainly based on the mouth type angle calculation. The shape of upper lip A and lower lip B will be separately processed, as shown in Fig. 6. It shows the method [25] of obtaining the mouth shapes of the corresponding moods and the matching the lip angles, as shown in Table 2.

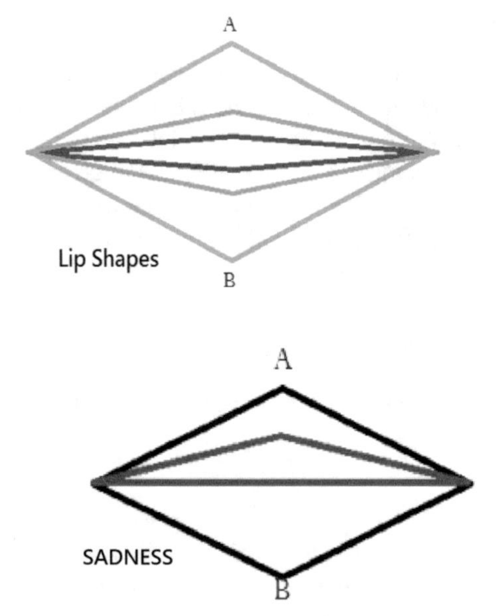

Fig. 6. Shapes of mouth for different emotions

Table 2. Criterion of emotions according to angles A, B

Emotion	Criterion
Neutral	A < −0.7 and B < −0.9
Happy	A > −0.75 and B < −0.9
Surprise	A > −0.3 and B > −0.8
Sad	B < −0.99

4 REBIS and Response Design Through Robot Motions

This study uses a Motion Enhanced Interactive eBook System [21] to serve as a medium for users to read e-books. E-Reader [19, 20], is a portable electronic device used to read e-books. This study will use an e-book reader to sense the user's emotional reactions. For example, one may want to get the user's feelings or emotional changes of reading a story in the book content, and then, using pre-designed human and robot interaction to achieve the purpose of two-way man-machine interaction. In other words, when the e-reader detects the user's emotions, some predesigned response, such as the traditional screen multimedia display, by the e-book authors can be performed. The proposed system has also combines the previous research results [21].

Designing the robot motion, mainly for the left and right hands and the lower body part, the left and right feet. And then, these designs will be a basis for further meaningful combinational actions, as shown in Fig. 7. For example, the posture of raising two hands high is actually a combination of two basic motions: left hand high and the right hand high simultaneously. However, considering the balance of the robot without falling, it is necessary to limit the movement of the left and right feet of the lower body of the robot not to a great extent.

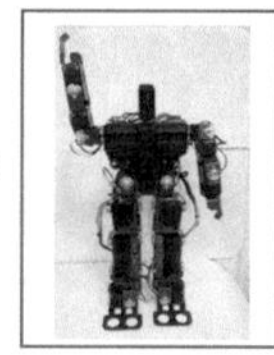

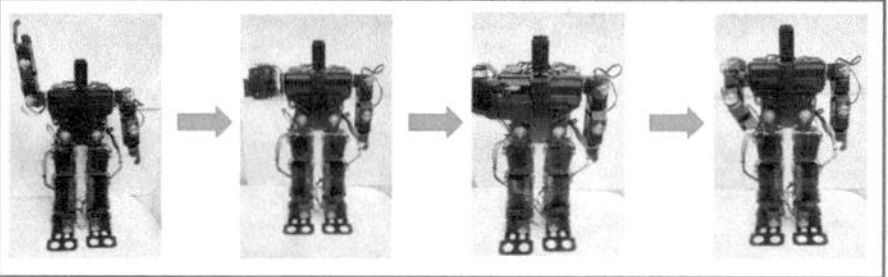

Fig. 7. Authoring a robot action

4.1 Robot Action Labels and RDSL

Robot Domain Specific Language (RDSL) is a way to describe the robot postures using predefined labels, so that programmers can simply design the robot actions, as shown in Table 3. This study refers to the RSDL architecture design as shown in Table 2.

Table 3. Tags of RDSL motion file

Label	Description
rm	Root of this document
name	Document name
rid	Robot ID
r-actno	Root of an action
basic	Basic posture name
cat	Posture attributes
buffer	Action command and time

4.2 Integrated Editing Module

As the user ends the design of the robot actions and emotion definitions, the integrated editing module will integrate the user's emotional information and robot action and then join the touch-type feedback mechanism. The Touch key feedback mechanism is the use of the robot sensor, the user can design the robot to select the action, such as: the system will determine user's mood and perform the corresponding robot action caused by the user mood. And, when the user touches the robot at the same time, the robot will respond to the user's feelings. These files will eventually be integrated into an integration file, as shown in Fig. 8.

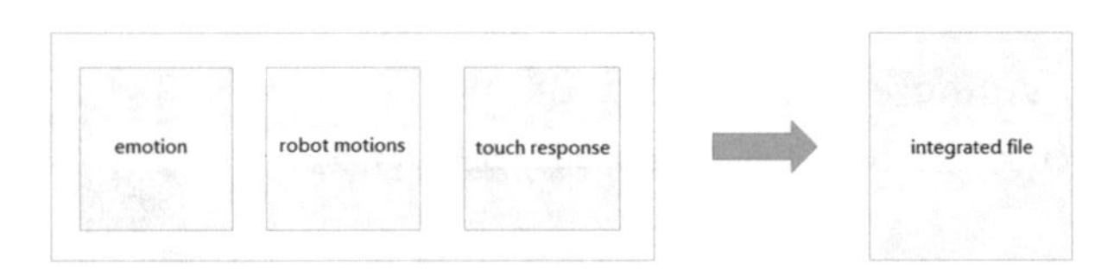

Fig. 8. Integrated file

5 Implementation

5.1 System Architecture

The proposed system architecture is divided into three layers: the application software layer, the middleware layer, and the robot control layer, as shown in Fig. 9. Kinect's emotional sensing, robot action design, feedback mechanism and communication with the intermediary layer are all designed in the application software layer; while the middleware layer is responsible for the robot command conversion and transmission. The robot control layer is responsible for the implementation of action instructions and touch Feedback reaction.

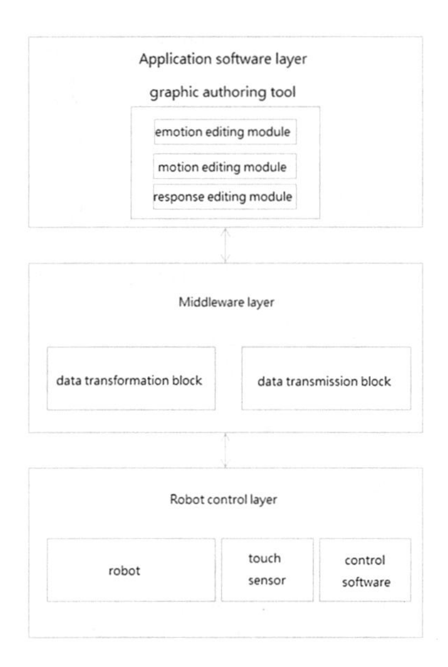

Fig. 9. System architecture of e-reader system

5.2 Application Software Layer

The application software layer is mainly to provide the user's editing function, as shown in Fig. 10. In this layer, Kinect sensor is used to sensing user emotional information; a graphical robot editing interface is used to allow users to design the robot action performance; and finally through the feedback function to integrate all the information. This layer is mainly divided into emotion editing function, action editing function, response editing function. They will be described in detail in the following chapters.

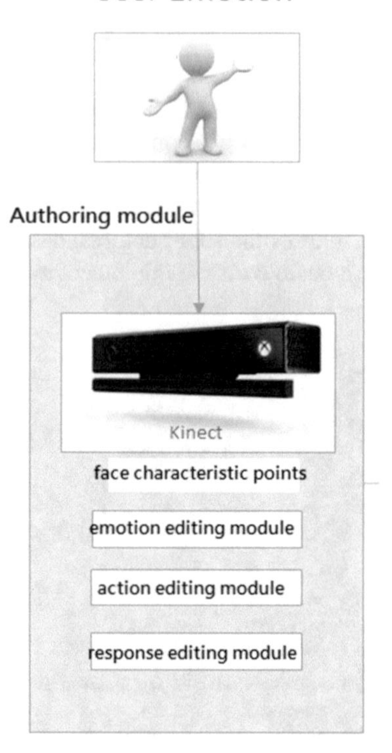

Fig. 10. Application software layer

5.2.1 User Interface and Emotion Editing Module

Emotion editing module interface is as shown in Fig. 11. The color screen shows the real time face sensing with the sixteen feature points, the user can then specify the emotional

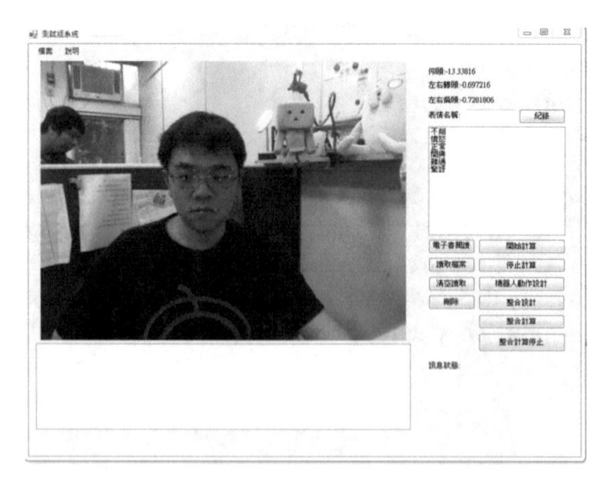

Fig. 11. Emotion definition module

name in the upper right corner of "expression name". The "Start calculation" function refers to the start of the operation of the user recorded by the emotional data. The system will show results after the resulting message displayed on the *ListBox*. In the bottom right of the window, the "Read file" function displays the mood definitions data of the previous user record on the screen, allowing the user to know the person's previously defined emotional face data message. The "Robot action design" function is to start the robot's action design, the "integration design" will collect emotional information, robot design information, as well as feedback function integration together. "Integration calculation" is to apply the integration design file for the final system integration.

5.2.2 Authoring Robot Actions Robot motion design module

The Robot motion design module interface is shown in Fig. 12. The robot motion design blocks located at the upper left corner are the left hand, right hand, and the feet respectively. The user can select the desired gestures through the right side part of the graphical interface. The robot's gestures are listed in their corresponding pictures. To design a whole robot action, a programmer should design separate motions for each stage. Robot programmers can therefore arrange the composed gesture for a stage by combining some robot gestures into a composed and desired action through selecting appropriate gesture pictures. Robot action programmer can also choose the desired motion speed for each action from the action design block below. A robot will therefore perform the predesigned actions stage by stage accordingly.

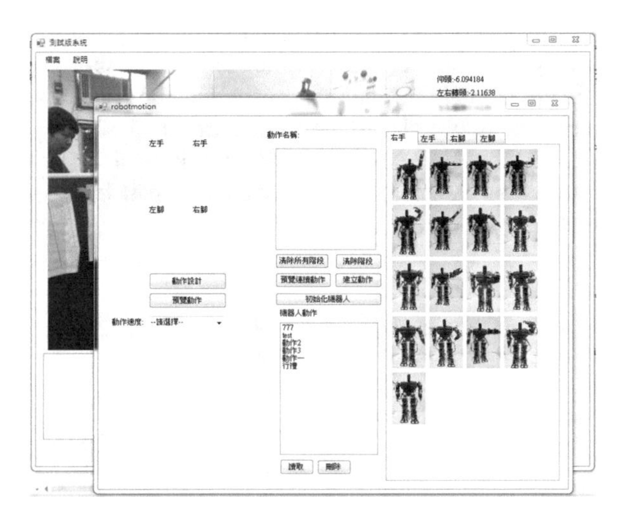

Fig. 12. Robot action design module

5.2.3 Response Design and Integration Module

The user interface for feedback editing is shown in Fig. 13. Feedback integration module can be implemented by joining the touch-type feedback to the robot action editing function that is responsible for reflecting the user's emotional information. The touch points of the robots are on the left and right hands as well as on both sides of the chest. Touch-type feedback function is mainly to achieve that not only looking but also touching in the interaction with the robot. So the user can design their own responding robot actions to the corresponding mood. For example: in defining the [sad mood] and the response, we can design the robot to wave its hands to get user's attention or other ways to appease the user's emotions. Another example is, when the user feels sad and touch the robot's right hand, the robot will be activated to perform some actions to try to please the user.

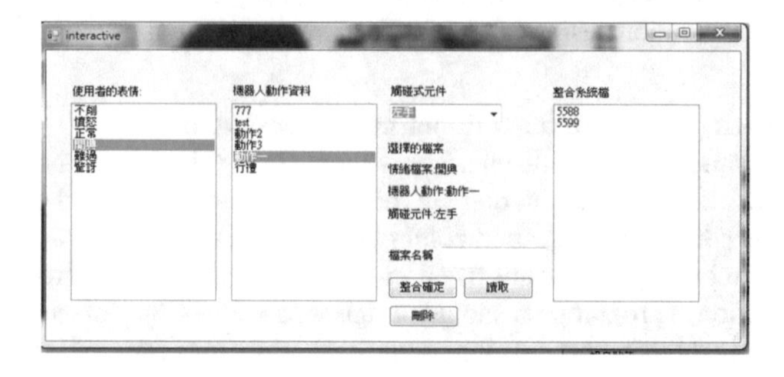

Fig. 13. Response design and integration module

5.3 e-Book Reader

The interface design of the e-book presented in this study is based on the e-book reading interface generated by the "Action Enhanced Interactive E-book System" that is previously completed in our laboratory. Emotion sensing functions is background executed in the e-book reader. The user designed interaction scenarios will be performed during user's reading and interacting with an eBook. As for the original feature of the interactive e-book system interaction with a robot performance, it will be inherited in the new system, as shown in Fig. 14.

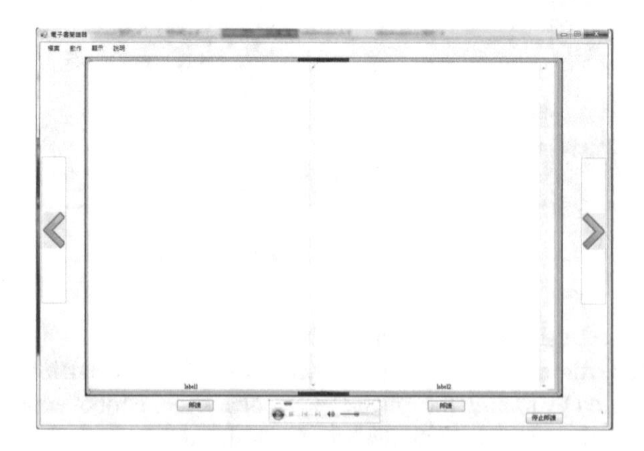

Fig. 14. e-Book reader

5.4 Middleware Layer

The main function of the middleware layer is to assist the process of conversion and transmission of robot commands, as shown in Fig. 15. When the application software layer transmits the commands to the data conversion block of the middleware layer, the system information from the application software layer is then converted into the appropriate instructions to the corresponding robot of a specific vendor. Finally, the commands to controlling an on-line robot through the data transmission block will be sent.

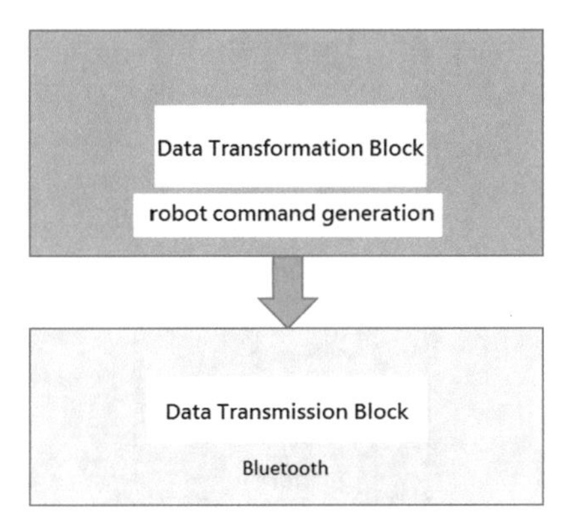

Fig. 15. Middleware layer

5.5 Robot Control Layer

The robot control layer performs command control on the robot device and is responsible for the robotic action design and the robotic command transmission after the integrated file is composed. Robot control software *innoBASICworkshop2* is designed based on the Basic language with the basic program syntax such as condition evaluation, loop, user-defined functions, and so on. In this study, we can carry out data transmission with the built-in wireless transmission function [22]. The robot control flow chart is shown in Fig. 16.

The program flow for robot control is described as follows:

Step1: Start the robot and initial the position of sixteen motors. Keep the standing posture.
Step2: Enter the loop waiting for entering the commands to the middleware layer, and start the feedback response mode. The feedback response commands are divided into three categories: action command, action time (seconds), and feedback response.
Step3: On receiving the message from the touch sensor, the system will update and trigger the robot interface action instructions.
Step4: Perform the actions.
Step5: If the program is disconnected, it will leave the loop and wait for the next connection to the system. Or, the robot will be off.

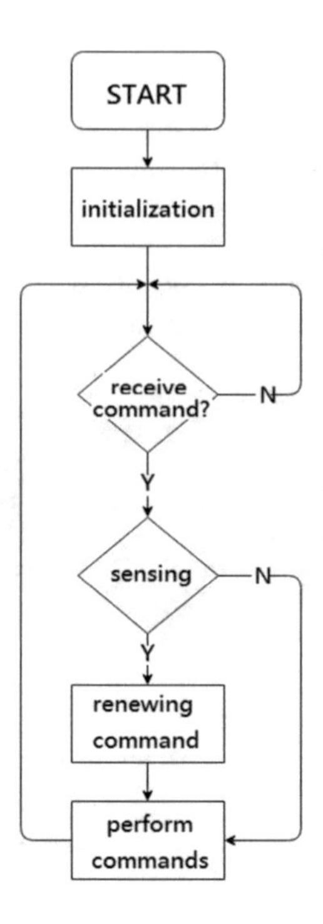

Fig. 16. Robot control program flow

6 Conclusions

In this study, the features of emotion detection and robot interaction are involved in an e-book reader, so that users can use a simple way to design their own emotional expression, and through the design of the robot to detect specific emotions to respond to action to increase reading eBooks interaction. In addition, the design of the response by touching the feedback function of the robot allows the user to respond differently to the different eBook content so as to achieve the goal of two-way interaction. The reader can thus integrate more with the situation in the eBook and then achieve a new kind of interaction between people and eBooks.

Acknowledgements. The authors would like to thank the Ministry of Science and Technology, Taiwan R.O.C. for the partly financial support (Grant numbers: MOST103-2221-E-035-051, MOST104-2221-E-035-008, and MOST105-2511-S-035-002-MY2) to this research.

References

1. Johal, W., Adam, C., Fiorino, H., Pesty, S., Jost, C., Duhaut, D.: Acceptability of a companion robot for children in daily life situations. In: IEEE Conference on Cognitive Infocommunications, Vietri sul Mare, pp. 31–36, (2014)
2. Zhang, S.-M., et al.: Research on the human computer interaction of E-learning. In: International Conference on Artificial Intelligence and Education (ICAIE) (2010)
3. Lee, K.M., Jung, Y., Kim, J., Kim, S.R.: Are physically embodied social agents better than disembodied social agents? The effects of physical embodiment, tactile interaction, and people's loneliness in human–robot interaction. Int. J. Hum Comput Stud. **64**, 962–973 (2006)
4. Sakamoto, D., Kanda, T., Ono, T., Ishiguro, H., Hagita, N.: Android as a telecommunication medium with a human-like presence. In: presented at the Proceedings of the ACM/IEEE International Conference on Human-Robot Interaction, Arlington, Virginia, USA (2007)
5. Chin, K.-Y.: IDML-based animated and physical pedagogical agents for computer-assisted learning. Ph.D. thesis, Department of Information Engineering and Computer Science, Feng Chia University, Taiwan (2011)
6. Li, K.-Y.: An IDML-based robot control mechanism. Master thesis, Department of Information Engineering and Computer Science, Feng Chia University, Taiwan (2009)
7. Manohar, V., Crandall, J.W.: Programming robots to express emotions: interaction paradigms, communication modalities, and context. IEEE Trans. Hum.-Mach. Syst. **44**(3), 362–373 (2014)
8. Lin, H.-A.: A study of the influences of e-book reading on tablet PC on the elementary school children's reading ability and attitude. Master thesis, Graduate Institute of Library & Information Studies, National Taiwan Normal University, Taiwan (2011)
9. Hung, K.-H.: Product and marketing strategy research of ebooks and e-readers: based on Amazon, Barnes & Noble, and Apple. Master thesis, Department of Business Administration, National Taipei University, Taiwan (2010)
10. 余宜芳、 葉靜瑜、 黃秀霜、 陳響亮,「電子書教學互動教具之設計研究–以國小 五年級國語教材為例」, *2010 臺灣網際網路研討會大會 (TANET)* (2010)
11. Zhang, Z.: Microsoft Kinect sensor and its effect. IEEE Multimedia Mag. **19**(2), 4–10 (2012)
12. Microsoft Kinect SDK, April 2014. http://www.microsoft.com/en-us/kinectforwindows/
13. Microsoft Kinect SDK: Programming with the Kinect for Windows SDK, April 2014
14. Getting Started with the Kinect for Windows SDK Beta from Microsoft Research, Kinect for Windows SDK beta. Programming Guide, pp. 19–20, July 2011
15. Zhou, M., Liang, L., Sun, J., Wang, Y.-S.: AAM based face tracking with temporal matching and face segmentation. In: IEEE Conference on Computer Vision and Pattern Recognition (CVPR), pp. 701–708 (2010)
16. Cootes, T., Edwards, G., Taylor, C.: Active appearance models. IEEE Trans. Pattern Anal. Mach. Intell. **23**(6), 681–685 (2001)
17. You, S.-D.: RDSL: a domain specific language for robot manipulation. Master thesis, Department of Information Engineering and Computer Science, Feng Chia University, Taiwan (2013)
18. XML Tutorial: https://www.w3schools.com/Xml/. Accessed July 2013
19. Lin, W.-J., Yueh, H.-P.: Examining college students' reading behaviors and needs for Ebook readers. J. Libr. Inf. Stud., pp. 113–142, December 2012
20. Wikipedia, e-book: http://en.wikipedia.org/wiki/E-book, December 2012
21. Lin, J.-M., Chiou, C.W., Lee, C.-Y., Hsiao, J.-R.: Supporting physical agents in an interactive e-book. In: The Ninth International Conference on Genetic and Evolutionary Computing (ICGEC 2015), pp. 243–252, Yangon, Myanmar 2015-08

22. Bluetooth Module: http://tinyurl.com/n7vm5yn, July 2013
23. Rahman, A.S.Md.M., Alam, K.M., Saddik, A.El.: A prototype haptic EBook system to support immersive remote reading in a smart space. In: IEEE International Workshop on Haptic Audio Visual Environments and Games (HAVE), pp. 61–84 (2011)
24. Alam, K.M., Rahman, A.S.Md.M., Saddik, A.El.: HE-Book: a prototype haptic interface for immersive E-Book reading. In: World Haptics Conference (WHC), pp. 21–24 (2011)
25. Lin, C.-C.: Development of automatic recognition system for facial expression. Master thesis, Institute of Automation Technology, National Taipei University of Technology, Taiwan (2006)
26. Ekman, P.: Emotions Revealed Recognizing Faces and Feelings to Improve Communication and Emotional Life, 2nd edn. Holt Paperbacks, New York (2007). ISBN 080507516X

On the Automatic Construction of Knowledge-Map from Handouts for MOOC Courses

Nen-Fu Huang(✉), Chia-An Lee, Yi-Wei Huang, Po-Wen Ou, How-Hsuan Hsu, So-Chen Chen, and Jian-Wei Tzengßer

Department of Computer Science, National Tsing Hua University, Hsinchu, Taiwan
nfhuang@cs.nthu.edu.tw, s104064522@m104.nthu.edu.tw
http://www.nthu.edu.tw/

Abstract. Massive open online courses (MOOCs) offer valuable opportunities for freedom in learning; however, many learners face cognitive overload and conceptual and navigational disorientation. In this study, we used handouts to automatically build domain-specific knowledge maps for MOOCs. We considered handouts as conceptual models created by teachers, and we performed text mining to extract keywords from MOOC handouts. Each knowlege map is based on the structure of the handouts, each consisting of an outline, title, and content. The findings suggest that using handouts to build knowledge maps is feasible.

Keywords: Knowledge maps · Learning styles · Massive open online courses · Open learning

1 Introduction

Massive open online courses (MOOCs) are open educational resources that are available to learners worldwide free of charge. MOOCs include high-quality instructional videos produced by professors from prestigious universities. Learners share their ideas and reflections in discussion forums and use an online exercise system to evaluate their learning outcomes. MOOCs have been adopted by prestigious educational institutions, such as Stanford, Harvard, and the Massachusetts Institute of Technology, as well as private organizations and individuals, such as Salman Khan. Examples of MOOCs include Coursera [1], edX [2], Udacity [3], and Khan Academy [4].

Although the self regulated learning structure of MOOCs offers considerable flexibility and learn material of interest, many learners face cognitive overload and conceptual and navigational disorientation [5]. Cognitive overload refers to learners being required to process more information than what the human memory can usually hold. This represents a significant challenge in learning

J.-S. Pan et al. (eds.), *Advances in Intelligent Information Hiding and Multimedia Signal Processing*, Smart Innovation, Systems and Technologies 81,
DOI 10.1007/978-3-319-63856-0_13

environments with large numbers of learners [9]. In addition, MOOCs pose other challenges that may affect learners' success and willingness to complete courses.

In recent years, knowledge maps have been used extensively in education and business. A knowledge map is a visual representation of knowledge originally developed by Holley and Dansereau [6]. Knowledge maps could improve learning when used to summarize information. Evidence suggests that creating or studying summaries enhances the ability to recall summarized ideas [7]. Studying maps rather than text passages assists in the recollection of central ideas and details [8].

Several studies have investigated the use of knowledge maps in open learning environments, and a number of authors have experimented with organizing and presenting learning materials in open learning environments and customizing materials to learners needs and preferences [8]. Knowledge maps facilitate the building schemata of learning concepts in learners memories, thereby assisting learners in the learning process [10]. The construction of knowledge maps requires help from domain experts, who are difficult to hire. Few studies have investigated the automatic construction of knowledge maps [11]. Several studies have attempted to automatically build domain knowledge maps for e-learning using text mining techniques [12].

This paper describes an innovative study combining knowledge maps and handouts from MOOCs. We regarded each handout as a conceptual model for a teacher. Many studies have briefly defined conceptual models as models created by professionals such as researchers, teachers, or engineers to facilitate the understanding and teaching of world systems and states of affairs [13]. The present study aimed to determine variations in accuracy between conceptual models and automatically constructed knowledge maps. This paper proposes a method for the automatic construction of knowledge maps and an example of how to implement the method by using MOOC handouts.

2 System Architecture

In this study, we designed and implemented a system for the automatic construction of knowledge maps incorporating a current MOOC platform. This section introduces the system architecture and data analysis module.

2.1 Layer Structure Overview

This system can be divided into the following four parts: a web server, user interface module, data analysis module, and data server (Fig. 1). The web server consists of a web application and a web application programming interface (API). It mainly handles requests from users, enabling them to upload course files and select keywords from handouts. The user interface module contains the analyzed results based on MOOC handouts and presented using knowledge maps. The data analysis module is a pure API server without a front end view that is responsible for collecting and analyzing course files.

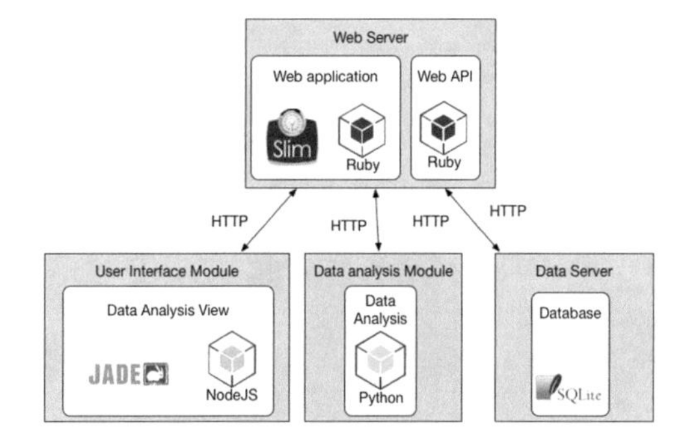

Fig. 1. System structure of knowledge map generator.

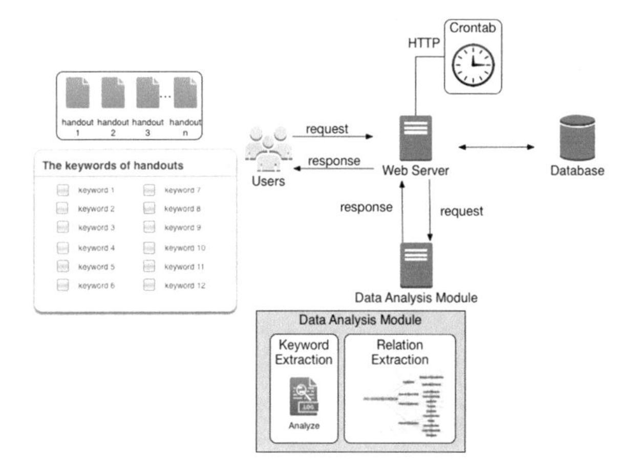

Fig. 2. System architecture for the web server and the data analysis module.

Figure 2 illustrates each part of the system architecture in detail. The system comprises two subsystems: a web server and data analysis server. The web server grants access to the system; a user requests desired websites through a browser. Moreover, the web server contains most of the web application logic and is responsible for front end presentation for the user. The data analysis module consists of keyword extraction and relation extraction. Keyword extraction is based on the term frequencyinverse document frequency (TFIDF) method [14], which provides numerical statistics to reflect how crucial a word is to a document in a collection or corpus [15].

2.2 Data Analysis Module

As previously mentioned, the data analysis module consists of keyword extraction and relation extraction (Fig. 3).

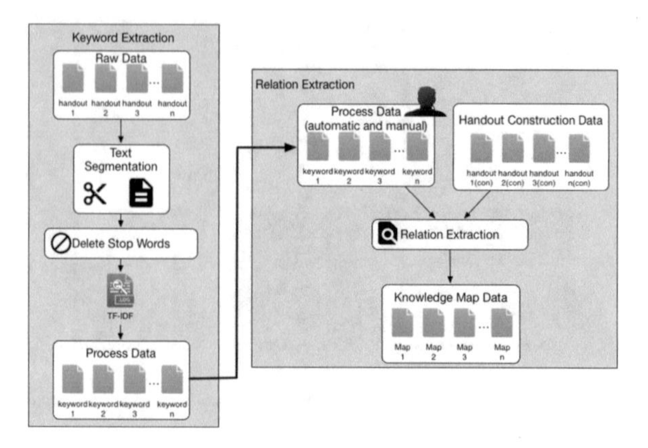

Fig. 3. Data analysis module for the keyword extraction and the relation extraction.

Keyword Extraction. In the first phase, keywords are extracted from course handouts. Before extraction, we must complete the following steps. First, we use the text segmentation method, which entails the division of written text into meaningful units such as words, sentences, and topics. Second, we delete stop words, which are words that are filtered out before or after the processing of natural language data (text) [16]. Once these preprocesses have been completed, we use the TFIDF algorithm to weight terms. Finally, we produce new files containing the keywords and their corresponding weights. The TFIDF algorithm is the product of the term frequency tf and inverse document frequency idf. We might count the term frequency, which refers to the number of times each term occurs in each document. The weight of a term in a document is simply proportional to the term frequency [17]. The inverse document frequency is a measure of how much information the word provides, or in other words, whether the term is common or rare across all documents. It is calculated based on the logarithmically scaled inverse fraction of the documents that contain the word.

$tfidf$ is calculated as follows [14]:

$$tfidf(t, d, D) = \frac{f_{t,d}}{\sum_k f_{k,d}} \cdot \log \frac{N}{|\{d \in D : t \in d\}|} \quad (1)$$

where:

$f_{t,d}$ = raw count (i.e., the number of times that term t occurs in document d)

N = total number of documents in the corpus $N = |D|$

$|\{d \in D : t \in d\}|$ = number of documents where the term t appears (i.e., $tf_{t,d} \neq 0$). A term being absent from the corpus leads to division by zero. Therefore, it is common to adjust the denominator to $1 + |\{d \in D : t \in d\}|$.

A high TFIDF weight is reached through a high term frequency in the given handouts and a low term frequency in the whole collection of documents. Hence, the weights tend to filter out common terms.

Relation Extraction. After we extract the keywords, the knowledge map user selects refined keywords from the list of highest ranked terms. When the refined keywords in the knowledge map have been determined, we can define relations. Each handout consists of an outline, title, and content. We must complete the following steps before extracting the relations. First, we filter out some punctuation marks, bullet points, and images. Second, we analyze the word size in every column and the sentence layers on every page. Finally, we produce analysis results and store files.

3 System Implementation

This section discusses the implementation of the proposed system, focusing on the functions of the data analysis module and user interface.

3.1 Data Analysis Module

The data analysis module is crucial in the proposed system. We use course handouts to determine the structure and relations of the module. In this subsection, we explain how the effective text and structure of a handout is extracted and how a knowledge map is generated.

Parsing PDF. First, we assume that the handouts are PDF files. We use this flexible tool to determine the structure bases of MOOC handouts. The layout analyzer returns objects for each page in the PDF document.

We extract the corresponding locations and font sizes based on objects. Because the tool does not have a hierarchy function, we designed a function to determine the hierarchy based on the corresponding locations. The hierarchy indicates which layer the sentence belongs to. In addition, we can determine which sentence is higher or lower than the sentence currently being analyzed.

Knowledge Map Generation. We use the refined keywords and parsing files of handouts to create knowledge maps and the name of the course chapter as the root of the knowledge map. The first layer is the outline. We collect the content of the same outline based on the parsing files of handouts. Because every content includes a title, the second layer is the title. The parsing files of handouts contain every sentence matched with layer numbers. Furthermore, we must determine keywords in sentences based on the refined keywords file. We can match keywords to layer numbers, and we can then regard the highest layer number as the third layer. In accordance with the rules, we create the remaining layers. Finally, we generate a knowledge map for each handout.

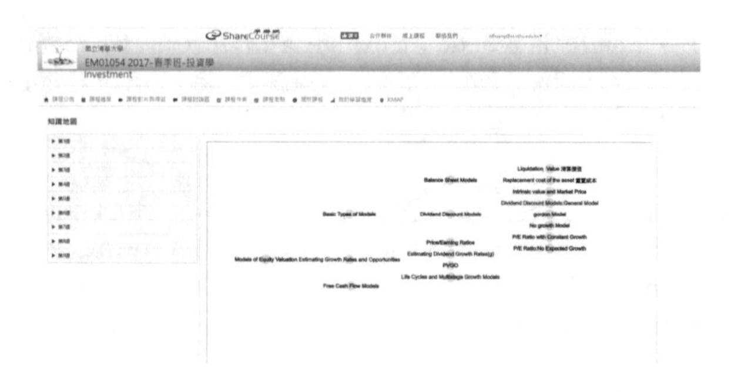

Fig. 4. Knowledge map for the week 3 of an "Investment" course.

3.2 User Interface

Knowledge Map Page. ShareCourse [18] is a well-known MOOC platform in Chinese-speaking countries. It provides courses online and in mobile applications. ShareCourse provides more than 1000 courses and has more than 60,000 learners registered online.

We added a tab beside the current learner progress page. When "KMAP" is clicked, the browser sends a request to the server to obtain information regarding which chapters have corresponding knowledge maps (Fig. 4). The request return data shows the vertical tabs of the "KMAP" page. The vertical tabs are clickable buttons and a learner can select the knowledge map they wish to view.

4 Results and Analysis

Because we sought to determine whether knowledge maps help teachers, a knowledge map was applied in one course on ShareCourse. This section presents an analysis of knowledge maps created manually and those created automatically, as well as an automatic knowledge map that was modified by an expert. Figures 5, 6 and 7 show knowledge maps for weeks 2, 3, and 4 of an "Investment" [19] course, respectively. The "Investment" course will understand the investment environment home and abroad. The course contents consist the presentation of investment evaluation tool and so on. First, we determined whether each map was created manually or automatically. Despite the high degree of

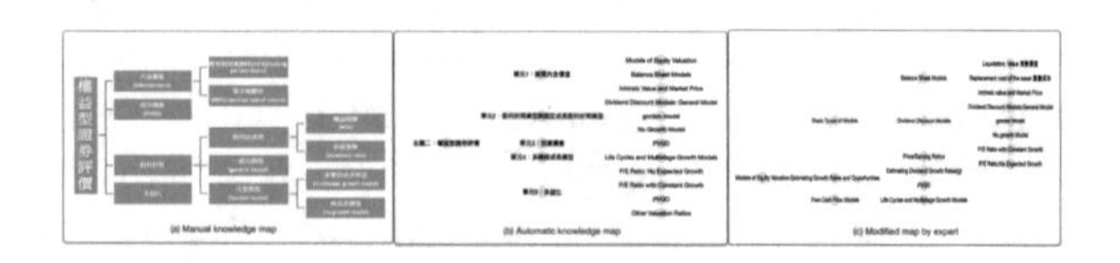

Fig. 5. Knowledge map comparison: Week 2 of the "Investment" course. (Left: Manual knowledge map; Middle: Automatic knowledge map; Right: Modified knowledge map by expert)

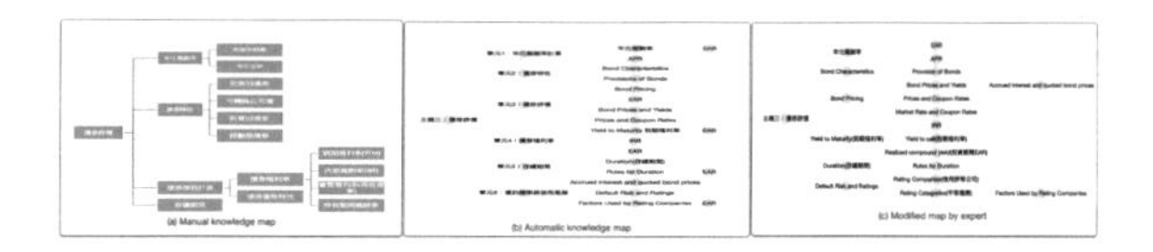

Fig. 6. Knowledge map comparison: Week 3 of the "Investment" course. (Left: Manual knowledge map; Middle: Automatic knowledge map; Right: Modified knowledge map by expert)

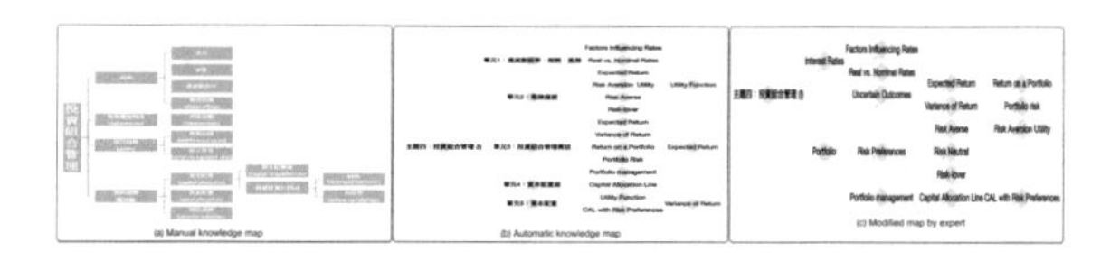

Fig. 7. Knowledge map comparison: Week 4 of the "Investment" course. (Left: Manual knowledge map; Middle: Automatic knowledge map; Right: Modified knowledge map by expert)

resemblance between the two types of map, only a short amount of time was required to obtain the relevant information from the automatic knowledge map. Subsequently, we examined the automatic knowledge map that was modified by an expert. The expert watched a video and modified the automatic knowledge map created by our system. A few concepts that were mentioned by the teacher in the video were not explained in the handouts. Because the teacher wanted to further explain the handout content, new concepts did not appear on the handouts. Although we lost some information, the expert could still add concepts to our system. Experts can use this flexible system to quickly create knowledge maps. With the correct keywords and accurate layer assignment, high-precision knowledge maps can be generated.

5 Conclusion and Future Works

According to the results of a meta-analysis study [8], knowledge maps effectively enhance knowledge retention by assisting learners in recalling central ideas and forming summaries based on acquired knowledge. However, studies have seldom investigated using handouts to build knowledge maps. The present study aimed to determine the difference in accuracy between conceptual models and automatic construction of knowledge maps. According to the result and analysis, the knowledge map generator is faster than manual knowledge map, moreover, it is very accurate. The findings suggest that using handouts to build knowledge maps is feasible.

Finally, we propose the following two directions for further study: (1) expansion of the data range to construct more robust knowledge maps for MOOCs; and (2) creation of a recommendation system to increase learning efficiency.

Acknowledgement. This study is supported by the Ministry of Science and Technology (MOST) of Taiwan under grant numbers MOST-105-2511-S-007-002-MY3 and MOST-105-2634-F-007-001.

References

1. Coursera. https://zh-tw.coursera.org/
2. edX. https://www.edx.org/
3. Udacity. https://www.udacity.com/
4. Khan Academy. https://www.khanacademy.org/
5. Wang, M., Peng, J., Cheng, B., Zhou, H., Liu, J.: Knowledge visualization for self-regulated learning. Educ. Technol. Soc. **14**, 28–42 (2011)
6. Holley, C.D., Dansereau, D.F.: Spatial Learning Strategies: Techniques, Applications, and Related Issues. Academic Press, New York (2014)
7. Foos, P.W.: The effect of variations in text summarization opportunities on test performance. J. Exper. Educ. **63**, 89–95 (1995)
8. Nesbit, J.C., Adesope, O.O.: Learning with concept and knowledge maps: a meta-analysis. Rev. Educ. Res. **76**, 413–448 (2006)
9. Fasihuddin, H.A., Skinner, G.D., Athauda, R.I.: Boosting the opportunities of open learning (MOOCs) through learning theories. GSTF J. Comput. (JoC) **3**, 112 (2013)
10. Fasihuddin, H., Skinner, G., Athauda, R.: Knowledge maps in open learning environments: an evaluation from learners perspectives. J. Inf. Technol. Appl. Educ. **4**, 18–29 (2015)
11. YongYue, C., HuoSong, X.: Research on the auto-construction methods of concept map. In: International Conference on Intelligent Human-Machine Systems and Cybernetics, IHMSC 2009, pp. 75–77. IEEE (2009)
12. Lee, J.H., Segev, A.: Knowledge maps for e-learning. Comput. Educ. **59**, 353–364 (2012)
13. Greca, I.M., Moreira, M.A.: Mental models, conceptual models, and modelling. Int. J. Sci. Educ. **22**, 1–11 (2000)
14. Salton, G., Buckley, C.: Term-weighting approaches in automatic text retrieval. Inf. Process. Manage. **24**, 513–523 (1988)
15. Leskovec, J., Rajaraman, A., Ullman, J.D.: Mining of Massive Datasets. Cambridge University Press, Cambridge (2014)
16. Flood, B.J.: Historical note: the start of a stop list at Biological Abstracts. J. Assoc. Inf. Sci. Technol. **50**, 1066 (1999)
17. Luhn, H.P.: A statistical approach to mechanized encoding and searching of literary information. IBM J. Res. Dev. **1**, 309–317 (1957)
18. ShareCourse. http://www.sharecourse.net
19. The "Investment" course on ShareCourse. http://www.sharecourse.net/sharecourse/course/view/courseInfo/987

Automated Music Composition Using Heart Rate Emotion Data

Chih-Fang Huang[1(✉)] and Yajun Cai[2]

[1] Department of Information Communications, Kainan University, No.1 Kainan Road, Luzhu, Taoyuan 33857, Taiwan
jeffh.me83g@gmail.com

[2] Master Program of Sound and Music Innovated Technologies, National Chiao Tung University, 1001 University Road, Hsinchu 300, Taiwan, ROC
sars40211@gmail.com

Abstract. This paper proposes an innovated way to compose music automatically according to the input of the heartbeat sensor, to generate music with the correspondent emotion states. The typical 2D emotion plane with arousal and valence (A-V) states are adapted into our system, to determine the generative music features. Algorithmic composition technique including Markov chain is used, with the emotion - music feature mapping method, to compose the desired correspondent music. The result show a pretty good success with various generative music, including sad, happy, joyful, and angry, and the heartbeat values show its good consistency for the correspondent emotion states finally.

Keywords: 2D emotion plane · Arousal · Valence · Algorithmic composition · Emotion - music feature mapping

1 Introduction

Algorithmic composition [1–3] is a long-term development issue since Lejaren Hiller's "Illiac Suite" in 1957 [4]. Nowadays, it is becoming more and more practical to perform the automated music composition using the idea and method of algorithmic composition with the new trends of technology such as genetic algorithm [5], and deep learning [6, 7].

Emotion features can be defined based a 2 dimensional plane in both arousal-valence axes [8, 9]. Several features mapping between music and emotion are discussed in the recent years, especially in the field of MIR (Music Information Retrieval) [10–12].

This research is mainly to adapt human's heart rate as the biofeedback input, to transfer the data to our proposed automated composition system. Various situations based on the heart rate data, can be correspondent to different generative music, to help the users to attain the goal of reviving energy or relieving tension. The 2 Dimensional A-V Plane is used for our proposed system to express the emotion states, and the feature mapping between emotion and music can be referred by [13, 14] with regression.

J.-S. Pan et al. (eds.), *Advances in Intelligent Information Hiding and Multimedia Signal Processing*, Smart Innovation, Systems and Technologies 81,
DOI 10.1007/978-3-319-63856-0_14

2 Method

The A-V 2D emotion plane (as shown in Fig. 1) is applied for our proposed algorithmic music composition, with Markov Chain [15, 16] based stochastic process to generate music pitches and chords automatically. Figure 2 shows the music generation method of the proposed system. A typical finger measuring heartbeat module, as shown in Fig. 3, is connected to Arduino microcontroller board for the user's biofeedback data, to obtain the correspondent emotion states in the A-V 2D emotion plane.

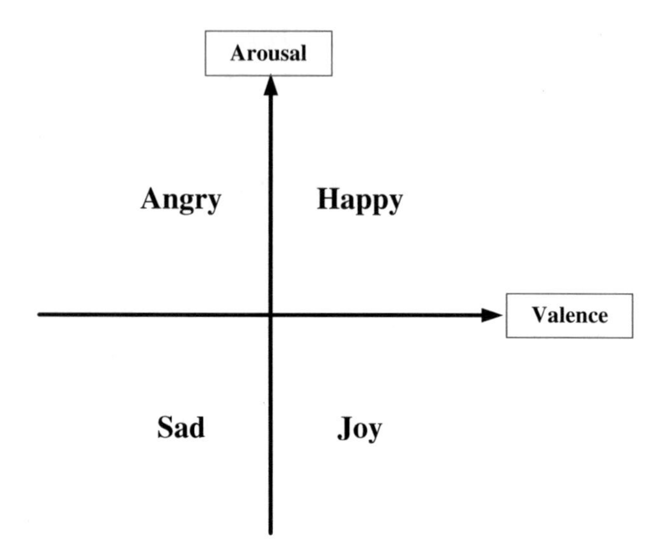

Fig. 1. 2D emotion plane

Fig. 2. The music generation method of the proposed system

According to music feature and physiological relation research [13, 14], the regression equations can be used to map the music features into the correspondent emotion Arousal-Valence states, therefore the music features can be converted from the 2D emotion plane A-V states too.

Fig. 3. The heart rate sensor and arduino microcontroller board

The experiment includes the following steps:

(1) Using the finger measuring heartbeat sensor to measure with the subject's right hand middle finger;
(2) Wearing headphones;
(3) Listening to music with the following sequence: (a) Pre-rest for 1 min; (b) Listening to the generative "Sad Music" for 3 min; (c) Rest for 1 min; (d) Listening to the generative "Happy Music" for 3 min; (e) Rest for 1 min; (f) Listening to the generative "Joyful Music" for 3 min; (g) Rest for 1 min; (h) Listening to the generative "Angry Music" for 3 min; (i) Post-rest for 1 min;

3 Result

Human heartbeat in the resting condition is usually in the range of 50 ~ 90 BPM [17]. According to the research [13, 14], heartbeat will be increased with the music changed to a faster tempo or modulated in a major mode; while the heartbeat going slower with the tempo decreased or the mode changed to a minor key. For the music tempo, there are seven levels with the middle value 3.5, and the following regression equation [13, 14] can be used:

$$\text{Tempo} = 1.2 + 0.1 \times \text{Valence} + 0.43 \times \text{Arousal} \tag{1}$$

When we need to revive our energy, the music tempo level can go up to 4.0, and keep watching the measured heartbeat to see if it is increased. If the heartbeat is increased, then the tempo level can be maintained to 4.0, if not, then the level should be increased to 4.5 or 5 until the heartbeat rate is 5% higher than the original.

If we need to relieve our tension, then the tempo level can be set to 3.0, and keep watching the measured heartbeat to see if it is decreased. If the heartbeat is decreased, then the tempo level can be maintained to 3.0, if not, then the level should be decreased to 2.5 or 2 until the heartbeat rate is 5% lower than the original.

In the other hand, generative music can be in either major or minor modes. When "riving energy" is needed, major mode is entered with the mode level of 7. On the contrary, if "relieving tension" is required, then minor mode is entered with the mode level of 0, with the regression equation [13, 14]:

$$\text{Mode} = 6.4 - 0.27 \times \text{Valence} - 0.71 \times \text{Arousal} + 0.11 \times \text{Valence} \times \text{Arousal} \quad (2)$$

With Tempo Eq. (1) and Mode Eq. (2), both valence and arousal variables can be solved to integrate with the 2D emotion A-V plane, to generate music automatically. The average heartbeat values correspondent to various emotion states, including Pre-Rest, Sad, Rest, Happy, Rest, Joyful, Rest, Angry, and Post-Rest, are shown in the blue area of Fig. 4. The raw data is also shown in the yellow area.

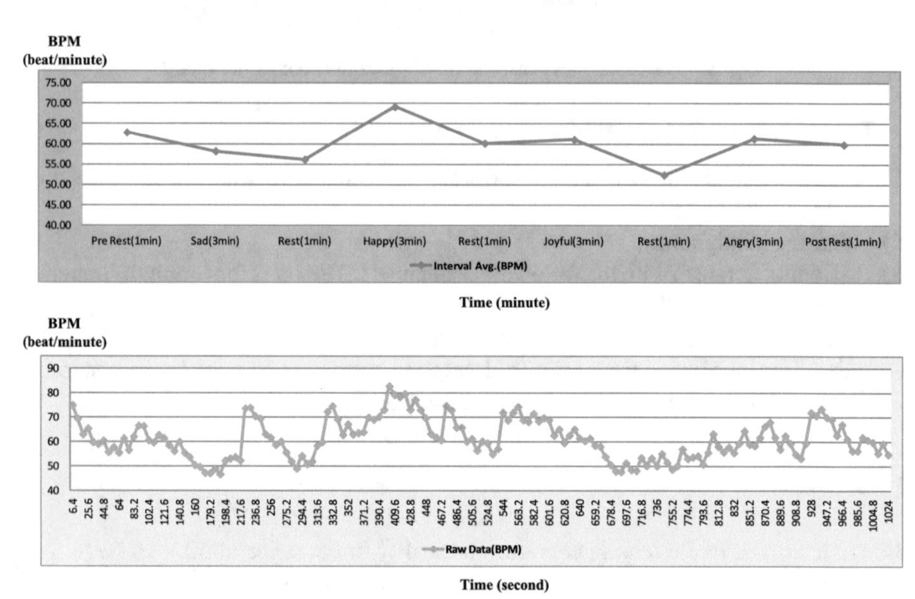

Fig. 4. The correspondent heartbeat data with various emotion states

Figure 5 shows the score of the generative Happy Music, while Fig. 6 shows the score of the generative Joyful Music.

Fig. 5. Generative happy music score

Fig. 6. Generative joyful music score

4 Conclusion

This proposed system shows a successful way to compose music with the input from heart rate sensor, with various emotion states automatically. The generative Happy Music shows the highest heartbeat 69.17 BPM, while the generative Sad Music shows its correspondent heartbeat 58.19 BMP. In the future, we would add more biosensors such as respiratory sensor, GSR (Galvanic Skin Response), etc., with IoT (Internet of Things) and wearing technologies, to integrate more music features, and make the system more convenient and precise with more music varieties.

Acknowledgement. The authors would like to appreciate the support from Ministry of Science and Technology projects of Taiwan: MOST 105-2410-H-424-008 and MOST 105-2218-E-007-031.

References

1. George, P., Wigginsm, G.: AI methods for algorithmic composition: A survey, a critical view and future prospects. In: AISB Symposium on Musical Creativity, Edinburgh, UK (1999)
2. Nierhaus, G.: Algorithmic Composition: Paradigms of Automated Music Generation. Springer, New York (2009)
3. Alpern, A.: Techniques for algorithmic composition of music 95, 120 (1995). On the web http://hamp.hampshire.edu/adaF92/algocomp/algocomp
4. Sandred, O., Laurson, M., Kuuskankare, M.: Revisiting the Illiac Suite–a rule-based approach to stochastic processes. Sonic Ideas/Ideas Sonicas 2, 42–46 (2009)
5. Biles, J.A.: GenJam: A genetic algorithm for generating jazz solos. In: ICMC, vol. 94 (1994)
6. Dubnov, S., et al.: Using machine-learning methods for musical style modeling. Computer **36**(10), 73–80 (2003)
7. Boulanger-Lewandowski, N., Bengio, Y., Vincent, P.: Modeling temporal dependencies in high-dimensional sequences: Application to polyphonic music generation and transcription (2012). arXiv preprint arXiv:1206.6392
8. Sun, K., et al.: An improved valence-arousal emotion space for video affective content representation and recognition. In: 2009 IEEE International Conference on Multimedia and Expo ICME 2009. IEEE (2009)
9. Bustamante, P.A., et al.: Recognition and regionalization of emotions in the arousal-valence plane. In: 2015 37th Annual International Conference of the IEEE Engineering in Medicine and Biology Society (EMBC). IEEE (2015)
10. Kim, Y.E., et al.: Music emotion recognition: A state of the art review. In: Proceedings of the ISMIR. (2010)
11. Barthet, M., Fazekas, G., Sandler, M.: Multidisciplinary perspectives on music emotion recognition: Implications for content and context-based models. In: Proceedings of the CMMR, pp. 492–507 (2012)
12. Laurier, C., et al.: Exploring relationships between audio features and emotion in music. In: 7th Triennial Conference of European Society for the Cognitive Sciences of Music ESCOM 2009 (2009)
13. Gomez, P., Danuser, B.: Affective and physiological responses to environmental noises and music. Int. J. Psychophysiol. **53**(2), 91–103 (2004)
14. Gomez, P., Danuser, B.: Relationships between musical structure and psychophysiological measures of emotion. Emotion **7**(2), 377–387 (2007). Washington DC
15. Jacob, B.L.: Algorithmic composition as a model of creativity. Organised Sound **1**(03), 157–165 (1996)
16. Bell, C.: Algorithmic music composition using dynamic Markov chains and genetic algorithms. J. Comput. Sci. Coll. **27**(2), 99–107 (2011)
17. Hoffmann, M.H.K., et al.: Noninvasive coronary angiography with 16–detector row CT: Effect of heart rate. Radiology **234**(1), 86–97 (2005)

Inter-vehicle Media Distribution for Driving Safety

Sheng-Zhi Huang, Chih-Lin Hu(✉), Ssuwei Chen, and Liangxing Guo

Department of Communication Engineering, National Central University, Taoyuan City 32001, Taiwan, R.O.C.
{shengzhi,liangxing}@g.ncu.edu.tw, clhu@ce.ncu.edu.tw, chenssuwei@gmail.com

Abstract. Regarding conventional driving scenarios, vehicles are isolated and self-secured without interaction although many vehicles may exist in a surrounding. Driving safety is thus secured by individual driving behavior; that says, drivers must take care of themselves. Whereas many technologies contribute to the rapid development of vehicular networking and applications, it can lead to better quality of driving experience. This paper proposes a system and method for inter-vehicle media content distribution for driving safety. The system aims to expand the range of a drivers' vision by sharing front-end views of neighbor vehicles on the road. In this system, we use the Wi-Fi facility to establish an inter-vehicle communication network, use the UPnP protocol to manage vehicles, and then use RTSP to deliver real-time video to other vehicles in a network. Thus, drivers can catch a broad view through the sharing services of front views. In addition, this system provides functions to record images, road features and geographical information, and then to transfer them onto a back-end database. The system will be able to notify traffic accidents and situations on the warming map, so the drivers can browse the information to recognize what happens in front of them before approaching to the area. Accordingly, we design a system architecture, develop a software prototype, and demonstrate practical scenarios.

Keywords: Driving safety · Media distribution · Inter-vehicle networks · Vehicular networks

1 Introduction

Traditional usages to learn traffic information are to listen radio broadcasting, browse traffic news on Web, and launch traffic report applications on mobile phones. These information services are beneficial to people who are going to drive on the way, but may be just informative somewhat to drivers who are always on the roads. Drivers care more about driving safety than traffic information, that is, what dangerous situations and threats in the surrounding of a driver. For example, in the process of driving, because of the different size of vehicles, there is a threatening event that a large vehicle blocks the sight of line of small vehicles nearby. Without clear sight of the road, drivers may fall in fatal threats because they cannot watch out the traffic road ahead and timely respond to dangers in front of them. Another example is that when drivers are not familiar

J.-S. Pan et al. (eds.), *Advances in Intelligent Information Hiding and Multimedia Signal Processing*, Smart Innovation, Systems and Technologies 81,
DOI 10.1007/978-3-319-63856-0_15

with routes and use the assistance of traffic navigation systems, they cannot always pay attention to the front view of the roads.

To boost up the traffic safety, the ordinary approach is based on Intelligent Transport Systems (ITS). Meanwhile, vehicular ad hoc networks (VANET) provide a vehicular computing environment, thanks to the advance of wireless communications, portable information devices, and network built-in vehicles, where new scenarios and services are burgeoning. From the prospective of service development in vehicular environments, vehicle- and driver-provided traffic information services will be flexible and time to market regardless of lengthy system integration and standardization processes.

In this paper, we design an inter-vehicle media distribution system for driving safety based on self-organized networks among neighbor vehicles on the roads. Ideally, we let neighbor vehicles communicate with each other, set up a vehicular network and share front views on the roads, thus expecting to reduce traffic accidents and elevate driving safety in a cooperative fashion. Our proposed system and methods have the following characteristics:

- Some vehicles in geographic proximity can build an inter-vehicle communication group through Wi-Fi technologies with drivers' willing to join or leave.
- A self-organized network is linked locally, and thus vehicles in a group can discover each other using the Universal Plug and Play (UPnP) protocols.
- Real-time video transmission is based on Real-Time Streaming Protocol (RTSP). Captured image and recorded video can be immediately shared to the members of a group.
- Captured image, recorded video, and road-segment information can be uploaded to a back-end database on the Clouds.
- Compilation of uploaded traffic information can result in a "warning map."
- The system can notify drivers of what happened in front of them, and vice versa, drivers can watch the map to recognize the situation in the neighborhood.

Hence, we design a system architecture, develop a prototype, and demonstrate practical scenarios. The result shows that the proposed system can help to solve the problem of the blind of sight using inter-vehicle cooperation in vehicular networks. Drivers can pro-actively sense dangerous situations and keep from possible traffic accidents.

The rest of this article is organized as follows. Section 2 presents related work. Section 3 describes the system design and development. Section 4 shows implementation and demonstration. Finally, the paper concludes in Sect. 5.

2 Related Work

Most data distribution technologies in vehicular environments generally arise from two research aspects, i.e., vehicle-to-vehicle (V2V) and vehicle-to-infrastructure (V2I). In addition to many layer-1 and layer 2 broadcasting schemes in V2V, [4] proposed a P2P mixed with V2V. This work used the concept of content dissemination and cache update (CDCU) and media-service counter (MSC). This work expected to solve the fairness problem as media sharing between V2V and V2I.

[5] adopted relay vehicles with more resource and capabilities. A relay node can provide network information to other vehicles inside its transmission range as defined by a cluster network. Assume that each vehicle owns sensors to detect its speed, GPS location, and moving direction. The packets to a relay vehicle will be further broadcast to the members of a cluster. [6] mentioned an information-sharing scheme for location-dependent data. They solved how to push or pull messages from vehicles to areas with different densities. Explicitly, it will be better to find cars in an area of high density to broadcast messages in urban areas. The messages in a high-density area will be broadcast fast and scattered efficiently, thus leading to the minimal cost of packet delivery. Compared with [6, 7] mentioned how to resolve packet collisions in a high-density area. When one vehicle wants to send packets, it will broadcast hello messages. Only other vehicles, which send ACK packet back, will be the candidates. Then, they used random network coding to improve the data throughput. This scheme can reduce the collisions by adjusting the frequency of hello messages.

Furthermore, when some vehicles inter-contact in a local area as the V2V context, our study can apply previous results [2, 3] to perform device and service discovery between vehicles. In [2], the UPnP protocol stack consists of six functional layers, including IP addressing, discovery, description, control, eventing and presentation. On top of TCP/IP layers, vehicles can discover other vehicles and advertise some services in a local network, In [3], a home gateway design was used to integrate P2P and home networks. Note that this role can play the similar functions to a relay nodes of a cluster on a V2V network. Thus, media content sharing can be facilitated by relay nodes that are capable of network bridging and media transformation.

To realize real-time media distribution in a network, prior studies mainly used the Real Time Streaming Protocol (RTSP) and H.264 as the encoding of the video. [8] discussed real-time multi-view video broadcasting to other vehicles and focused on maximizing the video quality. Highway vehicles are in the form of a platoon. Each vehicle will receive video streaming from road site units. By computing the distance from the vehicle to RSU, it will allocate the bandwidth and channel capacity. For multi-view video, users may switch to another view periodically. [9] used H.264 coding as considering the scenario that the quality of service (QoS) would worsen when vehicles play media streaming and move simultaneously. This work proposed that the fleet of vehicles can communicate with each other at the k-hop distance. Also, this work discussed how to adjust the bandwidth for each downlink to become an adaptive video streaming scheme.

Besides the connection in the V2V context, some recent works made connections outwards to the clouds. For example, [10] transferred the data on roads to be a traffic map in a 3D model. For example, the steep slope may lead to danger, so the color of the area in the warning map will be red as represented the danger. By analyzing the picture of roads, they rebuilt risk zones, called a "warning map."

3 System Model

We describe the design abstraction of the proposed system, as follows. The primary goal is to exempt from the problem of "blind vision of a driver" imposed by neighbor vehicles and roadside obstacles on roads. We think this problem can be resolved if any affected vehicles could get the visions from other vehicles in front of them.

3.1 Design Abstraction

To support V2V communications among vehicles, this system enables vehicles to build or join a vehicle's self-organized wireless network. Following TCP/IP conventions, the system employs the UPnP technology by which vehicles can discover others and their services in a local network. The system specifies a new type of vision sharing service. Vehicles can offer this service and distribute their real-time front visions in a request-response manner over UPnP communications. Drivers can thus promptly sense potential danger with other vehicles' shared visions. In addition, the system can remind drivers what just happened in front of them using traffic information sharing over a social-like driver community. The system can timely push location-dependent information to vehicles in any particular areas wherein some traffic accidents or events may occur. Furthermore, as a result of the history of traffic information, the system can analyze tendency of potential hazards. Thus, the system can offer warning information to have drivers' caution when they will go through some often-dangerous areas.

Accordingly, as Fig. 1 illustrates, we design the system which contains four subsystems: real-time video transmission, database management, warning map and voice control, as mentioned in respective subsections.

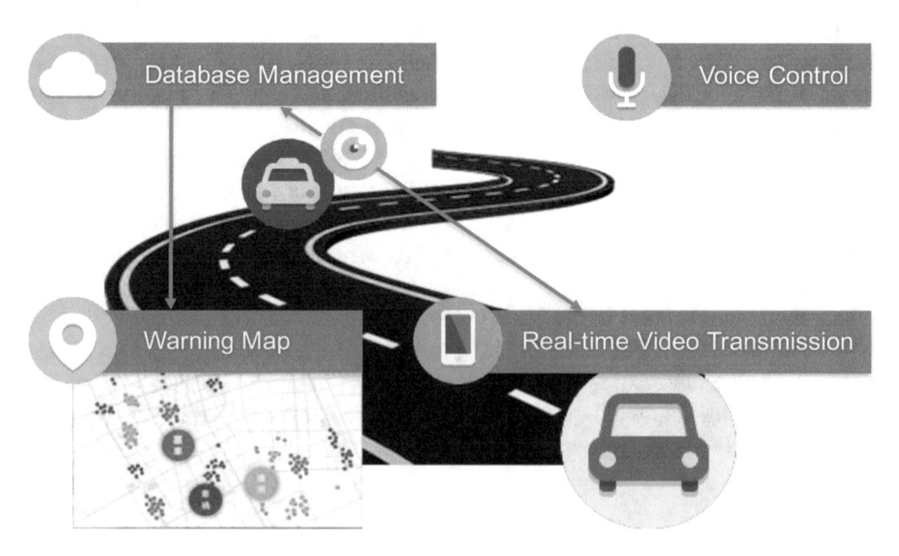

Fig. 1. System architecture and key components.

3.2 Real-Time Video Transmission

The system commences at the first step to connect vehicles in the surrounding. Each vehicle automatically detects the presence of any existing V2V networks that other vehicles self-organize with WiFi-enabled hotspots. Otherwise, it creates one and waits for neighbor vehicles' joining. Figure 2 depicts the action flow of an initial step.

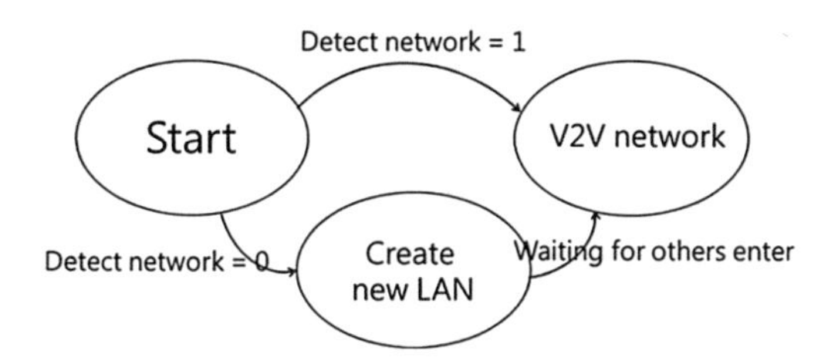

Fig. 2. Devices added into a V2V network

In a WiFi network, vehicles interact each other in light of standard HTTP/TCP/IP conventions. Vehicles are aware of others using UPnP-specific SSDP protocols. They send two types of service packets for inter-vehicle communications, as follows.

$$< \text{packet type-A} > : = < \text{IP address} >< \text{User name} >< \text{Service name} >$$
$$< \text{packet type-B} > : = < \text{IP address} >< \text{Service name} >< \text{Level of quality} >$$

There are two types of vehicles: one is to share real-time media, and the other is to select and watch media. Figure 3 shows, when a driver wants to share real-time captured view, it creates a specific WiFi hotspot, broadcasts packet type-A, and waits for someone connects. After another vehicle is connecting, it sends a packet type-B to inform the hosting vehicle of selecting quality level of data streaming. On the other hand, a driver can enter the selection mode to indicate which vehicle's front view. All above interactive messages follow the UPnP-specific formats. Thus, media format negotiation and transport will be conducted by IP-based connections.

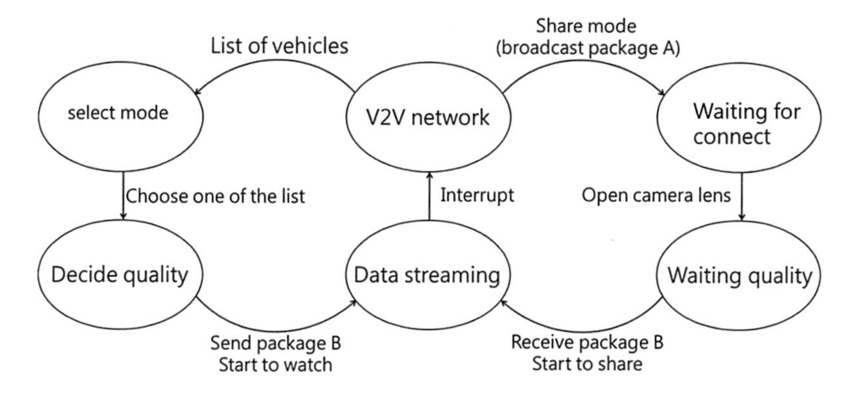

Fig. 3. Sharing real-time media between vehicles database management

The system maintains two types of information in a database. One is the information about roads, including GPS coordinates with events. Three events are defined in the system, including vehicle crash, traffic jam and illegal checking spots by polices. The other is driver-provided video content about events. As Fig. 4 shows, a driver can choose events to upload to the database. The system uses HTTP RESTful API to upload information, so data are formatted using JSON. An uploaded media object will be linked by a particular URL in reference to the event and object.

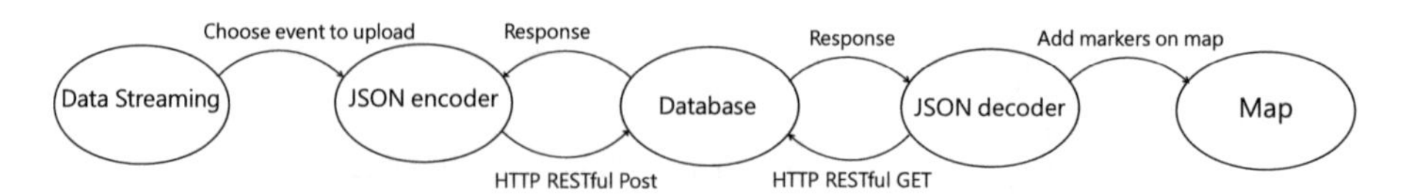

Fig. 4. Data uploaded to or downloaded from the database on the server

3.3 Warning Map

The vehicles can download traffic information from the server and accordingly generate a warning map. On the map, different kinds of events are labeled with colorful icons to catch drivers' caution. As Fig. 4 shows, the system uses HTTP GET to download traffic information. After decoding JSON, the system displays events at the location in a map.

3.4 Voice Control

In addition to manual operations, the system offers the voice control interface based on Google Speech API. Basically, the system transfers a driver's voice to string and compares it with certain control patterns. Two interfaces are supported to select menu items and to select which a vehicle will be connected to obtain shared media.

4 Demonstration

To realize real-time media distribution in a V2V network, we develop a software implementation with Android mobile phones. Drivers can run pre-installed mobile applications and place them in the front of vehicles. Figure 5 presents some snapshots.

The usage is that a driver first checks whether any V2V networks exist or not in its surrounding. If yes, a driver will select a network to connect. Otherwise, a driver can build a new network whose SSID is set as v2v-USERNAME. As shown in the right-up picture, there are four menu items of the screen from; from left to right they are "share real-time video," "select a video to watch," "update the map," and "upload video to the server." When a driver wants to share real-time front views, he/she will broadcast the IP address and license plate number of the vehicle to others in the same network by UPnP. On the other hand, when a driver wants to watch another view, he/she will select a vehicle indicated by the license plate number that can be input by typing or voice. In addition, a driver can adjust the video quality of media streaming over H.264 corresponding to the list of quality choices that the source vehicle can offer. After a media transport task is negotiated by UPnP, the content can be transmitted. Then, a driver can upload structured data to the server so as to record traffic events. As shown, from left to right, the events are "vehicle crash," "illegal driving behavior by the police," and "traffic jam." When a driver selectively touches a button, the information including the type of event, GPS coordinate, and username will be stored in the database. To bring better user experience, we use Google Map to show the history of road events, thereby resulting in a warning map. For example, red is for vehicle crash, yellow is for the spot of detecting illegal cases and blue is for traffic jam.

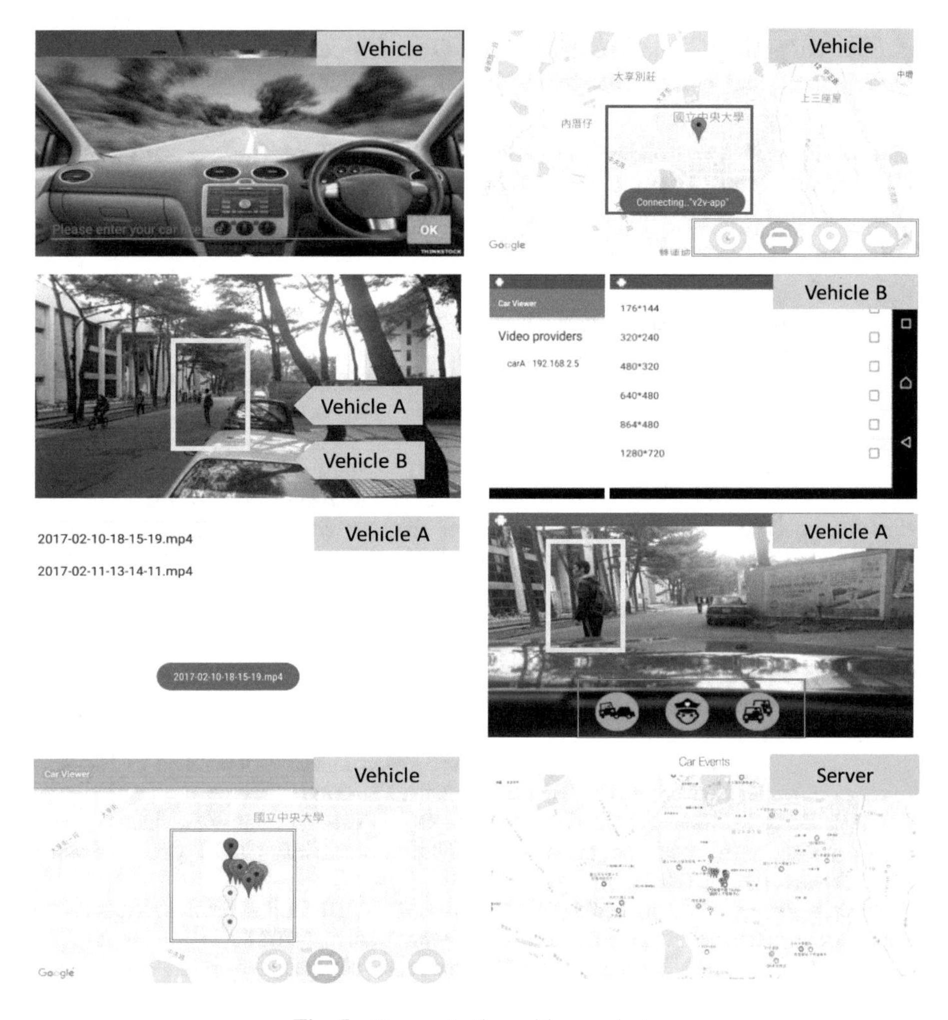

Fig. 5. Demonstration with snapshots

5 Conclusion

This paper presents a novel system design for inter-vehicle media distribution for driving safety on roads. This design can be deployed in a WiFi-enabled V2V network where neighbor vehicles can discover each other and access shared real-time media services over UPnP communications. In addition, this design includes a database on the Internet. Vehicles can upload traffic videos and events to the server, and the server can maintain the history of traffic information. Accordingly, a vehicle can generate a warning map in its surrounding by downloading location-dependent traffic information from the server. As a result of prototype implementation, this system is able to assist drivers in extending the view scope and avoiding dangerous driving threats on roads.

Acknowledgment. This work was supported in part by the Ministry of Science and Technology, Taiwan, under Contract MOST-105-2221-E-008-029-MY3.

References

1. Hartenstein, H., Laberteaux, K. (eds.): VANET: Vehicular Applications and Inter-Networking Technologies. Wiley, Chichester (2010)
2. Hu, C.-L., Liao, W.-S., Huang, Y.-J: Mobile media content sharing in UPnP-based home network environment. In: International Symposium on Applications and the Internet, vol. 8, pp. 1753–1769 (2008)
3. Hu, C.-L., Lin, H.-C., Hsu, Y.-F., Hsieh, B.-J.: A P2P-to-UPnP proxy gateway architecture for home multimedia content distribution. KSII Trans. Internet Inf. Syst. **6**, 405–424 (2012)
4. Huang, C.-M., Yang, C.-C., Lin, Y.-C.: An adaptive video streaming system over a cooperative fleet of vehicles using the mobile bandwidth aggregation approach. IEEE Syst. J. **10**, 568–579 (2016)
5. Chiu, K.-L., Huang, R.-H., Chen, Y.-S.: Cross-layer design vehicle-aided handover scheme in VANETs. Wirel. Commun. Mob. Comput. **9**, 916–928 (2009)
6. Okamoto, J., Ishihara, S.: Distributing location-dependent data in VANETs by guiding data traffic to high vehicle density areas. In: IEEE Vehicular Networking Conference, vol. 2, pp. 189–196 (2010)
7. Kusumine, N., Ishihara, S.: R2D2 V: RNC based regional data distribution on VANETs. In: IEEE Vehicular Networking Conference, vol. 2, pp. 271–278 (2010)
8. Liu, Z., Dong, M., Zhang, B., Ji, Y., Tanaka, Y.: RMV: Real-time multi-view video streaming in highway vehicle ad-hoc networks (VANETs). In: IEEE Global Communications Conference, pp. 1–6 (2016)
9. Zhou, L., Zhang, Y., Song, K., Jing, W., Vasilakos, A.-V.: Distributed media services in P2P-based vehicular networks. IEEE Trans. Veh. Technol. **60**, 692–703 (2011)
10. Arora, P., Corbin, D., Brennan, S.-N.: Variable-sensitivity road departure warning system based on static, mapped, near-road threats. In: IEEE Intelligent Vehicles Symposium, pp. 1217–1223 (2016)

High-Capacity ECG Steganography with Smart Offset Coefficients

Ching-Yu Yang[1(✉)] and Wen-Fong Wang[2]

[1] Department of Computer Science and Information Engineering, National Penghu University of Science and Technology, 300, Li-Ho Road, Makung, Penghu 880, Taiwan
chingyu@npu.edu.tw

[2] Department of Computer Science and Information Engineering, National Yunlin University of Science and Technology, Douliu, Yunlin 640, Taiwan
wwf@yuntech.edu.tw

Abstract. The authors present an economic way to hide patient diagnosis information in electrocardiogram (ECG) signal based on smart offset coefficient. Simulations indicated that hiding capacity is larger than existing techniques while the perceived quality is good. Moreover, the method is tolerant of the attacks such as inversion, translation, truncation, and Gaussian noise-addition attacks, which is rare in conventional ECG steganographic schemes. Since the privacy (or medical message) of the patients can be fast and effective embedded in ECG host by the proposed method, it is feasible for our method being employed in real-time applications.

Keywords: Data hiding · ECG steganography · Smart offset coefficients · Real-time applications

1 Introduction

Due to the ubiquitous broadband services, cloud computing, intelligent robots and internet of thing, it is convenient and easy for the individuals and organizations to share their resources and messages among one another. As we know, data can be eavesdropped and tampered with during transmission in the intranet (or internet) by the third parties (or adversaries). To protect private (or important) data, people often employ either encryption/decryption systems or data hiding techniques to achieve the goal. However, compared with the encryption/decryption systems, data hiding techniques have a merit of providing an economic way for the protection of secret message in multimedia or patient data such as blood pressure, glucose and temperature in biometric signal [1,2]. For example, an electrocardiogram (ECG) data is one of biometric signal being used by doctors to diagnose patient shape. Namely, an ECG signal provides a clear vision of the activities

J.-S. Pan et al. (eds.), *Advances in Intelligent Information Hiding and Multimedia Signal Processing*, Smart Innovation, Systems and Technologies 81,
DOI 10.1007/978-3-319-63856-0_16

of our heart. The cardiac cycle is characterized by five separate waves of deflections designated as P, Q, R, S and T [3]. Several ECG steganography approaches have been suggested to secure patient information [4–9]. Ibaida et al. [4] presented an ECG watermarking scheme for a wearable sensor-net health monitoring system. The payload was approximately 10,000 bits, and the method was tolerant of manipulations. Ibaida and Khalil [5] suggested an ECG steganographic approach via encryption and scrambling on the basis of discrete wavelet transform (DWT). The payload size was 2,531 bytes. Based on singular value transform (SVD) and DWT, Jero et al. [6] proposed an ECG steganography for protecting confidential data. Simulations indicated that the scheme had a good performance in percentage residual difference (PRD) and bit error rate (BER) when the high-high sub-band was used for hiding a 4,489-bit secret message. Based on integer wavelet domain (IWT), Yang and Lin [7] embedded data bits in an ECG host signal via a coefficient adjusting technique. Simulations indicated that the perceived quality was good with a high hiding-capacity. Further, Yang and Wang [8] proposed two hiding methods: lossy and reversible ECG steganographys for ECG signals. Simulations confirmed that the perceived quality generated by the lossy ECG steganography method was good, while hiding capacity was acceptable. In addition, the reversible one can not only hide secret messages but also completely restore the original ECG signal after bit extraction. In combination use of continuous ant colony optimization and DWT-SVD watermarking schemes, Jero et al. [9] presented a novel ECG steganography. The scaling factor in the quantization technique governs the tradeoff between perceived quality and robustness. Simulations indicated that the PRD was 0.007 with payload of size 1.77 kb.

From the above survey, we can see that either perceptual quality or payload size is not good enough. In this article, we present a better hiding technique to protect patient diagnosis information in ECG host signal. The remainder of this paper is organized as follows. Section 2 describes the procedures of bit embedding and bit extraction. Section 3 presents the simulation results, and Sect. 4 provides the conclusions of this study.

2 Proposed Method

To provide a high hiding-capacity, the proposed method embeds two data bits in a bundle of coefficients from the host ECG signal at a time. The details of bit embedding and bit extraction for our methods are specified in the following sections.

2.1 Bit Embedding

Let $H_j = \{s_{ji}\}_{i=0}^{n-1}$ be the j-th bundle of size n taken from a host ECG data, as shown in Fig. 1, with $n = 3$.

Below are the major steps of bit embedding for the proposed method.

Step 1. Input a bundle from the host ECG data. If the end of input is encountered, then proceed to Step 7.

S_{j0}	S_{j1}	S_{j2}

Fig. 1. Bundle of size 3.

Step 2. Input two data bits $b_m b_n$ from the secret message.
Step 3. If $b_m b_n = 00$, then call the procedure "HidingCase00" and go back to Step 1.
Step 4. If $b_m b_n = 01$, then call the procedure "HidingCase01" and go back to Step 1.
Step 5. If $b_m b_n = 10$, then call the procedure "HidingCase10" and go back to Step 1.
Step 6. If $b_m b_n = 11$, then call the procedure "HidingCase11" and go back to Step 1.
Step 7. Stop.

The procedure of "HidingCase00" and the corresponding coefficient-alignment are shown in Figs. 2(a)–(d). In Fig. 2(a), if the input data bits are "00" and both conditions $s_{j1} > s_{j0}$ and $s_{j1} > s_{j2}$ are satisfied, nothing is done

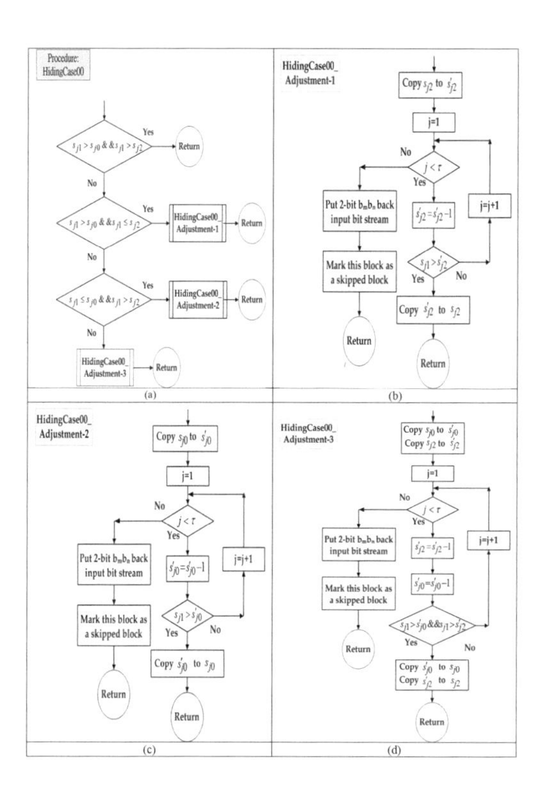

Fig. 2. Flow charts of HidingCase00. (a) Main procedure, (b) HidingCase00_Adjustment-1, (c) HidingCase00_Adjustment-2, and (d) HidingCase00_Adjustment-3.

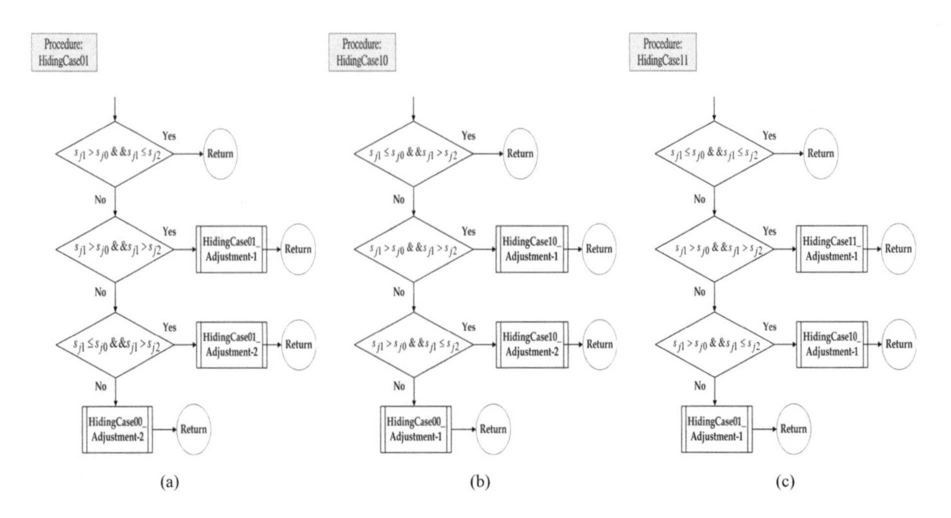

Fig. 3. Flow charts of the other three Main procedures. (a) HidingCase01, (b) HidingCase10, and (c) HidingCase11.

(which means that the block carries "00" data bits); otherwise, an adjustment policy is executed on the target coefficients s_{j0} and/or s_{j2} of the bundle (see Figs. 2(b)–(d)). Figures 3(a)–(d) show the main procedures of "HidingCase01," "HidingCase10," and "HidingCase11," respectively. Due to the limitation of the space, the sub-procedures such as "HidingCase01_Adjustment-x," "HidingCase10_Adjustment-x," and "HidingCase11_Adjustment-x" were omitted here.

Note that the control parameter τ used in our algorithm was determined by trial-and-error. An optimal SNR can be generated with the least number of skipped bundles. Notice as well a skipped bundle was rarely encountered in the proposed method during simulations. In fact, all input bits were directly (or undergone successful adjustment and) embedded in the host bundles of ECG with null skipped blocks.

2.2 Bit Extraction

The procedure of bit extraction for the proposed method is much simpler than that of bit embedding. The hidden message can be extracted according to the following steps:

Step 1. Input a bundle $\hat{H}_j$ from the marked ECG data. If the end of input is encountered, proceed to Step 6.

Step 2. If $\hat{H}_j$ is a skipped block, proceed to Step 1.

Step 3. If $\hat{s}_{j1} > \hat{s}_{j0}$ and $\hat{s}_{j1} > \hat{s}_{j2}$, are satisfied, then two data bits "00" are extracted, proceed to Step 1.

Step 4. If $\hat{s}_{j1} > \hat{s}_{j0}$ and $\hat{s}_{j1} \leq \hat{s}_{j2}$, then data bits "01" are extracted, proceed to Step 1.

Step 5. If $\hat{s}_{j1} \leq \hat{s}_{j0}$ and $\hat{s}_{j1} > \hat{s}_{j2}$, then data bits "10" are extracted, otherwise, "11" are extracted, and proceed to Step 1.
Step 6. Assemble all extracted bits and rebuild the secret message.
Step 7. Stop.

3 Experimental Results

To evaluate the performance of our proposed method, a set of 32 1D ECG host signal was taken from the Lead-1 in an ECG database, which obtained from the MIT-BIH arrhythmia database [10]. Each host data consists of 30,000 coefficients. A binary watermark was used as test data. The size of the bundle was 3 and each test data used various τ-value. Since two data bits can be embedded in each coefficient, the optimal payload for the proposed method is $(30,000 \div 3) \times 2 =$ 20,000 bits. Three objective measurements: signal-to-noise-ratio (SNR), mean absolute errors (MAE) and percentage residual difference (PRD), which used to assess the performance of the proposed method was given in Table 1. It can be seen from

Table 1. SNR/MAE/PRD performance of the proposed method with various τ, and SNR performance comparison with various methods.

ECG data sets	SNR/MAE/PRD/τ	Ref. [7]	Ref. [8]	Our method
102	45.44/1.32/0.0053/86	44.44	44.04	45.44
108	50.17/1.25/0.0031/46	40.03	49.2	50.17
109	45.06/1.86/0.0056/61	49.14	42.98	45.06
111	48.22/1.51/0.0039/34	43.42	46.43	48.22
112	46.59/1.28/0.0047/43	44.86	45.28	46.59
114	49.09/1.26/0.0035/60	48.05	47.53	49.09
117	45.81/1.39/0.0031/63	43.87	44.11	45.81
121	49.91/0.98/0.0032/33	48.4	48.25	49.91
124	43.54/1.49/0.0067/71	41.93	42.45	43.54
200	43.98/1.92/0.0063/93	42.18	42.14	43.98
201	47.62/1.25/0.0042/48	46.14	45.72	47.62
202	47.82/1.39/0.0041/53	46.33	46.34	47.82
205	44.84/1.34/0.0057/65	42.91	46.26	44.84
207	49.37/1.33/0.0034/45	47.8	46.66	49.37
221	44.18/1.75/0.0062/80	41.84	42.57	44.18
222	47.65/1.31/0.0041/69	46.49	46.29	47.65
231	41.67/1.82/0.0083/119	39.77	39.92	41.67
232	48.65/1.30/0.0037/58	46.96	46.63	48.65
Average	44.93/1.63/0.0060/–	44.70	45.16	46.65

Table 1 that the smaller the value of τ, the larger value of SNR with a small MAE value. Notice that no skipped block was used here during simulations. The SNR, MAE and PRD are defined as follows:

$$SNR = 10\log_{10}\frac{\sum_i {s_i}^2}{\sum_i (s_i - \hat{s}_i)^2}, \tag{1}$$

$$MAE = \frac{1}{N}\sum_{i=1}^{N}\left|s_i - \hat{s}_i\right|, \tag{2}$$

$$PRD = \sqrt{\frac{\sum_i (s_i - \hat{s}_i)^2}{\sum_i {s_i}^2}}, \tag{3}$$

respectively, where s_i and $\hat{s}_i$ are the values of the coefficients in original ECG and marked ECG. Performance comparison with Yang and Lin's technique [7] and Yang and Wang's approach [8] was listed in Table 1. It is obvious that the average SNR of the proposed method is the best among these compared methods. Notice that the payload provided by both two methods [7,8] was around 15,000 bits.

The marked ECG generated from ECG100, ECG114, ECG121, and ECG201 via the proposed method were depicted in Fig. 4. The perceived quality is quite good. No apparent distortion existed in the marked ECGs. Close observation of these marked ECGs (randomly selected in 10–12 s interval) were shown in Fig. 5. It can be seen that the marked signal introduced by the proposed methods (red line) was approximately similar to the original one (blue line), i.e., the distortion generated by the proposed methods was insignificant. Furthermore, the proposed

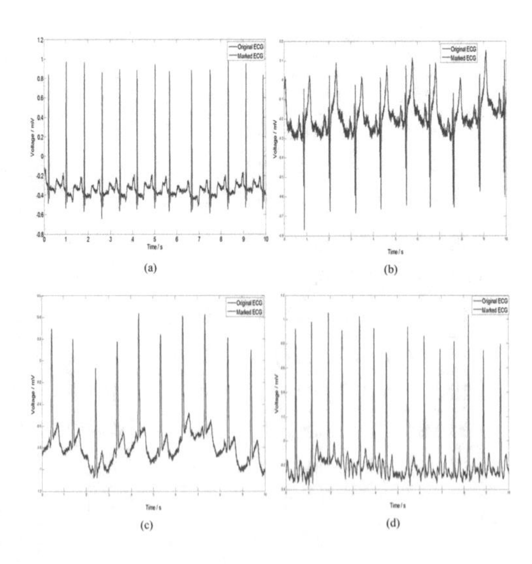

Fig. 4. Marked ECGs generated by the proposed method. (a) ECG100, (b) ECG114, (c) ECG121, and (d) ECG201.

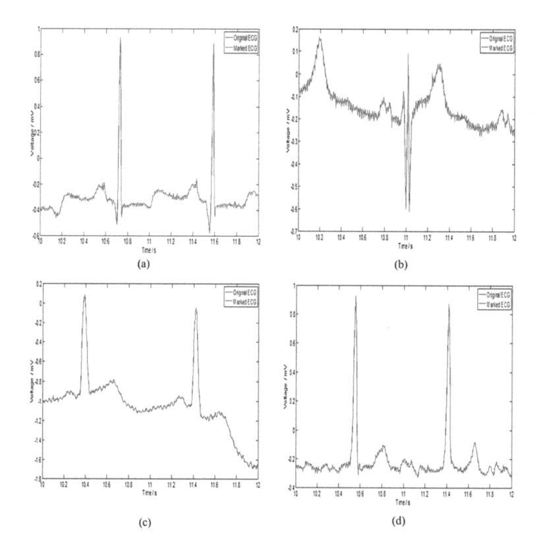

Fig. 5. Close observation of the host and marked ECGs generated by the proposed method. (a) ECG100, (b) ECG114, (c) ECG121, and (d) ECG201.

Attacks	Null-attack PRD = 0.0000	Inversion PRD = 1.8347	Truncation† PRD = 1.3111	Translation (+500) PRD = 0.0000	White-Gaussian noise (with SNR of 3 dB) PRD = 0.8921	White-Gaussian noise (with SNR of 2 dB) PRD = 0.9233	White-Gaussian noise (with SNR of 1 dB) PRD = 0.9719
Survived Watermarks							

† The last two bits of the marked data were truncated

Fig. 6. Examples of survived watermarks from the manipulations of marked ECG121.

method has a certain degree of robustness. Examples of survived watermarks from the manipulations of marked ECG121 (using $\tau = 33$) was given in Fig. 6. An extra binary logo of size 141×141 was used as a test watermark. The value of PRD equals 0 if a marked ECG were not being manipulated. In spite of the PRD of the survived watermarks which manipulated by "Inversion" and "Truncation" were larger than 1, it was recognized. Moreover, the marked ECG data under white-Gaussian noise attacks of three different signal strengths (with 1–3 dB) were demonstrated. Although the average PRD was about 0.9291, the extracted watermarks were recognizable. From Fig. 6 we can see that all extracted marked were identified.

4 Conclusion

Based on the smart offset coefficient, the patient data can be effectively hidden in ECG host signal. Experiments confirmed that the hiding capacity provided by the proposed method is larger than existing techniques while the perceived quality is good. In addition, the method is capable of resisting the attacks such as inversion, translation, truncation, and Gaussian noise-addition attacks, which is rare in conventional ECG steganography schemes. Because a secret message can be quickly embedded in ECG host via smart offset coefficients, the proposed method is suitable being utilized in real-time applications.

References

1. Cox, I.J., Miller, M.L., Bloom, J.A., Fridrich, J., Kalker, T.: Digital Watermarking and Steganography. Morgan Kaufmann, San Francisco (2008)
2. Eielinska, E., Mazurczyk, W., Szczypiorski, K.: Trends in steganography. Comm. ACM **57**, 86–95 (2014)
3. Skordalakis, E.: Syntactic ECG processing: a review. Pattern Recogn. **19**, 305–313 (1986)
4. Ibaida, A., Khalil, I., Schynde, R.V.: A low complexity high capacity ECG signal watermark for wearable sensor-net health monitoring system. In: Computing in Cardiology, vol. 38, pp. 393–396 (2011)
5. Ibaida, A., Khalil, I.: Wavelet-based ECG steganography for protecting patient confidential information in point-of-care systems. IEEE Trans. Biomed. Eng. **60**, 3322–3330 (2013)
6. Jero, S.E., Ramu, P., Ramakrishnan, S.: Discrete wavelet transform and singular value decomposition based ECG steganography for secured patient information transmission. J. Med. Syst. **38**, 132 (2014). doi:10.1007/s10916-014-0132-z
7. Yang, C.Y., Lin, K.T.: Hiding data in electrocardiogram based on IWT domain via simple coefficient adjustment. In: 4th International Conference on Annual Conference on Engineering and Information Technology, March 29–31, Kyoto, Japan (2016)
8. Yang, C.Y., Wang, W.F.: Effective electrocardiogram steganography based on coefficient alignment. J. Med. Syst. **40**, 1 (2016)
9. Jero, S.E., Ramu, P., Ramakrishnan, S.: Imperceptability-robustness tradeoff studies for ECG steganography using continuous ant colony optimization. Expert Syst. Appl. **49**, 123–135 (2016). doi:10.1007/s10916-015-0426-9
10. Moody, G.B., Mark, R.G.: The impact of the MIT-BIH arrhythmia database. IEEE Eng. Med. Biol. **20**, 45–50 (2001)

An Automatic People Counter in Stores Using a Low-Cost IoT Sensing Platform

Supatta Viriyavisuthisakul[1(✉)], Parinya Sanguansat[1], Satoshi Toriumi[2], Mikihara Hayashi[2], and Toshihiko Yamasaki[3]

[1] Panyapiwat Institute of Management, Nonthaburi, Thailand
{supattavir,parinysan}@pim.ac.th
[2] Future Standard Co., Ltd., Tokyo, Japan
{satoshi.toriumi,mikihisa.hayashi}@futurestandard.co.jp
[3] The University of Tokyo, Tokyo, Japan
yamasaki@hal.t.u-tokyo.ac.jp

Abstract. In this paper, we propose an automatic people counting system by using our low-cost Internet-of-Things (IoT) platform consisting of a single camera and Raspberry Pi. In this system, we count the number of moving people in bidirection by observing from a side view. Because the system can determine the height of the people, our system can be used to classify them into adults or children. This system is applied for no people overlapping problem in indoor environment only. The background subtraction and morphological operations are used to extract foreground objects from background images. The experimental results show proposed method can achieve 98% of people counting accuracy. It can also achieve 91% accuracy in adult/child classification. Although the algorithms for the people counting and classification are not novel, our technical contribution is that we have implemented them onto our IoT platform, whose cost is less than 100 US dollars. In addition, the images do not need be sent to the server, but all the image processing is done inside the device and only the results are uploaded to the server. This system can be applied to for customer behavior analysis or security.

Keywords: People counting · IoT sensing · Raspberry Pi

1 Introduction

People counting is widely used in service, security or data analysis. A gate counter is a primitive one but it is hard to install and not flexible. Image and video processing is applied to solve this problem. In [1], they proposed a method for people counting in monocular video sequences by setting up the camera four meters above the ground. Block-based background subtraction was introduced in which each pixel in sub block was modeled by a Gaussian distribution. It achieved 91% accuracy. The distance from camera to object was discussed in [2].

J.-S. Pan et al. (eds.), *Advances in Intelligent Information Hiding and Multimedia Signal Processing*, Smart Innovation, Systems and Technologies 81,
DOI 10.1007/978-3-319-63856-0_17

They tested the distance in three levels i.e. 3, 3.5 and 4 m. A camera was fixed at the wall of the building and used a line in counting. The experimental results showed that four meters could get the most accurate result similar to [1].

Detecting objects passing through a virtual line in an image is often used in counting the people [3]. This work could be used in both indoor and outdoor environments by using a single camera. The results showed the accuracy rate of 94.95% and low-cost processing at 100 frames per second. Moreover, using multiple lines for counting could be used to determine the direction of motion. In [4], they proposed people counting by using a single camera and prevented the miscalculation for abnormal walking. The results showed that the system could work well on the normal traffic and get high accuracy rate of 98%. But the accuracy will be dropped when objects are moving closely to each other or when the color of ground and clothes are similar to each other. Furthermore, illumination and shadow of the object are another issue that needs to be concerned.

In background subtraction, not only objects of interest are detected because of noise from light condition or environments. In [5], they found that a Gaussian Mixture Model (GMM)-based method [6] could work well in background subtraction and a morphological opening operation. The results showed that the MAE (Mean absolute error) was 1.15 and the MAP (Mean absolute percentage error) was 23%.

The light condition and environment are important factors that induce the system error. The morphological opening and Otsu's global threshold were applied in [7] to separate objects from the background for human tracking in surveillance system. The system was still not robust to environment noise such as light and shadow. In [8], the counter was applied on the bus that confronted with the different light conditions. There were three different light conditions that were strong daylight, side daylight (afternoon) and lamplight (night). They proposed an algorithm that consists of erosion and dilation for moving object extraction based on greedy algorithm. The result indicated that strong daylight got the best accuracy at 90% and mean accuracy was 87%. A method that deal with the variations of light and shadow was proposed in [9]. This method updated the background in real-time. This system achieved high accuracy rate at 97.7%. But the researchers reported that if it the location or position of the camera is changed, re-adjustment is required, which is not flexible.

In this paper, the people counting supports bi-directional movement and can classify a moving object into adult or child. This information is useful for marketing analysis, transportation or security. In addition, the system needs to be automatic, easy to install, low complexity and low-cost. Therefore, Raspberry Pi and normal webcam is applied with our proposed method.

Our IoT platform consists of real-time image acquisition, image processing and alert via LINE Notify. This system is operated on Raspberry Pi 3 model B with Raspbian version 8.0 (Jessie) and USB webcam (angle of view 60°). Our system was implemented in Python version 2.7.9 using OpenCV library version 3.1.0.

For real-time processing, our algorithm can operate at 8.5 frames per second for 320 × 240 video resolution, which a sufficient processing speed for this application. The source code for the people counter will be made public very soon and anyone can reproduce our system with low-cost.

2 Proposed Method

The proposed method is suitable for installing at the door to count the number adults or children, that enter and exit. Image acquisition is set as shown in Fig. 1. The camera is set 155 cm away from object. The height of an object is used to classify object into adult or child. If it is smaller than 130 cm, it will be counted as a child, otherwise it will be counted an adult. People will walk pass the camera one-by-one in both directions. The algorithm is rather simple to ensure that they can run in real-time in our IoT platform.

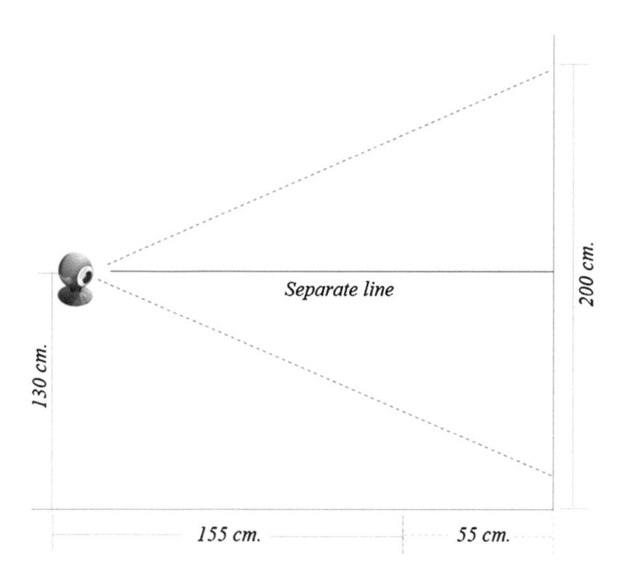

Fig. 1. Side view camera model

2.1 Background Subtraction

The background subtraction attempts to extract foreground objects from background. To make the people counting algorithm highly accurate, the background should be updated continuously. In this paper, K-nearest neighbors-based background subtractor [10] is applied.

2.2 Morphological Operation and Thresholding

Before the morphological operation, we apply Gaussian filter with 21×21 kernel size to reduce image noise and blinding object details. After that, we apply morphological opening operations because it can remove small objects and get the objects of interest. And then we convert the image to binary by thresholding at the half value of the intensity.

2.3 People Counting

Figure 2 shows a frame and anchor points for people counting. It can be used to count objects in bi-direction, which are from R_{in} to R_{out} or R_{out} to R_{in}, and classify objects into adult or child. In case of IN direction, when an object moving through line C^c_{in} to line C^c_{out} the object will be counted as IN direction. After that the height of the object is calculated. If it is higher than the separate line it will be counted an adult, otherwise it will be counted a child. The separate line is set at 0.35 of the height of the video resolution, which correspond to setting the threshold as 130 cm. When an object moving through line C^c_{out} to C^c_{in} the object will be counted as OUT direction.

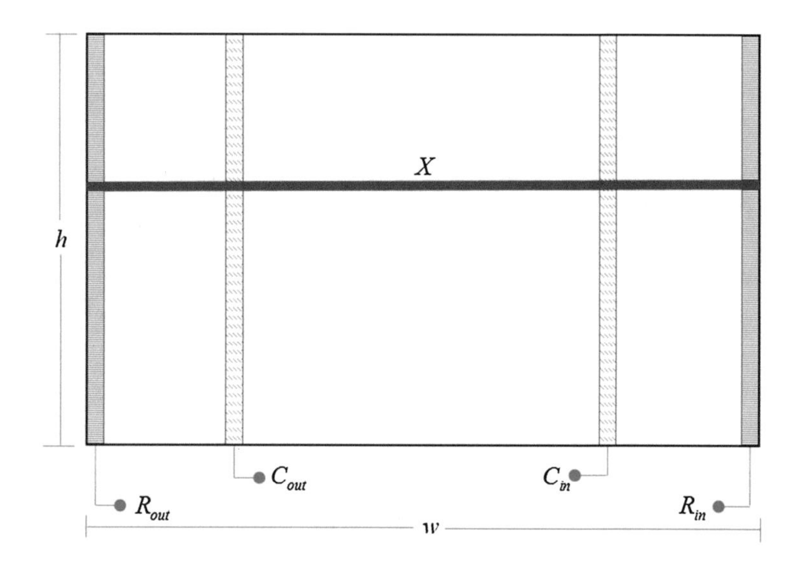

Fig. 2. Frame area for counting

All of parameters are shown in Table 1. The process is summarized in the pseudo code in Algorithm 1.

```
Data: Sequence of images {I_t}
Result: Numbers of adults in two directions N^a_in, N^a_out and numbers of children
        in two directions N^c_in, N^c_out
initialization;
Set N^a_in, N^a_out, N^c_in and N^c_out to 0;
Set P_in and P_out to the number of white pixels in their detected areas;
while not end of {I_t} do
    Read images I_t ;
    Apply background subtraction, Gaussian blur and morphological opening to
      I_t, respectively ;
    Convert I_t to a binary image B_t;
    Set C_in and C_in to the number of white pixels of B_t in their detected areas;
    if R_in > T_in or R_out > T_out then
        Set all booleans to FALSE ;
    end
    if C_in > T_in then
        Set d_in to TRUE ;
    end
    if C_out > T_out then
        Set d_out to TRUE ;
    end
    if X > T_a then
        Set d_a to TRUE ;
    end
    if P_in > T_in and d_out then
        if d_a then
            N^a_in <- N^a_in + 1;
        else
            N^c_in <- N^c_in + 1;
        end
        Set P_in and P_out to FALSE;
    end
    if P_out > T_out and d_in then
        if d_a then
            N^a_out <- N^a_out + 1;
        else
            N^c_out <- N^c_out + 1;
        end
        Set P_in and P_out to FALSE;
    end
end
```

Algorithm 1. People counter algorithm

3 Experimental Results

Our system was designed for side view camera, because from this view point the system can separate height of the objects. We setup single camera around 155 cm

Table 1. Parameters of people counting system

Parameter	Description
N^a_{in}	Number of adults for IN direction
N^a_{out}	Number of adults for OUT direction
N^c_{in}	Number of children for IN direction
N^c_{out}	Number of children for OUT direction
C_{in}	Detected area for IN direction
C_{out}	Detected area for OUT direction
P_{in}	Previous value of C_{in}
P_{out}	Previous value of C_{out}
X	Detected area for adult
R_{in}	Detected area for reset all booleans for IN direction
R_{out}	Detected area for reset all booleans for OUT direction
d_{in}	Boolean for C_{in}
d_{out}	Boolean for C_{out}
d_a	Boolean for X
T_{in}	Threshold for C_{in}
T_{out}	Threshold for C_{out}
T_a	Threshold for X

Table 2. Accuracy rate of people counting

Sequence	Direction	Counting accuracy	Classification accuracy
Video 1	IN	100%	92%
Video 2	OUT	100%	92%
Video 3	IN and OUT	95%	90%
Avg. accuracy		**98%**	**91%**

away from objects and separate line at 0.35 of height of the video resolution or 130 cm from the ground.

The experimental results are shown in Table 2, in which *counting accuracy* means the percent of the correct number of people that were counted by the system and *classification accuracy* means the percent of the correct number of people that were classified into adult or child. The system can count the number of people with high accuracy. We tested the performance of our system with 47 subjects in three different scenarios i.e. enter only in Video 1, exit only in Video 2, and both ways in Video 3. We can achieve 100% in counting accuracy at Video 1 and Video 2. For complex scenario as in Video 3 we got the counting accuracy at 95%. The average accuracy rate is 98%. For adult and child classification, its average accuracy rate is 91%. Figure 3a illustrates a sample of incorrect

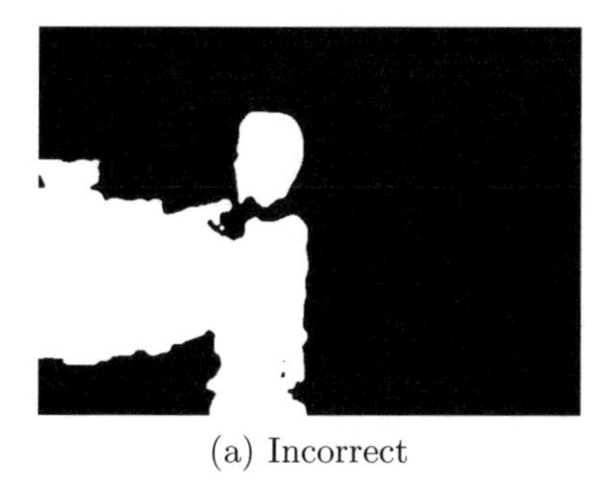
(a) Incorrect

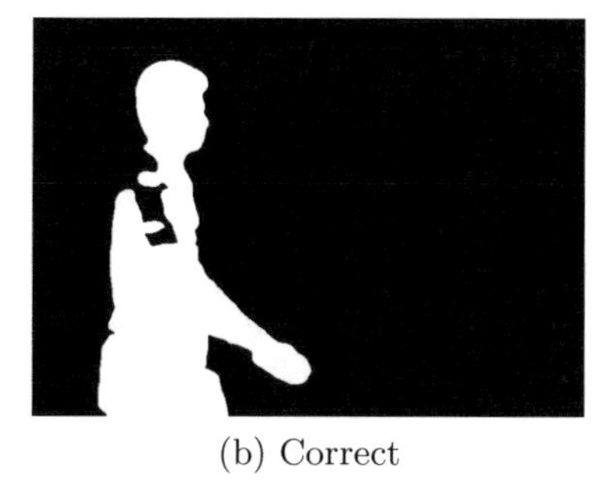
(b) Correct

Fig. 3. Examples of processed frame

classification that occurred because of the environment noise. A sample of correct one is demonstrated in Fig. 3b. The average processing speed is 8.5 frames per second. It can work smoothly in real-time processing.

Table 3 shows the confusion matrix of our system, the row is the actual values and the column is the classify results. This system can classify children perfectly and get some error for adult classification around 18%. This occurs due to environment noise and the variation of distance of objects from camera. The background subtraction cannot remove it in this case, thus the counting will be incorrect.

Table 3. Confusion matrix of adult-child classification

	Adult	Child
Adult	82%	18%
Child	0%	100%

4 Conclusions

An automatic people counter using a low-cost device such as a single webcam and Raspberry Pi for IoT platform is proposed in this paper. The side view camera is installed for detecting the height of object and it is easy to install. The algorithm based on image processing in real-time can perform on small computation unit such as Raspberry Pi. The background subtraction is used to extract foreground objects. Our system can count people passing frames IN or OUT directions and then classify them by their height. The system can be applied in many scenarios such as at entrance gate in shop, train station or other places which need the people pass through the gate one by one.

Acknowledgment. This is a collaboration project for joint internship program among Panyapiwat Institute of Management, Thailand, Future Standard Company and Yamasaki Lab of The University of Tokyo, Japan.

References

1. Xu, X.W., Wang, Z.Y., Liang, Y.H., Zhang, Y.Q.: A rapid method for passing people counting in monocular video sequences. In: 2007 International Conference on Machine Learning and Cybernetics, pp. 1657–1662. IEEE (2007)
2. Barandiaran, J., Murguia, B., Boto, F.: Real-time people counting using multiple lines. In: 2008 Proceedings of the Ninth International Workshop on Image Analysis for Multimedia Interactive Services, pp. 159–162. IEEE (2008)
3. Lee, K.Z., Tsai, L.W., Hung, P.C.: Fast people counting using sampled motion statistics. In: 2012 Proceedings of the Eighth International Conference on Intelligent Information Hiding and Multimedia Signal Processing, pp. 162–165. IEEE, July 2012
4. Cao, J., Sun, L., Odoom, M.G., Luan, F., Song, X.: Counting people by using a single camera without calibration. In: 2016 Chinese Control and Decision Conference (CCDC), pp. 2048–2051. IEEE, May 2016
5. Bamrungthai, P., Puengsawad, S.: Robust people counting using a region-based approach for a monocular vision system. In: 2015 International Conference on Science and Technology (TICST), pp. 309–312. IEEE, November 2015
6. KaewTraKulPong, P., Bowden, R.: An improved adaptive background mixture model for real-time tracking with shadow detection. In: Video-Based Surveillance Systems, pp. 135–144. Springer, Boston (2002)
7. Sahoo, A.K., Patnaik, S., Biswal, P.K., Sahani, A.K., Mohanta, P.B.: An efficient algorithm for human tracking in visual surveillance system. In: 2013 IEEE Second International Conference on Image Information Processing, ICIIP 2013, pp. 125–130. IEEE, December 2013
8. Perng, J.W., Wang, T.Y., Hsu, Y.W., Wu, B.F.: The design and implementation of a vision-based people counting system in buses. In: 2016 International Conference on System Science and Engineering (ICSSE), pp. 1–3. IEEE, July 2016
9. Yufeng, X., Qiuyu, Z., Baozhu, Z.: People counting system based on improved Gaussian background model. In: 2015 International Conference on Smart and Sustainable City and Big Data (ICSSC), p. 5. Institution of Engineering and Technology (2015)
10. Zivkovic, Z., van der Heijden, F.: Efficient adaptive density estimation per image pixel for the task of background subtraction. Pattern Recogn. Lett. **27**(7), 773–780 (2006)

Biomedical System Design and Applications

Determination of Coefficient of Thermal Expansion in High Power GaN-Based Light-Emitting Diodes via Optical Coherent Tomography

Ya-Ju Lee[1(✉)], Yung-Chi Yao[1], Yi-Kai Haung[1], and Meng-Tsan Tsai[2(✉)]

[1] Institute of Electro-Optical Science and Technology, National Taiwan Normal University, 88, Sec.4, Ting-Chou Road, Taipei 116, Taiwan
yajulee@ntnu.edu.tw

[2] Department of Electrical Engineering, Chang Gung University, 259, Wenhua 1st Rd, Taoyuan 333, Taiwan
mttsai@mail.cgu.edu.tw

Abstract. One of the most challenging issues when operating a high-power light-emitting diode (LED) is to conduct appropriate packaging materials for the reliable thermal management of the entire device. Generally, the considerable amount of heat produced around the junction area of the LED would transfers to the entire device, and causes thermal expansion in the packaging material. It induces strain inevitably, and hinders the output performances and possible applications of high-power LEDs. The coefficient of thermal expansion (CTE) is a physical quantity that indicates the expansion to which value a material will be upon heating. Therefore, as far as an advancement of thermal management is concerned, the quantitative and real-time determination of CTE of packaging materials becomes more important than ever since the demanding of high-power LEDs is increased in recent years. In this study, we measure the CTE of GaN-based (λ = 450 nm) high-power LED encapsulated with polystyrene resin by using optical coherent tomography (OCT). The displacement change between individual junctions of OCT image is directly observed and recorded to derive the CTE values of composed components of the LED device. The obtained instant CTE of polystyrene resin is estimated to be around $10 \times 10^{-5}/°C$, which is a spatial average value over the OCT scanning area of 10 μm × 10 μm. The OCT provides an alternative way to determine a real-time, non-destructive, and spatially resolved CTE values of the LED device, and that shows essential advantage over the typical CTE measurement techniques.

Keywords: Light-emitting diodes · Optical coherence tomography · Packaging materials

1 Introduction

In the past decade, the OCT has been widely adopted as an *in vivo* imaging modality that provides noninvasive, high speed, and 3-dimensional construction mainly in the

J.-S. Pan et al. (eds.), *Advances in Intelligent Information Hiding and Multimedia Signal Processing*, Smart Innovation, Systems and Technologies 81,
DOI 10.1007/978-3-319-63856-0_18

field of biological specimens such as gastroenterology, cardiology, dermatology, oral mucosa, and ophthalmology [1–3]. Typically, OCT can provide high resolutions of 1~10 μm in both axial and transverse directions, and the optical imaging depth can achieve ~2 mm. Yet, very few studies exerted OCT as the inspection tool on the semiconductor industry has been made [4–6], except for the biomedical applications mentioned above. Through the examination of 3D OCT images, the spatial distribution and change in sample structure can be identified, enabling application for optical inspection in many kinds of industrial products.

In this study, we examined the temperature-dependent and depth-resolved OCT images to determine the instantaneous CTE on the packaging materials of InGaN-based ($\lambda = 450$ nm), high-power LED. The distances between individual interfaces of OCT images is observed and recorded to derive the instantaneous CTE of the packaged LED device with different injected currents. The relationship between the junction temperature and the injected current is established through the forward voltage method. As a result, the measured instantaneous CTE of polystyrene resin varies from 5.86×10^{-5} to 14.10×10^{-5} °C^{-1} over a junction temperature range of 25–225 °C, and exhibits a uniform distribution with static root-mean-square value by an OCT scanning area of 200 μm × 200 μm. Most importantly, this work validates that the OCT can provide an alternative way to both directly and nondestructively determine the spatially resolved CTE on the packaged LED device, which offers essential advantages over traditional techniques for the CTE measurement.

2 Experiments

Figure 1(a) shows an image of packaged InGaN-based ($\lambda = 450$ nm), high-power LED used in this study. The LED chip with the dimension of 1 mm × 0.5 mm is mounted on the lead frame, and the anode and cathode of the LED chip are connected to the frame through gold wires. The lead frame is soldered on the printed circuit board (star shape) with a metal slug design for the supplement of injected currents. The inset of Fig. 1(a) shows the microscope top-view image focusing on the center of the packaged LED device. The wire bonding LED chip mounted on lead frame is clearly observed in the figure. In this study, the polystyrene resin is dispensed to encapsulate and form flat cavity geometry for the protection of the LED chip underneath. Such polymeric encapsulant generally exhibits high transparency, high refractive index, high temperature stability, and good hermeticity. Figure 1(b) shows the Raman spectrum of polystyrene resin excited by a 532-nm diode-pumped, solid-state laser. The repeating unit of chemical structure of polystyrene resin is also plotted in the figure. The polystyrene resin consists of a long-chain hydrocarbon wherein alternating carbon centers are attached to phenyl groups. A dominant peak associated with the vibration of aromatic carbon rings in the polystyrene resin appears at around 1000 cm^{-1}. In addition, two distinctive peaks assigned as low carbon-carbon (C-C) and high carbon-hydrogen (C-H) vibrations are clearly identified at around 600 cm^{-1} and 3000 cm^{-1}, respectively. We can also observe a stronger vibration of two carbon atoms with double bonds (C = C) than that of C-C single bond at a higher frequency regime of 1600 cm^{-1}. The Raman spectrum confirms that the encapsulant material of our LED device is primarily composed of polystyrene. Figure 1(c) plots the light-output power and the forward voltage versus the forward current for the packaged LED device. The turn-on voltage and series resistance of the packaged LED device estimated by the Shockley diode equation are around 2.66 V and 2.40 Ω, respectively, comparable to that of typical InGaN-based high-power LED chips. The LED's light output power increases gradually and is saturated at an approximately injected current of $I = 400$ mA, implying a considerable dissipation of electrical-input power was induced in the form of unwanted heat. The electroluminescence (EL) spectra of the packaged LED device [inserts of Fig. 1(c)] are blue-shifted slightly (from 454.2 nm to 453.1 nm) at lower injected currents of $I < 200$ mA due to the quantum-confined stark effect generally found in the InGaN-based LEDs. The EL spectrum becomes a pronounced red-shift (from 454.2 nm to 460.9 nm) at higher injected currents of $I > 300$ mA, validating that a considerable thermal heat was indeed induced and accumulated inside the LED chip, leading to severe thermal expansion in the packaged LED device.

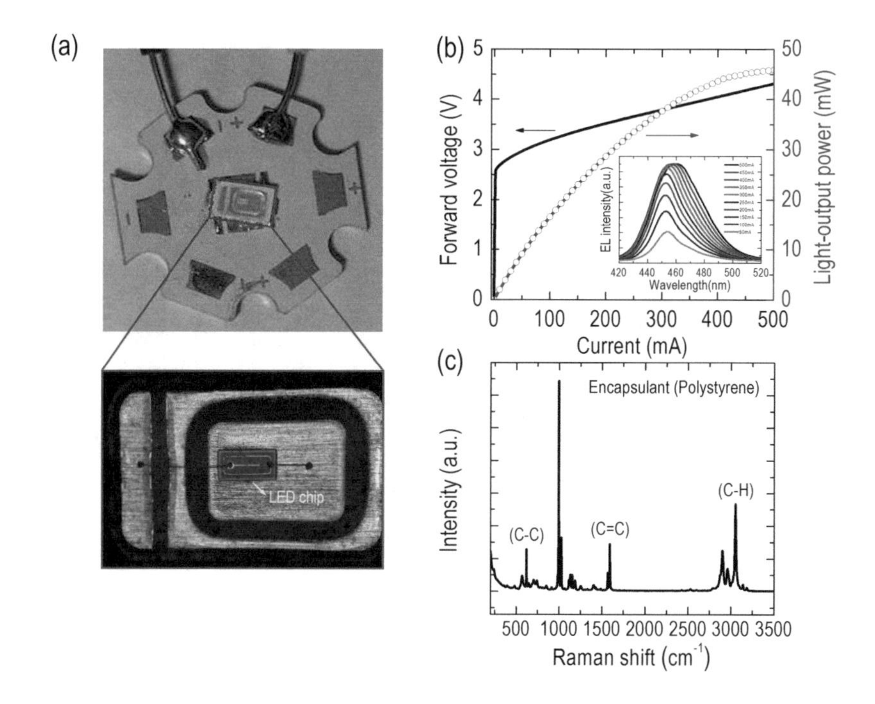

Fig. 1. (a) Photography of packaged InGaN-based (λ = 450 nm) high-power LED used in this study. Inset: Microscope top-view image focusing on the center of the packaged LED device. (b) Raman spectrum of polystyrene (PS) resin excited by a 532 nm diode-pumped, solid-state laser. Inset: Repeating unit of the chemical structure of polystyrene resin. (c) Light-output power and forward voltage versus forward current for the packaged LED device. Inset: EL spectra of the packaged LED device under different injected currents (from 50 to 500 mA).

Figure 2(a) depicts a schematic diagram of the SS-OCT system used in this study. The center wavelength of swept source (SSOCT-1060, AXSUN Technologies Inc., MA) is located at 1060 nm with a scanning spectral range of 100 nm. Then, the optical beam from the light source is split into the reference and sample arms by a fiber coupler. In the sample arm, a two-axis galvanometer (GVS302, Thorlabs Inc., NJ) is utilized to achieve the lateral and transverse scanning, and then, the optical beam is focused on the sample by a scanning lens (LSM02-BB, Thorlabs Inc., NJ). To reduce the dispersion resulting from the scanning lens in the sample arm, a dispersion compensator is inserted in the reference arm. Finally, the return signal from both arms is received by a balanced detector (PDB460C, Thorlabs Inc., NJ) and digitized by a high-speed digitizer (ATS-9350, Alazar Technologies Inc., QC, Canada). The corresponding axial and transverse resolutions of the developed OCT system are approximately 4 m in the sample and 7 m, respectively. With the developed OCT system, 2D or 3D scanning can be performed. The physical scanning range covers a square area of 4 mm × 4 mm and the imaging depth can achieve 2 ~ 3 mm, which depends on the optical properties of sample. The scan rate of light source is 100 kHz, which corresponds to a frame rate of 100 frames/s. Figure 2(b) shows a photography for the SS-OCT system. Figure 2(c) shows a cross-sectional 3D OCT image of the packaged LED device without injecting current.

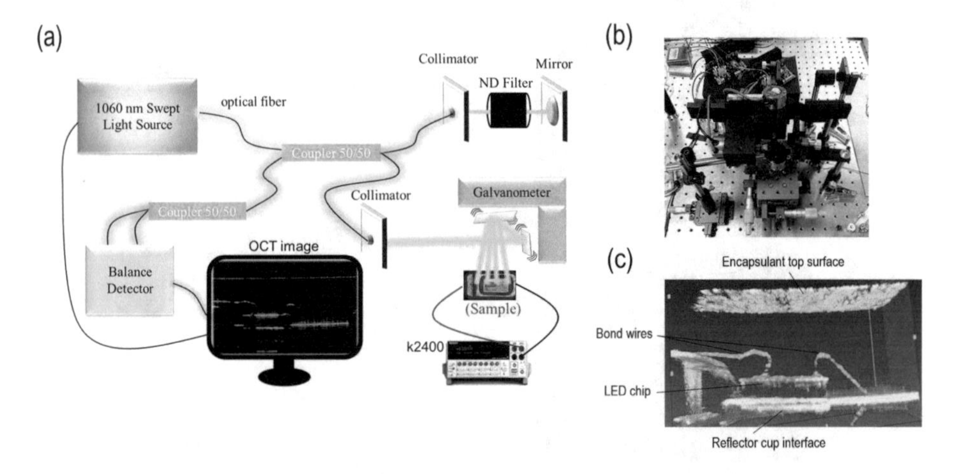

Fig. 2. (a) Schematic diagram and (b) photography of the SS-OCT system. (c) Reconstructed 3D OCT images of the packaged LED device without injecting current.

3 Results and Discussions

Figure 3(a) presents thickness variance versus injected current, derived by the depth-resolved OCT signal curve of the package LED device, for polystyrene resin layer. The thickness of polystyrene resin is sensitive to the variance of injected currents and increases linearly with the increase in injected currents. Figure 3(b) shows $\varepsilon_{thermal}$ and α_I of the polystyrene resin against the junction temperature of the packaged LED device. The measured α_I is varied from 5.86×10^{-5} to 14.10×10^{-5} °C^{-1} over a junction temperature range of 25–225 °C, agreeing well with the previously reported values in the literature. This suggests the adoption of polystyrene resin with such high CTE may not be appropriate in terms of packaging applications of high-power LEDs, and the development of advanced polymeric composites with reduced CTE is necessary. Most importantly, through the layer-by-layer inspection of OCT images over a range of junction temperature, our study provides a direct and promising way to quantitatively determine the CTE with high accuracy for the constitute elements of the packaged LED device.

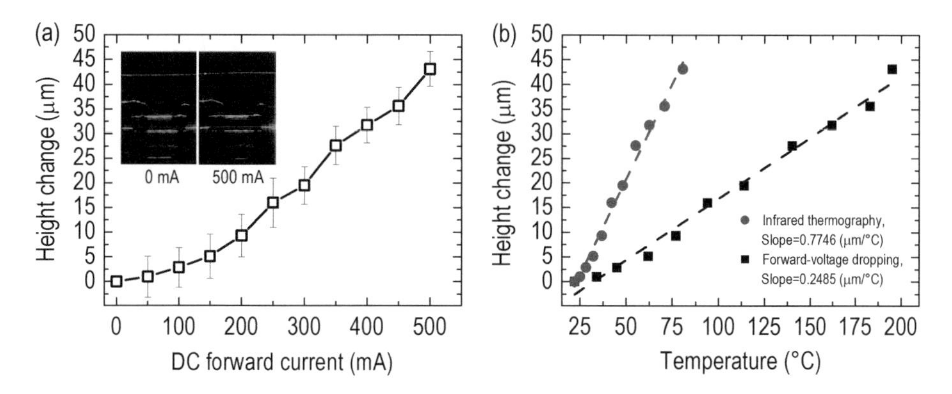

Fig. 3. (a) Thickness variance versus injected current for polystyrene resin (black squares) and sapphire substrate (blue squares). (b) Thermal strain ($\varepsilon_{thermal}$) and instantaneous coefficient of thermal expansion (α_I) versus junction temperature of the packaged LED device for the polystyrene resin.

4 Conclusion

Benefiting from the advantageous abilities such as nondestructive inspection, real-time visualization, layer-by-layer tomography, and three-dimensional reconstruction, the OCT was used to determine the CTE of InGaN-based high-power LED encapsulated with polystyrene resin as the packaging material. As compared to the LED's sapphire substrate, the thickness of polystyrene resin is sensitive to the variance of injected currents and increases linearly with the increasing of injected currents. The derived instantaneous CTE of polystyrene resin varies from 5.86×10^{-5} to 14.10×10^{-5} °C^{-1} over a junction temperature range of 25–225 °C and exhibits a uniform distribution with static root-mean-square value over a scanning area of 200 μm × 200 μm. We believe the revealed scheme associated with OCT will contribute essential improvement in terms of inspection efficacy and measurement accuracy over traditional techniques for the CTE measurement.

Acknowledgement. The authors would like thank the founding support from Ministry of Science and Technology (Contract. No. MOST 103–2112–M–003–008–MY3).

References

1. Adler, D.C., Chen, Y., Huber, R., Schmitt, J., Connolly, J., Fujimoto, J.G.: Three-dimensional endomicroscopy using optical coherence tomography. Nat. Photon. **1**, 709–716 (2007)
2. Tsai, M.T., Lee, H.C., Lee, C.K., Yu, C.H., Chen, H.M., Chiang, C.P., Chang, C.C., Wang, Y.M., Yang, C.C.: Effective indicators for diagnosis of oral cancer using optical coherence tomography. Opt. Express **16**, 15847–15862 (2008)
3. Campbell, J.P., Zhang, M., Hwang, T.S., Bailey, S.T., Wilson, D.J., Jia, Y., Huang, D.: Detailed vascular anatomy of the human retina by projection-resolved optical coherence tomography angiography. Sci. Rep. **7**, 42201 (2017)
4. Kim, S.H., Kim, J.H., Kang, S.W.: Nondestructive defect inspection for LCDs using optical coherence tomography. Displays **32**, 325–329 (2011)
5. Tsai, M.T., Chang, F.Y., Yao, Y.C., Mei, J., Lee, Y.J.: Optical inspection of solar cells with phase-sensitive optical coherence tomography. Sol. Energ. Mat. Sol. Cells **136**, 193–199 (2015)
6. Prykari, T., Czajkowski, J., Alarousu, E., Myllyla, R.: Optical coherence tomography as an accurate inspection and quality evaluation technique in paper industry. Opt. Rev. **17**, 218–222 (2010)

Compression-Efficient Reversible Data Hiding in Zero Quantized Coefficients of JPEG Images

Jen-Chun Chang, Yu-Hsien Lee, and Hsin-Lung Wu(✉)

Department of Computer Science and Information Engineering,
National Taipei University, New Taipei City, Taiwan
{jcchang,hsinlung}@mail.ntpu.edu.tw, minerva1218@gmail.com

Abstract. In this paper, we study reversible data hiding schemes for JPEG images and we propose a method which is constructed by embedding secret data into zero quantized coefficients. In the proposed method, consecutive zero quantized coefficients in a fixed set of entries of DCT block are used to embed data in the following way. First, the algorithm selects several positions from this fixed set according to the first part of the secret message and modifies selected zero coefficients by small nonzero values according to the second part of the secret message. Experimental results show that our proposed methods have higher embedding capacity and obtain lower ratios between the increased file size and payload than previous methods.

Keywords: Reversible data hiding · JPEG images · Zero quantized coefficients · Compression-efficient

1 Introduction

Reversible data hiding is a technique which can embed a secret message into an image such that, from the stego image, one can recover losslessly the original image after the whole embedded message is extracted. This novel technique has found many applications in the area of medical image processing, military, and forensics where it is required to restore the original image without any distortion. Generally speaking, known reversible data hiding techniques focus on three domains: spatial, transformation, and compression domains. Among these domains, only a few data hiding schemes are proposed for the compression domain. Compressed images draw a lot of attention recently when people use internet as a major transmission tool. Compressed images become popular since they save much storage space and improve the transmission speed on the internet. Among many compression methods, JPEG is a popular compressed format used on many platforms.

There are many reversible data hiding schemes developed for JPEG images. The method of Chang et al. [1] losslessly embeds secret data into some selected diagonal arrays of the quantized DCT blocks in JPEG images. The method of

J.-S. Pan et al. (eds.), *Advances in Intelligent Information Hiding and Multimedia Signal Processing*, Smart Innovation, Systems and Technologies 81,
DOI 10.1007/978-3-319-63856-0_19

Xuan et al. [7] shifts the quantized DCT coefficient histogram and embeds secret data based on histogram pairs. Based on the method of Xuan et al., the adaptive method of Sakai et al. [5] detects whether a block is suitable for embedding data or not. The scheme of Liao and Zhang [3] embeds secret data into a selected zero coefficient from those last remaining zero quantized coefficients in each DCT block. The method of Wang et al. [6] modifies both the quantization table and quantized DCT coefficients to embed data bits. The scheme of Nikolaidis [4] embeds secret data by modifying fixed consecutive zero quantized coefficients in each DCT block where embedding and extraction procedures are performed without the need to preprocess the whole image. The scheme of Huang et al. [2] uses the block-selection strategy to embed secret data into quantized coefficients with values 1 and -1 by using the histogram-shifting technique.

Existing reversible data hiding schemes for JPEG images can be classified according to quantization tables used in data hiding schemes. Some schemes such as [6] use the modified quantization tables to generate the stego JPEG images. For this approach, the advantage is the good stego-image quality and large embedding capacity but the disadvantage is that the increased file size of the generated JPEG file is large. On the other hand, most existing schemes such as [1–5, 7] use the standard quantization table to generate the stego JPEG images. The advantage is that the increased file size of the generated marked JPEG file is small but the disadvantage is the degradation of stego-image quality. These schemes can be further divided into two groups based on quantized coefficients used for embedding data. Reversible data hiding schemes [1, 3–5, 7] embed secret data into zero quantized coefficients. The advantage is that embedding capacity is quite large but the ratio between the increased file size and the payload is also large. On the contrary, schemes such as [2] embed secret data into nonzero quantized coefficients. In this approach, the ratio between the increased file size and the payload is small but the maximum embedding capacity is small.

In this paper, a new reversible data hiding scheme for JPEG images by embedding secret data into zero quantized coefficients is proposed. The proposed scheme obtains high embedding capacity and acceptable stego-image quality. Moreover, in the proposed scheme, the ratio between the increased file size and the payload is smaller than those reversible data hiding schemes which embed secret data into zero quantized coefficients. In addition, as the scheme of Nikolaidis [4], our proposed algorithm requires only a single pass on the quantized coefficient values during embedding data into the JPEG image.

The rest of this paper is organized as follows. In Sect. 2, the JPEG compression procedure is briefly reviewed. In Sect. 3, we describe our proposed reversible data hiding scheme in detail. Experimental results are presented and explained in Sect. 4. Finally, a conclusion is given in Sect. 5.

2 Background on JPEG Compression Procedure

In this section, we review the standard JPEG compression procedure. Figure 1(a) shows the flowchart of the JPEG compression. Based on JPEG standard, a given

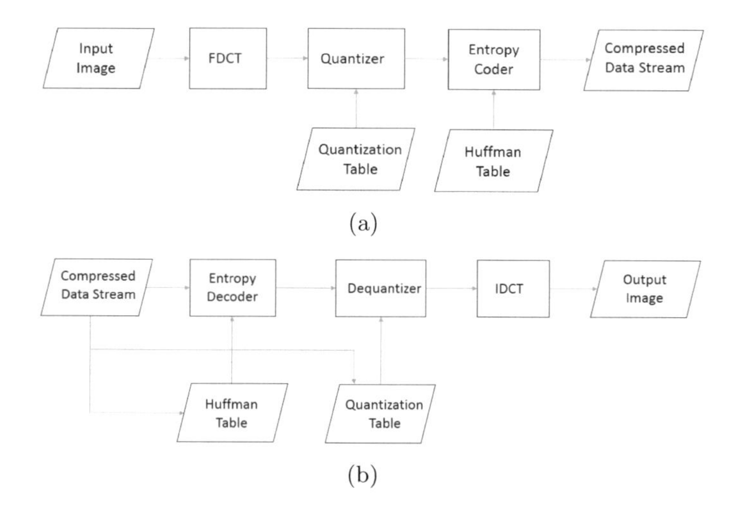

Fig. 1. (a) The flowchart of the JPEG compression and (b) the flowchart of the JPEG decompression

512×512 image is divided into 4096 non-overlapping 8×8 blocks. By using the discrete cosine transformation, each block is transformed into a DCT-block where each entry is called a DCT coefficient. In each block, DCT-coefficients are classified into two groups: AC coefficients and DC coefficients. The top-left corner entry in each block is the DC coefficient. The remaining 63 coefficients in each block are the AC coefficients. After quantizing by a quantization table with fixed quality factor, each DCT-coefficient is converted into a quantized DCT-coefficient. In Fig. 2(a), the standard quantization table with quality factor 50 is shown. Quantized DC coefficients are then encoded with the Huffman code after transforming the actual values into difference values. Since many quantized AC coefficients are zero, quantized AC coefficients in each block are efficiently encoded with the run-length coding in a fixed zig-zag order shown in Fig. 2(b).

The flowchart of the JPEG decompression procedure is shown in Fig. 1(b). The procedure consists of entropy decoding, dequantization, and IDCT. First, the decompression procedure gets the quantized DCT coefficients and the defined quantization table. Then, for each block and for each position in the block, it calculates the multiplication of the quantized DCT coefficient and the corresponding entry in the quantization table. The resulting values are approximations of the original DCT coefficients. Finally, the procedure applies IDCT to the approximated DCT coefficients to obtain the decompressed image.

3 Embedding Data into Zero Quantized Coefficients

In this section, we propose a high-capacity reversible data hiding scheme for JPEG images by embedding data into zero quantized coefficients. In order to obtain high embedding capacity, it is a good way to embed data into zero quantized coefficients since most quantized coefficients are zero quantized coefficients.

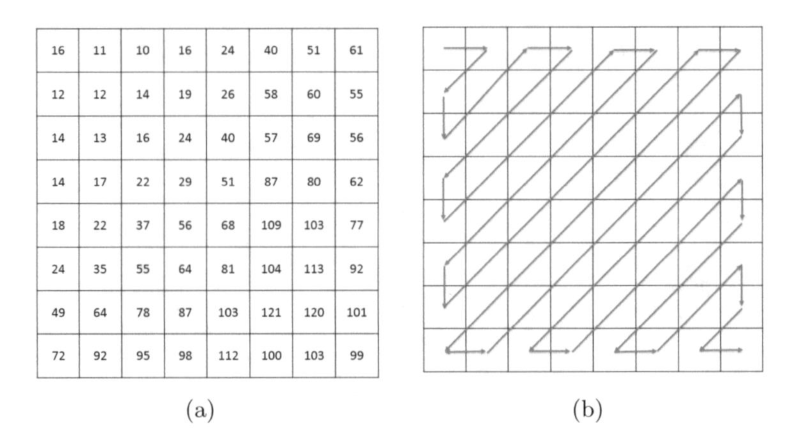

16	11	10	16	24	40	51	61
12	12	14	19	26	58	60	55
14	13	16	24	40	57	69	56
14	17	22	29	51	87	80	62
18	22	37	56	68	109	103	77
24	35	55	64	81	104	113	92
49	64	78	87	103	121	120	101
72	92	95	98	112	100	103	99

Fig. 2. (a) Standard quantization table (QF = 50) and (b) Zig-zag scanning order in a single block.

Another issue is to reduce the increased file size of the generated stego-JPEG image. Based on JPEG standard, modifying zero quantized coefficients by values 1 and -1 would not increase the file size too much but cause low embedding capacity. So in order to obtain high embedding capacity, we must modify zero quantized coefficients by values different from 1 and -1. To avoid increasing file size too much, the proposed algorithm chooses some positions of zero quantized coefficients to be embedded according to the to-be-embedded secret data. By using this technique, our proposed method obtains good performance on the ratio between increased file size and payload. Next, we describe our embedding algorithm in detail.

3.1 Data Embedding Procedure

Let t, ℓ be integers such that $t + \ell < 64$ and k be an integer such that $k \leq \ell$. We give the detail of the embedding algorithm called $P(n, \ell, k)$ in the following.

$P(t, \ell, k)$ **Reversible Data Embedding Scheme**:
Input: A secret data w and a JPEG image J
The watermarked JPEG image J_w is obtained as follows:

1. Divide $w = ((w_{1,1}, w_{1,2}), \ldots, (w_{i,1}, w_{i,2}), \ldots, (w_{m,1}, w_{m,2}))$ where each
$$(w_{i,1}, w_{i,2}) \in \{1, 2, \ldots, \binom{\ell}{k}\} \times \{-3, -2, -1, 1, 2, 3\}^k.$$
2. For each block, check whether the coefficients from the $(t+1)$-st entry to the $(t + \ell)$-th entry in the zig-zag order are all zero. Such a block is called an embeddable block.

3. For the i-th embeddable block, the algorithm chooses a k-element subset S_i from $\{t+1, t+2, \ldots, t+\ell\}$ according to the data digit $w_{i,1}$ where $w_{i,1}$ represents a unique k-element subset S_i of $\{t+1, t+2, \ldots, t+\ell\}$. After selecting the subset S_i, the algorithm embeds $w_{i,2}$ into zero quantized coefficients C_{t+j}'s for $t+j \in S_i$ as follows. For $1 \le j \le k$,

$$\widetilde{C}_{t+j} = C_{t+j} + (w_{i,2})_j$$

   where $(w_{i,2})_j$ denotes the j-th digit of $w_{i,2}$.
4. For those non-embeddable block, if $C_{t+j} \in \{-3, -2, -1, 1, 2, 3\}$ for each $1 \le j \le \ell$, then record the block index in a location map L.
5. In the JPEG header, write the location map L in it.
6. Apply the entropy coder to the modified quantized DCT coefficients to generate the stego JPEG image J_w.

3.2 Data Extracting Procedure

The extraction algorithm is described in the following.

Proposed Data Extraction Scheme of $P(t, \ell, k)$:

Input: A stego JPEG image J_w.
The restored JPEG image J and the secret data w are obtained as follows:

1. By using the location map L written in the JPEG header, the extraction algorithm can identify embeddable blocks.
2. For the i-th embeddable block, compute the subset S_i by checking $t+j \in S_i$ if and only if $\tilde{C}_{t+j} \neq 0$ for $1 \le j \le \ell$. After determining S_i, the data digit $w_{i,1}$ is retrieved. Moreover, it is easily seen that $(w_{i,2})_j = \tilde{C}_{t+j}$ for $1 \le j \le \ell$. After retrieving $w_{i,2}$, set $\tilde{C}_{t+j} = 0$ for $1 \le j \le \ell$.
3. Apply the entropy coder to the restored quantized DCT coefficients to generate the original JPEG image J.

4 Experimental Results

In this section, we show our experimental results. For the embedding algorithm $P(t, \ell, k)$, we set $(t, \ell, k) = (9, 10, 2)$ or $(t, \ell, k) = (9, 10, 3)$ in our experiments. We use *Proposed*(2) and *Proposed*(3) to denote $P(9, 10, 2)$ and $P(9, 10, 3)$, respectively. Here, we consider three aspects: the stego JPEG image quality (PSNR), embedding capacity (payload), and the ratio between the increased file size and payload (rip). To measure the stego-image quality, we use the well-known peak signal-to-noise ratio (PSNR) between the original decompressed image and the decompressed stego JPEG image. The test images are 512×512 JPEG images (Lena, Peppers, and Airplane) constructed by quantization tables with quality factor 50 for our experiment and the secret data is generated randomly since the encrypted secret data looks like a random string. Furthermore, we also consider

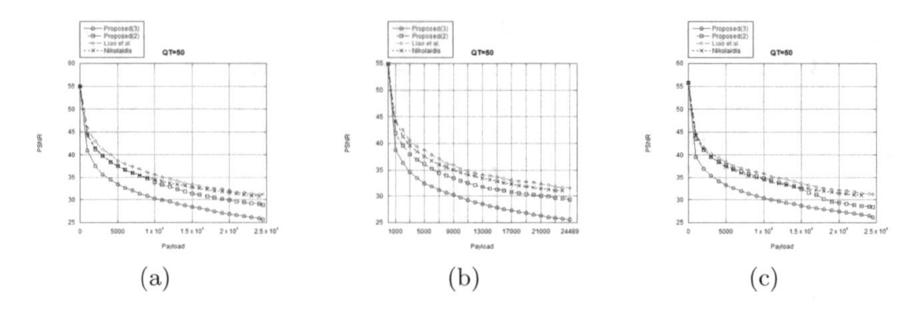

Fig. 3. PSNR versus payload for JPEG images with QF = 50 (a) Lena (b) Peppers (c) Airplane

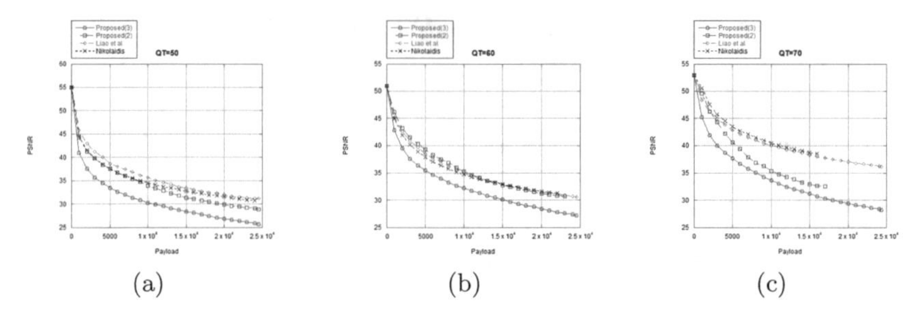

Fig. 4. PSNR versus payload for Lena image (a) QF = 50 (b) QF = 60 (c) QF = 70

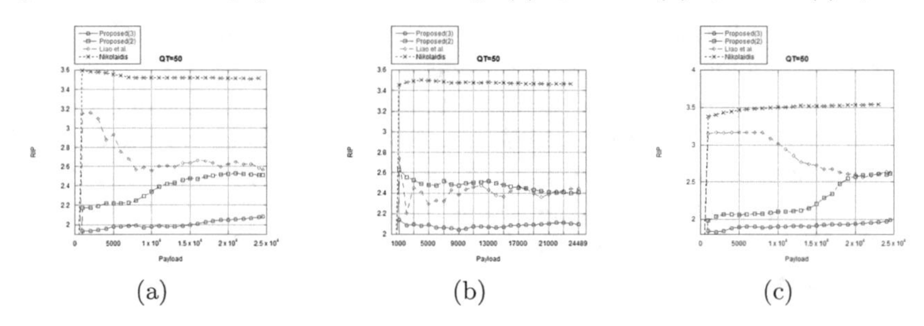

Fig. 5. The ratio between increased file size and payload for JPEG images with QF = 50 (a) Lena (b) Peppers (c) Airplane

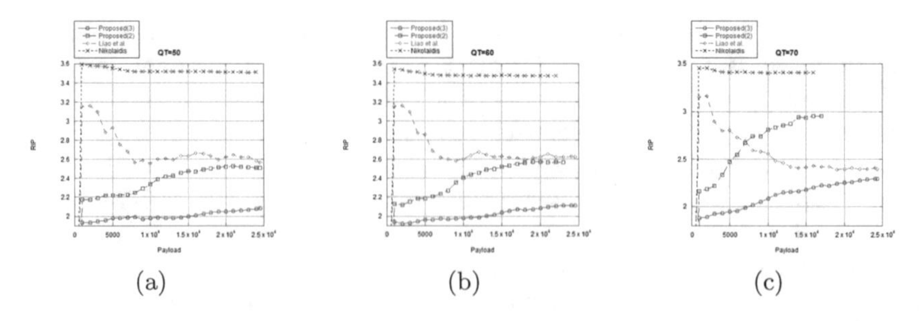

Fig. 6. The ratio between increased file size and payload for Lena image (a) QF = 50 (b) QF = 60 (c) QF = 70

Table 1. Payload (bits) of Lena image with QF = 50, 60, and 70.

QF	50	60	70
[3]	24517	24517	24281
[4]	23760	21330	15480
[2]	14087	15860	19450
Proposed(2)	25332	22741	16504
Proposed(3)	41964	37672	27340

Table 2. Payload (bits) of test images with QF = 50.

QF	Lena	Peppers	Airplane
[3]	24517	24489	24521
[4]	23760	23390	22580
[2]	14087	14299	14575
Proposed(2)	25332	24937	24074
Proposed(3)	41964	41310	39880

512×512 JPEG Lena images constructed by quantization tables with quality factors $60, 70$ for our experiment.

We compare our embedding algorithms with the method of Liao et al. [3] and the method of Nikolaidis [4] since these two methods are zero-coefficient-based embedding algorithms. Figure 3 shows the PSNR against payload for test images with quality factor 50. In addition, Fig. 4 shows the PSNR against payload for Lena images with quality factors 50, 60, and 70. It is expected that our proposed algorithms obtain lower PSNR values since our embedding algorithms modify many zero quantized coefficients by the quantized values larger than 1 or lower than -1. Next, Fig. 5 shows the ratio between the increased file size and payload for test images with quality factor 50. Figure 6 shows the ratio between the increased file size and payload for Lena images with quality factors 50, 60, and 70. It is shown that ratios obtained by our proposed methods $P(9, 10, 2)$ and $P(9, 10, 3)$ are much lower than ratios obtained by the method of Liao et al. [3] and the method of Nikolaidis [4]. In average, the ratio obtained by $P(9, 10, 3)$ is approximately 2 while the ratio obtained in [4] is approximately 3.5 and the ratio obtained in [3] is approximately 2.5. This gives evidence that our proposed methods are compression-efficient.

In Table 1, we show the maximum embedding capacity for Lena image with quality factors 50, 60, and 70. We compare our methods with methods in [2–4]. One can see that our proposed method $P(9, 10, 3)$ obtains higher embedding capacity than other methods. In Table 2, we show the maximum embedding capacity for test images with quality factor 50. One can see that our proposed method $P(9, 10, 3)$ also has higher embedding capacity than other methods.

5 Conclusion

In this paper, we construct a zero-coefficient-based reversible data hiding scheme for JPEG images which modifies zero quantized coefficients during embedding data. Our proposed method embeds data by modifying the fixed all-zero coefficient subsequence in each embeddable block. First, the embedding method selects several positions from the all-zero coefficient subsequence according to one part of the secret message. Then, the proposed method modifies zero coefficients in these chosen positions by small non-zero values according to the remaining part of the secret message. Experimental results show that our proposed methods outperform previous methods on maximum embedding capacities and ratios between the increased file size and payload.

References

1. Chang, C.-C., Lin, C.-C., Tseng, C.-S., Tai, W.-L.: Reversible hiding in DCT-based compressed images. Inf. Sci. **177**(13), 2768–2786 (2007)
2. Huang, F., Qu, X., Kim, H.J., Huang, J.: Reversible data hiding in JPEG images. IEEE Trans. Circuits Syst. Video Technol. **26**(9), 1610–1621 (2016)
3. Liao, G., Zhang, X.: A reversible hiding for JPEG images. Adv. Mater. Res. **433–440**, 4615–4620 (2012)
4. Nikolaidis, A.: Reversible data hiding in JPEG images utilising zero quantised coefficients. IET Image Proc. **9**(7), 560–568 (2015)
5. Sakai, H., Kuribayashi, M., Morii, M.: Adaptive reversible data hiding for JPEG images. In: Proceedings of the International Symposium on Information Theory and its Applications, pp. 1–6 (2008)
6. Wang, K., Lu, Z.-M., Hu, Y.-J.: A high capacity lossless data hiding schee for JPEG images. J. Syst. Softw. **86**, 1965–1975 (2013)
7. Xuan, G.R., Shi, Y.Q., Ni, Z.C., Chai, P.Q., Cui, X., Tong, X.F.: Reversible data hiding for JPEG images based on histogram pairs. In: Proceedings of the International Conference on Image Analysis and Recognition, pp. 715–727 (2007)

Using Optical Coherence Tomography to Identify of Oral Mucosae with 3D-Printing Probe

Ying-Dan Chen[1,2], Cheng-Yu Lee[2], Trung Nguyen Hoang[2], Yen-Li Wang[3], Ya-Ju Lee[4], and Meng-Tsan Tsai[2,5(✉)]

[1] School of Information and Electronic Engineering, Zhejiang Gongshang University, Hangzhou 310018, China

[2] Department of Electrical Engineering, Chang Gung University, Taoyuan 33302, Taiwan
mttsai@mail.cgu.edu.tw

[3] Department of Periodontics, Chang Gung Memorial Hospital, Taoyuan, Taiwan

[4] Institute of Electro-Optical Science and Technology, National Taiwan Normal University, Taipei, Taiwan

[5] Department of Dermatology, Chang Gung Memorial Hospital, Linkou, Taipei, Taiwan

Abstract. Biomedical materials have different optical properties (e.g. absorption) for different wavelengths. Therefore, probing biomedical materials with multiple wavelengths not only can get in-depth understanding of the detected biomedical materials but also can differentiate the detected biomedical materials. To achieve this purpose, in this conference paper, we present our initial results of building up a portable multiple-wavelength biomedical sensing system. At this initial phase, we assembled this kind of system with multiple wavelengths of light sources and photodetectors and preliminarily tested the absorbance of the glucose solutions with different concentrations. The result shows good linearity of the absorbance of the glucose solution with concentration. In addition, we also measured the absorbance of the glucose solutions using a broadband white light source and a spectrometer. This result also exhibits linearity but different slop of absorbance with glucose concentration, which confirms the linearity result obtained from the built sensing system. The difference of slop maybe relates to the difference of optical design between these two systems.

Keywords: Multiple wavelengths · Biomedical sensing · Glucose concentration

1 Introduction

Oral cancer is not a common disease for low-risk group, but is the fifth in the world's top ten cancer and estimated new cancer cases and deaths will be 45,780 and 8,650 in the US in 2015 [1]. Particularly for under-privileged groups in the developed or developing countries, the oral cancer has been one of the most important health burdens in terms of their prevalence, severity and associated healthcare cost. The main etiologies of oral squamous cell carcinoma (OSCC), which is the most common type of oral carcinoma, in Taiwan are areca quid (AQ) chewing, cigarette smoking, and alcohol consumption. Apart from low public awareness of the oral cancer, current visual examination by

J.-S. Pan et al. (eds.), *Advances in Intelligent Information Hiding and Multimedia Signal Processing*, Smart Innovation, Systems and Technologies 81,
DOI 10.1007/978-3-319-63856-0_20

treatment modalities including radiotherapy or a combination of surgery and chemotherapy or radiotherapy have limited opportunity to manage the oral cancer at early stage because of subjective criteria of the examiners. In Taiwan, there are two million people who habitually chew AQs [2]; approximated 80% of all oral cancer deaths are associated with this habit. The five-year survival rate for oral cancer patients in Taiwan is very low. Accordingly, necessity for an effective diagnostic tool has been on high demand for prognosis, diagnosis and early-treatment of the oral cancer and precancer.

A majority of oral cancers are found to develop from oral premalignant lesions such as leukoplakia, erythroplakia, erythroleukoplakia, dysplasia, and carcinoma in situ. In addition, these oral premalignant lesions are not homogeneous lesions, i.e., some part of the lesion may show only hyperkeratosis and acanthosis, while others may show epithelial dysplasia, carcinoma in situ or invasive carcinoma. Therefore, one lesion may need multiple biopsies to avoid misdiagnosis of most severe part of the lesion.

In the words of the European Journal of Cancer Prevention, "Early diagnosis and treatment is the key to improved patient survival" [3]. Due to many diseases of the oral cavity are accompanied by changes in tissue structure, the ability to perform high resolution diagnostic imaging is therefore crucial to the detection and treatment of these diseases. Conventional X-ray radiography has too large resolution, suitable for characterizing macroscopic structural changes in teeth. Ultrasound (US) is limited by the wavelength of the source, and sub-millimeter clinical resolutions can be achieved only under optimal conditions. The increased image acquisition times necessary to achieve sub-millimeter resolution in Magnetic Resonance Imaging (MRI), as well as its physical footprint and expense, make it impractical in routine clinical dentistry setting. The recent development of optical coherence tomography (OCT) has attracted much attention as a noninvasive optical modality in the realm of biomedical imaging because of its micrometer resolution and appreciable imaging depth, high speed, and high sensitivity and made it possible to overcome these limitations by using interferometric cross-correlation techniques to detect the coherent backscattered components of short coherence length light [4–6]. OCT has become one of the best ways to help us to choose the most appropriate site for diagnoses of oral cancer and precancer [7–9].

In this paper, we make a report about two-dimensional (2D) imaging of structure and three-dimensional (3D) imaging of microvessel within oral cavity tissues in vivo using a high-speed swept-source optical coherence tomography (SS-OCT) at 1310 nm with a designed probe by using 3D printing technique. Volumetric structural OCT images of the inner tissues of oral cavities are acquired with a field of view of 2 mm 2 mm. A type of detachable probe attachment is devised and applied to the port of the imaging probe of OCT system that easy to access to deeper cavity tissue. Blood perfusion is mapped with OCT-based correlation microangiography from 3D structural OCT images, in which a novel vessel extraction algorithm is used to decouple dynamic light scattering signals, due to moving blood cells, from the background scattering signals due to static tissue elements. Characteristic tissue anatomy and microvessel architectures of various cavity tissue regions of healthy human volunteers are identified with the 3D OCT images. The initial finding suggests that the proposed method may be engineered into a promising tool for discovering and treating oral cancer and precancer.

2 Experiments

Figure 1 demonstrates the flow chart of probe design. First, we should get lost in thinking probe design, and then purchasing optical components from Thorlab which will use in probe. What's more, we devise a scanning probe by SolidWorks and print it by 3D printing. In addition, to facilitate the access to the tissues in the oral mucosa, we design a specialized L-shaped and a line-shape probe with three achromatic doublets, as shown in Fig. 2. And use the ZEMAX software to design the optical path showing in Fig. 3. It is quite convenient for us to probe any area of oral mucosa. The length of the line-shaped probe is around 110 mm. If not succeed in assembling parts and scanning 2D images, we should return to the mechanical design by SolidWorks or optical design by ZEMAX. When all is well done, we can scan the oral cavity tissue. Figure 4 shows the full view of oral scanning probe.

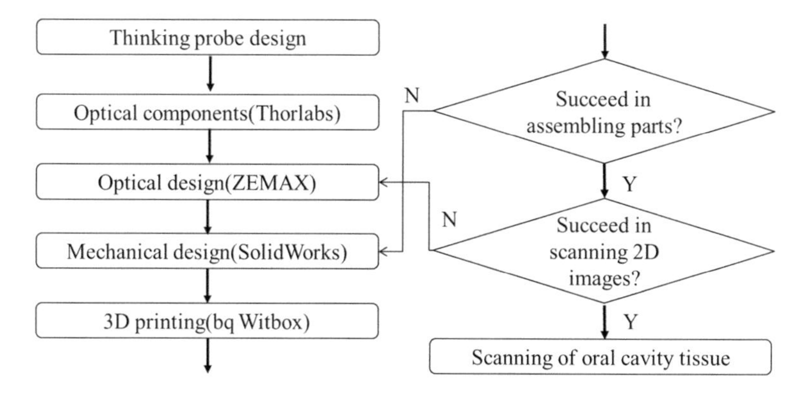

Fig. 1. Flow chart of probe design

Fig. 2. The L-shaped scanning probe. The length is around 110 mm and the width is around 11 mm.

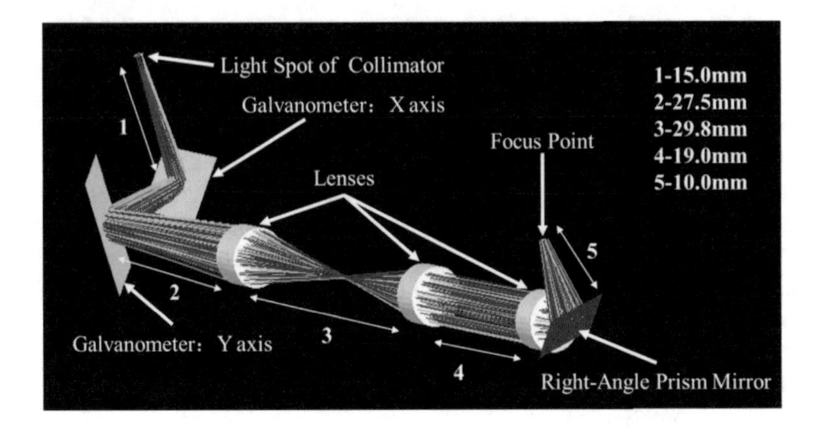

Fig. 3. Optical design of scanning probe

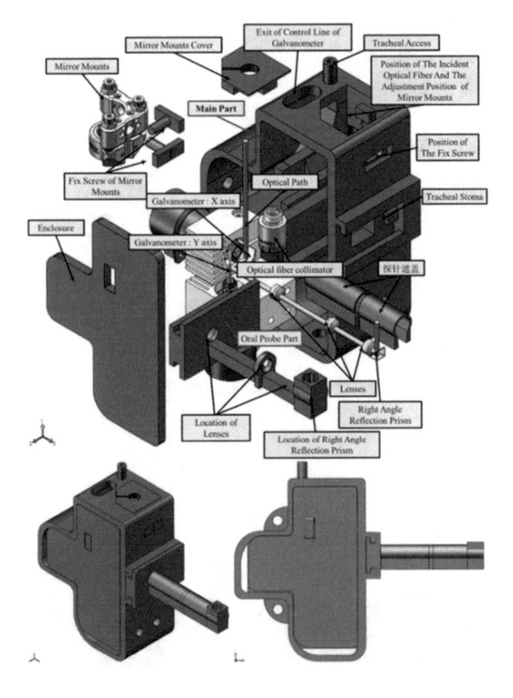

Fig. 4. Full view of oral scanning probe

As shown in Fig. 5, we use a fiber-based OCT system to perform OCT imaging of oral mucosa. The optical source power incident on the tissue was 20 mW at a wavelength of 1310 nm and by divided the source light into two beams and directing one beam to the sample and the other identical beam to a reference mirror whose location is accurately known. Light returning from the sample and from the reference mirror is recombined at a detector where the interference between the two beams is registered. The OCT system used in this study acquired images with 2 mm × 2 mm.

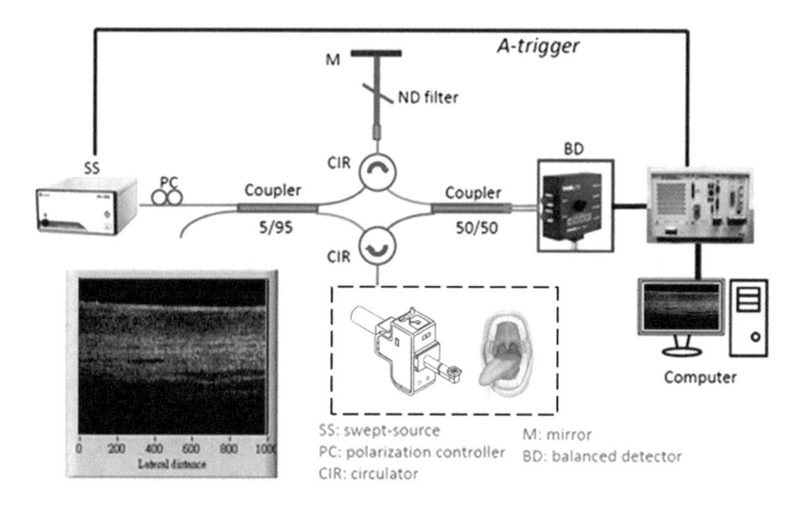

Fig. 5. Schematic of OCT system for oral mucosa

For vascular mapping of the oral cavity tissues, we adopted a method that utilizes the algorithms of correlation mapping OCT (cmOCT) where the vessel extraction is achieved with the combination of two angiograms obtained from each algorithm with the same OCT data set. The first stage in the cmOCT algorithm is determination of the correlation between two OCT frames. This is calculated by cross correlating a grid frame A (I_A) to the same grid from frame B (I_B) using Eq. (1) [4]. 3D OCT imaging was performed on the oral cavity tissue region of 2 mm (X) × 2 mm (Y) using scanning of the galvanometric scnners. The post data processing was performed with a laboratory-software written by a MATLAB language (R2010b, MathWorks, Inc).

$$cmOCT(x,y) = \sum_{p=0}^{M}\sum_{q=0}^{N}\frac{\left[I_A(x+p,y+q)-\overline{I_A(x,y)}\right]\left[I_B(x+p,y+q)-\overline{I_B(x,y)}\right]}{\sqrt{\left[I_A(x+p,y+q)-\overline{I_A(x,y)}\right]^2+\left[I_B(x+p,y+q)-\overline{I_B(x,y)}\right]^2}} \quad (1)$$

Where M and N is the grid size and $\bar{I}$ is the grids mean value. This grid is then shifted across the entire XY image and a 2D correlation map is generated. The resulting correlation map contains values on the range of 0 ± 1 indicating weak correlation and strong correlation respectively.

3 Results and Discussions

The oral mucosa can be divided into three types: the masticatory mucosa (gingival and hard palate mucosa), the lining mucosa (alveolar, soft palate, labial, and buccal mucosa, as well as the mucosa of the mouth floor and the ventral surface of the tongue), and the specialized mucosa (lips, dorsum of the tongue). The two-dimensional images of the vermilion border of lower lips, lower lip mucosa, and right cheek mucosa, tongue abdomen are shown as follows. The last image in each oral cavity tissue describes typical labial anatomical features; an epithelial layer (EP) and an underlying lamina propria (LP).

4 Conclusion

In this paper, we make a report about two-dimensional (2D) imaging of structure and three-dimensional (3D) imaging of microvessel within oral cavity tissues in vivo using a high-speed swept-source optical coherence tomography (SS-OCT) at 1310 nm with a designed probe by using 3D printing technique. Volumetric structural OCT images of the inner tissues of oral cavities are acquired with a field of view of 2 mm 2 mm. A type of detachable probe attachment is devised and applied to the port of the imaging probe of OCT system that easy to access to deeper cavity tissue. Blood perfusion is mapped with OCT-based correlation microangiography from 3D structural OCT images, in which a novel vessel extraction algorithm is used to decouple dynamic light scattering signals, due to moving blood cells, from the background scattering signals due to static tissue elements. Characteristic tissue anatomy and microvessel architectures of various cavity tissue regions of healthy human volunteers are identified with the 3D OCT images. The initial finding suggests that the proposed method may be engineered into a promising tool for discovering and treating oral cancer and precancer.

References

1. Reta, N., Michelmore, A., Saint, C., Prieto-Simon, B., Voelcker, N.H.: Porous silicon membrane-modified electrodes for label-free voltammetric detection of MS2 bacteriophage. Biosens. Bioelectron. **80**, 47–53 (2016)
2. Syshchyk, O., Skryshevsky, V.A., Soldatkin, O.O., Soldatkin, A.P.: Enzyme biosensor systems based on porous silicon photoluminescence for detection of glucose, urea and heavy metals. Biosens. Bioelectron. **66**, 89–94 (2015)
3. http://omlc.org/spectra/hemoglobin/index.html
4. http://www.andor.com/learning-academy/spectral-response-of-glucose-spectral-response-within-optical-window-of-tissue
5. Kamasahayam, S., Haindavi, S., Kavala, B., Chowdhury, S.R.: Non invasive estimation of blood glucose using near infra red spectroscopy and double regression analysis. In: Seventh International Conference on Sensing Technology, pp. 627–631 (2013)
6. McNichols, R.J., Cote, G.L.: Optical glucose sensing in biological fluids: an overview. J. Biomed. Optics **5**, 5–16 (2000)
7. Hecht, E.: Optics, 4th edn. Addison Wesley, Reading (2001)

Novel Approach of Respiratory Sound Monitoring Under Motion

Yan-Di Wang[1], Chun-Hui Liu[2], Ren-Yi Jiang[2], Bor-Shing Lin[3], and Bor-Shyh Lin[2(✉)]

[1] Institute of Photonic System, National Chiao Tung University, Tainan 711, Taiwan, R.O.C.
[2] Institute of Imaging and Biomedical Photonics, National Chiao Tung University, Tainan 711, Taiwan, R.O.C.
borshyhlin@gmail.com
[3] Department of Computer Science and Information Engineering, National Taipei University, New Taipei 237, Taiwan, R.O.C.

Abstract. Electronic stethoscope system is the most frequently used approach to collect respiratory sound to evaluate the lung function of patients or investigate various kinds of lung diseases of in recent years. However, current electronic stethoscope systems are just suitable for measuring respiratory sounds under static state. Under motion, the respiratory sounds are easily affected by the vibration of the human body. Moreover, it is also inconvenient to wear the conventional electronic stethoscope system. In order to improve the above issues, a novel wireless respiratory sound recording system was proposed to collect respiratory sound under motion. Here, a wireless and wearable respiratory sound recording device was designed to collect respiratory sound wirelessly. It is also easy to wear and monitor respiratory sound under motion due to its advantages of small volume and wireless transmission. Moreover, the technique of adaptive filter was also applied to enhance the noisy respiratory sound from single channel trial. From the experimental results, the noisy respiratory sound can be effectively improved by the proposed adaptive filter. Therefore, the proposed system exactly contains the potential of being a good assisting tool for lung diseases and may be applied in the applications of lung and sports medicine in the future.

Keywords: Respiratory sound · Vibration of human body · Wireless transmission · Adaptive filter

1 Introduction

According to the statistics of World Health Organization (WHO), more than half (54%) of the 56.4 million deaths worldwide in 2015 were due to the top ten causes, and the third, fourth, fifth causes of death in the top ten are breathing-related diseases [1]. Therefore, how to effectively detect potentially lung diseases becomes important in clinical. Abnormal respiratory sounds (such as crackles, rhonchus, and wheezing sounds) are often considered to be symptoms of chronic respiratory disease [2], such as chronic obstructive pulmonary disease (COPD), chronic bronchitis, and bronchial asthma, etc.

J.-S. Pan et al. (eds.), *Advances in Intelligent Information Hiding and Multimedia Signal Processing*, Smart Innovation, Systems and Technologies 81,
DOI 10.1007/978-3-319-63856-0_21

The traditional diagnosis method of respiratory or lung diseases is using the method of stethoscope auscultation. However, the state of lung function is usually determined by the physician with their subjective factors or clinical experiences, it may easily affect the evaluation of lung function due to this non-quantitative approach. Recently, the development of electronic stethoscope system became more mature, and they are widely applied in clinical.

In 2006, Bor-Shyh Lin et al. used an electronic system to recognize wheeze, it not only is low-cost but also has a high performance [3]. In 2016, Bor-Shyh Lin et al. proposed a technique of speech recognition to detect wheezes [4], and an electronic stethoscope system to monitor real-time wheezes [5], these approaches can detect wheezes simultaneously. Moreover, the approaches to detect sounds under different noise conditions were proposed [6, 7].

However, the above studies focus on the investigation of respiratory sound under static state. Under motion, the respiratory sounds are easily affected by the vibration of the human body. Moreover, it is also inconvenient to wear the conventional electronic stethoscope system.

In order to improve the above issue, a novel approach of respiratory sound monitoring under motion was proposed to evaluate lung function under motion in this study. Here, a wireless and wearable respiratory sound recording device was designed to collect respiratory sound wirelessly. It mainly consists of a wireless respiratory sound acquisition module and a wearable mechanical design. It is also easy to wear and monitor respiratory sound under motion due to its advantages of small volume and wireless transmission. Moreover, the technique of adaptive filter was also applied to enhance the noisy respiratory sound from single channel trial. The reference signal related to the clean respiratory sound could be extracted from the noisy respiratory sound signal, and then applied in the adaptive filter [8–11] to enhance the noisy respiratory sounds. Finally, the performance of enhancing noisy respiratory sound under motion was also investigated in this study. From the experimental results, the noisy respiratory sound can be effectively improved by the proposed approach.

2 Enabling Wireless Respiratory Sound Monitoring

The proposed wireless respiratory sound recording system mainly contains a wireless and wearable respiratory sound recording device and a host system. Here, the wireless and wearable respiratory sound recording device mainly consists of a wireless respiratory sound acquisition module and a wearable mechanical design. Here, the wireless respiratory sound acquisition module, embedded in the wearable mechanical design, was designed to monitor respiratory sound wirelessly under motion. The wearable mechanical design was designed to provide a convenient and reliable measuring procedure. The proposed wireless respiratory sound recording device can be worn easily and comfortably for long-term monitoring, and transmits respiratory data to the host system wirelessly. In the host system, a respiratory sound monitoring program is developed to monitor and record real-time respiratory sounds. An adaptive filter is also built in the respiratory sound monitoring program to effectively improve the influence of human body vibration in recording respiratory sounds.

2.1 Wireless Respiratory Sound Acquisition Module

The block diagram of the wireless respiratory sound acquisition module includes an acoustic sensor, a front-end amplifier circuits, a microphone driving circuit, a microprocessor, and a wireless transmission circuit. The acoustic sensor consists of a stethoscope bell and a microphone. Here, the respiratory sound will be collected by the stethoscope bell and then will be transferred to the electrical signal by the microphone. In order to eliminate the variability of the power source, the microphone driving circuit is designed to provide a stable reference voltage. Then, the acquired respiratory sound signal will be amplified and filtered by the front-end amplifier circuits. Next, an analog-to-digital converter in the microprocessor will digitize the amplified respiratory sound signal with 2 k Hz sampling rate. Finally, the processed respiratory sound signal will be transmitted to the host system by the wireless transmission circuit.

2.2 Wearable Mechanical Design

The proposed wearable mechanical design mainly consists of a shoulder brace, an elastic band, and an area of Velcro. Here, Velcro is used to fix the proposed wireless respiratory sound acquisition module on the shoulder brace. Moreover, the length of the elastic band can be adjusted to suitably fit the user's chest contour. Based on the above design, the proposed system can be worn and used easily, and it provides a convenient and reliable measuring procedure.

2.3 System Software Design

In this study, a commercial laptop is used as the platform of the host system. Here, Window 10 is used as the basic operation system of the host system. The respiratory sound monitoring program is designed by using Microsoft C#, and can provide the functions of real-time monitoring and recording respiratory sound signal.

2.4 Algorithm of Adaptive Filter for Human Body Vibration Cancellation

Under motion, monitoring respiratory sound is easily affected by various kinds of noise, in particular, human body vibration. In order to effectively improve the above issue, an adaptive filter is designed to eliminate the interference of human body vibration and enhance the respiratory sound from the noisy respiratory sound. An adaptive filter is a commonly used method which can self-regulate and learn to adapt according to the changes of signal or noise features. The basic scheme of the adaptive filter used in this study is shown in Fig. 1. Compared with the feature of respiratory sounds, the frequency band of human body vibration is lower. Therefore, the reference signal related to the clean respiratory sound will first be extracted from single-channel noisy respiratory sound by using a high-pass filter. The extracted reference signal will then be used as the reference input of the adaptive filter. The recorded noise respiratory sound will be used as the primary input of the adaptive filter. Next, the noisy respiratory sound signal will be adjusted by the adaptive filter until the error, which the difference between the filter output and the primary input signal, is minimized. Finally, the estimated clean respiratory sound will be obtained by the filter output.

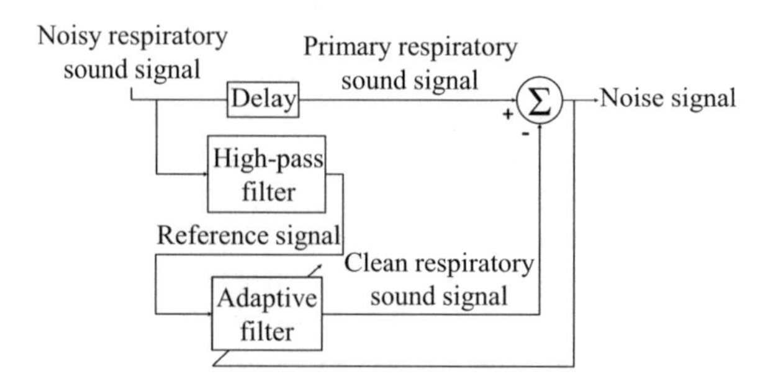

Fig. 1. Basic scheme of proposed adaptive filter for human body vibration cancellation.

3 Results

3.1 Simulated Performance of Enhancing Noisy Respiratory Sound

The performance of the proposed approach on enhancing the noisy respiratory sound is first simulated and investigated in this section. Here, the noisy respiratory sounds are generated by mix the pre-recorded respiratory sound with the pre-recorded human body vibration noise or the white Gaussian noise.

Figures 2 and 3 show the simulated results of enhancing noisy respiratory sounds with different frequency bands Gaussian noise (0 Hz – 150 Hz, and 150 Hz – 800 Hz

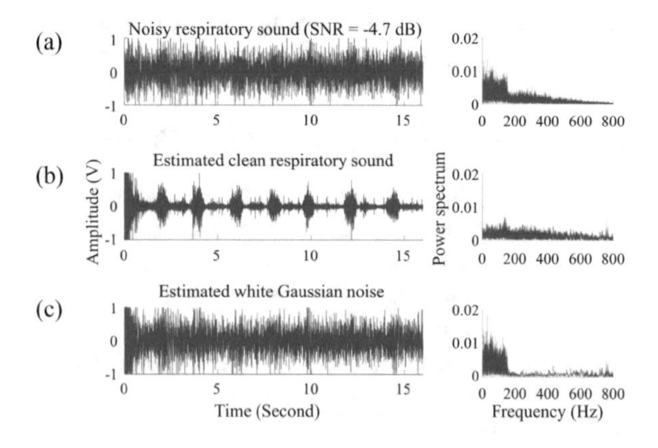

Fig. 2. (a) Noisy respiratory sound (SNR = −4.7 dB), (b) estimated clean respiratory sound, (c) estimated white Gaussian noise, and their frequency spectra.

respectively) and different signal-to-noise ratio (SNR; −4.7, and −4.3 dB respectively), and their frequency spectra. Here, mean square error (MSE) is used to validate the performance of the proposed algorithm. The values of MSE for the noisy respiratory sound and the estimated clean respiratory sound in Figs. 2 and 3 are (0.1144, 0.0239), and (0.107, 0.0265) respectively.

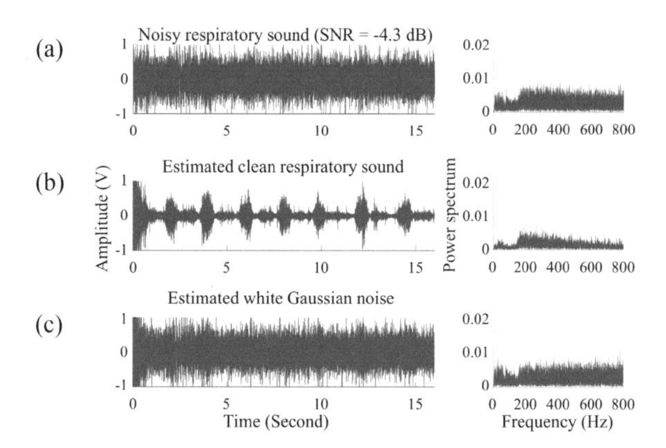

Fig. 3. (a) Noisy respiratory sound (SNR = −4.3 dB), (b) estimated clean respiratory sound, (c) estimated white Gaussian noise, and their frequency spectra.

Figure 4(a), (b), and (c) show the noisy respiratory sound with human body vibration noise, the estimated clean respiratory sound, the estimated human body vibration noise, and their frequency spectra respectively. Here, the SNR of the noisy respiratory sound is set to −3.9 dB. The MSE values of the noisy respiratory sound and the estimated clean respiratory sound are 0.0155 and 0.0021 respectively.

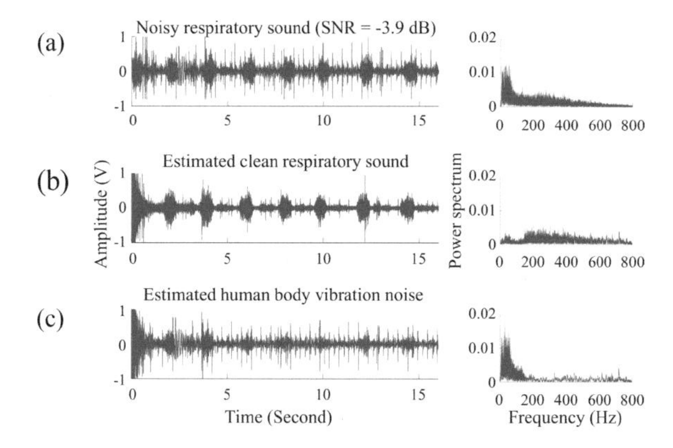

Fig. 4. (a) Noisy respiratory sound (SNR = −3.9 dB), (b) estimated clean respiratory sound, (c) estimated human body vibration noise, and their frequency spectra.

3.2 Performance of Enhancing Real Noisy Respiratory Sound

In this section, the performance of the proposed approach on real-time enhancing real noisy respiratory sound is investigated. The real noisy respiratory sound is recorded by the proposed wireless respiratory sound acquisition module during walking. The length of the trail of noisy respiratory sound is about 16 s. Figure 5(a), (b), and (c) show the real noisy respiratory sound with human body vibration noise, the estimated clean respiratory sound, and the estimated human body vibration noise respectively. It showed that the respiratory sound can exactly enhanced and the human body vibration noise was effectively eliminated.

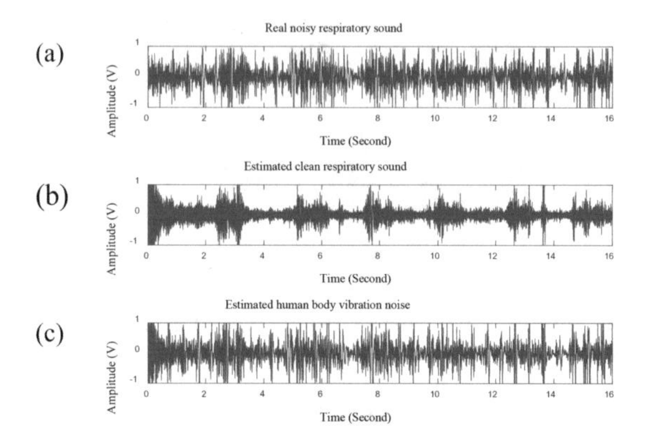

Fig. 5. (a) Real noisy respiratory sound, (b) estimated clean respiratory sound, and (c) estimated human body vibration noise.

4 Discussion

In this study, the wireless and wearable respiratory sound recording device is designed to measure respiratory sounds under motion. Here the proposed adaptive filtering algorithm could enhance the respiratory sound and eliminate the noise of human body vibration from the noisy respiratory sound. From the experimental results in Figs. 2 and 3, the noisy respiratory sound could be effectively enhanced, and their MSE also obviously improved (from 0.1144 to 0.0239 and from 0.107 to 0.0265). However, from the results of frequency spectra, it indicated that the frequency spectra of respiratory sound and Gaussian noise in Fig. 2 can be effectively split. Although the noisy respiratory sound in Fig. 3 could be also enhanced, its frequency spectrum seems to result in a little distortion. This may be explained by that the frequency band overlap between the reference signal of adaptive filter and the Gaussian noise increased, and this also affected the performance of enhancing noisy respiratory sound. From the experimental results in Fig. 4, it showed that the noise of the pre-recorded human body vibration could be effectively eliminated to enhance the noisy respiratory sound. The main frequency feature of human body vibration might be lower than the frequency band of the respiratory sound, and therefore, the reference signal related to the clean respiratory sound could be easily extracted to apply in the adaptive filter to enhance the noisy respiratory sound. For real-time enhancing noisy respiratory sound (Fig. 5), the noise of human body vibration could be exactly eliminated by the proposed approach.

5 Conclusions

In this study, a wearable and wireless respiratory sound monitoring system is proposed to monitor the respiratory under motion. Here, the wireless respiratory sound acquisition module embedded in the wearable mechanical design is developed to collect the respiratory sound under motion. Moreover, the technique of adaptive filter was also applied in enhancing respiratory sounds and eliminating human body vibration noise from single-channel noisy respiratory sound. The simulated results showed that the value of MSE of noisy respiratory sound could be exactly improved, and the frequency spectra of the enhanced respiratory sounds are similar to that of the clean respiratory sound. Therefore, the proposed system could effectively provide a good performance of monitoring respiratory sound under motion to evaluate the lung function, and might be applied in the researches of lung diseases or sports medicine in the future.

References

1. World Health Organization (WHO): The top 10 causes of death, January 2017. http://www.who.int/mediacentre/factsheets/fs310/en/
2. Moretz, C., Zhou, Y., Dhamane, A.D., Burslem, K., Saverno, K., Jain, G., Devercelli, G., Kaila, S., Ellis, J.J., Hernandez, G., Renda, A.: Development and validation of a predictive model to identify individuals likely to have undiagnosed chronic obstructive pulmonary disease using an administrative claims database. J. Managed Care Spec. Pharm. **21**(12), 1149–1159 (2015)
3. Lin, B.-S., Lin, B.-S., Wu, H.-D., Chong, F.-C., Chen, S.-J.: Wheeze recognition based on 2D bilateral filtering of spectrogram. Biomed. Eng.-Appl. Basis Commun. **18**(3), 128–137 (2006)
4. Lin, B.-S., Lin, B.-S.: Automatic wheezing detection using speech recognition technique. J. Med. Biol. Eng. **36**(4), 545–554 (2016)
5. Li, S.-H., Lin, B.-S., Tsai, C.-H., Yang, C.-T., Lin, B.-S.: Design of wearable breathing sound monitoring system for real-time wheeze detection. Sensors **17**(171), 1–15 (2017)
6. Sheu, M.-J., Lin, P.-Y., Chen, J.-Y., Lee, C.-C., Lin, B.-S.: Higher-order-statistics-based fractal dimension for noisy bowel sound detection. IEEE Signal Process. Lett. **22**(7), 789–793 (2015)
7. Lin, B.-S., Sheu, M.-J., Chuang, C.-C., Tseng, K.-C., Chen, J.-Y.: Enhancing bowel sounds by using a higher order statistics-based radial basis function network. IEEE J. Biomed. Health Inform. **17**(3), 675–680 (2013)
8. Widrow, B., Glover, J.R., McCool, J.M., Kaunitz, J., Williams, C.S., Hearn, R.H., Zeidler Jr., J.R., Dong, E., Goodlin, R.C.: Adaptive noise cancelling: principles and applications. Proc. IEEE **63**(12), 1692–1716 (1975)
9. Dhiman, J., Ahmad, S., Gulia, K.: Comparison between adaptive filter algorithms (LMS, NLMS and RLS). Int. J. Sci. Eng. Technol. Res. (IJSETR) **2**(5), 1100–1103 (2013)

10. Gnitecki, J., Hossain, I., Pasterkamp, H., Moussavi, Z.: Qualitative and quantitative evaluation of heart sound reduction from lung sound recordings. IEEE Trans. Biomed. Eng. **52**(10), 1788–1792 (2005)
11. Yang, C., Tavassolian, N.: Motion noise cancellation in seismocardiogrpahic monitoring of moving subjects. In: IEEE Biomedical Circuits and Systems Conference, December 2015

A General Auto-Alignment Algorithm for Three-Degree Freedom Stage by Local Inverse Information with Regression Method

Yu-Min Hung(✉) and Yao-Chin Wang

Department of Biomedical Engineering, HunagKang University,
No. 1018, Sec. 6, Taiwan Boulevard, Shalu District, Taichung City 43302, Taiwan (R.O.C.)
ymhung@sunreise.hk.edu.tw

Abstract. This paper demonstrated a general algorithm to carry out precision alignment with UVW or XXY and XYθ stages using two cameras without rotation center information. Image processing with template match was implemented to search for fiducial mark under the field of view of a camera as the positioning tool. We established the relationship between cameras and motion stage by calibration procedure with inverse regression method. We finished the substrate's alignment parallel to the mechanical axis in 2–4 times iterations and applied this algorithm in sixth-generation LCD prober system successfully.

Keywords: UVW axis · Auto alignment · Precision positioning

1 Introduction

To increase the production efficiency, automatic systems play a very important role in industry applications and manufacturing. In addition, to promote the high throughputs, the more precision assembly could be executed using vision servo system to acquire stable and high quality to replace human labors. A precision alignment function provides the possibilities to a product in a process or assembly to enhance production quality especially size downing conditions. A mechanism with precision movement, rotation mechanism and feedback system is usually implemented to construct this function. With computation efficiency increasing, vision system also provides a notable function in alignment, inspection, and quality checking in semiconductor, liquid crystal display (LCD), automobile, food industry and more applications.

In a precision mechanism, in generally, there are two types displacement and rotation stage in the industry applications; which are XYθ and UVW or XXY type respectively. The size of UVW type is compact and thinner than XYθ type but limited in displacement and orientation range. In a lot of applications, the substrate just only adjusts tiny displacement and orientation, i.e., print circuit board (PCB), surface mounting technology (SMT), IC bounding, solar cells and so on, to align the substrate to parallel the system. Based on these reasons, an UVW type stage is very suitable for precision alignment functions. Nian and Tarng [1, 2] proposed a new type UVW type stage with

J.-S. Pan et al. (eds.), *Advances in Intelligent Information Hiding and Multimedia Signal Processing*, Smart Innovation, Systems and Technologies 81,
DOI 10.1007/978-3-319-63856-0_22

specifically designed rotation center to archive ±5 um positioning with image processing technique. The algorithm of auto-alignment was developed to correspond to their design mechanism and image processing with locking of circle shape fiducial mark to define object center. Lee et al. [3] proposed a novel XXY stage with the kinematic equation to derivate the alignment algorithm for the specific mechanism, and finish ±7 um position accuracy.

The In general, UVW or XXY type axis motion stage is difficult to define the rotation center to derivate the kinematic equation of the mechanism. In this paper, we proposed a general alignment algorithm to derivate displacement and rotation for two CCD cameras system not only for UVW stage but also suitable for XYθ stage. In this study, we successfully applied this algorithm to the UVW stage in our laboratory and transferred this algorithm to the sixth-generation LCD prober equipment achieving precision alignment function.

2 Alignment Algorithm

2.1 Modeling

This section discusses the ideal conditions the difference between two cameras with stage orientation. In general, the motion stage provides the movement and rotation function. Figure 1 shows the layout of rotation stage with two cameras. In generally, the coordinates of image center could be expressed following Eqs. (1) and (2):

$$\begin{aligned} x_{ccd1} &= x_0 + r_{ccd1} \cos \alpha_1 \\ y_{ccd1} &= y_0 + r_{ccd1} \sin \alpha_1 \end{aligned} \tag{1}$$

$$\begin{aligned} x_{ccd2} &= x_0 + r_{ccd2} \cos \alpha_2 \\ y_{ccd2} &= y_0 + r_{ccd2} \sin \alpha_2 \end{aligned} \tag{2}$$

where x_o and y_o denotes the rotation center of motion stage, α_1 and α_2 denotes the angle camera center to global coordinate with right-hand principal, r_{ccd1} and r_{ccd2} denotes the distance of two cameras to rotation center. L denotes the length of two cameras. Herein, we give a small rotation angle $\Delta\theta$ by motion stage; the new positions are given as following equations:

$$\begin{aligned} x'_{ccd1} &= x_0 + r_{ccd1} \cos(\alpha_1 + \Delta\theta) \\ y'_{ccd1} &= y_0 + r_{ccd2} \sin(\alpha_1 + \Delta\theta) \end{aligned} \tag{3}$$

$$\begin{aligned} x'_{ccd2} &= x_0 + r_{ccd2} \cos(\alpha_2 + \Delta\theta) \\ y'_{ccd2} &= y_0 + r_{ccd2} \sin(\alpha_2 + \Delta\theta) \end{aligned} \tag{4}$$

where x'_{ccd1}, y'_{ccd1}, x'_{ccd2} and y'_{ccd2} denotes the new position coordinates after rotation $\Delta\theta$. The difference of reference and new position in X direction could be expressed as following Eqs. (5) and (6):

$$\delta x_{ccd1} = x'_{ccd1} - x_{ccd1} = 2r_{ccd1}(\sin \alpha_1 \cos \frac{\Delta\theta}{2} + \cos \alpha_1 \sin \frac{\Delta\theta}{2}) \sin \frac{-\Delta\theta}{2} \tag{5}$$

$$\delta x_{ccd2} = x'_{ccd2} - x_{ccd2} = 2r_{ccd2}(\sin \alpha_2 \cos \frac{\Delta\theta}{2} + \cos \alpha_2 \sin \frac{\Delta\theta}{2}) \sin \frac{-\Delta\theta}{2} \tag{6}$$

where δx_{ccd1} and δx_{ccd2} denotes the offset in X direction with camera 1 and camera 2 respectively. The difference of two cameras with a rotation angle $\Delta\theta$ in X direction could be expressed as follow equation with Taylor series expansion when $\Delta\theta$ is small enough.

$$\delta x_{ccd2} - \delta x_{ccd1} \cong -L_y\Delta\theta - \frac{1}{2}L_x\Delta\theta^2 + \frac{1}{6}L_y\Delta\theta^3 - \frac{1}{24}L_x\Delta\theta^4 + \cdots \tag{7}$$

where L_x and L_y denote the distance between two cameras in X and Y direction respectively. Similarly, we could obtain the relationship between two cameras in Y direction shown in Eq. (8):

$$\delta y_{ccd2} - \delta y_{ccd1} \cong L_x\Delta\theta - \frac{1}{2}L_y\Delta\theta^2 + \frac{1}{6}L_x\Delta\theta^3 + \frac{1}{24}L_y\Delta\theta^4 + \cdots \tag{8}$$

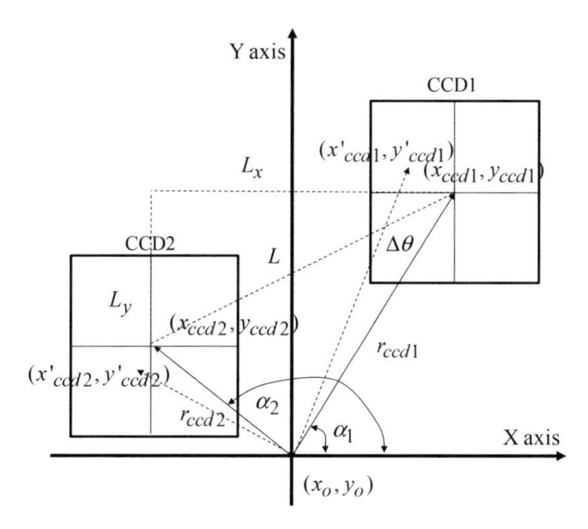

Fig. 1. The layout of motion stage and two cameras.

The summation displacement of two cameras after $\Delta\theta$ rotation in X and Y direction could also be expressed as following:

$$\delta x_{ccd2} + \delta x_{ccd1} \cong S_y\Delta\theta + \frac{1}{2}S_x\Delta\theta^2 - \frac{1}{6}S_y\Delta\theta^3 + \cdots \tag{9}$$

$$\delta y_{ccd2} + \delta y_{ccd1} \cong S_x \Delta\theta - \frac{1}{2} S_y \Delta\theta^2 + \frac{1}{6} S_x \Delta\theta^3 + \cdots \tag{10}$$

where S_x and S_y denotes the summation and subtraction of two cameras in X and Y respectively. From the modeling in this chapter, we could find the value of summation and subtraction of two cameras is a polynomial function of orientation. The relationship could be constructed in the procedure of calibration.

2.2 Algorithm

Figure 2 shows the relationship of reference fiducial mark and new fiducial mark under the camera field view. The difference between reference pattern and new pattern position is δx_{ccd1} and δy_{ccd1} in CCD1 and δx_{ccd2} and δy_{ccd2} in CCD2 respectively. Herein we assume that the substrate moves with rigid body movement and rotation denoting as (Δx, Δy, $\Delta\theta$) which is need to solved. The Eqs. (11) and (12) describes the relationship pattern offset in X direction under two cameras. In these two equations, where $\delta x_{ccd1}(\Delta\theta)$ denotes the displacement in X direction contributed from substrate's orientation under CCD1 view. Where $\delta x_{ccd2}(\Delta\theta)$ denotes displacement in X direction contributed from substrate's orientation under CCD2 view.

$$\delta x_{ccd1} = \Delta x + \delta x_{ccd1}(\Delta\theta) \tag{11}$$

$$\delta x_{ccd2} = \Delta x + \delta x_{ccd2}(\Delta\theta) \tag{12}$$

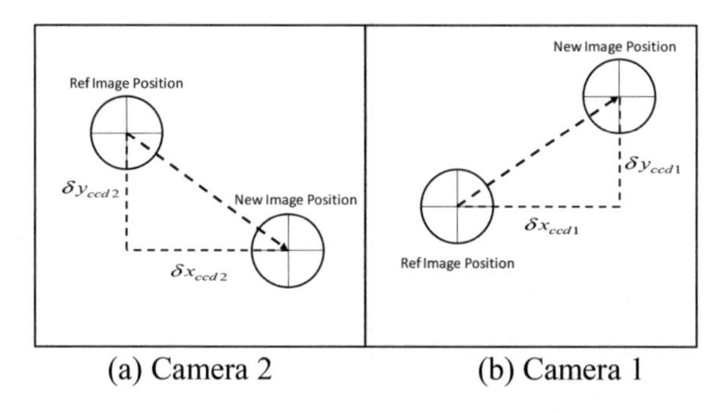

Fig. 2. The relationship between new image positions to reference image under two cameras.

Similarly, the relationship between two cameras and substrate in Y direction could be described in Eqs. (13) and (14):

$$\delta y_{ccd1} = \Delta y + \delta y_{ccd1}(\Delta\theta) \tag{13}$$

$$\delta y_{ccd2} = \Delta y + \delta y_{ccd2}(\Delta\theta) \tag{14}$$

where $\delta y_{ccd1}(\Delta\theta)$ and $\delta y_{ccd2}(\Delta\theta)$ denotes the Y direction movement contributed from substrate's after orientation $\Delta\theta$. We could obtain Eqs. (15) and (16) by operating Eq. (12) minus Eq. (12) and Eq. (14) minus Eq. (13):

$$\delta x_{ccd2} - \delta x_{ccd1} = \delta x_{ccd2}(\Delta\theta) - \delta x_{ccd1}(\Delta\theta) = f_x(\Delta\theta) \tag{15}$$

$$\delta y_{ccd2} - \delta y_{ccd1} = \delta y_{ccd2}(\Delta\theta) - \delta y_{ccd1}(\Delta\theta) = f_y(\Delta\theta) \tag{16}$$

The rigid displacement of substrate will be eliminated in subtraction operation. The movement Δx and Δy of substrate could be solved by Eqs. (17) and (18), if we could know Eqs. (15) and (16).

$$\Delta x = \frac{\delta x_{ccd1} + \delta x_{ccd2} - \delta x_{ccd1}(\Delta\theta) - \delta x_{ccd2}(\Delta\theta)}{2} \tag{17}$$

$$\Delta y = \frac{\delta y_{ccd1} + \delta y_{ccd2} - \delta_{ccd1} y(\Delta\theta) - \delta y_{ccd2}(\Delta\theta)}{2} \tag{18}$$

In an alignment process, δx_{ccd1}, δx_{ccd2}, δy_{ccd1} and δy_{ccd2} are known values by an image processing technology. The orientation of substrate could be interpolated through inversing $f_x(\Delta\theta)$ and $f_y(\Delta\theta)$ which established in calibration procedure. Then $\delta x_{ccd1}(\Delta\theta) + \delta x_{ccd2}(\Delta\theta)$ and $\delta y_{ccd1}(\Delta\theta) + \delta y_{ccd2}(\Delta\theta)$ could be interpolated from calibration procedure.

3 Experimental Setup

We setup the UVW axes motion stage to implement the movement and orientation, shown in Fig. 3(a). Three servo motors are used to control the movement of UVW stage. Moving X travel, it just only controls X-axis servo motor. Moving Y travel, it needs to control Y1 and Y2 axes motors simultaneously same direction and distance. If want to move orientation, the X motor moves the same distance with Y1 and Y2 servo motor,

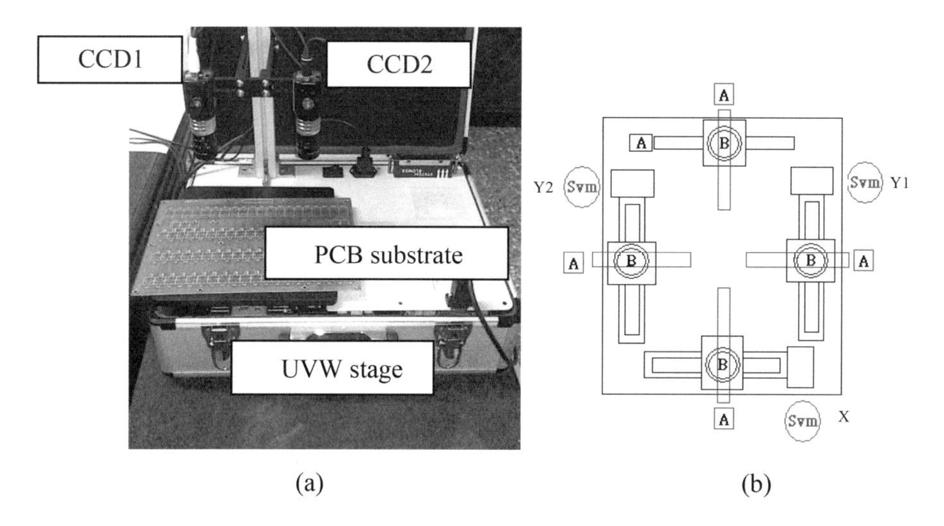

Fig. 3. (a) The hardware of UVW axis with two camera alignment system, (b) the schematic drawing of UVW axis stage

but Y1 and Y2 are opposite of its direction, shown in Fig. 3(b). In this figure, a PCB substrate lies on platform of UVW axis stage. The image capture system which includes two cameras (WATEC), embedded 35 mm TV lens, connected to PC with image grabbing frame card (Adlink RTV24 Frame Grabber Card). In this study, the Matrox Image Library (MIL) [4] was implemented in the unknown pattern searching and fiducial mark location by pattern machine. The normalized cross correlation has been expensively used in machine vision for industry inspection or alignment, especially in unknown image definition and registration.

4 Results and Application

4.1 Camera Calibration

Camera calibration is an important procedure to corresponding pixel dimension to physic dimension. To simplify the hardware installation of cameras, calibration procedure will find the orientation angle of cameras. In the calibration procedure, we also construct the relationship for two cameras in $\delta x_{ccd2}(\Delta\theta) - \delta x_{ccd1}(\Delta\theta)$, $\delta y_{ccd2}(\Delta\theta) - \delta y_{ccd1}(\Delta\theta)$, $\delta x_{ccd2}(\Delta\theta) + \delta x_{ccd1}(\Delta\theta)$ and $\delta y_{ccd2}(\Delta\theta) - \delta y_{ccd1}(\Delta\theta)$. Figure 4 shows the calibration procedure. After calibration, we could obtain the pixel size and orientation of cameras, shown in Table 1. Basically, the orientation of camera value is nearly by zero. In this case, orientation of camera is not equal to zero, means chip sensor of cameras are not parallel to mechanical axis. Figure 5 shows summation and subtraction of two camera's offset with respect to stage orientation. We could observe $\delta y_{ccd2}(\Delta\theta) - \delta y_{ccd1}(\Delta\theta)$ and $\delta x_{ccd2}(\Delta\theta) - \delta x_{ccd1}(\Delta\theta)$ is linear with respect to rotation angle in our experimental setup due to two cameras locating two side of rotating center. In the case, we establish the polynomial function with order two by least square method.

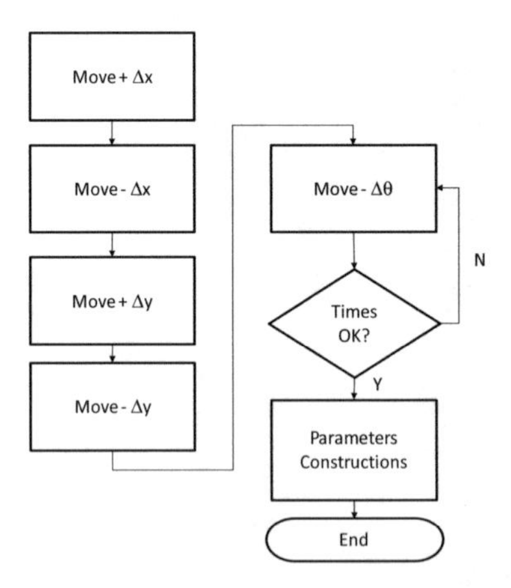

Fig. 4. The procedure of camera calibration

Table 1. Calibration data of two cameras

	X pixel size (nm)	Y pixel size (nm)	Orientation (φ, deg)
CCD1	30528	30768	1.949
CCD2	−30545	−30750	1.326

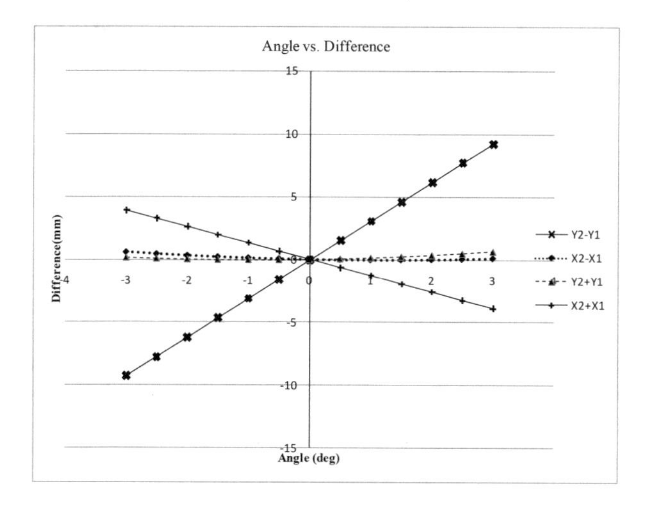

Fig. 5. The relationships of orientation with difference of two cameras applications

4.2 Application

Figure 6 shows six generations LCD Prober system which alignment system's function constructed by Y and θ axis motor and two cameras mounted in gantry system is to achieve precise and automatic contact position for panel testing. Figure 7(a) shows the fiducial mark under the field of view. Applied the inverse regression method algorithm, the fiducial marks was in the same line in Y direction, but still have offset in X direction due to mechanism of LCD prober, shown in Fig. 7(b).

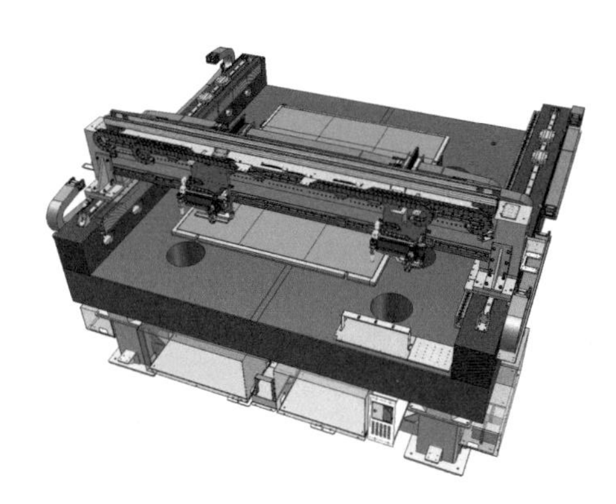

Fig. 6. The 3D drawing of six generation LCD prober

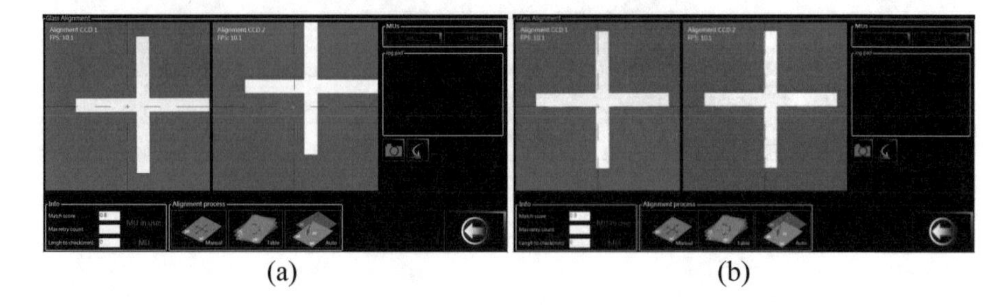

Fig. 7. (a) Un-aligned fiducial marks under two CCD's view (b) alignment result applied with local regression method

5 Conclusion

We proposed a general algorithm for three axis stage in alignment without center information of rotation center and applied in UVW and XYθ stage. Through camera calibration procedure, we established the relationship between orientation and difference of offset of two cameras with polynomial function. By pattern matching, we obtain the offset of two cameras. These data derivate the substrate's $(\Delta x, \Delta y, \Delta \theta)$ from pre-defined regression function in 2–3 times iterations.

References

1. Nian, C.Y., Tarng, Y.S.: An auto-alignment vision system with three-axis motion control mechanism. Int. J. Adv. Manuf. Technol. **26**, 1121–1131 (2005)
2. Nian, C.Y., Chuang, S.F., Tarng, Y.S.: A new algorithm for a three-axis auto-alignment system using vision inspection. J. Mater. Process. Technol. **171**, 319–329 (2006)
3. Lee, H.-W., Liu, C.-H., Chiu, Y.-Y., Fang, T.-H.: Design and control of an optical alignment system using a parallel XXY stage and four CCDs for micro pattern alignment. In: Proceedings of the DTIP 2012, pp. 3–17 (2012)
4. Matrox Image Library User Manual. Matrox Image

Initial Phase of Building up a Portable Multiple-Wavelength Biomedical Sensing System

Yen-Lin Yeh[1], Zu-Po Yang[1(✉)], and Yao-Chin Wang[2(✉)]

[1] Institute of Photonic System, National Chiao Tung University, Tainan, Taiwan
zupoyang@nctu.edu.tw
[2] Department of Biomedical Engineering, Hungkuang University, Taichung, Taiwan
autherkyn@gmail.com

Abstract. Biomedical materials have different optical properties (e.g. absorption) for different wavelengths. Therefore, probing biomedical materials with multiple wavelengths not only can get in-depth understanding of the detected biomedical materials but also can differentiate the detected biomedical materials. To achieve this purpose, in this conference paper, we present our initial results of building up a portable multiple-wavelength biomedical sensing system. At this initial phase, we assembled this kind of system with multiple wavelengths of light sources and photodetectors and preliminarily tested the absorbance of the glucose solutions with different concentrations. The result shows good linearity of the absorbance of the glucose solution with concentration. In addition, we also measured the absorbance of the glucose solutions using a broadband white light source and a spectrometer. This result also exhibits linearity but different slop of absorbance with glucose concentration, which confirms the linearity result obtained from the built sensing system. The difference of slop maybe relates to the difference of optical design between these two systems.

Keywords: Multiple wavelengths · Biomedical sensing · Glucose concentration

1 Introduction

Biomedical materials detection is of increasing importance and has wide applications, e.g. pathological examination, industrial chemical detection, personal physiological monitoring, etc. [1, 2]. There are different strategies, e.g. electrical-base and optical-base measurements, have been proposed to detect biomedical materials. The optical detection means utilizes the optical properties (e.g. absorption) of detected biomedical materials. For instance, as shown in Fig. 1(a), the hemoglobin exhibits different characteristic of optical absorption [3] for oxy- or deoxy-states. Comparing with its deoxy-state (deoxy-hemoglobin, Hb), oxy-hemoglobin (hemoglobin bound to oxygen, HbO2) has stronger absorbance in the near infrared regime (wavelength longer than ~ 800 nm) but weaker absorbance in the spectral regime between ~ 600 nm and ~ 800 nm. And both show similar value of absorbance for wavelength shorter than ~ 600 nm. Therefore,

J.-S. Pan et al. (eds.), *Advances in Intelligent Information Hiding and Multimedia Signal Processing*, Smart Innovation, Systems and Technologies 81,
DOI 10.1007/978-3-319-63856-0_23

probing hemoglobin with multiple wavelengths can differentiate its state (oxy- or deoxy-states). For glucose, it has relatively stronger absorption than water for some near infrared ranges, e.g. ~ 900–1000 nm (Fig. 1(b)) [4–6], so that this spectral and other near-infrared ranges can be used to sensing glucose. Conventional optical-base measurements typically use single or single-band wavelength light source and photodetector, which will limit capability of the differentiation of biomedical materials. For example, both oxy-hemoglobin and deoxy-hemoglobin absorb the lights in the range of 900 nm-1000 nm although oxy-hemoglobin has stronger absorption. Hence, for the mixture biomedical materials of oxy-hemoglobin and deoxy-hemoglobin, it is difficult to quantitatively extract the individual absorption of oxy-hemoglobin and deoxy-hemoglobin by using single or single band wavelength. However, this issue can possible be solved by using multiple (band) wavelengths. Besides the detectability, portability is another consideration for modern biomedical sensing systems. It is even better if the sensing system can be integrated as a chip. In this conference paper, we present our preliminary results of building up a portable multiple-wavelength biomedical sensing system. At this initial phase, we preliminary tested the glucose solution with different concentrations. The result shows the good linearity of absorbance with the glucose concentration. In addition, we also measured the absorbance using a broad band white light source and a spectrometer, which also exhibits good linearity with the glucose concentration but has different slop from the built multiple-wavelength sensing system. The cause of slop difference is still under investigation but maybe related to difference of optical design.

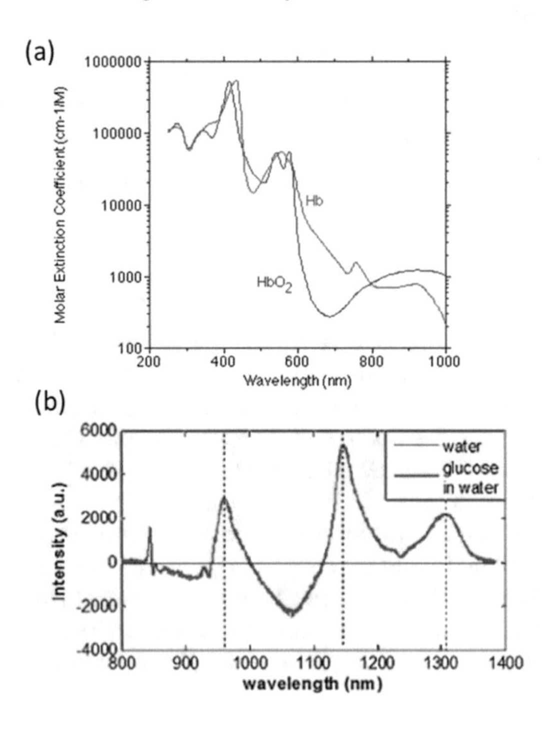

Fig. 1. (a) Molar extinction coefficient of oxy-hemoglobin (Hb) and deoxy-hemoglobin (HbO) [2]. (b) Absorption spectrum of glucose [3].

2 Experiments

Figure 2(a) shows the schematic of the preliminary design of our multiple-wavelength biomedical sensing system. The light source, located on the top, is composed of a multiple-wavelength light-emitting-diode (LED) array with emission colors of blue, green, red, and near infrared. Their emission spectra are plotted in Fig. 3(a). The bottom is of a detector array denoting as B, G, R, and IR with (response range, peak wavelength) of (400–540 nm, 460 nm), (480–600 nm, 540 nm), (590–720 nm, 660) and (880–1050 nm, 940 nm), respectively, which are obtained from the datasheet of detectors and replotted in Fig. 3(b). So that, the response ranges of individual photodetectors basically match the emission spectra of the corresponding LEDs. Both of LED and photodetector arrays are controlled an Arduino circuit board (model ArduinoUNORev3). Each individual color of LED is turned on (0.5 s) in sequence of near infrared, red, green, and blue and only the single of the corresponding photodetector is read during the LED is on. Sample, which is glucose solution here, is placed in between of the LED and photodetector arrays. The distance *d* between sample and detector is varied (here are 7 and 4.2 cm) to examined the effect of optical design (will be discussed later). In addition, we also built up another system, as shown in Fig. 2(b), to double check the built multiple-wavelength biomedical sensing system. The multiple-wavelength LED array is replaced by a broad band light source (tungsten lamp) with emission light coupled into a fiber. After passes through the sample, the light was collected by a fiber through a collimated lens and was then sent into an Ocean Optics spectrometer. To preliminary test the built multiple-wavelength biomedical sensing system, different concentrations of glucose solutions were used as samples, which were prepared by added different weights of glucose into deionized (DI) water of fixed volume (20 ml). The pure DI water is used as a reference sample. The optical density (*OD*) of glucose solution was calculated using

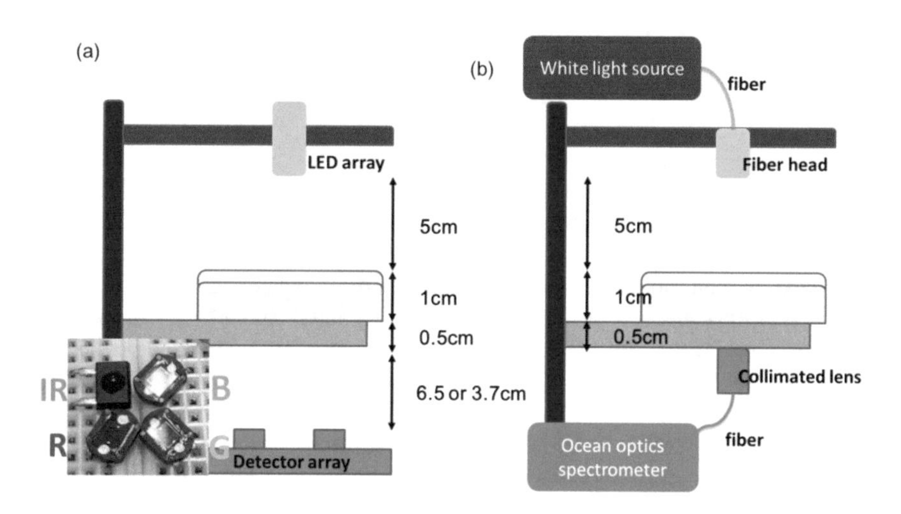

Fig. 2. (a) schematic of the preliminary design of our multiple-wavelength biomedical sensing system. (b) Another biomedical system with a white light source and a spectrometer.

$OD = logI_0 - logI$, where I_0 and I are the transmitted light intensity of DI water and glucose solution, respectively.

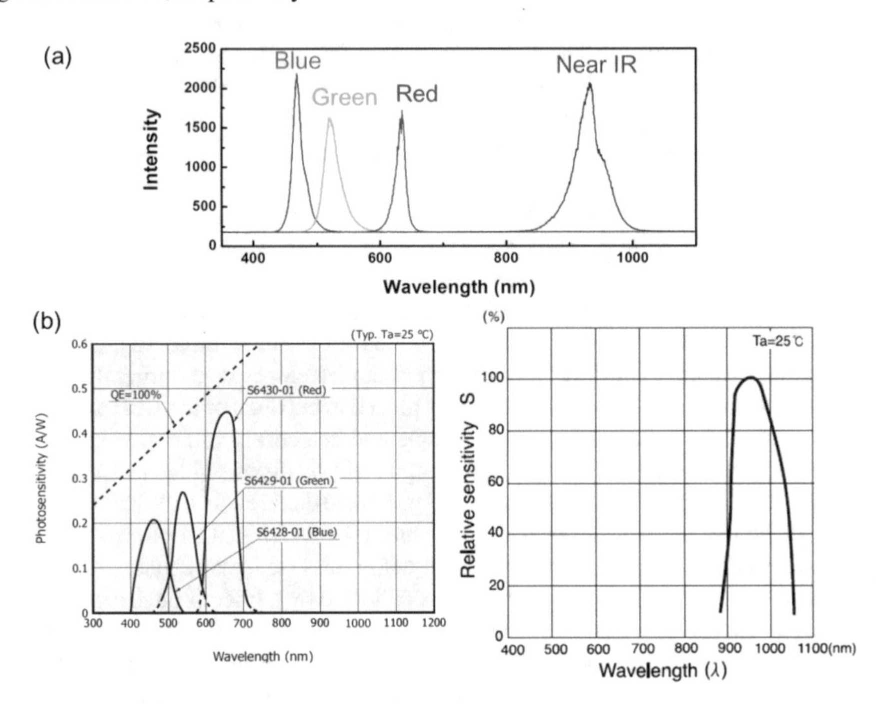

Fig. 3. (a) Emission spectra of blue, green, red, and near infrared LEDs. (b) Response spectra of photodetectors for B, G, R, and IR.

3 Results and Discussions

Figure 4(a) shows the measured *OD* (for d = 7 cm) of different glucose concentrations (different glucose weights) for blue, green, red, and near infrared wavelengths. The *OD*s for all different wavelengths shows good linearity with the glucose concentration. The slop decreases from 0.0468 to 0.0133 as wavelength increases from blue to near infrared. As shown in Fig. 4(b), the *OD* for different wavelengths (438.2, 566.24, 688.32 and 988.02 nm) measured by an Ocean Optics spectrometer also exhibits good linearity with glucose concentration. However, the slops for the wavelengths of blue (438.2 nm), green (566.24 nm), and red (688.32 nm) have nearly identical slop and are larger than the slop for near infrared (988.02). Rather than calculate *OD* using single wavelength, the *OD*s were also calculated by the integrated spectral ranges of 415.23 ~ 520.43 nm, 495.38 ~ 585.67 nm, 615.4 ~ 705.65 nm and 900.2 ~ 1049.75 nm for the corresponding photodetectors of B, G, R, and IR (Fig. 4(d)). The result is similar to the result obtained from single wavelength. This observed phenomenon is much different from the observation by the built multiple-wavelength sensing system. To figure out this issue, we decreased the distance *d* from 7 cm to 4.2 cm and re-measured the *OD* of different glucose concentration by the built multiple-wavelength sensing system. The result (Fig. 4(c)) shows that *OD*s still show good linearity for different blue, green, red, and near infrared wavelengths. But the slops for blue, green, and red wavelengths now are nearly the same and are larger than that for near infrared wavelength, which is consistent with the results obtained by the spectrometer system. In addition, the slop for near infrared is basically no change for different values of *d*. This observation is associated with the effect of light scattering since the *OD* we calculated includes absorption as well as light scattering. This argument is based on the two points: (1) for d = 7 cm, the slop decreases as wavelength increases which agree with the decrease of light scattering with increase of wavelength [7]; (2) glucose has certain absorption for near infrared regime but no absorption for the spectral range between blue to red. However, for blue, green, and red lights, the measured *OD*s are larger than that for near infrared. This result also indicates that the light scattering could be used as another freedom to differentiate biomedical materials but more investigations still need to be done. Although, by shortening the distance *d*, the observation by the built multiple-wavelength sensing system is similar to the phenomenon observed by the spectrometer system, the slop of linearity extracted by this sensing system for all wavelengths is still smaller than that obtained by the spectrum system. This difference of slop maybe relates to the difference of optical design between these two systems but more investigations have to be done to confirm this speculation.

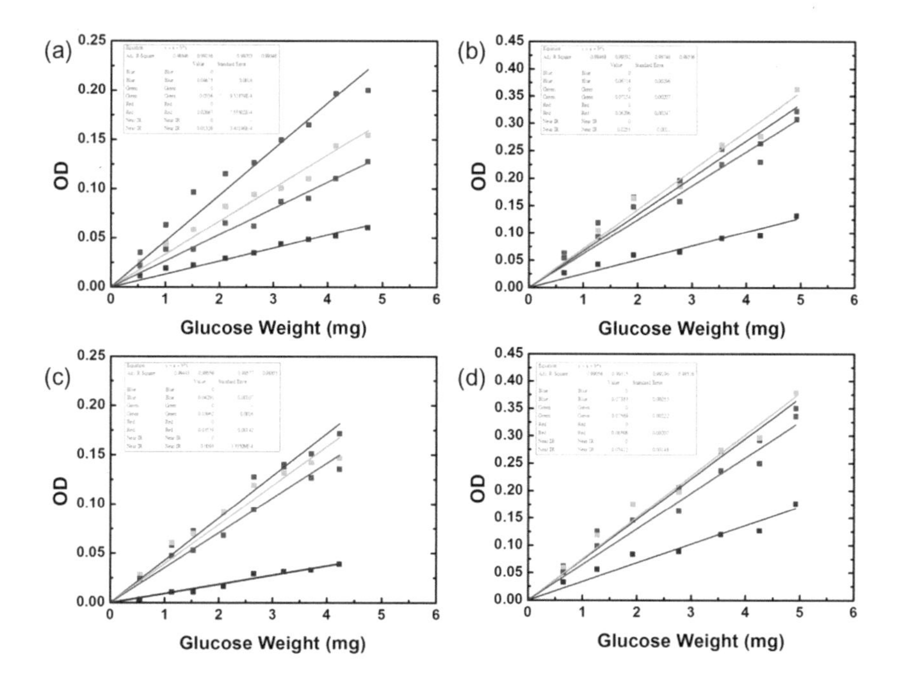

Fig. 4. OD vs glucose weight for different colors (blue, green, red, near IR) measured by (a) the multiple-wavelength sensing system with d = 7 cm, (b) the spectrometer system (single wavelength), (c) the multiple-wavelength sensing system with d = 4.2 cm, and (d) the spectrometer system (integrated spectral range).

4 Conclusion

We tried to build up a portable multiple-wavelength biomedical sensing system by assembling multiple colors of LEDs and photodetector. To inspect our preliminary build-up system, we measured the optical density of glucose solution with different concentrations, which exhibits good linearity with the glucose concentration. We also find that the scattering light may affect the measurement and probably can be serve as another freedom to differentiate biomedical materials. It has to be noted that by replace the Arduino circuit board with wireless function, the built system can become a portable wireless sensing system. Also the size of the build-up system can be shrunk further.

Acknowledgement. The authors would like thank the founding support from Ministry of Science and Technology (Contract. No. MOST 105-2221-E-009-073)

References

1. Reta, N., Michelmore, A., Saint, C., Prieto-Simon, B., Voelcker, N.H.: Porous silicon membrane-modified electrodes for label-free voltammetric detection of MS2 bacteriophage. Biosens. Bioelectron. **80**, 47–53 (2016)
2. Syshchyk, O., Skryshevsky, V.A., Soldatkin, O.O., Soldatkin, A.P.: Enzyme biosensor systems based on porous silicon photoluminescence for detection of glucose, urea and heavy metals. Biosens. Bioelectron. **66**, 89–94 (2015)
3. Obtained from. http://omlc.org/spectra/hemoglobin/index.html
4. http://www.andor.com/learning-academy/spectral-response-of-glucose-spectral-response-within-optical-window-of-tissue
5. Kamasahayam, S., Haindavi, S., Kavala, B., Chowdhury, S.R.: Non invasive estimation of blood glucose using near infra red spectroscopy and double regression analysis. In: Seventh International Conference on Sensing Technology, pp. 627–631 (2013)
6. McNichols, R.J., Cote, G.L.: Optical glucose sensing in biological fluids: an overview. J. Biomed. Optics **5**, 5–16 (2000)
7. Hecht, E.: Optics, 4th edn. Addison Wesley, Reading (2001)

Emerging Techniques and Its Applications

On the Security of a Certificateless Public Key Encryption with Keyword Search

Tsu-Yang Wu[1,2(✉)], Chao Meng[3], Chien-Ming Chen[3], King-Hang Wang[4], and Jeng-Shyang Pan[1,2]

[1] Fujian Provincial Key Lab of Big Data Mining and Apllications, Fujian University of Technology, Fuzhou 350118, China
wutsuyang@gmail.com, jengshyangpan@gmail.com
[2] National Demonstration Center for Experimental Electronic Information and Electrical Technology Education, Fujian University of Technology, Fuzhou 350118, China
[3] Harbin Institute of Technology Shenzhen Graduate School, Shenzhen 518055, China
171521532@qq.com, chienming.taiwan@gmail.com
[4] Department of Computer Science and Engineering, The Hong Kong University of Science and Technology, Clear Water Bay, Kowloon, Hong Kong
kevinw@cse.ust.hk

Abstract. Public key encryption with keyword search (PEKS) is one of searchable encryption mechanisms. It not only provides user to retrieve ciphertext by keyword but also protects the confidentiality of keyword. In the past, many PEKS schemes based on different cryptosystems were proposed. Recently. Zheng et al. proposed a certificateless based PEKS scheme called CLKS. In this paper, we show that Zheng et al.'s CLKS scheme has some security flaw, i.e. their scheme suffered from an off-line keyword guessing attack.

Keywords: Public key encryption with keyword search · Certificateless · Off-line keyword guessing attack · Cryptanalysis

1 Introduction

In order to solve the problem of searching an encrypted data, public key encryption with keyword search (PEKS) is proposed for the first time by Boneh et al. [3] in 2004. This method provides not only the confidentiality of keyword but also quickly retrieving ciphertext by keyword. Hence, PEKS has a better security properties and widely applied to real environments such as mail system and cloud storage system [10]. Afterwards, many variants of PEKS have been proposed such that supporting single keyword search [2,6,10], supporting conjunctive keyword search [5,17,22], supporting fuzzy keyword search [13,32], supporting multi-user environments [17,19,24], supporting keyword updating functionality [20], and supporting ranking keyword search [7,15,24].

J.-S. Pan et al. (eds.), *Advances in Intelligent Information Hiding and Multimedia Signal Processing*, Smart Innovation, Systems and Technologies 81,
DOI 10.1007/978-3-319-63856-0_24

However, due to the relatively limited number of keywords, numerous PEKS schemes are vulnerable to off-line keyword guessing attacks [8,14,26–28,34]. The process of off-line keyword guessing attacks can be divided into two steps. In the first step, attacker guesses an appropriate keyword kw. Then, it verifies the guessed keyword kw whether related to captured trapdoor by some equation and public parameters. If true, this attack is successful. Otherwise, return to the first step and repeat the process until all possible keywords are tried. In order to overcome the mentioned off-line keyword guessing attacks, Rhee et al. [25] defined a new security requirement called "trapdoor indistinguishability" and showed that if a PEKS scheme provides trapdoor indistinguishability, then the scheme is secure against off-line keyword guessing attacks. In 2014, Wu et al. [30] defined a new security model and requirements for PEKS in the ID-based public key cryptosystems [4,9,31]. Their scheme is demonstrated to resist off-line keyword guessing attacks by formal security proof.

Certificateless public key cryptosystem is introduced by Al-Riyami et al. [1] in 2003 to solve the key escrow problem in the ID-based cryptosystems. Later, several cryptographic schemes and protocols based on this cryptosystem were proposed such as encryption scheme [11,12], signature scheme [16,18], and two-party key agreement protocol [21]. Recently, Peng et al. [23] and Zheng et al. [36] proposed certificateless based PEKS schemes, respectively. However, Wu et al. [28] had shown that Peng et al.'s scheme is insecure against an off-line keyword guessing attack. In this paper, we first revisit Zheng et al.'s scheme and show that it is also insecure against an off-line keyword guessing attack. Concretely, an external attacker or server can launch this attack.

2 Review of Zheng et al.'s CLKS Scheme

2.1 Bilinear Map

Let G_1 and G_2 be two groups with large prime p. we define a bilinear map e by $e : G_1 \times G_1 \rightarrow G_2$ which satisfies the following properties:

1. Bilinear. For any a, $b \in \mathbb{Z}_p^*$ and g_1, $g_2 \in G_1$, $e(g_1^a, g_2^b) = e(g_1, g_2)^{ab}$.
2. Non-degenerate. For any identity $1_{G_1} \in G_1$, $e(1_{G_1}, 1_{G_1})$ is also an identity of G_2.
3. Computable. There exist several algorithms to compute e.

For the details of bilinear map, readers can refer to [4,9,29,31,33,35] for a full descriptions.

2.2 A Brief Review

Zheng et al.'s CLKS scheme [36] consists of the following seven algorithms: *Setup*, *Partial Private Key Generation*, *Private Key Generation*, *Public Key Generation*, *CLKS*, *Trapdoor Generation*, and *Test* as shown in following Fig. 1.

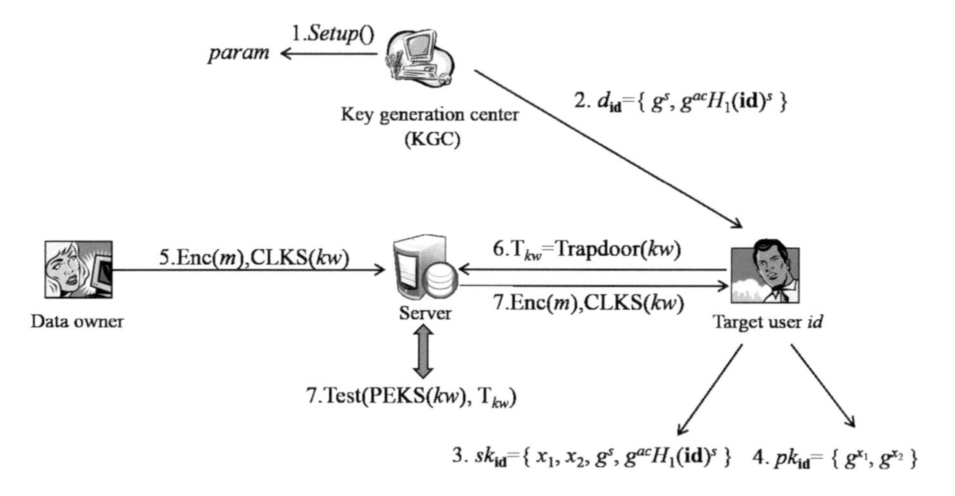

Fig. 1. Overview of Zheng et al.'s CLKS scheme

1. *Setup.* Key generation center (KGC) runs this algorithm to initialize the system and generates some system parameters as follows.
   (a) Inputting a security parameter 1^l, it generates a bilinear map $e : G_1 \times G_1 \rightarrow G_2$.
   (b) Randomly selecting a, b, $c \in \mathbb{Z}_p^*$ and then computing g^a, g^b, and g^c, where g is a generator of G_1.
   (c) Randomly select a vector $(u, u_1, \ldots, u_n) \in G_1^n$ and define a hash function H_1 by
   $$H_1(\mathbf{id}) = u \prod_{j=1}^{n} u_j^{id_j},$$
   where $\mathbf{id} = id_1||id_2|| \cdots ||id_n$ is an n-bit value.
   (d) Choosing another hash function $H_2 : \{0,1\}^* \rightarrow \mathbb{Z}_p^*$ and public parameters $param$ are defined as $\{e, G_1, G_2, p, g, g^a, g^b, g^c, H_1, H_2, u, u_1, \ldots, u_n\}$.
2. *Partial Private Key Generation.* The KGC runs this algorithm to generate a user $\mathbf{id}$'s partial private key $d_{\mathbf{id}}$ as follows.
   (a) Selecting a random value $s \in \mathbb{Z}_p^*$ and computing g^s and $g^{ac} \cdot H_1(\mathbf{id})^s$.
   (b) The partial private key $d_{\mathbf{id}}$ is defined as $d_{\mathbf{id}} = (g^s, g^{ac} H_1(\mathbf{id})^s)$ and then the KGC sends it to user $\mathbf{id}$ via a secure channel.
3. *Private Key Generation.* User $\mathbf{id}$ runs this algorithm to generate her/his private key $sk_{\mathbf{id}}$ as follows.
   (a) Randomly selecting x_1, $x_2 \in \mathbb{Z}_p^*$.
   (b) The private key $sk_{\mathbf{id}}$ is defined as $sk_{\mathbf{id}} = (x_1, x_2, g^s, g^{ac} H_1(\mathbf{id})^s))$.
4. *Public Key Generation.* User $\mathbf{id}$ runs this algorithm to generate her/his public key $pk_{\mathbf{id}}$ as follows.
   (a) The user $\mathbf{id}$ inputs the own private key (x_1, x_2) and computes g^{x_1} and g^{x_2}.

(b) The public key $pk_{\mathbf{id}}$ is defined as $pk_{\mathbf{id}} = (g^{x_1}, g^{x_2})$.

5. *CLKS.* Assume that data owner (DO) wants to share his data to user **id**. Then, she/he first selects a keyword kw according to the shared data and runs this algorithm to encrypt kw as follows.
   (a) The DO selects two random values r_1 and $r_2 \in \mathbb{Z}_p^*$ and computing
      i. $C_1 = (g^c \cdot g^{x_2})^{r_1} = g^{(c+x_2)r_1}$,
      ii. $C_2 = (g^a \cdot g^{x_1})^{(r_1+r_2)} \cdot (g^b)^{H_2(kw)\cdot r_1} = g^{(a+x_1)(r_1+r_2)} \cdot g^{bH_2(kw)r_1}$,
      iii. $C_3 = g^{r_2}$,
      iv. $C_4 = H_1(\mathbf{id})^{r_2}$.
   (b) The CLKS ciphertext of keyword kw is defined as $\mathcal{C} = (C_1, C_2, C_3, C_4)$.
6. *Trapdoor Generation.* In order to retrieve a CLKS ciphertext of keyword kw, user **id** runs this algorithm to generate a trapdoor T_{kw} as follows.
   (a) Selecting a random number r.
   (b) Using private key $sk_{\mathbf{id}}$ and keyword kw to compute
      i. $T_1 = (g^s)^r$,
      ii. $T_2 = [(g^a)^{x_2} \cdot (g^c)^{x_1} \cdot (g)^{x_1 \cdot x_2}]^r \cdot [g^{ac} H_1(\mathbf{id})^s]^r = g^{(a+x_1)(c+x_2)r} \cdot H_1(\mathbf{id})^{sr}$,
      iii. $T_3 = (g^c \cdot g^{x_2})^r = g^{(c+x_2)r}$,
      iv. $T_4 = (g^a \cdot g^{x_1})^r \cdot (g^b)^{H_2(kw)\cdot r} = g^{(a+x_1)r} \cdot g^{bH_2(kw)r}$.
   (c) The trapdoor of keyword kw is set by $T_{kw} = (T_1, T_2, T_3, T_4)$.
7. *Test.* In order to find out user **id** wants to retrieve, server runs this algorithm to match the trapdoor T_{kw} sent by the user and a CLKS ciphertext $\mathcal{C}$ as follows:
   (a) Computing $\alpha = e(C_3, T_2)/e(C_4, T_1)$.
   (b) Verifying $e(C_2, T_3) = \alpha \cdot e(C_1, T_4)$.

3 Drawbacks of Zheng et al.'s CLKS Scheme

In Zheng et al.'s CLKS scheme, we know that if user wants to retrieve some ciphertext $\mathcal{C}$ with keyword kw, then she/he must send a trapdoor T_{kw} to server. Here, we demonstrate a off-line keyword guessing attack while an external attacker captures T_{kw} or server receives T_{kw}. The detail steps are as follows:

1. To guess an appropriate keyword kw'.
2. To verify $e(g^a \cdot g^{x_1} \cdot (g^b)^{H_2(kw')}, T_3) = e(g^c \cdot g^{x_2}, T_4)$.

Here, we demonstrate the correctness:

$$\begin{aligned}
\text{Left hand side} &= e(g^a \cdot g^{x_1} \cdot g^{bH_2(kw')}, g^{(c+x_2)r}) \\
&= e(g^{a+x_1} \cdot g^{bH_2(kw')}, g^{(c+x_2)r}) \\
&= e(g,g)^{(a+x_1)(c+x_2)r} \cdot e(g,g)^{b(c+x_2)H_2(kw')r}.
\end{aligned}$$

$$\begin{aligned}
\text{Right hand side} &= e(g^c \cdot g^{x_2}, g^{(a+x_1)r} \cdot g^{bH_2(kw)r}) \\
&= e(g^{c+x_2}, g^{(a+x_1)r} \cdot g^{bH_2(kw)r}) \\
&= e(g,g)^{(a+x_1)(c+x_2)r} \cdot e(g,g)^{b(c+x_2)H_2(kw)r}.
\end{aligned}$$

4 Conclusions

In this paper, we have demonstrated an external attacker or server can launch an off-line keyword guessing attack on Zheng et al.'s CLKS scheme. In the future, we will propose a improvement based on their scheme with formal security proof.

Acknowledgments. The authors would thank anonymous referees for a valuable comments and suggestions. The work of Chien-Ming Chen was supported in part by the Project NSFC (National Natural Science Foundation of China) under Grant number 61402135 and in part by Shenzhen Technical Project under Grant number JCYJ20150513151706574.

References

1. Al-Riyami, S.S., Paterson, K.G.: Certificateless public key cryptography. In: International Conference on the Theory and Application of Cryptology and Information Security, pp. 452–473. Springer (2003)
2. Baek, J., Safavi-Naini, R., Susilo, W.: Public key encryption with keyword search revisited. In: Computational science and its applications-ICCSA 2008, pp. 1249–1259 (2008)
3. Boneh, D., Di Crescenzo, G., Ostrovsky, R., Persiano, G.: Public key encryption with keyword search. In: International Conference on the Theory and Applications of Cryptographic Techniques, pp. 506–522. Springer (2004)
4. Boneh, D., Franklin, M.: Identity-based encryption from the weil pairing. In: Annual International Cryptology Conference, pp. 213–229. Springer (2001)
5. Boneh, D., Waters, B.: Conjunctive, subset, and range queries on encrypted data. In: Theory of Cryptography Conference, pp. 535–554. Springer (2007)
6. Buccafurri, F., Lax, G., Sahu, R.A., Saraswat, V.: Practical and secure integrated PKE+PEKS with keyword privacy. In: 2015 12th International Joint Conference on e-Business and Telecommunications (ICETE), vol. 4, pp. 448–453. IEEE (2015)
7. Buyrukbilen, S., Bakiras, S.: Privacy-preserving ranked search on public-key encrypted data. In: 2013 IEEE International Conference on Embedded and Ubiquitous Computing (HPCC_EUC), 2013 IEEE 10th International Conference on High Performance Computing and Communications, pp. 165–174. IEEE (2013)
8. Byun, J.W., Rhee, H.S., Park, H.A., Lee, D.H.: Off-line keyword guessing attacks on recent keyword search schemes over encrypted data. In: Workshop on Secure Data Management, pp. 75–83. Springer (2006)
9. Chen, L., Cheng, Z., Smart, N.P.: Identity-based key agreement protocols from pairings. Int. J. Inf. Secur. **6**(4), 213–241 (2007)
10. Cheng, L., Jin, Z., Wen, O., Zhang, H.: A novel privacy preserving keyword searching for cloud storage. In: 2013 Eleventh Annual International Conference on Privacy, Security and Trust (PST), pp. 77–81. IEEE (2013)
11. Cheng, Z., Chen, L., Ling, L., Comley, R.: General and efficient certificateless public key encryption constructions. In: International Conference on Pairing-Based Cryptography, pp. 83–107. Springer (2007)
12. Dent, A.W., Libert, B., Paterson, K.G.: Certificateless encryption schemes strongly secure in the standard model. In: International Workshop on Public Key Cryptography, pp. 344–359. Springer (2008)

13. He, T., Ma, W.: An effective fuzzy keyword search scheme in cloud computing. In: 2013 5th International Conference on Intelligent Networking and Collaborative Systems (INCoS), pp. 786–789. IEEE (2013)
14. Hu, C., Liu, P.: An enhanced searchable public key encryption scheme with a designated tester and its extensions. J. Comput. **7**(3), 716–723 (2012)
15. Hu, C., Liu, P.: Public key encryption with ranked multi-keyword search. In: 2013 5th International Conference on Intelligent Networking and Collaborative Systems (INCoS), pp. 109–113. IEEE (2013)
16. Huang, X., Susilo, W., Mu, Y., Zhang, F.: On the security of certificateless signature schemes from asiacrypt 2003. In: International Conference on Cryptology and Network Security, pp. 13–25. Springer (2005)
17. Hwang, Y.H., Lee, P.J.: Public key encryption with conjunctive keyword search and its extension to a multi-user system. In: International Conference on Pairing-Based Cryptography, pp. 2–22. Springer (2007)
18. Li, X., Chen, K., Sun, L.: Certificateless signature and proxy signature schemes from bilinear pairings. Lith. Math. J. **45**(1), 76–83 (2005)
19. Li, Z., Zhao, M., Jiang, H., Xu, Q.: Multi-user searchable encryption with a designated server. Ann. Telecommun. (2017). doi:10.1007/s12243-017-0571-x
20. Liang, K., Su, C., Chen, J., Liu, J.K.: Efficient multi-function data sharing and searching mechanism for cloud-based encrypted data. In: Proceedings of the 11th ACM on Asia Conference on Computer and Communications Security, pp. 83–94. ACM (2016)
21. Lippold, G., Boyd, C., Nieto, J.G.: Strongly secure certificateless key agreement. In: International Conference on Pairing-Based Cryptography, pp. 206–230. Springer (2009)
22. Park, D.J., Kim, K., Lee, P.J.: Public key encryption with conjunctive field keyword search. In: International Workshop on Information Security Applications, pp. 73–86. Springer (2004)
23. Peng, Y., Cui, J., Peng, C., Ying, Z.: Certificateless public key encryption with keyword search. China Commun. **11**(11), 100–113 (2014)
24. Rane, D.D., Ghorpade, V.: Multi-user multi-keyword privacy preserving ranked based search over encrypted cloud data. In: 2015 International Conference on Pervasive Computing (ICPC), pp. 1–4. IEEE (2015)
25. Rhee, H.S., Park, J.H., Susilo, W., Lee, D.H.: Trapdoor security in a searchable public-key encryption scheme with a designated tester. J. Syst. Softw. **83**(5), 763–771 (2010)
26. Rhee, H.S., Susilo, W., Kim, H.J.: Secure searchable public key encryption scheme against keyword guessing attacks. IEICE Electron. Express **6**(5), 237–243 (2009)
27. Wang, B., Chen, T., Jeng, F.: Security improvement against malicious server's attackfor a dPEKS scheme. Int. J. Inf. Educ. Technol. **1**(4), 350 (2011)
28. Wu, T.Y., Meng, F., Chen, C.M., Liu, S., Pan, J.S.: On the security of a certificateless searchable public key encryption scheme. In: International Conference on Genetic and Evolutionary Computing, pp. 113–119. Springer (2016)
29. Wu, T.Y., Tsai, T.T., Tseng, Y.M.: A revocable id-based signcryption scheme. J. Inf. Hiding Multimed. Signal Process. **3**(3), 240–251 (2012)
30. Wu, T.Y., Tsai, T.T., Tseng, Y.M.: Efficient searchable id-based encryption with a designated server. Ann. Telecommun. annales des télécommunications 69(7–8), 391–402 (2014)
31. Wu, T.Y., Tseng, Y.M.: An id-based mutual authentication and key exchange protocol for low-power mobile devices. Comput. J. **53**(7), 1062–1070 (2010)

32. Xu, P., Jin, H., Wu, Q., Wang, W.: Public-key encryption with fuzzy keyword search: A provably secure scheme under keyword guessing attack. IEEE Trans. Comput. **62**(11), 2266–2277 (2013)
33. Xu, Y., Zhong, H., Cui, J.: An improved identity-based multi-proxy multi-signature scheme. J. Inf. Hiding Multimed. Signal Process. **7**(2), 343–351 (2016)
34. Yau, W.C., Phan, R.C.W., Heng, S.H., Goi, B.M.: Keyword guessing attacks on secure searchable public key encryption schemes with a designated tester. Int. J. Comput. Math. **90**(12), 2581–2587 (2013)
35. Yin, S.L., Li, H., Liu, J.: A new provable secure certificateless aggregate signcryption scheme. J. Inf. Hiding Multimed. Signal Process. **7**(6), 1274–1281 (2016)
36. Zheng, Q., Li, X., Azgin, A.: CLKS: Certificateless keyword search on encrypted data. In: International Conference on Network and System Security, pp. 239–253. Springer (2015)

Efficient Mining of High Average-Utility Itemsets with Multiple Thresholds

Tsu-Yang Wu[1,2], Jerry Chun-Wei Lin[3(✉)], and Shifeng Ren[3]

[1] Fujian Provincial Key Laboratory of Big Data Mining and Applications, Fujian University of Technology, Fuzhou 350118, China
wutsuyang@gmail.com

[2] National Demonstration Center for Experimental Electronic Information and Electrical Technology Education, Fujian University of Technology, Fuzhou, China

[3] School of Computer Science and Technology, Harbin Institute of Technology Shenzhen Graduate School, Shenzhen, China
jerrylin@ieee.org, renshifeng@stmail.hitsz.edu.cn

Abstract. In this paper, we propose an efficient algorithm to discover HAUIs based on the compact average-utility list structure. A tighter upper-bound model is used to instead of the traditional *auub* model used in HAUIM to lower the upper-bound value. Three pruning strategies are also respectively developed to facilitate mining performance of HAUIM. Experiments show that the proposed algorithm outperforms the state-of-the-art HAUIM-MMAU algorithm in terms of runtime and memory usage.

Keywords: Data mining · High average-utility itemsets · List structure · Multiple thresholds

1 Introduction and Background

The main purpose of data mining techniques is to reveal interesting, important and useful information from databases according to different requirements and applications [1,5]. Apriori [1] is the well-known algorithm for mining association rules (ARs) from databases. The association-rule mining only considers the occurrence frequency of items in database, thus it is not efficient to mine the interesting patterns under this limitation.

High-utility itemset mining (HUIM) [8,13] was presented to mine the set of high-utility itemset. Several algorithms [2,3,9] were respectively presented to mine the high-utility itemsets (HUIs) based on the TWU model [8]. However, the above studies suffer from an important limitation that each discovered itemset is only measured by a single minimum high-utility threshold. High-utility itemset mining with multiple minimum high-utility threshold (HUIM-MMU) framework [10] was presented to assign different thresholds for different items. Although the HUIM can reveal more information than that of the association-rule mining, the HUIM still suffers, however, from the intrinsic drawback that

J.-S. Pan et al. (eds.), *Advances in Intelligent Information Hiding and Multimedia Signal Processing*, Smart Innovation, Systems and Technologies 81,
DOI 10.1007/978-3-319-63856-0_25

the longer itemset has the greater utility. The high average-utility itemset mining (HAUIM) was proposed [6] by considering the size of the itemset to provide another measurement for evaluating the utility of the itemset based on the developed average-utility upper-bound (*auub*) model [6]. The first algorithm called high average-utility itemset mining with multiple minimum average-utility thresholds (HAUIM-MMAU) [11] was presented to mine the HAUIs under multiple minimum average-utility thresholds based on the Apriori-like approach.

In this paper, we present an algorithm with an efficient list-based structure to mine the HAUIs without candidate generation and multiple database scans. An efficient upper-bound model and three efficient pruning strategies are also developed to reduce the search space. Extensive experiments are conducted on two real-world datasets to show that the proposed algorithm has better performance in terms of runtime and memory usage.

2 Preliminaries and Problem Statement

Let $I = \{i_1, i_2, \ldots, i_m\}$ be a finite set with m distinct items in the database D. Each item i_j in a transaction T_q, has its purchase quantity (defined as a positive integer), and denoted as $q(i_j, T_q)$. A profit table $ptable = \{p(i_1), p(i_2), \ldots, p(i_m)\}$ shows the profit value of each item i_j. A set of k distinct items $X = \{i_1, i_2, \ldots, i_k\}$ such that $X \subseteq I$ is said to be a k-itemset, in which k is the length of the itemset. An itemset X is considered to be contained in a transaction T_q if $X \subseteq T_q$.

Definition 1 (Average-utility of an item in a transaction). The average-utility of an item (i_j) in a transaction T_q is denoted as $au(i_j)$, and defined as:

$$au(i_j, T_q) = \frac{q(i_j, T_q) \times p(i_j)}{1} = \frac{u(i_j, T_q)}{1}. \tag{1}$$

Definition 2 (Average-utility of an itemset in a transaction). The average-utility of a k-itemset X in a transaction T_q is denoted as $au(X, T_q)$, and defined as:

$$au(X, T_q) = \frac{\sum_{i_j \in X \wedge X \subseteq T_q} u(i_j, T_q)}{|X| = k}. \tag{2}$$

Definition 3 (Average-utility of an itemset in *D*). The average-utility of an itemset X in a database D is denoted as $au(X)$, and defined as:

$$au(X) = \sum_{X \subseteq T_q \wedge T_q \in D} au(X, T_q). \tag{3}$$

Definition 4 (Multiple minimum high average-utility count). The minimum high average-utility count of an item i_j in the database D is denoted as $mau(i_j)$. A *MAU-Table* indicates the set of minimum high average-utility count of each item in D, which defined as:

$$MAU\text{-}Table = \{mau(i_1), mau(i_2), \ldots, mau(i_r)\}. \tag{4}$$

Definition 5 (Least minimum high average-utility count, LMAU). The least minimum high average-utility count is the minimum mau value of each item in D, which is denoted as $LMAU$ and defined as:

$$LMAU = min\{mau(i_1), mau(i_2), \ldots, mau(i_r)\}. \tag{5}$$

Definition 6 (Minimum high average-utility count of an itemset). The minimum high average-utility count of a k-itemset X in D is denoted as $mau(X)$, and defined as:

$$mau(X) = \frac{\sum_{i_j \in X} mau(i_j)}{|X|(=k)} \tag{6}$$

Definition 7 (Transaction-maximum utility, *tmu*). The transaction-maximum utility of a transaction T_q is denoted as $tmu(T_q)$, and defined as:

$$tmu(T_q) = max\{u(i_j, T_q)|i_j \subseteq T_q\}. \tag{7}$$

Definition 8 (Average-utility upper bound of an itemset in *D*, *auub*). The average-utility upper-bound of an itemset X in a database D is denoted as $auub(X)$, and defined as:

$$auub(X) = \sum_{X \subseteq T_q \wedge T_q \in D} tmu(X, T_q). \tag{8}$$

Definition 9 (High average-utility upper-bound itemset, *HAUUBI*). An itemset X is called a high average-utility upper-bound itemset (HAUUBI) if its average-utility upper-bound is no less than the minimum high average-utility count, which is defined as:

$$HAUUBI \leftarrow \{X|auub(X) \geq mau(X)\}. \tag{9}$$

Problem Statement: An itemset X is considered as a HAUI iff its average-utility is no less than the minimum high average-utility count, that is:

$$HAUI \leftarrow \{X|au(X) \geq mau(X)\}. \tag{10}$$

3 Proposed Framework and Pruning Strategies

In the past, an Apriori-like HAUIM-MMAU [11] algorithm was first proposed to discover all HAUIs with multiple minimum high average-utility counts. To improve mining performance, a more efficient algorithm with multiple minimum high average-utility count is proposed in this paper to efficiently discover HAUIs without candidate generation and multiple database scans. A sorted enumeration tree with the tighter upper-bound model are respectively presented to speed up mining performance and limit search space for mining HAUIs. Details are given below.

3.1 Sorted Enumeration Tree and a Tighter Upper-Bound Model

The previous works [6,12] based on the transaction-maximum utility downward closure (TMUDC) property does not hold under multiple minimum average-utility counts. Thus, a sorting strategy is developed here and used in the enumeration tree for exploring the HAUIs in the search space.

Definition 10 (Sorting strategy). All items in the *MAU-Table* are sorted according to their *mau*-ascending order as $\prec$.

Based on the sorting strategy property, the completeness and correctness to explore the promising itemsets in the enumeration tree for mining the HAUIs can thus be obtained [10]. Besides, the processed item is extended with the items after it according to the $\prec$ in the enumeration tree. This process can help reduce the *tmu* value of the processing transaction and obtain lower upper-bound value of the itemset within it.

Definition 11 (Irrelevant itemsets in the database). The irrelevant itemset indicates that its *auub* value does not satisfy the condition against the minimum high average-utility count, which cannot be extended for generating the promising itemsets.

Definition 12 (Revised transaction-maximum utility, *rmu*). The revised transaction-maximum utility of an itemset X in transaction T'_q is denoted as $rmu(X, T'_q)$, and defined as:

$$rmu(X, T'_q) = max\{u(i_1, T_q), u(i_2, T_q), \ldots, u(i_{|T_q|}, T_q)\}, \tag{11}$$

where $X \prec i_1 \prec i_2 \prec, \ldots, i_{|T_q|}$, $T'_q \subseteq T_q$, and $T_q \backslash X = T'_q$.

Definition 13 (Revised tighter upper-bound model, *rtub*). The revised tighter upper-bound of an itemset X is denoted as $rtub(X)$, and defined as:

$$rtub(X) = \sum_{X \subseteq T'_q \in D} rmu(X, T'_q). \tag{12}$$

Definition 14 (High revised tighter upper-bound itemset, HRTUBI). An itemset X is called a high revised tighter upper-bound itemset (HRTUBI) if its revised tighter upper-bound value is no less than the minimum high average-utility count, which defined as:

$$HRTUBI \leftarrow \{X | rtub(X) \geq mau(X)\} \tag{13}$$

To facilitate the efficiently for removing the irrelevant items, a compact average-utility (CAU)-list structure is then designed to avoid the multiple database scans.

Definition 15 (Compact average-utility (CAU)-list). The proposed compact average-utility (CAU)-list records the relevant information of itemset (X) and the transactions T_q containing (X). Each transaction containing X is represented by an entry which has three fields: (1) the transaction id of (X) in T_q (*tid*); (2) the utility of (X) in T_q (*iutil*); (3) the revised maximal utility of (X) in T_q (*rmu*).

3.2 Proposed Algorithm and Pruning Strategies

From the above definitions, we can obtain that if $auub(X) < LMAU$, the supersets of X will not be the HAUIs. Based on this property, the first pruning strategy can be obtained as follows.

Pruning strategy 1. (Remove unpromising items from database, RUI): The unpromising itemset, such that $auub(i_j) < LMAU$, is removed from the database to obtain lower transaction-maximum utility (*tmu*) of each transaction, which can be used to limit the search space but would not affect final discovered HAUIs.

Since the downward closure property holds for the developed *rtub* model, it is easily to early prune the unpromising itemsets. To efficiently check the k-itemsets ($k \geq 2$), the estimated average-utility matrix is developed as follows.

Definition 16 (Estimated average-utility matrix). The structure is a compact matrix in which the element is the form of $< I_x, I_y, auub >$, and I_x, I_y are co-occurrence items.

Pruning strategy 2. (Estimated average-utility matrix pruning strategy): If the *auub* of a 2-itemset in the designed matrix is less than the *LMAU*, any supersets of it will not be HAUIs, and can be directly pruned.

Recently, a HUP-Miner [7] algorithm was proposed to improve mining performance of HAUIs. The LA-Prune provides a tighter utility upper-bound value for each itemset used in join operation while generating the new itemsets and its utility list. Here, we further extend the *auub* and *rtub* model to the LA-Prune strategy [7] for handling the HAUIM under multiple minimum high average-utility counts.

Pruning Strategy 3 (LA-Prune strategy): Let P_x and P_y be the two itemsets, the join operation is unnecessary to be performed if any of the following equations is less than the minimum high average-utility count while performing the join operation:

$$LAPA(P_x) = auub(P_x) - \sum_{T_q \in tids(P_x) \wedge T_q \notin tids(P_y), T_q \in D} tmu(T_q). \quad (14)$$

$$LAPR(P_x) = rtub(P_x) - \sum_{T_q \in tids(P_x) \wedge T_q \notin tids(P_y), T_q \in D} rmu(T_q). \quad (15)$$

With the designed tighter upper-bound model and three pruning strategies, the proposed more efficient algorithm is described in Algorithm 1, and details of the **Search** algorithm is shown in Algorithm 2.

4 Experimental Evaluation

In this section, the performance of the proposed algorithm is compared to the state-of-art HAUIM-MMAU algorithm [11]. Experiments were conducted on two

Algorithm 1. Proposed algorithm

Input: D, a transactional database; *ptable*, a profit table; *MAU-table* = $\{mau(i_1), mau(i_2), \ldots, mau(i_r)\}$, user-specified multiple minimum high average-utility counts.
Output: The set of high average-utility itemsets, HAUIs

```
1 calculate auub of each item;
2 LMAU ← min{mau(i_1), ..., mau(i_r)};
3 if auub(i_j) < LMAU then
4     remove i_j from D;
5 re-calculate auub of each remaining item in D;
6 sort items in the transactions in mau-ascending order;
7 Search(,I*.AULs, matrix, MAU-table, LMAU);
8 return HAUIs;
```

Algorithm 2. Search Algorithm

Input: X, an itemset; *extnesionOfX*, a set of CAU-list of all 1-items in $I^*.AULs$; *matrix*, an estimated average-utility matrix; *LMAU*, a least minimum high average-utility count.
Output: The set of high average-utility itemsets, HAUIs.

```
1  for each X_a ∈ extnesionOfX do
2      if sum(X_a.iutils)/|X_a| ≥ mau(X_a) then
3          HAUIs ← HAUIs ∪ X_a;
4      if rtub(X_a) ≥ mau(X_a) then
5          extensionOfX_a ← φ;
6          for each X_b ∈ extnesionOfX such that a ≺ b do
7              if matrix(X_a, X_b) ≥ LMAU then
8                  X_ab = X_a ∪ X_b;
9                  X_ab ← Constrcut (X, X_a, X_b, LMAU);
10                 if X_ab.AULs ∉ null then
11                     extensionOfX_a ← extensionOfX_a ∪ X._ab.AUL;
12         Search(X_a, extensionOfX_a, matrix, MAU-table, LMAU);
13 return HAUIs;
```

real-world [4] datasets. A simulation model [10] was developed to generate the quantity and unit profit of items in transactions for all datasets. A log-normal distribution was used to randomly assign quantities in the [1, 5] interval and item profit values in the interval [1, 1000]. To generate the *mau* value of each item [10] in our proposed algorithm, the following equation is used to automatically set the *mau* value of each item as:

$$mau(i_j) = max\{\beta \times p(i_j), glmau\}, \tag{16}$$

where β is a constant value, which is used to multiply the profit of item $p(i_j)$. The global least minimum average utility named *glmau*, is a user-defined parameter.

4.1 Runtime

We first compare the runtime of all algorithms to evaluate their efficiency with various *glmau* and a fixed β is randomly generated within an interval.

From Fig. 1, it can be observed that the proposed algorithm performs well than the state-of-the-art HAUIM-MMAU algorithm and the designed *rtub* model is more efficient than traditional *auub* model. For example in Fig. 1(a), when *glmau* is set to 9,000K, the runtime of HAUIM-MMAU for discovering the complete HAUIs requires 500 s while the proposed algorithm only needs less than 24 s. The proposed algorithm also utilizes the CAU-list structure to represent the database as the vertical one and explores the promising itemsets in the enumeration tree, which is more efficient than the level-wise HAUIM-MMAU approach.

4.2 Memory Usage

The memory usage is measured by means of Java API and the peak memory usage of compared algorithms for different datasets is recorded as the final memory usage. From Fig. 2, it can be observed that the proposed algorithm requires less memory than that the state-of-the-art HAUIM-MMAU algorithm. Thus, the multiple database scans of the proposed algorithm to keep the promising itemsets for later mining progress is unnecessary. Therefore, the proposed algorithm outperforms the HAUIM-MMAU algorithm.

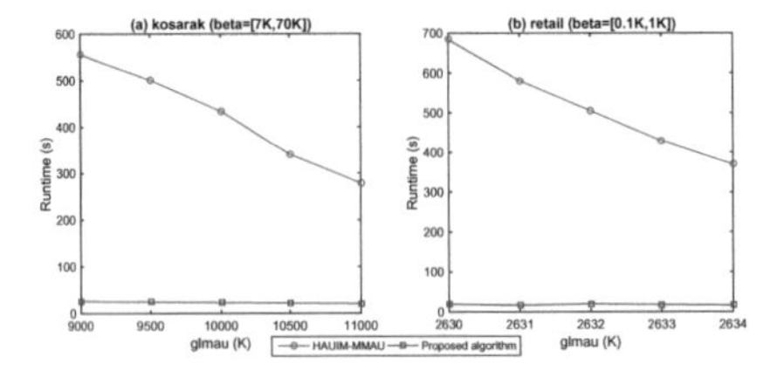

Fig. 1. Runtime comparisons.

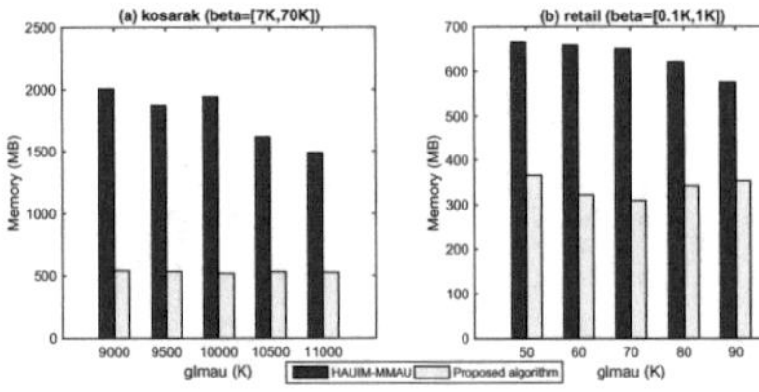

Fig. 2. Memory usage comparisons.

5 Conclusion

In this paper, we first present an efficient algorithm to discover the high average-utility itemsets with multiple minimum high average-utility counts. A novel CAU-list structure and a sorted enumeration tree are respectively developed to mine HAUIs. A tighter upper-bound model is also designed to lower the over-estimated value of the promising itemsets compared to the traditional *auub* model. Three pruning strategies are then proposed based on the proposed upper-bound model to further limit the search space for mining HAUIs.

Acknowledgments. This research was partially supported by the National Natural Science Foundation of China (NSFC) under grant No. 61503092, by the Research on the Technical Platform of Rural Cultural Tourism Planning Basing on Digital Media under grant 2017A020220011, and by the Tencent Project under grant CCF-Tencent IAGR20160115.

References

1. Agrawal, R., Srikant, R.: Fast algorithms for mining association rules. In: International Conference on Very Large Data Bases, pp. 487–499 (1994)
2. Ahmed, C.F., Tanbeer, S.K., Jeong, B.S., Lee, Y.K.: Efficient tree structures for high utility pattern mining in incremental databases. IEEE Trans. Knowl. Data Eng. **21**(12), 1708–1721 (2009)
3. Erwin, A., Gopalan, R.P., Achuthan, N.R.: Efficient mining of high utility itemsets from large datasets. In: Pacific-Asia Conference on Knowledge Discovery and Data Mining, pp. 554–561 (2008)
4. Fournier-Viger, P., Lin, J.C.W., Gomariz, A., Gueniche, T., Soltani, A., Deng, Z., Lam, H.T.: The SPMF open-source data mining library version 2. In: Joint European Conference on Machine Learning and Knowledge Discovery in Databases, pp. 36–40 (2016)
5. Fournier-Viger, P., Lin, J.C.W., Kiran, R.U., Koh, Y.S., Thomas, R.: A survey of sequential pattern mining. Data Sci. Pattern Recogn. **1**(1), 54–77 (2017)
6. Hong, T.P., Lee, C.H., Wang, S.L.: Effective utility mining with the measure of average utility. Expert Syst. Appl. **38**(7), 8259–8265 (2011)
7. Krishnamoorthy, S.: Pruning strategies for mining high utility itemsets. Expert Syst. Appl. **42**(5), 2371–2381 (2015)
8. Liu, Y., Liao, W., Choudhary, A.: A fast high utility itemsets mining algorithm. In: International Workshop on Utility-based Data Mining, pp. 90–99 (2005)
9. Liu, J., Wang, K., Fung, B.C.M.: Mining high utility patterns in one phase without generating candidates. IEEE Trans. Knowl. Data Eng. **28**(5), 1245–1257 (2016)
10. Lin, J.C.W., Gan, W., Fournier-viger, P., Hong, T.P., Zhang, J.: Efficient mining of high-utility itemsets using multiple minimum utility thresholds. Knowl.-Based Syst. **113**, 100–115 (2016)
11. Lin, J.C.W., Li, T., Fournier-Viger, P., Hong, T.P., Su, J.H.: Efficient mining of high average-utility itemsets with multiple minimum thresholds. In: Industrial Conference on Data Mining, vol. 9728, pp. 14–28 (2016)
12. Lin, C.W., Hong, T.P., Lu, W.H.: Efficiently mining high average utility itemsets with a tree structure. In: International Conference on Intelligent Information and Database Systems, pp. 131–139 (2010)
13. Yao, H., Hamilton, H.J., Butz, C.J.: A foundational approach to mining itemset utilities from databases. In: SIAM International Conference on Data Mining, pp. 482–486 (2004)

Cryptanalysis of an Anonymous Mutual Authentication Scheme for Secure Inter-device Communication in Mobile Networks

Tsu-Yang Wu[1,2], Weicheng Fang[3], Chien-Ming Chen[3(✉)], and Guangjie Wang[3]

[1] Fujian Provincial Key Laboratory of Big Data Mining and Applications, Fujian University of Technology, Fuzhou 350118, China
wutsuyang@gmail.com

[2] National Demonstration Center for Experimental Electronic Information and Electrical Technology Education, Fujian University of Technology, Fuzhou 350118, China

[3] Harbin Institute of Technology Shenzhen Graduate School, Shenzhen, China
626558837@qq.com, chienming.taiwan@gmail.com, wangguangjie115@foxmail.com

Abstract. Anonymous authentication allows one entity to be authenticated by the other without revealing the identity information. In mobile networks, mobile devices communicate with each other to exchange resources. To achieve anonymous mutual authentication, the devices are anonymously authenticated under the trusted server. Recently, Chung et al. proposed a efficient anonymous mutual authentication scheme for inter-device communication using only low-cost functions, such as hash functions and exclusive-or operations. However, we find that their protocol does not preserve user's privacy in terms of untraceability. Also, their protocol is vulnerable to a denial of service attack and a user impersonation attack.

Keywords: Anonymity · Mutual authentication · Privacy · Mobile network

1 Introduction

Anonymous authentication is an important cryptographic technique to protect privacy. While it seems paradoxical to combine both anonymity with authentication together, such technique has been extensively studied in the recent literatures. Some of them focused on the anonymous authentication in a client-server architecture, where clients authenticate themselves to the server without revealing their identities [1,2]. The others proposed a scheme for some various network systems such as peer-to-peer systems [3–5].

Authentication protocols [6–10] preserving user anonymity are also desirable in mobile networks [11–13]. Under a three-party setting, a trusted third party, or a trusted server, is introduced into the mobile networks to help the other two parties authenticate each other anonymously, and establish a session key for

J.-S. Pan et al. (eds.), *Advances in Intelligent Information Hiding and Multimedia Signal Processing*, Smart Innovation, Systems and Technologies 81,
DOI 10.1007/978-3-319-63856-0_26

direct communication after successful authentication. In mobile networks, anonymous authentication allows authenticated mobile devices to access resources from local service provider or other mobile devices, to whom their identities remain anonymous. Under such situations, it is natural and convenient to introduce the trusted third party. Typically, mobile devices get registered at its home agent, also known as the trusted third party. When roaming around foreign places, the resources located at foreign agents or nearby devices will be available. Thus, to protect mobile devices' privacy, their identities are hidden when communicating with foreign agents or nearby devices.

Recently, several anonymous authentication protocols designed for mobile networks are proposed. These protocols are proposed and enhanced to be efficiently employed for roaming users in mobile networks to anonymously access resources from local foreign agents. In 2015, Shin et al. [14] improved an authentication scheme with only low-cost functions, such as hash functions and exclusive-or operations. However, Farash et al. [15] demonstrated that their scheme does not protect user's privacy in terms of untraceablility, and thus proposed a new protocol. To achieve secure inter-device communication in mobile networks, Chung et al. [16] proposed an efficient anonymous mutual authentication scheme that allows registered devices authenticate and communicate with each other, both anonymously and directly. Unfortunately, we find that their scheme does not guarantee untraceability neither. A passive attacker can easily identify the protocol runs initiated by the same user, thus breaking user's anonymity. Also, Chung et al.'s protocol is vulnerable to a denial of service attack and a user impersonation attack.

In this paper, we reconsider Chung et al.'s protocol, and find that their protocol does not provide untraceability and security against know attacks. The paper is organized as follows. In next section, we review the protocol proposed by Chung et al.. In Sect. 3, our cryptanalysis of the protocol shows the attacks in details. Finally, we conclude the paper in Sect. 4.

2 Review of Chung et al.'s Protocol

In this section, we describe Chung et al.'s protocol, which consists of three phases: registration, authentication and session key establishment, and renewal. There are two types of entities involved in the protocol: the mobile devices and the trusted server. The notations used in the protocol are listed in Table 1.

2.1 Registration Phase

In the registration phase, a mobile device M get registered using its chosen identity ID_M. First, it sends ID_M to the server R through a secure channel. After checking the validity of its identity, R retrieves its secret key x, chooses a random number x_M, and computes a virtual identity for the mobile device as

$$VID_M = h(ID_M \| x_M \| x). \tag{1}$$

Table 1. The notations used in Chung et al.'s protocol

Notations	Descriptions
M, M_i	The entity of an mobile device i
R	The entity of the trusted server
ID_X	The identity of an entity X
x	The secret key of the server
x_X	The shared secret between X and the server
$h(\cdot)$	Secure hash function
$\|\|$	Concatenation operation
$\oplus$	Exclusive-or (XOR) operation

Then, R stores $[VID_M, ID_M, x_M]$ in its database and send back $[VID_M, x_M, h(\cdot)]$. Finally, M store these values securely.

2.2 Authentication and Session Key Establishment Phase

In the authentication and session key establishment phase, two mobile devices, M_1, M_2, authenticate each other anonymously with the help of the server R. The following descriptions explain this phase in steps.

1. M_1 first generates two nonces n_{M_1} and r_{M_1} and retrieves $[VID_{M_1}, x_{M_1}, h(\cdot)]$ with input ID_{M_1}. Then, it computes

$$SID_{M_1} = VID_{M_1} \oplus h(h(x)\|n_{M_1}), \tag{2}$$

$$V_1 = r_{M_1} \oplus h(x_{M_1}\|n_{M_1}), \tag{3}$$

$$H_1 = h(x_{M_1}\|SID_{M_1}\|V_1\|n_{M_1}), \tag{4}$$

and sends $m_1 = \{SID_{M_1}, V_1, H_1, n_{M_1}, ID_R\}$ to the nearby mobile device.

2. When M_2 receives an authentication request from another mobile device, it generates a nonce n_{M_2}, retrieves $[VID_{M_2}, x_{M_2}, h(\cdot)]$ with input ID_{M_2}, and computes

$$SID_{M_2} = VID_{M_2} \oplus h(h(x)\|n_{M_2}), \tag{5}$$

$$H_2 = h(x_{M_2}\|SID_{M_2}\|V_2\|n_{M_2}). \tag{6}$$

Then, it forward m_1 together with the computed values to R as $m_2 = \{SID_{M_1}, V_1, H_1, n_{M_1}, ID_R, SID_{M_2}, H_2, n_{M_2}\}$.

3. When R receives m_2 from a mobile device, it first recovers the virtual identity as follows:

$$VID'_{M_1} = SID_{M_1} \oplus h((h(x)\|n_{M_1}), \tag{7}$$

$$VID'_{M_2} = SID_{M_2} \oplus h((h(x)\|n_{M_2}). \tag{8}$$

Then, R retrieves the records $[VID_{M_1}, ID_{M_1}, x_{M_1}]$ and $[VID_{M_2}, ID_{M_2}, x_{M_2}]$ from the database. If the records do not exist, R terminates the session.

Otherwise, it checks H_1 and H_2 respectively using the obtained values. If both are valid, R authenticates both mobile devices. Next, it generates a nonce n_R and computes

$$r'_{M_1} = V_1 \oplus h(x_{M_1} \| n_{M_1}), \tag{9}$$

$$V_2 = r'_{M_1} \oplus h(x_{M_2} \| n_{M_2}), \tag{10}$$

$$V_3 = h(x_{M_2} \| VID_{M_1} \| n_{M_2}), \tag{11}$$

$$V_4 = h(x_{M_1} \| VID_{M_2} \| n_{M_1}), \tag{12}$$

$$H_3 = h(x_{M_2} \| V_2 \| V_3 \| V_4 \| n_R), \tag{13}$$

and sends $m_3 = \{V_2, V_3, V_4, H_3, n_R\}$ back to M_2.

4. When M_2 receive the response from R, it computes the required values to check H_3. Also, it recovers M_1's virtual identity as (7), and check V_3 accordingly. If both are valid, it recovers the nonce r'_{M_1} as (9), generates a new nonce r_{M_2}, and computes

$$SK = h(h(x) \| r_{M_1} \| r_{M_2} \| VID_{M_1} \| VID_{M_2}), \tag{14}$$

$$V_5 = r_{M_2} \oplus h(h(x) \| r_{M_1}), \tag{15}$$

$$V_6 = h(SK \| n_{M_1}), \tag{16}$$

$$H_4 = h(r_{M_1} \| SID_{M_2} \| V_4 \| V_5 \| V_6 \| n_{M_2}). \tag{17}$$

Then, it sends $m_4 = \{SID_{M_2}, V_4, V_5, V_6, H_4, n_{M_2}\}$ back to M_1.

5. When M_1 receive the response from the nearby mobile device, it computes the required values to check H_4. Also, it recovers M_2's virtual identity as (8), and M_2's new nonce as

$$r'_{M_2} = V_5 \oplus h(h(x) \| n_{M_2}). \tag{18}$$

Then, it checks V_4 and V_6 accordingly. If all checking procedures return valid, it computes

$$V_7 = h(SK \| n_{M_2}), \tag{19}$$

and sends $m_5 = \{V_7\}$ to M_2.

6. When M_2 receives the response from M_1, it checks the validity of V_7. If it is correct, the session key SK is established between M_1 and M_2.

2.3 Renewal Phase

In the renewal phase, a mobile device M renews its virtual identity through the server R. The following descriptions explain this phase in steps.

1. M first generates a nonce n_M and retrieves $[VID_M, x_M, h(\cdot)]$ with input ID_M. Then, it computes similarly as (5) and (6):

$$SID_M = VID_M \oplus h(h(x) \| n_M), \tag{20}$$

$$H_5 = h(x_M \| SID_M \| n_M), \tag{21}$$

and sends $m_6 = \{SID_M, H_5, n_M\}$ to R.

2. When R receives a renewal request from M, it recovers the virtual identity similarly as (7) or (8):

$$VID'_M = SID_M \oplus h(h(x) \| n_M), \tag{22}$$

and retrieves the records $[VID_M, ID_M, x_M]$ from the database. If the record does not exist, or it fails to check H_5, R terminates the session immediately. Otherwise, it generates a nonce n_R and a new shared secret x^*_M, computes the new virtual identity for M as

$$VID^*_M = h(ID_M \| x^*_M \| x), \tag{23}$$

and update the database accordingly. Then, R sends back $m_7 = \{SID'_M, V_8, H_6, n_R\}$ to M, where

$$V_8 = x^*_M \oplus h(x_M \| n_M), \tag{24}$$

$$SID'_M = VID^*_M \oplus h(h(x) \| x^*_M), \tag{25}$$

$$H_6 = h(x_M \| SID'_M \| V_8 \| n_R). \tag{26}$$

3. When M receives response from R, it first checks H_6. If it is incorrect, M terminates the renewal phase immediately. Otherwise, it computes the new shared secret and virtual identity as

$$x^*_M = V_8 \oplus h(x_M \| n_M), \tag{27}$$

$$VID^*_M = SID'_M \oplus h(h(x) \| x^*_M). \tag{28}$$

Then, R renews the stored values as $[VID^*_M, x^*_M, h(\cdot)]$, and will use the new virtual identity for later logins.

3 Cryptanalysis of Chung et al.'s Protocol

In this section, we analyze Chung et al.'s protocol in terms of security against known attacks. We assume that an attacker obtains a registered mobile device in all attacks. The analysis shows that their protocol does not protect user's privacy in terms of untraceability. Also an acitve attacker can launch a denial of service (DoS) attack and a user impersonation attack.

3.1 No Provision of Untraceability

In this attack, a passive attacker can trace the mobile devices in different protocol runs. First, the attacker E extracts $h(x)$ from the registered mobile device. Then, E starts to eavesdrop messages comming from other mobile devices in the authentication and session key establishment phase, and captures m_1 and m_2 continuously. According to (2) and (5), the virtual identities can be correctly computed by

$$VID_{M_i} = SID_{M_i} \oplus h(h(x) \| n_{M_i}). \tag{29}$$

Thus, E can easily keeps the track of the mobile devices through the same virtual identity.

In Chung et al.'s protocol, the renewal phase allows the mobile device holder to renew the virtual identity if it is revealed. As a countermeasure to resist such attack launched by E, one may think of a simple invocation of a renewal phase immediately after an authentication and session key establishment phase. However, this incurs extra communication overhead. Furthermore, it still cannot resist the attacks discussed in the following subsections.

3.2 Denial of Service Attack

In this attack, an active attacker can force the trusted server to deny a valid authentication request coming from a registered mobile device. The following descriptions explain the attack in steps.

1. An active attacker E first eavesdrops a successful protocol run Π_{Auth} involving a mobile device as M_2, and captures m_2 in the authentication and session key establishment phase.
2. Then, E extracts $m_2' = \{SID_{M_2}, H_2, n_{M_2}\}$ from m_2, and initiates a fresh renewal phase by replacing m_6 with m_2'.

In the renewal phase, the trusted server R will accept the renewal request. This is because (1) m_2' and m_6 consists of the same components of values, namely a masked virtual identity SID_M, a hashed value H_i and a nonce n_M, and (2) m_2' has passed the verification of H_2 in Π_{Auth}, and then it must pass the verification of H_5 in the renewal phase. As soon as R updates its database with the new generated virtual identity and the shared secret, M_2 has no chance to login again until either R or M_2 discover such an attack.

It is essential to prevent the denial of service attacks on those schemes with an on-line renewal phase. Although the values stored in the mobile device and those in the server's database are desynchronized by the attacker, the server may take measures to avoid desynchronization. It can keep the most recently renewed virtual identity as well as the values before the renewal. However, such enhancement incurs computation overhead since the server may query the database twice to find the correct identity for the mobile device. Furthermore, it does not resist a user impersonation attack described in the next subsection.

3.3 User Impersonation Attack

In this attack, an active attacker can successfully impersonate as another registered mobile device without knowing its real identity. The attack extends from the denial of service attack in Subsect. 3.3. The following descriptions explain the attack in steps.

1. An active attacker E initiates an authentication and session key establishment phase with target mobile device M_2. Let Π_{auth} denotes this protocol run. Then, E captures m_3 to obtain V_2 in Π_{auth}, and computes

$$H_E = V_2 \oplus r_{M_1}. \tag{30}$$

The value of H_E should be equal to $h(x_{M_2} \| n_{M_2})$ according to (5).
2. Next, E follows the steps described in the denial of service attack in Subsect. 3.2. Let Π_{new} denotes the protocol run of the renewal phase.
3. In Π_{new}, E also intercepts m_7 to obtain V_8 and SID'_M. To further obtain the renewed virtual identity and the shared secret stored in R's database, E computes

$$x_{M_E} = V_8 \oplus H_E, \tag{31}$$

and derives VID_{M_E} similarly as (28) using the computed x_{M_E}. Both values should be equal to the renewed ones stored in the database according to (24) and (25).
4. To impersonate M_2, E always retrieves $[VID_{M_E}, x_{M_E}, h(x)]$ to compute required values in the authentication and session key establishment phase or the renewal phase.

The user impersonation attack is based on the denial of service attack discussed in Subsect. 3.2. So, the active attacker E can continue to impersonate M_2 until someone figures out the attack.

4 Conclusions

In this paper, we reconsider Chung et al.'s anonymous mutual authentication protocol. Their protocol aims to provide secure authentication and preserve user's privacy for inter-device communication in mobile networks. However, we find that their protocol does not guarantee user's privacy in terms of untraceability. Also, we demonstrate that the protocol is vulnerable to a denial of service attack and a user impersonation attack.

Acknowledgement. The work of Chien-Ming Chen was supported in part by the Project NSFC (National Natural Science Foundation of China) under Grant number 61402135 and in part by Shenzhen Strategic Emerging Industries Program under Grants No. ZDSY20120613125016389.

References

1. Zhang, Z., Yang, K., Hu, X., Wang, Y.: Practical anonymous password authentication and tls with anonymous client authentication. In: Proceedings of the 2016 ACM SIGSAC Conference on Computer and Communications Security, pp. 1179–1191. ACM (2016)
2. Shin, S., Kobara, K.: Simple anonymous password-based authenticated key exchange (sapake), reconsidered. IEICE Trans. Fundam. Electron. Commun. Comput. Sci. **100**, 639–652 (2017)
3. Tsang, P.P., Smith, S.W.: Ppaa: peer-to-peer anonymous authentication. In: International Conference on Applied Cryptography and Network Security, pp. 55–74. Springer (2008)
4. Lu, L., Han, J., Liu, Y., Hu, L., Huai, J.P., Ni, L., Ma, J.: Pseudo trust: Zero-knowledge authentication in anonymous p2ps. IEEE Trans. Parallel Distrib. Syst. **19**(10), 1325–1337 (2008)

5. Wang, F., Xu, Y., Zhang, H., Zhang, Y., Zhu, L.: 2flip: a two-factor lightweight privacy-preserving authentication scheme for vanet. IEEE Trans. Veh. Technol. **65**(2), 896–911 (2016)
6. Chen, C.M., Li, C.T., Liu, S., Wu, T.Y., Pan, J.S.: A provable secure private data delegation scheme for mountaineering events in emergency system. IEEE Access **5**, 3410–3422 (2017)
7. Chen, C.M., Fang, W., Wang, K.H., Wu, T.Y.: Comments on an improved secure and efficient password and chaos-based two-party key agreement protocol. Nonlinear Dyn. **87**, 1–3 (2016)
8. Chen, C.M., Xu, L., Wu, T.Y., Li, C.R.: On the security of a chaotic maps-based three-party authenticated key agreement protocol. J. Netw. Intell. **2**, 61–65 (2016)
9. Chen, C.M., Wang, K.H., Wu, T.Y., Pan, J.S., Sun, H.M.: A scalable transitive human-verifiable authentication protocol for mobile devices. IEEE Trans. Inf. Forensics Secur. **8**(8), 1318–1330 (2013)
10. Chen, C.M., Chen, S.M., Zheng, X., Yan, L., Wang, H., Sun, H.M.: Pitfalls in an ecc-based lightweight authentication protocol for low-cost rfid. J. Inf. Hiding Multimedia Sig. Process. **5**(4), 642–648 (2014)
11. Zhao, D., Peng, H., Li, L., Yang, Y.: A secure and effective anonymous authentication scheme for roaming service in global mobility networks. Wirel. Pers. Commun. **78**(1), 247–269 (2014)
12. Gope, P., Hwang, T.: Enhanced secure mutual authentication and key agreement scheme preserving user anonymity in global mobile networks. Wirel. Pers. Commun. **82**(4), 2231–2245 (2015)
13. Wang, E.K., Cao, Z., Wu, T.Y., Chen, C.M.: Mapmp: a mutual authentication protocol for mobile payment. J. Inf. Hiding Multimedia Sig. Process. **6**(4), 697–707 (2015)
14. Shin, S., Yeh, H., Kim, K.: An efficient secure authentication scheme with user anonymity for roaming user in ubiquitous networks. Peer-to-peer Netw. Appl. **8**(4), 674–683 (2015)
15. Farash, M.S., Chaudhry, S.A., Heydari, M., Sadough, S., Mohammad, S., Kumari, S., Khan, M.K.: A lightweight anonymous authentication scheme for consumer roaming in ubiquitous networks with provable security. Int. J. Commun. Syst. (2015)
16. Chung, Y., Choi, S., Won, D.: Anonymous mutual authentication scheme for secure inter-device communication in mobile networks. In: International Conference on Computational Science and Its Applications, pp. 289–301. Springer (2016)

DCT-Based Compressed Image with Reversibility Using Modified Quantization

Chi-Yao Weng[1], Cheng-Ta Huang[2](✉), and Hung-Wei Kao[1]

[1] Department of Computer Science, National Pingtung University, Pingtung 900, Taiwan
[2] Department of Information Management, Oriental Institute of Technology, New Taipei City 220, Taiwan
cthuang@mail.oit.edu.tw

Abstract. This paper presents a lossless data hiding scheme for concealing message in each block of discrete cosine transform (DCT) coefficients. Our proposed improves the Chang et al.'s scheme for obtaining higher performance in terms of embedding capacity and image quality. In our scheme, two neighbor coefficients which are near to zero as an group are selected in each block, and then, hiding the message into each group. The coefficient is modified by one unit; therefore, the stego-image can keep the high image quality. Experimental results show that the proposed scheme not only provides better performance than that of previous works but also achieves coefficients reversibility.

Keywords: Lossless data hiding · Discrete Cosine Transform (DCT) · Coefficient modification · Reversibility

1 Introduction

For decades, many researchers have developed information-hiding schemes [1–7] to protect the integrity and copyright for digital multimedia. In general, these methods embed the copyright information/secret data into the raw images and generate a stego images or a watermarked images. According to the embedding domain, information hiding methods can be categorized into three main domains: methods in the spatial domain [1–3], methods in the compression domain [4, 5] and methods in the frequency domain [6, 7].

Information hiding methods with reversibility [8–12] in the spatial domain are the most intuitive that are designed to modify the pixel values directly, such as histogram-based, least significant bit (LSB), difference expansion (DE), pixel value difference (PVD), etc. In 2016, Dadgostar and Afsari proposed a steganography method based on fuzzy edge detection and modified LSB [1]. A novel mechanism is used for ensuring the secret message can be extracted without any distortion in the fuzzy edge detection area. In 2017, Arham et al. proposed a multiple layer data hiding method based on difference expansion of quad [2]. To increase capacity and visual quality, the method can be improved by applying improved reduced difference expansion. A PVD-based

J.-S. Pan et al. (eds.), *Advances in Intelligent Information Hiding and Multimedia Signal Processing*, Smart Innovation, Systems and Technologies 81,
DOI 10.1007/978-3-319-63856-0_27

data hiding method was proposed by Shen and Huang in 2015 [3]. Shen and Huang adopted the concept of PVD and employ two pixels as an embedding unit.

In the compression domain, cover media are usually stored in compressed formats, such as joint photographic experts group (JPEG), vector quantization (VQ), search ordering coding (SOC), etc. In 2017, a reversible data hiding scheme for encrypted JPGE was proposed by Chang et al. [4]. The scheme is constructed with a reserving-room-before-encryption manner. The separable scheme for encrypted JPEG has good performance on image quality and embedding capacity. In 2015, Tu and Wang proposed a reversible VQ data hiding method with high payload based on referred frequency [5]. The method can embeds m bits of secret data in each VQ index, where $m >= 1$.

Information hiding methods in the frequency domain use a transformation function to embed secret data, such as discrete cosine transform (DCT), discrete wavelet transform (DWT), fast fourier transform (FFT), etc. In 2014, Lin proposed a DCT information hiding method based on the ability of human vision [6]. The concept of decomposition of images is used in the method to improve the image quality of the stego image. In 2007, Chang et al. proposed a reversible data hiding method based on DCT approach [7]. The secret data of Chang et al.'s method are embedded by two successive zero coefficients of the medium-frequency components in each block. In order to keep the image quality, the method modifies the quantization table in the embedding phase.

In this paper, the reversible DCT-based data hiding method using modified quantization method is proposed. The proposed method discovers the aspect that many DCT-quantized coefficients are near to zero. We apply the coefficient shifting to obtain high image quality. Experimental results demonstrate that the proposed scheme can achieve both high image quality and large embedding capacity in the stego-image. The remainder of this paper is organized as follows: In Sect. 2, the related works are introduced. The proposed DCT-based information hiding method is presented in Sect. 3 and then the experimental results are given in Sect. 4. Finally, conclusions are presented in Sect. 5.

2 Related Works

In this subsection, we will review the reversible data hiding based on DCT-based compressed image proposed by Chang et al. [7] and reversible data hiding using histogram-shifting introduced by Ni et al. [8].

2.1 Chang et al.'s Reversible Data Hiding in DCT-Based [7]

Chang et al. proposed a reversible data hiding scheme in DCT-based compressed image using adjusted the quantized coefficients. In this method, an image should be first divided into non-overlapping blocks sized with 8×8 pixels. Then, perform the operation of DCT Transform to each block for obtaining DCT coefficients, and quantize the DCT coefficients for all blocks. After that, run the embedding procedure for concealing secret data.

Before hiding the data, the DCT coefficient should be defined into several sets $R_i(i = 1, 2, \ldots, 9)$, the example of set delimitation is shown in Fig. 1. Each set contains 4 to 7 coefficients according to set delimitation.

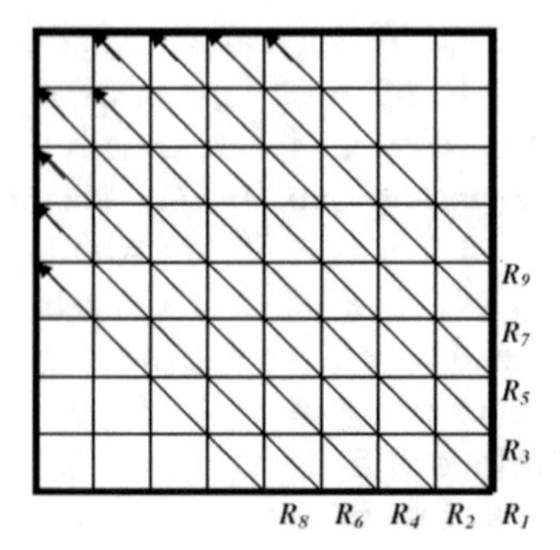

Fig. 1. The example of sets for data embedding

Hide the secret message into each set following the estimation rule. If $b >= 2$, set can hide one bit into. Where b indicates the length of consecutive zero in by scanning order from high frequency coefficient to lowest frequency coefficients. Apply the parameters tz_{i1} and tz_{i2} to record the recover information. The tz_{i1} means the zero value of the lowest frequency of set R_i, and tz_{i2} represents the lower right component of tz_{i1}, respectively.

The secret message is hidden into set R_i by modifying the value of tz_{i2}, the modifying equation is give as Eq. (1)

$$tz_{i2} = \begin{cases} 0, & \text{if } s_i = 0 \\ 1 \text{ or } -1, & \text{if } s_i = 1 \end{cases} \tag{1}$$

where 0 or -1 is selected by randomization.

2.2 Ni et al.'s Histogram Shifting Scheme [8]

The Histogram shifting approach is proposed by Ni et al. in 2006. In their approach, first of all, they applied the statistic method to generate the distribution of pixel values. The pixel distribution is depicted in Fig. 2. Next, find out the hiding information (P, Z) of peak point and zero point from this distribution. Here, peak point indicates that the pixel value has the maximum number. In other words, the zero points means that the

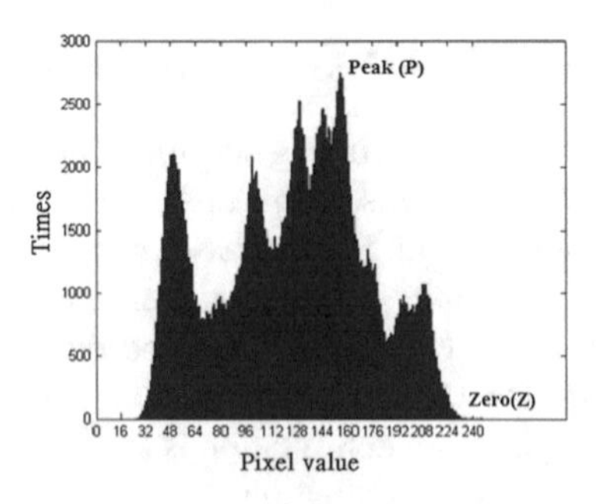

Fig. 2. The example of pixel distribution using Lena as cover image

pixel value has zero number. The zero points will not be occurred in a special pixel distribution. In this case, pixel value is the minimum number treated as the zero points.

After determining the hiding information, the range of peak point and zero point is selected, the secret bitstreams are hidden into the selected range. Before the data concealing into the peak point, all the pixel values x falling into the range of [P + 1, Z] or [Z, P − 1] are shifting by one unit. The shifted function is given as Eq. (2).

$$\begin{cases} x' = x + 1, \text{ } if \text{ } Z > P \text{ } and \text{ } x \in [P + 1, Z] \\ x' = x - 1, \text{ } if \text{ } Z < P \text{ } and \text{ } x \in [Z, P - 1] \end{cases} \tag{2}$$

Following up the hiding strategy, for each pixel values x, hide the secret data into x. The strategy is shown as below.

$$\begin{cases} x' = x + 1, \text{ if } Z > P \\ x' = x - 1, \text{ if } Z < P \end{cases} \tag{3}$$

Finally, all the pixels have been visited, then, output the scanned pixel as the stego pixel (called as stego image). The receiver can perform the data extraction and pixel recovering when receiver holds a pair-data (P, Z).

3 Our Proposed Method

In this section, we introduce our reversible data hiding scheme based on modified DCT coefficient. Our proposed method improves the Chang et al.'s approach. In Chang et al.'s hiding method, one bit is hidden into a set of R_i when b >= 2. However, Chang et al.'s manner has the drawback since consecutive coefficients are not existed or less then two in all the set of R_i, the data hiding procedure will be skipped, leading to fewer hiding capacity. In order to improve the hiding capacity, we adjust the searching order for set delimitations, and gather two consecutive coefficients as a group. We apply the histogram-shifting method to hide the secret data. In addition, we used two side histogram-shifting to pursue high embedding capacity. The detail of our data embedding algorithm and extracting algorithm is shown as below.

3.1 Data Embedding Procedure

Input: Cover image CI with size $M \times N$, secret message $SMsg = b_1 b_2 b_3 \ldots b_n$, $b_i \in [1, 0]$.

Output: Stego-coefficient SC, two pair information (LP, LZ) and (RP, RZ).

Step 1: The Cover image CI is divided into non-overlapping blocks with 8 × 8 pixels. Then, perform the 2-Dimensional DCT to transform each block into 8 × 8 block of DCT coefficient. Scan each coefficient and select two coefficients (GC_i, GC_{i+1}) as a group.

Step 2: Gather all the groups and generated a 2-D distribution, H(groups).

Step 3: Find two side-information (*LP*, *LZ*) and (*RP*, *RZ*) from the group histogram distribution. Where the pair information of (*LP*, *LZ*) is picked in the left-side of the peak point. In the same way, the pair information (*RP*, *RZ*) is picked by the right-side of the peak point.

Step 4: Shift all group's coefficient GC_{i+1} by 1 unit according to following function.

$$GC'_{i+1} = \begin{cases} GC_{i+1} + 1, \text{ if } RP < RZ \text{ and } GC_{i+1} \in [RP + 1, RZ] \\ GC_{i+1} - 1, \text{ if } LZ < LP \text{ and } GC_{i+1} \in [LZ, LP - 1] \end{cases} \quad (4)$$

Step 5: Scan all the GC_{i+1} and fetch each secret bits from secret message *SMsg*, and then, hide the bitstring into while GC_{i+1} equals to the peak values *RP* and *LP*. The data embedding function is shown as below.

$$GC'_{i+1} = \begin{matrix} GC_{i+1} + 1, \text{ if } GC_{i+1} = RP \text{ and } b_i = 1 \\ GC_{i+1}, \quad \text{if } GC_{i+1} = RP \text{ and } b_i = 0 \\ GC_{i+1}, \quad \text{if } GC_{i+1} = LP \text{ and } b_i = 0 \\ GC_{i+1} - 1, \text{ if } GC_{i+1} = LP \text{ and } b_i = 1 \end{matrix} \quad (5)$$

Step 6: All the secret messages *SMsg* have been concealed or all group's coefficients GC_{i+1} have been visited, then, obtain the new GC'_{i+1} called as Stego-coefficient *SC*. Recover the *SC* to be a stego-image *SI* according the inverse of DCT-transform and quantization.

3.2 Message Extracting Procedure

The secret message can be extracted from a DCT-based stego image. The process is the inverse of the data embedding procedure. Given an $M \times N$ stego image performs DCT-transform and quantization. Two pairs information (*LP*, *LZ*) and (*RP*, *RZ*) are also needed in extracting procedure. The data extracting and pixel recovering are described in detail as below.

Step 1: Input a stego image *SI*, and divided *SI* into non-overlapping blocks with 8×8 pixels. Perform the DCT-transform and quantization, then, obtain the DCT coefficients *SC*.

Step 2: Scan each block and select two coefficients (GC'_i, GC'_{i+1}) as a group.

Step 3: For each block, run the following sub-steps to extract the message and recover the coefficients.

Step 3.1: Shifting operation

$$GC_{i+1} = \begin{cases} GC'_{i+1} - 1, \text{if } RP < RZ \text{ and } GC'_{i+1} \in [RP + 1, RZ] \\ GC'_{i+1} + 1, \text{if } LZ < LP \text{ and } GC'_{i+1} \in [LZ, LP - 1] \end{cases} \quad (6)$$

Step 3.2: Message extracting and pixel recovering

(a) The extracted message b_i equals to "1" according to the following function.

$$GC_{i+1} = \begin{cases} GC'_{i+1} - 1, \text{if } GC'_{i+1} = RP + 1 \\ GC'_{i+1} + 1, \text{if } GC'_{i+1} = LP - 1 \end{cases} \quad (7)$$

(b) The extracted message b_i equals to "0" according to the following function.

$$GC_{i+1} = \begin{cases} GC'_{i+1}, \text{if } GC'_{i+1} = RP \\ GC'_{i+1}, \text{if } GC'_{i+1} = LP \end{cases} \quad (8)$$

(c) If GC'_{i+1} does not satisfy with the request of Eq. (7) or Eq. (8), GC'_{i+1} remains the same value, $GC_{i+1} = GC'_{i+1}$.

Step 4: Repeat Step 3 until all blocks are processed. The secret data will be completely extracted, and the coefficients are successfully recovered.

4 Experiment Results

In section, we present the performance of our approach. Six gray-level images with sized 512 × 512, including Lena, Boat, Airplane, Pepper, Baboon, and Zelda. We used the program of Matlab to be a simulation tool, and secret bitstring in our experiment is generated by the random number generator. The visual image quality between the cover image and stego image is estimated by commonly measure function peak-signal-to-noise-ratio (PSNR). The measure functions are given as below.

$$PSNR = 10 \times \log_{10}(255^2/MSE) \quad (8)$$

$$MSE = 1/MH \sum_{i=1}^{M} \sum_{j=1}^{H} (I(i,j) - I'(i,j))^2 \quad (9)$$

where W and H are defined as the width and height of the image. $I(i,j)$ and $I'(i,j)$ indicates the values of cover image and stego image, respectively. The Max is the maximum value of cover image. Notably, the gray-scale image is our tested image, therefore, the Max value in gray-scale is $255(=2^8)$ (Fig. 3).

This paper is to improve the approach of Chang et al. We compare our experimental results with the Chang et al.'s DCT-based scheme, the comparison result is demonstrated in Table 1. From this comparison, in average, it is obvious to know that our scheme has better performance in terms of high embedding capacity and good image quality than that of Chang et al.'s methods. But the tested image of Baboon has not good result. We analyzed and inspected the coefficient distribution of Baboon image, we discovered a phenomenon that the coefficient distribution is randomly scattered and has bad grouping outcomes, leading to lower embedding capacity. In future work, we will recheck the coefficient distribution to develop more complicated hiding strategy for pursuing high efficient performances.

Fig. 3. Six cover images with size 512 × 512. (a) Lean; (b) Baboon; (c) Airplane; (d) Boat; (e) Zelda; (f) Pepper.

Table 1. The performance comparison with Chang et al.'s approach

	Chang et al.'s approach		Our proposed approach	
	Embedding capacity (L = 3)	PSNR	Embedding capacity (L = 3)	PSNR
Lean	12,288	35.15	28,364	42.42
Baboon	12,288	31.34	6,708	48.17
Airplane	12,288	35.22	27,694	45.73
Boat	12,288	34.92	15,564	45.72
Zelda	12,288	38.04	22,504	43.05
Pepper	12,288	36.32	20,191	48.15
Average	12,288	35.17	20,170	45.54

5 Conclusion

In this paper, based on DCT-based image, an efficient reversible data hiding scheme is proposed. Our proposed scheme improve the Chang et al.'s DCT-based scheme using grouping coefficients manner to generate more circumstance for seeking high performances. The two side histogram shifting approach is adopted for pursuing large embedding capacity. Experimental results further show that our proposed scheme can provides better stego-image quality and high embedding capacity then that of Chang et al. scheme especially for Baboon tested image. Our future work has to recheck the coefficient distribution and readjust our hiding method to produce a new scheme with complicated tactics for pursuing high efficient performances.

Acknowledgment. This work was supported in part by Ministry of Science of Technology, Taiwan, Under Contract MOST 105-2221-E-153-010.

References

1. Dadogstar, H., Afsari, F.: Image steganography based on interval-valued intuitionistic fuzzy edge detection and modified LSB. J. Inf. Secur. Appl. **30**, 94–104 (2016)
2. Arham, A., Nugroho, H.A., Adji, T.B.: Multiple layer data hiding scheme based on difference expansion of quad. Sig. Process. **137**, 52–62 (2017)
3. Shen, S.Y., Huang, L.H.: A data hiding scheme using pixel value differencing and improving exploiting modification directions. Comput. Secur. **48**, 131–141 (2015)
4. Chang, J.C., Lu, Y.Z., Wu, H.L.: A separable reversible data hiding scheme for encrypted JPEG bitstreams. Sig. Process. **133**, 135–143 (2017)
5. Tu, T.Y., Wang, C.H.: Reversible data hiding with high payload based on referred frequency for VQ compressed codes index. Sig. Process. **108**, 278–287 (2015)
6. Lin, Y.K.: A data hiding scheme based upon DCT coefficient modification. Comput. Stand. Interfaces **36**, 855–862 (2014)
7. Chang, C.C., Lin, C.C., Tseng, C.S., Tai, W.L.: Reversible hiding in DCT-based compressed images. Inf. Sci. **177**, 2768–2786 (2007)
8. Ni, Z.C., Shi, Y.Q., Ansari, N., Su, W.: Reversible data hiding. IEEE Trans. Circuits Syst. Video Technol. **16**, 354–362 (2006)
9. Yang, C.H., Tsai, M.H.: Improving histogram-based reversible data hiding by interleaving predictions. IET Image Proc. **4**, 223–234 (2010)
10. He, W.G., Cai, J., Zhou, K., Xiong, G.Q.: Efficient PVO-based reversible data hiding using multistage blocking and prediction accuracy matrix. J. Vis. Commun. Image Represent. **46**, 58–69 (2017)
11. Ou, B., Li, X.L., Wang, J.W., Peng, F.: High-fidelity reversible data hiding based on geodesic path and pairwise prediction-error expansion. Neurocomputing **226**, 23–34 (2017)
12. Xu, D.W., Wang, R.D.: Separable and error-free reversible data hiding in encrypted image. Sig. Process. **123**, 9–21 (2016)

Soft Computing and Its Application

Studying the Influence of Tourism Flow on Foreign Exchange Rate by IABC and Time-Series Models

Pei-Wei Tsai[1], Zhi-Sheng Chen[2], Xingsi Xue[3,4], and Jui-Fang Chang[2](✉)

[1] Department of Computer Science and Software Engineering, Swinburne University of Technology, Hawthorn, VIC 3122, Australia
ptsai@swin.edu.au

[2] Department of International Business, National Kaohsiung University of Applied Sciences, 807 Kaohsiung, Taiwan
1104346114@gm.kuas.edu.tw, rose@kuas.edu.tw

[3] College of Information Sciences and Engineering, Fujian University of Technology, Fuzhou 350118, Fujian Province, China

[4] Fujian Provincial Key Laboratory of Big Data Mining and Applications, Fujian University of Technology, Fuzhou 350118, Fujian Province, China
xxs@fjut.edu.cn

Abstract. In this study, we focus on analysing the relationship between the foreign exchange rate and the international tourism flow. Three foreign exchange rate forecasting models including GARCH(1,1), EGARCH(1,1), and the IABC forecasting model based on the computational intelligence are employed to produce the forecasting results. The Mean Absolute Percentage Error (MAPE) is selected to be the evaluation criterion for comparing the forecasting results of these models. The experiments contain the USD/NTD foreign exchange rate and the inbound international tourism flows in years of 2009 to 2010. The experimental results reveal that adding the international tourism flow as the new reference in the forecasting process has the positive contribution to the foreign exchange rate forecasting results.

Keywords: GARCH · EGARCH · IABC · Rate forecasting · Tourism flow

1 Introduction

Along with the rapid growths and developments of the global economy and convenient international public transportation, tourism industries are gradually taken as an important determination of the economic growth for the reason that it is one of the major activities in the forex. To our knowledge, many researchers have indicated that the number of the tourism flow is closely relevant to the foreign exchange rate [9]. On the other hand, the foreign exchange rate is also

J.-S. Pan et al. (eds.), *Advances in Intelligent Information Hiding and Multimedia Signal Processing*, Smart Innovation, Systems and Technologies 81,
DOI 10.1007/978-3-319-63856-0_28

frequently used in the discussion of the tourism demand model and it is with the explanatory ability to estimate the international flows [10]. When discussing the forecasting models in the finance expertise, the Time-Series analysis models are the most common type among many different forecasting models. For instance, the ARCH model, which is developed by Engle (1982) [1], indicates the conditional heteroscedastic would be influenced by the square of earlier stage error term, and simultaneously, the distribution of conditional heteroscedastic error term naturally fits in with the normal distribution. Bollerslev (1986) [2] utilised the ARCH model as the basis and introduced the GARCH model by expanding the ARCH model. He suggested that conditional heteroscedastic error term would not only be influenced by the square of the error term in the earlier stage but also the past conditional heteroscedastic in the earlier stages. When adopting the GARCH model in the real-world applications, the GARCH(1,1) model is usually employed. Moreover, the Exponential GARCH model (EGARCH) was developed by Nelson (1991) [3]. The EGARCH model is frequently used to evaluate the influence of different issues on markets.

On the other hand, the swarm intelligence methodology raises with a remarkable speed along with the development of the computer science. The huge computing power provided by the computers in recent days provides the scientists many different possibilities in developing new methods for solving problems in engineering, finance, and management. Unlike the Time-Series models, the algorithms in the swarm intelligence category are developed based on creature's collective behaviours and tiny intelligence existing in Mother Nature. For example, Tsai et al. (2009) [4] proposed the Interactive Artificial Bee Colony (IABC) by improving the searching ability of the conventional ABC algorithm (Karaboga, 2005) [11]. The IABC utilises the concept of the universal gravitation to enhance the interaction between the artificial agents. The swarm intelligence algorithms not only can be used to solve problems in optimisation but also can be utilised to construct the model for forecasting. In 2016, Tsai et al. utilise the swarm intelligence algorithm to produce the foreign exchange rate forecasting result by reading eleven macroeconomic variables as the input for the decision-making [5]. Although the forecasting result is satisfactory, the potential of information from other data, which may be contributed to the foreign exchange rate forecasting remains unstudied. Starting from filling up the vacancy, the tourism flow is taken into considering as a new input variable in this study to investigate whether it provides a positive contribution to the foreign exchange rate forecasting. The rest of this paper is structured as follows: a brief review of the related works is presented in Sect. 2, the experiment design is discussed in Sect. 3, the experimental results and discussions are given in Sect. 4, and the conclusion is made in Sect. 5.

2 Related Works

Researches on the foreign exchange rate related topics are always popular in the finance fields. For example, Ito claims three findings of the behaviors in the Japanese Yen to US dollar exchanging market: the first is that the market participants are heterogeneous, the seconds finding shows that the phenomenon of violating the rational expectation hypothesis can be found in many institutions, and the last is the forecasting with long horizons presents less Japanese Yen appreciation than the forecasting with short horizons [6]. His research findings provide the observational findings of the foreign exchange rate market in Japan. In 1990, Baillie proposes a multivariate generalised ARCH approach for modelling the risk in forwarding foreign exchange rate markets [7]. In his method, the 30-day forward rate forecast errors from using the weekly data are derived by the time series process. This work provides the proofed conclusion on that the risk premium is a linear function of the conditional variances. Diebold et al. provide a framework for both evaluating and improving the multivariate density forecasts in 1999 [8]. Their work provides the method for the application in the multivariate high-frequency exchange rate density forecasting. A density forecast calibration technique under the predefined conditions comes with their method for improving the deficient density forecasts. Although the findings and proposed methods in the literature mentioned above are valuable, all of them are focusing on the risk and the forecasting analysis produced by the time-series methods. The investigation on the forecasting achieved by other models and the input data still remain vacant. In 2015, Tsai et al. treat the foreign exchange rate forecasting in a different way by investigating indexes in the macroeconomic and other potential elements, which may contribute to the foreign exchange rate in an indirect way [5]. In their work, ten variables from the macroeconomic are co-aligned with the Consumer Confidence Index (CCI) as the input data for feeding to the swarm intelligence based forecasting model. The experimental results indicate that the swarm intelligence based forecasting model with the newly involved variables present higher accuracy than the time-series models.

3 Experiment Design

Inspired by Tsai et al.'s research outcome, it gives us the idea to discover what other elements can be possibly contributed to the foreign exchange rate forecasting. Thus, we employ the inbound international tourism flow in Taiwan as one of the input variables to the IABC based forecasting model in this work. The variables utilised in this work are listed in Table 1.

Table 1. Input variables for the IABC forecasting model.

Variables	Variables
Exchange Rate (NTD/USD)	Consumer Price Index
Commercial Paper Rate	Federal Fund Rate
Balance of Trade	Foreign Investment
Stock Return	M1
M1B	Consumer Confidence Index
Monitoring Indicator	Tourism Flow

In this work, twelve variables are applied as the input to the IABC based foreign exchange rate forecasting model. All data used in the experiment is collected from the TEJ database in Taiwan. To test the effect caused by the newly involved input variable, the experimental data covering the period in January 1st in 2009 to December 31st in 2010 is categorised in 2 sets on the annual basis.

The collected variables need to be preprocessed before feeding into the forecasting model. The preprocessing operations include the Pearson's correlation coefficient test, the autocorrelation examination, and the unit root test. The series procedures of preprocessing help us to examine whether the input variables fit the normality and the stationary criteria. The procedures of preprocessing require the steps listed as follows:

1. Conducting Pearson's correlation coefficient test. As given in Eq. (1), the tourism flow is the new variable to examine whether it has any relevance with exchange rates.

$$r_{XY} = \frac{COV_{XY}}{S_X S_Y} \tag{1}$$

where X and Y are in linear correlation, and the amount is in the range between minus 1 and positive 1.

2. The JB test is a method, which helps us to determine whether the examined data is correspondence with the normal distribution status. It is an essential condition required by the time-series models. The formula of JB test is given in Eq. (2):

$$JB = \frac{T-n}{6}\left[S^2 + 0.25 \times (K-3)^2\right] \tag{2}$$

where S stands for skewness and K denotes the kurtosis.

3. For correlation problems, this research conducted Lijun-Box Q tests to examine whether variables are self-correlated. The formula is given as follows:

$$Q(p) = n(n+2)\sum_{k=1}^{p}\frac{1}{n-k\rho_k^2} \sim \chi^2(p) \tag{3}$$

where n represents the samples and k stands for lag order. The statistical results are capable of fitting in the chi-square distribution, which means the degree of freedom is zero.

4. In Time Series, the data can be classified in to the stationary and the non-stationary categories. If the data is in the non-stationary category, the regression analysis will be a spurious regression, which means it is less reliable than data in the stationary category. Thus, the unit root test is the essential process to make sure the data is in the stationary category. The unit root tests used in this research are ADF test and PP test.
5. Utilising GARCH, EGARCH, and IABC models to produce the foreign exchange rate forecasting results. The GARCH model elaborates the phenomena of data in the time-series by the equations listed as follows:

$$y_t \mid \Omega_t \sim N(x_t\alpha, \sigma^2) \tag{4}$$

$$\varepsilon_t = y_t - x_t \alpha \quad (5)$$

$$\sigma_t^2 = \alpha_0 + \sum_{i=1}^{q} \alpha_i \varepsilon_{t-i}^2 + \sum_{j=1}^{p} \beta_j \sigma_{t-j}^2, \text{ subject to } \alpha_i \geq 0,\ \beta_j \geq 0 \quad (6)$$

where ε_{t-i}^2 represents the function of squared residues in the past q period, and σ_t^2 denotes the conditional variance in the past p period, p and q are both GARCH levels.

The EGARCH model equations are showing as follows:

$$y_t = x_t b + \varepsilon_t \quad (7)$$

$$\varepsilon_t \mid \Omega_{t-1} \sim N(0, \sigma^2) \quad (8)$$

$$\ln(\sigma_t^2) = \alpha_0 + \sum_{i=1}^{q} \left[\alpha_i \left(\left| \frac{\varepsilon_{t-i}}{\sigma_{t-i}} \right| - E\left| \frac{\varepsilon_{t-i}}{\sigma_{t-i}} \right| + \gamma \frac{\varepsilon_{t-i}}{\sigma_{t-i}} \right) \right] + \sum_{j=1}^{p} \beta_j \ln(\sigma_{t-j}^2) \quad (9)$$

where β_j is the accumulated stats of β_{j-1}, α_i stands for the parameter impacted by α_{i-1}, γ means the parameter in the condition of asymmetry deviation from last period to current period, and $\frac{\varepsilon_{t-i}}{\sigma_{t-i}}$ indicates the normal residuals. The object function in the IABC model for the training and the forecasting phases is a bit different. In the training phase, the object function is listed in Eq. (10):

$$\min f(W) = \sum_{i=1}^{n} \left| \left(\sum_{d=1}^{D} w_d \times v_{t,d} \right) - R_{real,t} \right| \quad (10)$$

where $f(W)$ represents the fitness value, $w = (w_1, w_2, \cdots, w_d)$ stands for the corresponding weights of the referenced inputs, D is the total amount of referenced days, v indicates the input variables, n is the total number of inputs, and $R_{real,t}$ stands for the actual exchange rate on day t. In the forecasting phase, the trained weights are adopted into Eq. (11) to produce the forecasting result:

$$R_{pd,D+1} = \sum_{d=2}^{D+1} w_d \times v_{t,d} \quad (11)$$

where $R_{pd,D+1}$ stands for the forecasted exchange rate.

6. Calculating the MAPE values of all forecasting models. The MAPE is utilised as the evaluation criterion to compare the forecasting abilities of all forecasting models. The forecasting model with lower MAPE value implies that it presents a more accurate forecasting ability. The formula is as follows:

$$MAPE = \frac{1}{n} \sum_{t=1}^{n} \frac{|\hat{S}_t - S_t|}{S_t} \times 100\% \quad (12)$$

where $\hat{S}_t$ represents the predicted exchange rate in period t and S_t indicates the actual exchange rate in period t, and n denotes the amount records of data.

4 Experiments and Experimental Results

As mentioned in the above section, two sets of experiments are taken in place for testing the forecasting accuracy of different models. The input variables contain ten indexes from the microeconomics, the Consumer Confidence Index (CCI), the Monitoring Indicator, and the inbound Tourism flow from over the world to Taiwan. The forecasting period covers two years in 2009 to 2010. The exchange rate forecasting outcomes are produced by GARCH(1,1), EGARCH(1,1), and IABC models, respectively. To compare the forecasting accuracy of these models, the MAPE values on a monthly basis is utilised as the measurement.

The MAPE value reveals the forecasting error between the forecasted value and the actual value. Thus, in the ideal case, the MAPE value is zero; otherwise, the forecasting model with larger MAPE value implies that the forecasting accuracy of the model is lower. We utilise the sliding window strategy on a 30-day historical data referencing period to compose the forecasting result for all these three models. The IABC forecasting model contains 16 populations with 120 iterations, the universal gravity effect consideration is set to 4 employed bees, and 30 runs are executed with different random seeds. The MAPE values of the experimental results are given in Figs. 1 and 2:

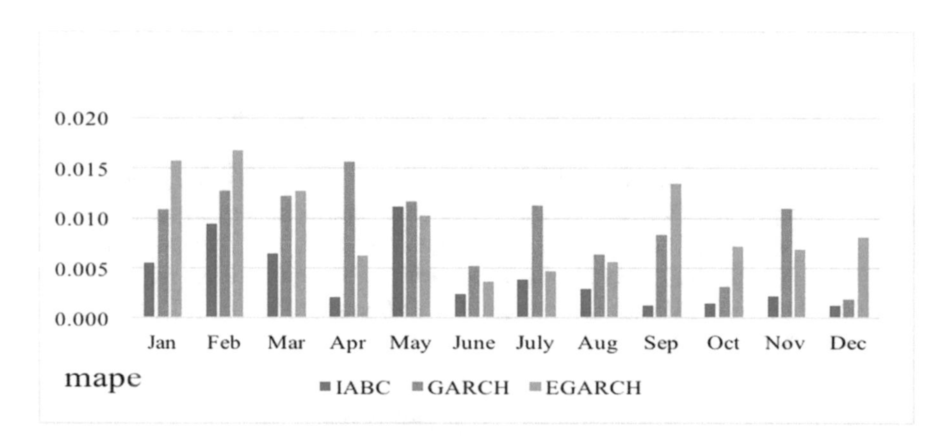

Fig. 1. Monthly MAPE value obtained by GARCH(1,1), EGARCH(1,1), and IABC in 2009.

As the results shown in Fig. 1, the IABC forecasting model presents the lowest MAPE value in all months except that its MAPE value is slightly greater than the MAPE value produced by EGARCH(1,1) in May.

In Fig. 2, the IABC model presents the best among all in 66.67% over the whole year; it presents a greater MAPE value than EGARCH(1,1) in October and November, and it presents 16.67% chance to have the highest MAPE value over the whole year. The experimental results are summarised in Table 2 with the average MAPE value and the standard deviation for all forecasting models tested in the experiments.

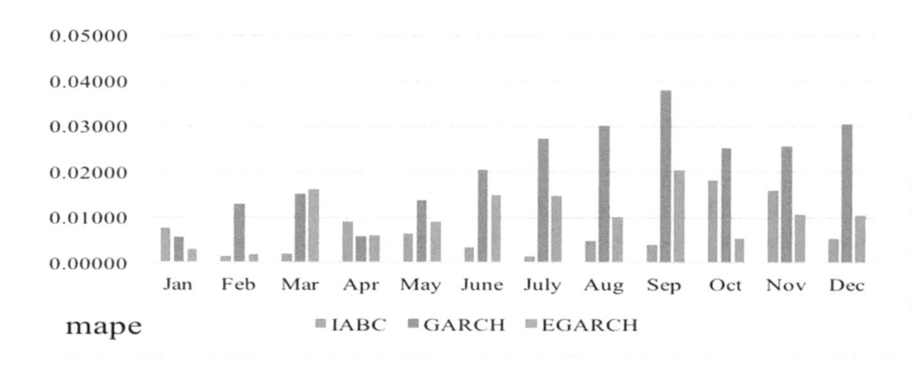

Fig. 2. Monthly MAPE value obtained by GARCH(1,1), EGARCH(1,1), and IABC in 2010.

Table 2. The before-and-after of adding Monitoring Indicator from 2009 to 2015.

	Average MAPE value	Standard deviation
GARCH(1,1)	0.01504	0.00826
EGARCH(1,1)	0.00976	0.00071
IABC	0.00424	0.00014

5 Conclusions and Future Works

In this paper, the inbound international tourism flow in Taiwan is included to be the new input variable with other known variables in the foreign exchange rate forecasting process. Three existing forecasting models including GARCH(1,1), EGARCH(1,1), and IABC forecasting model. The experimental results indicate that by considering the inbound international tourism flow as the new variable in the IABC model presents higher forecasting accuracy in most cases comparing to the time series models. Nevertheless, the IABC model, sometimes, is suffered from trapping in the local optimum and cannot deliver the accurate forecasting result in some time periods. The stability of the IABC model will be discussed and improved in the future work.

Acknowledgement. This work is funded by the Key Project of Fujian Provincial Education Bureau (JA15323).

References

1. Engle, R.F.: Autoregressive conditional heteroskedasticity with estimates of the variance of united kindom inflation. Econometrica **50**(4), 987–1007 (1982)
2. Bollerslev, T.R.: Generalized autoregressive conditional heteroskedasticity. J. Econometr. **31**(3), 307–327 (1986)

3. Nelson, D.B.: Conditional heteroskedasticity in asset returns: a new approach. Economietrica **59**(2), 347–370 (1991)
4. Tsai, P.-W., Khurram, M.K., Pan, J.-S., Liao, B.-Y.: Interactive artificial bee colony supported passive continuous authentication system. IEEE Syst. J. **8**(2), 395–405 (2014)
5. Tsai, P.-W., Liu, C.-H., Liao, L.-C., Chang, J.-F.: Using consumer confidence index in the foreign exchange rate forecasting. In: The 11th International Conference on Intelligent Information Hiding and Multimedia Signal Processing (IIH-MSP), pp. 360–363. IEEE Press, Adelaide (2015)
6. Ito, T.: Foreign exchange rate expectations: micro survey data. Am. Econ. Assoc. **80**(3), 434–449 (1990)
7. Baillie, R.T.: A multivariate generalized ARCH approach to modeling risk premia in forward foreign exchange rate markets. J. Int. Money Financ. **9**(3), 309–324 (1990)
8. Diebold, F.X., Hahn, J., Tay, A.S.: Multivariate density forecast evaluation and calibration in financial risk management: high-frequency returns on foreign exchange. Rev. Econ. Stat. **81**(4), 661–673 (1999)
9. World Travel & Tourism Council, Exchange Rate Trends and Travel & Tourism Performance, August 2016. https://www.wttc.org/-/media/files/reports/special-and-periodic-reports/exchange-rates-august-2016/exchange-rates-and-tt-performance.pdf
10. World Travel & Tourism Council, The Impact of Exchange Rates on Travel & Tourism, August 2016. https://www.wttc.org/-/media/files/reports/special-and-periodic-reports/exchange-rates-august-2016/wttc-exchange-rate-analysis-august-2016.pdf
11. Karaboga, D.: An idea based on honey bee swarm for numerical optimization. Technical report-TR06, 2005 (2005)

A New Solution Method for a Class of Fuzzy Random Bilevel Programming Problems

Aihong Ren[1(✉)] and Xingsi Xue[2,3]

[1] Department of Mathematics, Baoji University of Arts and Sciences, Baoji 721013, China
raih2003@hotmail.com
[2] College of Information Science and Engineering, Fujian University of Technology, Fuzhou, Fujian, China
[3] Fujian Provincial Key Laboratory of Big Data Mining and Applications, Fujian University of Technology, Fuzhou, Fujian, China

Abstract. This paper investigates a kind of bilevel programming with fuzzy random variable coefficients in both objective functions and the right hand side of constraints. On the basis of the notion of Er-expected value of fuzzy random variable, the upper and lower level objective functions can be replaced with their corresponding Er-expected values. In terms of probability over defuzzified operator, fuzzy stochastic constraints can be converted into the equivalent forms. Based on these, the fuzzy random bilevel programming problem can be transformed into its deterministic one. Then we suggest differential evolution algorithm to solve the final crisp problem. Finally, a numerical example is given to illustrate the proposed method.

Keywords: Bilevel programming · Fuzzy random variable · Er-expected value of fuzzy random variable · Differential evolution algorithm

1 Introduction

Bilevel programming is a complicated mathematical technique for modeling and handling hierarchical decision processes consisting in two levels. Over the past decades, bilevel programming has been widely used in electricity markets [1], transport network design [2,3], principal-agent problems [4], and others. In the light of these extensive and diverse applications, many researchers have also made great contributions to theoretical developments and solution approaches for such a kind of problem [5–7]. The interested reader is referred to good textbooks on the subject [8–10]. It is worthwhile to note that all the coefficients in a conventional bilevel programming problem are usually assumed as crisp values. However, the coefficient values in most real-world bilevel decision making are often imprecise and uncertain. Therefore, it is necessary to

J.-S. Pan et al. (eds.), *Advances in Intelligent Information Hiding and Multimedia Signal Processing*, Smart Innovation, Systems and Technologies 81,
DOI 10.1007/978-3-319-63856-0_29

build uncertain bilevel optimization models for meeting the demands of practical hierarchical decision making situations under various uncertain environments.

In recent ten years, lots of research efforts have mainly centered on two types of uncertain bilevel programming problems, namely the fuzzy bilevel programming problem and the stochastic bilevel programming problem, respectively. For example, Zhang, Lu and Dillon [11] proposed fuzzy Kuhn-Tucker approach, fuzzy branch-and-bound approach and fuzzy Kth-best approach through fuzzy set techniques to deal with fuzzy linear bilevel programming problems. Gao et al. [12] developed a λ–cut and goal programming-based algorithm to handle fuzzy linear multiple objective bilevel programming problems. Besides, Sakawa and Katagiri [13] used the concept of chance constrained programming and the fractile criterion optimization model to convert bilevel stochastic linear programming problems into deterministic ones, and then introduced interactive fuzzy programming to find satisfactory solutions. Yano [14] suggested a probability maximization model and a fractile optimization model to solve hierarchical multiobjective stochastic linear programming problems. It is worth mentioning that the above two kinds of uncertain bilevel programming in many applications are effective in dealing with only one type of uncertainty.

Nevertheless, in practical bilevel decision-making situations, sometimes fuzziness and randomness exist simultaneously in various input parameters. Fuzzy random variable, first introduced by Kwakernaak [15], is a powerful tool to express this hybrid type of uncertainty. In recent years, the fuzzy random bilevel programming problem with the aid of the concept of fuzzy random variable is getting attention by many researchers. Nevertheless, research on solution strategies of this kind of problem is relatively little because of NP-hardness of bilevel programming and complexities of hybrid uncertainties. In some recent studies, most solution methodologies of the fuzzy random bilevel programming problem focus on converting the problem into the corresponding equivalent crisp form through all kinds of defuzzifying and derandomizing approaches. Sakawa and Katagiri [16] applied level sets and fractile criterion to transform a fuzzy random bilevel linear programming into a deterministic one, and obtained a Stackelberg solution. Sakawa and Katagiri [17] developed an interactive fuzzy programming through level sets based probability maximization to cope with fuzzy random cooperative two-level linear programming. In addition, Ren and Wang [18] presented a computational method based on level sets and interval programming technique to solve bilevel linear programming with fuzzy random variable coefficients and obtained optimistic Stackelberg solutions.

This study aims to develop a new solution approach to deal with the fuzzy random bilevel programming problem in which all coefficients in both objective functions and the right hand side of constraints are fuzzy random variable. Making use of the concept of Er-expected value of fuzzy random variable, we substitute Er-expected values of the upper and lower level objective functions for these two objective functions. Meanwhile, equivalent transformations for fuzzy stochastic constraints are performed based on probability over defuzzified operator. As a result, the fuzzy random bilevel programming problem can be converted into a deterministic one dealt with by differential evolution algorithm. Finally, we provide a numerical example to illustrate the proposed method.

The rest of the paper is organized as follows: Sect. 2 gives the basic definitions and results throughout this paper. Section 3 describes the fuzzy random bilevel programming model. In Sect. 4, the solution approach by combing Er-expected value of fuzzy random variable and probability over defuzzified operator is proposed. Section 5 provides a numerical example to illustrate the proposed approach. Finally, Sect. 6 summarizes the paper.

2 Preliminaries

In this section, the definitions of Yager index of fuzzy number, fuzzy random variable, and Er-expected value of fuzzy random variable are reviewed.

Definition 1 ([19]). The Yager index of a fuzzy number $\tilde{c}$ can be defined as:

$$I(\tilde{c}) = \frac{1}{2}\int_0^1 [c_\alpha^- + c_\alpha^+]d\alpha,$$

where $[c_\alpha^-, c_\alpha^+]$ is the α-level set of $\tilde{c}$.

Definition 2 ([20]). Let (Ω, A, P) be a probability space, where Ω is a sample space, A is a $\sigma-$field and P is a probability measure. Then a fuzzy random variable on this probability space, denoted by $\tilde{\bar{c}}$, is a fuzzy set-valued mapping:

$$\tilde{\bar{c}} : \Omega \to F_0(R), \omega \to \tilde{c}(\omega)$$

such that for any Borel set B of R and for every $\alpha \in [0, 1]$,

$$\{\omega \in \Omega | \tilde{c}_\omega^\alpha \subset B\} \in A,$$

where $F_0(R)$ is the set of fuzzy numbers with compact supports in R, and $\tilde{c}_\omega^\alpha$ represents the α-level set of the fuzzy set $\tilde{c}(\omega)$.

For any $\alpha \in [0, 1]$, we denote the α-level set of a fuzzy random variable $\tilde{\bar{c}}$ by $\tilde{\bar{c}}_\alpha(\omega) = \{x | \mu_{\tilde{\bar{c}}(\omega)}(x) \geq \alpha\} = [c_\alpha^-(\omega), c_\alpha^+(\omega)]$.

Definition 3 ([21]). The Er-expected value of a fuzzy random variable $\tilde{\bar{c}}$ denoted by $Er(\tilde{\bar{c}})$ is defined as follows:

$$Er(\tilde{\bar{c}}) = \frac{1}{2}\int_0^1 [E(c_\alpha^-) + E(c_\alpha^+)]d\alpha.$$

where $E(c_\alpha^-)$ and $E(c_\alpha^+)$ represent expected values of $c_\alpha^-(\omega)$ and $c_\alpha^+(\omega)$.

Corollary 1 ([21]). Let $\tilde{\bar{c}}_1$ and $\tilde{\bar{c}}_2$ be two fuzzy random variables, then the three relations are defined respectively as follows:

(1) $\tilde{\bar{c}}_1 = \tilde{\bar{c}}_2 \ iff \ Er(\tilde{\bar{c}}_1) = Er(\tilde{\bar{c}}_2)$;
(2) $\tilde{\bar{c}}_1 \leq \tilde{\bar{c}}_2 \ iff \ Er(\tilde{\bar{c}}_1) \leq Er(\tilde{\bar{c}}_2)$;
(3) $\tilde{\bar{c}}_1 \geq \tilde{\bar{c}}_2 \ iff \ Er(\tilde{\bar{c}}_1) \geq Er(\tilde{\bar{c}}_2)$.

3 Problem Formulation

Consider the following fuzzy random bilevel linear programming problem in which fuzzy random variable coefficients are not only in the upper and lower level objective functions but also in the right hand side of constraints:

$$\begin{cases} \min\limits_{x} & \tilde{\bar{c}}_1 x + \tilde{\bar{d}}_1 y \\ & \text{where } y \text{ solves} \\ \min\limits_{y} & \tilde{\bar{c}}_2 x + \tilde{\bar{d}}_2 y \\ \text{s.t.} & a_{i1}x + a_{i2}y \leq \tilde{\bar{b}}_i, i = 1, 2, \cdots, s, \\ & x \geq 0, y \geq 0, \end{cases} \tag{1}$$

where x is an $n-$dimensional upper level decision vector and y is an $m-$dimensional lower level decision vector, $\tilde{\bar{c}}_l = (\tilde{\bar{c}}_{l1}, \tilde{\bar{c}}_{l2}, \ldots, \tilde{\bar{c}}_{ln})$, $l = 1, 2$ are $n-$dimensional fuzzy random vectors, $\tilde{\bar{d}}_l = (\tilde{\bar{d}}_{l1}, \tilde{\bar{d}}_{l2}, \ldots, \tilde{\bar{d}}_{lm})$ are $m-$dimensional fuzzy random vectors, $\tilde{\bar{b}}_i$, $i = 1, 2, \cdots, s$ are fuzzy random variables, a_{i1} and a_{i2} are $n-$dimensional and $m-$dimensional crisp vectors, respectively.

Clearly, model (1) is ill-defined owing to hybrid uncertainty. To cope with this type of problem, it is necessary to develop a reasonable alternative way to convert the problem into its deterministic equivalent problem.

4 Methodology

In this section, we adopt the concept of the scalar expected value of fuzzy random variable to model both objective functions and apply probability over defuzzifying operation to deal with fuzzy stochastic constraints, and then transform problem (1) into its equivalent crisp form. Furthermore, differential evolution algorithm is suggested to solve the final transformed problem.

4.1 Equivalent Deterministic Bilevel Model

In view of the fact that each coefficient in upper and lower level objective functions of problem (1) is a fuzzy random variable, these two objective functions are also fuzzy random variables by Zadeh's extension principle. As we all know, expected value of fuzzy random variable is one of the most frequently used methods to perform the defuzzifying and derandomizing processes at the same time. So far there have been several types of definitions of the expected value of a fuzzy random variable [22,23]. However, the computation process of some of the existing definitions is rather complicated and time consuming. For ease of calculation, we adopt the concept of the scalar expected value of fuzzy random variable introduced by Eshghi and Nematian [21] to handle the upper and lower objective functions. In this way, the upper and lower level objective functions of problem (1) are replaced with their own Er-expected values as follows:

$$Er[\tilde{\bar{c}}_l x + \tilde{\bar{d}}_l y], l = 1, 2.$$

Next, we deal with fuzzy stochastic constraints of problem (1). In order to transform fuzzy stochastic constraints into their deterministic ones, probability over defuzzified fuzzy quantities introduced by Aiche et al. [24] is employed. Considering that the Yager ranking indices technique has the linear and additive properties, we use this method as the defuzifying operation. As a result, fuzzy-stochastic constraints of problem (1) can be written as the following constraints by combining probability with Yager ranking indices method:

$$P_r\{\omega | a_{i1}x + a_{i2}y \leq I(\tilde{\bar{b}}_i)\} \geq \eta_i, i = 1, 2, \cdots, s,$$

where P_r represents probability, $I(\tilde{\bar{b}}_i)$ denotes the Yager-index of $\tilde{\bar{b}}_i$, and η_i is a satisfying level of the $i-$th constraint.

Based on the above these discussions, problem (1) can be converted into the following problem:

$$\begin{cases} \min\limits_{x} & Er[\tilde{\bar{c}}_1 x + \tilde{\bar{d}}_1 y] \\ & \text{where } y \text{ solves} \\ \min\limits_{y} & Er[\tilde{\bar{c}}_2 x + \tilde{\bar{d}}_2 y] \\ \text{s.t.} & Pr\{\omega | a_{i1}x + a_{i2}y \leq I(\tilde{\bar{b}}_i)\} \geq \eta_i, i = 1, 2, \cdots, s, \\ & x \geq 0, y \geq 0. \end{cases} \tag{2}$$

For simplicity, here we assume that all fuzzy random variable coefficients of problem (2) are taken as triangular forms with their mean values being normally distributed random variables. We denote $\tilde{\bar{c}}_l = (\tilde{\bar{c}}_{l1}, \tilde{\bar{c}}_{l2}, \ldots, \tilde{\bar{c}}_{ln})$ with all their elements $\tilde{\bar{c}}_{lk}(\omega) = (\bar{c}_{lk}(\omega), \alpha^c_{lk}, \beta^c_{lk})$ with $\bar{c}_{lk}(\omega) \sim N(m^c_{lk}, (\sigma^c_{lk})^2)$, $k = 1, 2, \cdots, n$, $\tilde{\bar{d}}_l = (\tilde{\bar{d}}_{l1}, \tilde{\bar{d}}_{l2}, \ldots, \tilde{\bar{d}}_{lm})$ with all their elements $\tilde{\bar{d}}_{lp}(\omega) = (\bar{d}_{lp}(\omega), \alpha^d_{lp}, \beta^d_{lp})$ with $\bar{d}_{lp}(\omega) \sim N(m^d_{lp}, (\sigma^d_{lp})^2)$, $p = 1, 2, \cdots, m$, and $\tilde{\bar{b}}_i(\omega) = (\bar{b}_i(\omega), \alpha^b_i, \beta^b_i)$ with $\bar{b}_i(\omega) \sim N(m^b_i, (\sigma^b_i)^2)$, $i = 1, 2, \cdots, s$.

By Definition 1, the Yager-index of $\tilde{\bar{b}}_i$ can be calculated as:

$$I(\tilde{\bar{b}}_i) = \bar{b}_i + \frac{\beta^b_i - \alpha^b_i}{4}.$$

Obviously, $I(\tilde{\bar{b}}_i)$ is a random variable. Then the $i-$th constraint of problem (2) can be represented as follows

$$\begin{aligned} & Pr\{\omega | a_{i1}x + a_{i2}y \leq I(\tilde{\bar{b}}_i)\} \geq \eta_i \\ \Leftrightarrow & Pr\{\omega | a_{i1}x + a_{i2}y \leq \bar{b}_i + \tfrac{\beta^b_i - \alpha^b_i}{4}\} \geq \eta_i \\ \Leftrightarrow & a_{i1}x + a_{i2}y - \tfrac{\beta^b_i - \alpha^b_i}{4} - m^b_i \leq \Phi^{-1}(1 - \eta_i)\sigma^b_i. \end{aligned}$$

Besides, according to Definition 3, the Er-expected values of the upper and lower level objective functions of problem (2) can be rewritten as:

$$Er[\tilde{\bar{c}}_l x + \tilde{\bar{d}}_l y] = Er(\tilde{\bar{c}}_l)x + Er(\tilde{\bar{d}}_l)y = [m^c_l + \frac{\beta^c_l - \alpha^c_l}{4}]x + [m^d_l + \frac{\beta^d_l - \alpha^d_l}{4}]y, l = 1, 2.$$

Thereby, problem (2) can be converted into the following problem:

$$\begin{cases} \min\limits_{x} & [m_1^c + \frac{\beta_1^c - \alpha_1^c}{4}]x + [m_1^d + \frac{\beta_1^d - \alpha_1^d}{4}]y \\ & \text{where } y \text{ solves} \\ \min\limits_{y} & [m_2^c + \frac{\beta_2^c - \alpha_2^c}{4}]x + [m_2^d + \frac{\beta_2^d - \alpha_2^d}{4}]y \\ \text{s.t.} & a_{i1}x + a_{i2}y - \frac{\beta_i^b - \alpha_i^b}{4} - m_i^b \leq \Phi^{-1}(1-\eta_i)\sigma_i^b, i = 1, 2, \cdots, s, \\ & x \geq 0, y \geq 0. \end{cases} \tag{3}$$

Clearly, model (3) is already deterministic bilevel programming problem.

4.2 Differential Evolution Algorithm

Up to now all kinds of solution approaches including traditional methods and heuristic algorithms have been developed to deal with different types of bilevel programming. Particularly, various heuristic algorithms are increasingly applied due to their simple, efficient and robust characteristics in the recent years. In this paper, differential evolution algorithm is employed to handle the transformed problem (3).

Differential evolution algorithm (DE) is a population-based evolutionary algorithm designed to cope with continuous optimization problems [25]. The main four phases of this way include initialization, mutation, crossover and selection. It is noted here that differential evolution algorithm unlike most of other evolutionary algorithms utilizes the linear combination of the difference vector between two individuals and a third individual to produce a new candidate solution in the mutation phase.

In our work, a solution approach which is the combination of differential evolution algorithm and linear programming technique to cope with problem (2). Specifically, we adopt differential evolution algorithm to solve the upper level optimization problem, and then apply linear programming technique to deal with the lower level programming problem for each given upper level variable.

5 Numerical Example

In order to illustrate the feasibility of the proposed method, we consider the following fuzzy random bilevel linear programming problem [26]:

$$\begin{cases} \min\limits_{x_1} & \tilde{\bar{c}}_{11}x_1 + \tilde{\bar{c}}_{12}x_2 + \tilde{\bar{c}}_{13}x_3 \\ & \text{where } (x_2, x_3) \text{ solves} \\ \min\limits_{x_2, x_3} & \tilde{\bar{c}}_{21}x_1 + \tilde{\bar{c}}_{22}x_2 + \tilde{\bar{c}}_{23}x_3 \\ \text{s. t.} & 2x_1 + 5x_2 + x_3 \leq \tilde{\bar{b}}_1, \\ & 5x_1 + 3x_2 + 2x_3 \leq \tilde{\bar{b}}_2, \\ & x_1 + 2x_2 + 4x_3 \leq \tilde{\bar{b}}_3, \\ & x_1 \geq 0, x_2 \geq 0, x_3 \geq 0, \end{cases} \tag{4}$$

Table 1. Values of fuzzy random variable coefficients

$\tilde{\bar{c}}_{11} = (\bar{c}_{11}, 1, 1)$ with $\bar{c}_{11} \sim N(-4, 1^2)$	$\tilde{\bar{c}}_{12} = (\bar{c}_{12}, 1.5, 1.5)$ with $\bar{c}_{12} \sim N(-2, 1.2^2)$
$\tilde{\bar{c}}_{13} = (\bar{c}_{13}, 1, 1)$ with $\bar{c}_{13} \sim N(-6, 1.1^2)$	$\tilde{\bar{c}}_{21} = (\bar{c}_{21}, 1.5, 1.5)$ with $\bar{c}_{21} \sim N(3, 1^2)$
$\tilde{\bar{c}}_{22} = (\bar{c}_{22}, 1, 1)$ with $\bar{c}_{22} \sim N(1, 1.2^2)$	$\tilde{\bar{c}}_{23} = (\bar{c}_{23}, 1.5, 1.5)$ with $\bar{c}_{23} \sim N(-4.8, 1.1^2)$
$\tilde{\bar{b}}_1 = (\bar{b}_1, 10, 10)$ with $\bar{b}_1 \sim N(120, 3^2)$	$\tilde{\bar{b}}_2 = (\bar{b}_2, 8, 8)$ with $\bar{b}_2 \sim N(115, 2^2)$
$\tilde{\bar{b}}_3 = (\bar{b}_3, 5, 5)$ with $\bar{b}_3 \sim N(100, 1^2)$	

where all coefficients in both objective functions and constraints are triangular fuzzy random variables. Tables 1 shows the values of all these coefficients.

From model (3), the above problem can be equivalently converted into the following problem:

$$\begin{cases} \min\limits_{x_1} & -4x_1 - 2x_2 - 6x_3 \\ & \text{where } (x_2, x_3) \text{ solves} \\ \min\limits_{x_2, x_3} & 3x_1 + x_2 - 4.8x_3 \\ \text{s. t.} & 2x_1 + 5x_2 + x_3 - 120 \le 3\Phi^{-1}(1 - \eta_1), \\ & 5x_1 + 3x_2 + 2x_3 - 115 \le 2\Phi^{-1}(1 - \eta_2), \\ & x_1 + 2x_2 + 4x_3 - 100 \le \Phi^{-1}(1 - \eta_3), \\ & x_1 \ge 0, x_2 \ge 0, x_3 \ge 0, \end{cases} \tag{5}$$

Next, we use differential evolution algorithm to solve the above problem (5). We set $\eta_1 = \eta_2 = \eta_3 = 0.9$, then $\Phi^{-1}(1 - \eta_1) = \Phi^{-1}(1 - \eta_2) = \Phi^{-1}(1 - \eta_3) = -1.28$. In addition, we set the parameters associated with differential evolution algorithm as follows: population size $N = 30$, maximum number of iterations $M = 50$, scaling factor $F = 0.8$, crossover rate $CR = 0.6$.

After solving problem (5), the following results can be obtained: $(x_1^*, x_2^*, x_3^*) = (14.0160, 0, 21.1760)$ with the corresponding upper level objective function value -183.1200.

For this example, Ren and Wang [26] suggested an interval programming approach based on the α–level set to find the best and worst optimal values. In this way, the best and worst optimal solutions are obtained at (14.4, 0, 21.5) and (13.4, 0, 20.7). Corresponding to these two optimal solutions, the upper level objective values are –186.60 and –177.8000. Obviously, the optimal objective function value obtained by our proposed method is falling between the two objective function values found by the existing method. In essence, the two optimal solutions obtained in [26] are two extreme cases, which is more inclined to theoretical significance than practical value. In view of this, the result obtained by our approach is valid in some practical problems.

6 Conclusion

This paper focuses on a type of bilevel programming with fuzzy random variable coefficients in both objective functions and the right-hand side of constraints.

To cope with such a problem, the concept of the scalar expected value of fuzzy random variable is applied to model both objective functions and probability over defuzzifying operation is used to handle fuzzy stochastic constraints, and then the fuzzy random bilevel programming problem can be converted into its deterministic bilevel programming. Subsequently, we suggest differential evolution algorithm to solve the transformed problem. Finally, we provide a numerical example to illustrate the proposed method. In future, we will apply the proposed approach to some bilevel practical problems.

Acknowledgements. This work was supported by the National Natural Science Foundation of China (Grant No.61602010), Natural Science Basic Research Plan in Shaanxi Province of China (Grant No.2017JQ6046) and Science Foundation of Baoji University of Arts and Sciences (Grant No.ZK16049).

References

1. Zhang, G.Q., Gao, Y., Lu, J.: Competitive strategic bidding optimization in electricity markets using bilevel programming and swarm technique. IEEE Trans. Industr. Electron. **58**(6), 2138–2146 (2011)
2. Gzara, F.: A cutting plane approach for bilevel hazardous material transport network design. Oper. Res. Lett. **41**(1), 40–46 (2013)
3. Fontaine, P., Minner, S.: Benders decomposition for discrete-continuous linear bilevel problems with application to traffic network design. Transp. Res. Part B Methodol. **70**, 163–172 (2014)
4. Cecchini, M., Ecker, J., Kupferschmid, M., Leitch, R.: Solving nonlinear principal-agent problems using bilevel programming. Eur. J. Oper. Res. **230**(2), 364–373 (2013)
5. Dempe, S.: Annotated bibliography on bilevel programming and mathematical programs with equilibrium constraints. Optimization **52**(3), 333–359 (2003)
6. Colson, B., Marcotte, P., Savard, G.: Bilevel programming a survey. 4OR **3**(2), 87–107 (2005)
7. Colson, B., Marcotte, P., Savard, G.: An overview of bilevel optimization. Ann. Oper. Res. **153**(1), 235–256 (2007)
8. Bard, J.F.: Practical Bilevel Optimization: Algorithms and Applications. Kluwer Academic Publishers, Dordrecht, Boston, London (1998)
9. Dempe, S.: Foundations of Bilevel Programming. Kluwer Academic Publishers, Dordrecht, Boston, London (2002)
10. Dempe, S., Kalashnikov, V., Pérez-Valdés, G.A., Kalashnykova, N.: Bilevel Programming Problems: Theory, Algorithms and Applications to Energy Networks. Kluwer Academic Publishers, Springer, Berlin (2015)
11. Zhang, G.Q., Lu, J., Dillon, T.: Fuzzy linear bilevel optimization: solution concepts, approaches and applications. Stud. Fuzziness Soft Comput. **215**, 351–379 (2007)
12. Gao, Y., Zhang, G.Q., Ma, J., Lu, J.: A λ-cut and goal-programming-based algorithm for fuzzy-linear multiple-objective bilevel optimization. IEEE Trans. Fuzzy Syst. **18**(1), 1–13 (2010)
13. Sakawa, M., Katagiri, H.: Interactive fuzzy programming based on fractile criterion optimization model for two-level stochastic linear programming problems. Cybern. Syst. **41**(7), 508–521 (2010)

14. Yano, H.: Hierarchical Multiobjective stochastic linear programming problems considering both probability maximization and fractile optimization. IAENG Int. J. Appl. Mathe. **42**(2), 91–98 (2012)
15. Kwakernaak, H.: Fuzzy random variables-I. definitions and theorems. Inf. Sci. **15**(1), 1–29 (1978)
16. Sakawa, M., Katagiri, H.: Stackelberg solutions for fuzzy random two-level linear programming through level sets and fractile criterion optimization. CEJOR **20**, 101–117 (2012)
17. Sakawa, M., Katagiri, H.: Interactive fuzzy random cooperative two-level linear programming through level sets based probability maximization. Expert Syst. Appl. **40**, 1400–1406 (2013)
18. Ren, A., Wang, Y.P.: Optimistic Stackelberg solutions to bilevel linear programming with fuzzy random variable coefficients. Knowl.-Based Syst. **67**, 206–217 (2014)
19. Yager, R.: A procedure for ordering fuzzy subsets of the unit interval. Inf. Sci. **24**, 143–161 (1981)
20. Luhandjula, M.K.: Fuzziness and randomness in an optimization framework. Fuzzy Sets Syst. **77**, 291–297 (1996)
21. Eshghi, K., Nematian, J.: Special classes of mathematical programming models with fuzzy random variables. J. Intell. Fuzzy Syst. **19**(2), 131–140 (2008)
22. Wang, G.Y., Zhong, Q.: Linear programming with fuzzy random variable coefficients. Fuzzy Sets Syst. **57**(3), 295–311 (1993)
23. Liu, Y.K., Liu, B.: A class of fuzzy random optimization: expected value models. Inf. Sci. **155**, 89–102 (2003)
24. Aiche, F., Abbas, M., Dubois, D.: Chance-constrained programming with fuzzy stochastic coefficients. Fuzzy Optim. Decis. Making **12**, 125–152 (2013)
25. Storn, R., Price, K.: Differential evolution - a simple and efficient heuristic for global optimization over continuous spaces. J. Global Optim. **11**(4), 341–359 (1997)
26. Ren, A.H., Wang, Y.P.: An interval programming approach for bilevel linear programming problem with fuzzy random coefficients. In: 2013 IEEE Congress on Evolutionary Computation (CEC2013), pp. 462–469 (2013)

A New Decomposition Many-Objective Evolutionary Algorithm Based on - Efficiency Order Dominance

Guo Xiaofang(✉)

School of Science, Xi'an Technological University, Xi'an, Shaanxi, China
gxfang1981@126.com

Abstract. Decomposition-based evolutionary algorithms are promising for handling many objective optimization problems with more than three objectives in the past decade. In the proposed algorithm, we develop a new dominance relation based on - efficiency order dominance (MOEA/D-εEOD) in each subproblem to realize the selection and update of the individuals. Besides, a dynamic adaptive weight vector generation method is proposed, which is able to dynamically adjust the weight vector setting according to the current distribution of the non-dominated solution set. The proposed algorithm has been tested extensively on six widely used benchmark problems, and an extensive comparison indicates that the proposed algorithm offers competitive advantages in convergence and diversity.

Keywords: Decomposition · ε-dominance efficiency order rank · Many-objective

1 Introduction

In the past decades, multi-objective optimization problems (MOPs) have been applied to the fields of engineering and management [1]. With the optimization problems becoming more and more complex, the number of objectives increases to 4 or more, and optimization problems in dealing with more than three objectives are named as many-objective problems [2] (MAPs). In MOPs, the optimal solution is aiming to find a set of Pareto optimal solutions, which can be converged as close to the target Pareto front (PF) as possible in the search space and be distributed as evenly as possible. As an evolutionary algorithm (EA) works with a population of individuals, it has the potential to achieve these two goals of MOPs, and actually, multi-objective evolutionary algorithms (MOEAs) have been successfully utilized to obtain the Pareto optimal solution set with two or three objective problems.

However, when dealing with many-objective problems which have more than three objectives, the search ability of MOEAs, such as NSGA-II and SPEA2, may degrade. The main reason is that the MOEAs rely primarily on Pareto ranking to guide the search, which can only bring a little selection pressure for many-objective problems [3]. In particular, as the number of objectives increases, solutions have a greater chance of becoming incomparable by using Pareto dominance relation. As a result, there is no

J.-S. Pan et al. (eds.), *Advances in Intelligent Information Hiding and Multimedia Signal Processing*, Smart Innovation, Systems and Technologies 81,
DOI 10.1007/978-3-319-63856-0_30

quantitative distinction between solutions, which lead to the loss of the selection pressure and the algorithm cannot drive the population toward the Pareto front efficiently. Furthermore, the number of points required to approximate the Pareto front increases exponentially with the growing number of objectives, which makes it hard to capture the whole Pareto front in limited computation resources.

Zhang and Li [4] proposed the multi-objective evolutionary algorithm based on decomposition (MOEA/D), which decomposes the original problem into many single-objective scale optimization problems and optimizes all the scalar problems simultaneously using the evolutionary algorithm. MOEA/D can easily determine the superiority or inferiority of solutions through scalarizing function, and it has achieved great success in MAPs.

Nevertheless, MOEA/D still suffers from several shortcomings [5]. In each sub-problem corresponding to a specific weight vector, whether or not a new solution replaces an old one is completely determined by their scalarizing function values during the update process, and this ignores the actual position in the objective space of the new solution locates on. In some cases, such replacement can make a severe loss of the diversity of population. In the proposed algorithm, we develop a new dominance relation based on -efficiency order dominance in each sub-problem to realize the selection and update of the individuals. Besides, a dynamic adaptive weight vector generation method is proposed, which is able to dynamically adjust the weight vector setting according to the current distribution of the non-dominated solution set.

The rest of this paper is organized as follows. Section 2 introduces the characteristics of the MOEA/D. Section 3 presents the details of the proposed algorithm. Section 4 makes a comparison between our proposal and the state-of-the-art algorithms. Finally, conclusions are drawn in Sect. 5.

2 Backgrounds

2.1 The Idea of MOEA/D

MOEA/D is a decomposition based MOEA, which is proposed by Qingfu Zhang. It combines the evolutionary algorithm with the traditional mathematical programming method, and decomposes a multi-objective optimization problem into a number of scalar optimization sub-problems via a scalar function, and optimizes them simultaneously using evolutionary algorithm. In MOEA/D, Tchebycheff approach is used to decompose a multi-objective optimization problem into N sub-problems, and the objective function of the j-th (j = 1,2,...N) sub-problem is as follows.

$$g^{te}(x \mid \lambda^j, z^*) = \max_{1 \leq i \leq m} \{ \lambda_i^j \mid f_i(x) - z_i^* \mid \} \tag{1}$$

where $\lambda^j = (\lambda_1^j, \ldots \lambda_m^j)^T$ is a weight vector and $z^* = (z_1^*, \ldots z_m^*)^T$ is a reference point. For each Pareto optimal point x^*, there exists a weight vector $\lambda^j = (\lambda_1^j, \ldots \lambda_m^j)^T$ such that x^* is the optimal solution of problem (1) and each optimal solution of (1) is a Pareto optimal solution of problem. Therefore, one is able to obtain different Pareto optimal solutions by altering the weight vector, and different weighted vectors will direct the search towards different regions of the objective space. In MOEA/D, the population is composed of the best solution found so far for each sub-problem.

2.2 ε -Dominance

Definition 1 (ε-dominance)
The concept of ε-dominance is introduced in [6], and it is a kind of relaxed form of Pareto dominance. It acts as an archiving strategy to ensure both properties of convergence towards the Pareto optimal sets and properties of diversity among the solutions found. The idea is to use a set of boxes to cover the Pareto front, where the size of such boxes is defined by a user defined parameter, called ε. Within each box, it is only allowed a single non-dominated solution to be retained, and the definition is described as follows.

For any two solutions a and b in the decision space, a ε-dominance b ($a \prec_\varepsilon b$), if and only if

$$\forall i\, B_i(a) \le B_i(b),\ \exists j\, B_j(a) < B_j(b),\ 1 \le i,j \le m \tag{2}$$

where $B_i(a) = \left\lfloor (f_i(a) - f_i^{\min}) / \varepsilon_i \right\rfloor$, $1 \le i,j \le m$, and it denoted the partition in the objective space for a. $f_i^{\min}$ is the assumed minimum value of the i-th objective and ε_i is the accepted tolerance value in the i-th objective.

2.3 Generalized Decomposition

The generalized decomposition method [7] is a new weight vector generation strategy proposed by Giagkiozis in 2014, and it can generate weight vector dynamically in order to guide the search according to the prior knowledge of users.

Definition 2 (Generalized Decomposition)
In MOEA/D, the objective function corresponding to the j-th (j = 1,2,…N) sub-problem (1) can be rewritten as the equivalent form [28], as shown in the formula (3).

$$\begin{aligned} &\min_x \| \lambda \circ | f(x) - z^* | \|_\infty, \\ &\sum_{i=1}^{m} \lambda_i = 1, \\ &\lambda_i \ge 0, \quad \forall i \in \{1, 2, \ldots m\} \end{aligned} \tag{3}$$

where $\circ$ is the Hadamard product of vector λ and vector $|f(x) - z^*|$, and $\| \|_\infty$ represents he infinite norm of the vector. In formula (3), given a certain weight vector, the optimal solution of a single objective optimization problem (3) corresponding to the weight vector is a non-dominated solution for the multi-objective optimization problem. In turn, given an optimal solution x^* of the problem (3), we want to find a unique weight vector

$\tilde{\lambda}$ in the convex set of weights, satisfying $\left\| \tilde{\lambda} \circ |F(x^*) - z^*| \right\|_\infty \leq \| \lambda \circ |F(x^*) - z^*| \|_\infty$, and $\tilde{\lambda}$ is the optimal weight vector of the given solution x^*. Therefore, if a set of initial non dominated solution is known in advance, the optimal weight vector set corresponding to the given non dominated solution set can be calculated according to formula (3), which will guide the search and adjust the setting the weight vector dynamically according to the shape of Pareto front.

3 The Algorithm MOEA/D-εEOD

3.1 Motivation of the Proposed Algorithm

In order to obtain a uniformly distributed optimal solution set on the Pareto front of target MAPs according to the geometric shape of the Pareto front, we have made two significant changes with the original MOEA/D. Firstly, we develop a new dominance relation, called ε-efficiency order dominance instead of scalarizing function to realize selection and update operation, which can increase the selective pressure and improve the distribution of individuals in each sub-problem. Besides, through analyzing the geometric relationship between weight vectors and their corresponding non-dominated solutions, we develop a novel adaptive weight vector generation method, which will adjust the setting of weight vector dynamically according to the current distribution of the non-dominated solutions.

3.2 ε-Efficiency Order Dominance

In the proposed algorithm, in order to overcome the shortcoming of the loss of diversity in sub-population with the use of scalarizing function, we develop a new dominance relation to evaluate the quality of non-dominated solutions. On one hand, a more stringent dominance relation, considering the efficient of order k, is introduced to realize the selection and update of the individuals. On the other hand, we borrow the concept of ε-dominance to maintain the diversity of Pareto optimal solutions in the newly proposed dominance relation, and the new dominance relation is named ε efficiency order dominance.

Definition 3 (ε-Efficiency Order Dominance).
For a given MOP, x_A, x_B are two non-dominated solutions in original objective set $\{f_1(x), f_2(x), \cdots, f_m(x)\}$. Consider all possible k-element subsets of the m given objectives in multi-objective problem $(1 \leq k \leq m)$, denoted by ρ_k, and $\rho_k = \{s_k | s_k \subseteq \{f_1, \cdots f_m\} \wedge |s_k| = k\}$.

Given a fixed value of k and $\boldsymbol{\varepsilon}$, we compare the $\boldsymbol{\varepsilon}$ dominance relation with x_A and x_B in all possible k-element subsets ρ_k, and count the number of k-element subsets satisfying $x_A\varepsilon$-dominance x_B, $x_B\varepsilon$-dominance x_A, and x_A and x_B are not compared with ε-dominance relation with k-element objective subset, denoted by $n_b(x_A, x_B \mid \varepsilon, \rho_k)$, $n_w(x_A, x_B \mid \varepsilon, \rho_k)$, $n_e(x_A, x_B \mid \varepsilon, \rho_k)$ respectively.

$$n_b(x_A, x_B|\varepsilon, \rho_k) = \{|\{s_k\}||x_A \prec_{\varepsilon}^{s_k} x_{B,} \forall s_k \in \rho_k\} \tag{4}$$

$$n_w(x_A, x_B|\varepsilon, \rho_k) = \{|\{s_k\}||x_B \prec_{\varepsilon}^{s_k} x_A, \forall s_k \in \rho_k\}; \tag{5}$$

$$n_e(x_A, x_B|\varepsilon, \rho_k) = \{|\{s_k\}||\neg(x_A \prec_{\varepsilon}^{s_k} x_B) \wedge \neg(x_B \prec_{\varepsilon}^{s_k} x_A), \forall s_k \in \rho_k\} \tag{6}$$

where $x_A \prec_{\varepsilon}^{s_k} x_B$ denote $x_A\varepsilon$-dominance x_B in one of the k-element objective subset $s_k \in \rho_k$.

For two non-dominated solutions x_A, x_B in original objective set, if x_A and x_B satisfy $n_b(x_A, x_B \mid \varepsilon, \rho_k) + n_e(x_A, x_B \mid \varepsilon, \rho_k) > n_w(x_A, x_B \mid \varepsilon, \rho_k)$, $x_A\varepsilon$ efficient dominance x_B with order k.

3.3 Main Flowchart

In the proposed algorithm, we divide population into N sub-population according to the weight vectors, and each individual is assigned to the sub-population to which the nearest weight vector belongs. In the first phase, the uniform design method is adopted to generate the initial weight vector, so that the objective space can be searched evenly in the initial stage. In the second phase, we use generalized decomposition to inversely obtain the optimal weight vector corresponding to the each non-dominated solution, for the purpose of adjusting the setting of weight vector according to the geometric shape of the Pareto front. Subsequently, in according to the shape and the property of the Pareto front, we use the formula $u^i = \alpha_1 \omega^i + \alpha_2 \delta^i \ (i = 1, 2)$ to obtain the newly adaptive weight vector μ^1, μ^2. By changing the value of the parameter α_1 and α_2, we can adjust the degree to the newly adaptive weight vector affected by the current Pareto optimal solutions set. Here, the smaller the value of α_1, the greater impact will be on the process of adapting the weight vectors by the current shape of the Pareto front.

Detailed description of the proposed algorithm is shown in Algorithm 1.

4 Simulation Results

4.1 Benchmark Problems and Parameter Settings

In the experimental study, we select 3widely used test problems DTLZ1 [8] to compare the proposed MOEA/D-εEOD with MOEA/D-M2 M [9], UMOEA/D [10] and NSGA2-CE [3]. Meanwhile, in order to investigate the capability of MOEA/D-εEOD for solving MOPs whose PFs are of complex shapes and of degenerate Pareto front, we also use the test cases F1, and DTLZ5(I,M) for our empirical studies. For DTLZ5 (I, M), where symbol I represents the actual number of objectives in Pareto front and symbol M denotes the original number of objectives.

Algorithm1. MOEA/D--εEOD

Step 1 **Initialization**

1.1 Generate the initial weight vector set $w^1, w^2, \ldots, w^N$ by applying the uniform design method;

1.2 Initialize the population POP at random with the size of $N \times K$, $x_1, \ldots x_{N \times K}$;

1.3 Calculate the ideal point $Z^* = (Z_1^*, \ldots Z_m^*)$ of the current population, where $Z_i^* = \min(f_i(x^1), \ldots f_i(x^N))$ $i = 1, 2, \ldots N$.

1.4 Divide the initial population POP into N subpopulations, and each subpopulation contains K individuals, e.g. the individuals of the $i-th$ sub-population is the set $\{x_{i(K-1)+1}, \ldots x_{iK}\}$. The discrimination of individuals is based on the proposed ε efficient order dominance relation and the best individual is referred to as x_{iK}.

1.5 Calculate the neighborhood list of the $i-th$ weight vector as $B_i = \{i_1, \ldots, i_T\}$.

Step 2 **Evolution**

For $i = 1$ to N do

2.1 For the $i-th$ sub-problem, select two individuals x_{i1}, x_{i2} from the neighbor sub-problem $B_i = \{i_1, \ldots, i_T\}$, and combine the best individual x_{iK} with x_{i1}, x_{i2} to execute crossover and mutation operator and obtain the new individual y';

2.2 Update: Update the ideal point with the new individual y', and find out the sub-population where y' located in. If y' lies in the $l-th$ sub-population, we compare the ε efficient order dominance relation between y' and the best individual x_{lK}. If x_{lK} better than y', delete y'; if y' better than x_{lK}, delete one of the individual from the $l-th$ sub-population, and add y' to the $l-th$ sub-population;

2.3 Update the neighborhood list of the $i-th$ sub-problem;

Step 3 **Adaptive Weight Adjustment**

IF $\mathrm{mod}(gen, iter) == 0$ then

Generate the new weight vector W. Execute evolutionary algorithm for several generations with the current weight vector set $w^1, w^2, \ldots, w^N$, and obtain a non-dominated solution set. Adopt the method of generalized decomposition method to generate the optimal weight vector δ^i corresponding to each non-dominated solution x^i according to formula (3) and get $\delta^1, \delta^2, \ldots, \delta^N$, and through the formula $u^i = \alpha_1 \omega^i + \alpha_2 \delta^i$ $(i = 1, 2, \ldots N)$, we can obtain the weight vector $\mu^1, \mu^2, \ldots, \mu^N$. Set $w^i := \mu^i, (i = 1, 2, \ldots N)$, and go to Step2.

Step 4 **Stopping criterion:**

If the stopping criterion is met, stop and output the best individual in each sub-problem; else set $gen = gen + 1$, go to step2.

Here, the IGD (Inverted Generational Distance) index is used to measure the convergence and distribution of obtained Pareto optimal solutions. For the sake of fairness, the four algorithms, i.e. MOEA/D-εEOD, MOEA/D-M2 M, UMOEA/D and NSGA2-CE, adopt the same evolution operator for each test function, which are summarized as follows. The simulated binary crossover and polynomial mutation are used, and the probability of crossover and mutation are respectively 0.9 and 1/d, where d is the number of decision variables; the distribution indexes for crossover and mutation are 10 and 20, respectively. The number of weight vectors is set to 200, and the size of neighborhood of each weight vector is set to 20. In each sub-population, the number of individuals is set to 5. The stopping condition of an algorithm is a predefined number of generations, which is set to 1000 in this work, and the results of each algorithm are the average value of 20 times independent runs on each test problem. The value of ε and $iter$ are set to 0.01 and 100, and the adaptive weight parameter $[\alpha_1, \alpha_2]$ is set to [0.5, 0.5] in DTLZ5(I,M) and [0.01, 0.99] in other test cases.

4.2 Experimental Studies on MOEA/D-εEOD and Comparison

This part of experiment is designed to study the effectiveness of MOEA/D-εEOD on different types of MAPs. Table 1 presents the mean and standard deviation of the IGD-metric values of the final non-dominated solutions obtained by MOEA/D-εEOD, MOEA/D-M2 M, UMOEA/D and NSGA2-GE over 20 independent runs, where the best metric values are highlighted in bold face. The t-test with significance level $\gamma = 0.05$ is used to study the significance of difference between the results achieved by the MOEA/D-εEOD and other algorithms for comparison, and the results are also shown in Table 1. The symbols +, −, = in each column respectively represent the performance of the proposed MOEA/D-εEOD is better than, worse than and equal to the compared algorithms.

Table 1. Statistic IGD-metric values obtained by AWVD-MOEAD and the other algorithms on 6 benchmark problems with different number of objectives

Problem	Obj	MOEA/D-*εEOD*		MOEA/D-M2M			UMOEA/D			NSGAII-CE		
		mean	std	mean	std	t	mean	std	t	mean	std	t
DTLZ1	5	**0.0405**	**0.0029**	0.0642	0.0022	+	0.0692	0.0026	+	0.1359	0.0190	+
	10	**0.1169**	**0.0314**	0.1299	0.0085	+	0.1739	0.0169	+	0.1209	0.0221	+
	15	**0.0642**	**0.0096**	0.1573	0.0112	+	0.2118	0.0032	+	0.1379	0.0332	+
	20	**0.1404**	**0.0397**	0.3627	0.0602	+	0.3467	0.0029	+	0.1652	0.0374	+
F1(5)	5	**0.0859**	**0.0020**	0.1490	0.0035	+	0.1772	0.0047	+	1.1512	0.1068	+
F1(10)	10	**0.2771**	**0.0553**	0.3552	0.0374	+	0.7074	0.0145	+	1.3108	0.1522	+
F1(15)	15	**0.4004**	**0.0751**	0.5113	0.0096	+	0.7445	0.0014	+	0.7438	0.0027	+
F2(5)	5	0.1453	0.0009	**0.1250**	**0.0066**	−	0.1049	0.0027	+	0.4860	0.1008	+
F2(10)	10	0.4603	0.0135	**0.2347**	**0.0239**	−	0.4083	0.0071	+	0.5325	0.0446	+
F2(15)	15	0.5037	0.0816	**0.3780**	**0.0572**	−	0.4429	0.0027	+	0.4887	0.0588	+
DTLZ5(3,5)	5	**0.0608**	**0.0033**	0.0720	0.0039	+	0.2896	0.0598	+	0.3245	0.0167	+
DTLZ5(2,10)	10	**0.1743**	**0.0156**	0.2062	0.0028	+	0.5643+	0.0855	+	0.7512	0.1132	+

From the experimental results, we can see that MOEA/D-εEOD shows better performance than the other three EMO algorithms in DTLZ1, F1 and DTLZ5(I,M) with all 5 to 20 objective cases.

5 Conclusion

In this paper, we focus on a new many objective evolutionary algorithm, which combines dominance and decomposition-based approaches. In particular, we propose a MOEA/D-εEOD, which uses an adaptive weight vector design method that can adjust the weight vector setting dynamically according to the current distribution of the non-dominated solution set. Furthermore, we develop a new dominance relation ε efficient order instead of scalarizing function to realize selection and update operation in each sub-problem, which can increase the selective pressure and improve the distribution of individuals. The performance of MOEA/D-εEOD has been investigated on a set of benchmark problems with up to 20 objectives. The empirical results demonstrate that the proposed MOEA/D-εEOD is able to find a well-converged and well-distributed set of points for most test cases.

Acknowledgement. This work was supported by the special scientific research project fund of education department of Shaanxi Province (16JK1381) and the National Natural Science Foundations of China (No. 61472297). The authors would like to thank the anonymous reviewers for their valuable comments and suggestions.

References

1. Purshouse, R.C.: Evolutionary many-objective optimization: An exploratory analysis. In: The 2003 Congress on Evolutionary Computation, pp. 2066–2073 (2003)
2. Ishibuchi, H., Tsukamoto, N., Nojima, Y.: Evolutionary many-objective optimization: A short review. In: The Congress of Evolutionary Algorithm, pp. 2424–2431 (2008)
3. Dai, C., Wang, Y., Ye, M.: A new evolutionary algorithm based on contraction method for many-objective optimization problems. Appl. Math. Comput. **245**, 191–205 (2014)
4. Zhang, Q., Li, H.: MOEA/D: A multi-objective evolutionary algorithm based on decomposition. IEEE Trans. Evol. Comput. **11**(6), 712–731 (2007)
5. Yutao, Qi: MOEA/D with adaptive weight adjustment. Evol. Comput. **22**(2), 231–264 (2014)
6. Hernández-Díaz, A.G., Santana-Quintero, L.V., Coello, C.C.A.: Pareto-adaptive epsilon-dominance. Evol. Comput. 15(4), 493–517 (2007)
7. Giagkiozis, I., Purshouse, R.C., Fleming, P.J.: Generalized decomposition and cross entropy methods for many-objective optimization. Inf. Sci. **282**, 363–387 (2014)
8. Deb, K., Thiele, L., Laumanns, M., et al.: Scalable multi-objective optimization test problems. In: IEEE Congress on Evolutionary Computation, pp. 825–830 (2002)
9. Liu, H., Gu, F., Zhang, Q.: Decomposition of a multi-objective optimization problems into a number of simple multi-objective sub-problems. IEEE Trans. Evol. Comput. **18**(3), 450–455 (2014)
10. Tan, Y., Jiao, Y., Li, H.: MOEA/D + uniform design A new version of MOEA/D for optimization problems with many objectives. Comput. Oper. Res. **40**(6), 1648–1660 (2013)

A Large Scale Multi-objective Ontology Matching Framework

Xingsi Xue[1,2(✉)] and Aihong Ren[3]

[1] College of Information Science and Engineering,
Fujian University of Technology, Fuzhou, Fujian, China
jack8375@gmail.com

[2] Fujian Provincial Key Laboratory of Big Data Mining and Applications,
Fujian University of Technology, Fuzhou, Fujian, China

[3] Department of Mathematics, Baoji University of Arts and Sciences,
Baoji, Shaanxi, China

Abstract. Multi-Objective Evolutionary Algorithm (MOEA) is emerging as a state-of-the-art methodology to solve the ontology meta-matching problem. However, the huge search scale of large scale ontology matching problem stops MOEA based ontology matching technology from correctly and completely identifying the semantic correspondences. To this end, in this paper, a large scale multi-objective ontology matching framework is proposed, which works with three sequential steps: (1) partition the large scale ontologies into similar ontology segment pairs; (2) utilize MOEA to match the similar ontology segments in parallel; (3) select the representative ontology segment alignments, which are further aggregated to obtain the final ontology alignment. In addition, a novel multi-objective model is also constructed for ontology matching problem and the MOEA and entity similarity measure that could be used in this framework are also recommended. The experimental result shows the effectiveness of our proposal.

Keywords: Multi-Objective Evolutionary Algorithm · Large scale ontology matching · Ontology partition

1 Introduction

Since its beginning, Evolutionary Algorithm (EA) is appearing as the most suitable methodology to address the ontology matching problem [5]. But most EA based ontology matching technologies use a single objective to evaluate the alignment quality during the generation process, even though a suitable computation of parameters could be better performed by evaluating the right compromise among different objectives involved in the matching process. Thus, Multi-Objective Evolutionary Algorithm (MOEA) is emerging as a state-of-the-art methodology to solve the ontology meta-matching problem. However, the huge search scale of large scale ontology matching problem stops MOEA based

J.-S. Pan et al. (eds.), *Advances in Intelligent Information Hiding and Multimedia Signal Processing*, Smart Innovation, Systems and Technologies 81,
DOI 10.1007/978-3-319-63856-0_31

ontology matching technology from correctly and completely identifying the semantic correspondences. To this end, in this paper, a large scale multi-objective ontology matching framework is presented, which utilizes the ontology partition algorithm to reduce the search space of MOEA and works without the reference alignment.

The rest of the paper is organized as follows: Sect. 2 formulates the multi-objective ontology matching problem and the profile-based similarity measure; Sect. 3 presents the details of large scale multi-objective ontology matching framework; Sect. 4 shows the experimental result; and Sect. 5 draws the conclusions.

2 Preliminaries

2.1 Multi-objective Ontology Matching

In this work, an ontology is defined as three-tuple (C, P, I) [12], where C, P, I are respectively referred to the set of classes, properties and instances. In addition, an ontology alignment A between two ontologies is a correspondence set and each correspondence inside is a 3-tuples (e, e', n), where e and e' are respectively the entities of two ontologies, n is the similarity score between them, and the relation of the correspondence is the equivalence. Supposing one entity in source ontology is matched with only one entity in target ontology and vice versa, based on the observations that the more correspondences found and the higher mean similarity score of the alignment is, the better the alignment quality is [1], the multi-objective optimal model of ontology matching problem can be defined as follows:

$$\begin{cases} max \quad (MF(X), avg(X)) \\ s.t. \quad X = (x_1, x_2, \cdots, x_{|O_1|})^T \\ \qquad\quad x_i \in \{1, 2, \cdots, |O_2|\}, i = 1, 2, \cdots, |O_1| \end{cases} \tag{1}$$

where $MF()$ and $avg()$ are respectively the functions of calculating the alignment X's MatchFmeasure [10] and the average similarity score of X, $|O_1|$ and $|O_2|$ respectively represent the cardinalities of two ontologies O_1 and O_2, and $x_i, i = 1, 2, \cdots, |O_1|$ represents the ith correspondence.

2.2 Profile-Based Similarity Measure

The foundation of ontology matching technology is the similarity measure of ontology entities [4], which is a function that measures the degree to which objects are similar to one another. In this framework, an profile-based similarity measure is recommended to calculate the entities' similarity values. To be specific, first, for each ontology entity, a profile is constructed for it by collecting the information from itself and all its ascendants. Then, the similarity of two entities is measured based on the similarity of their profiles. With respect to the details, please see also [11].

3 Large Scale Multi-objective Ontology Matching Framework

Figure 1 shows the proposed framework, and as can be seen from the figure, our proposal consists of three phases:

- ontology partition phase aims to reduce the search space of ontology matching process by partitioning two ontologies into several similar ontology segments, and the hereafter matching process only needs to be carried out in similar ontology segments,
- MOEA based ontology matching phase utilizes MOEA to automatically determine the non-dominated ontology segment alignments in parallel,
- alignment aggregation phase selects the representative segment alignments and aggregates them into a final alignment.

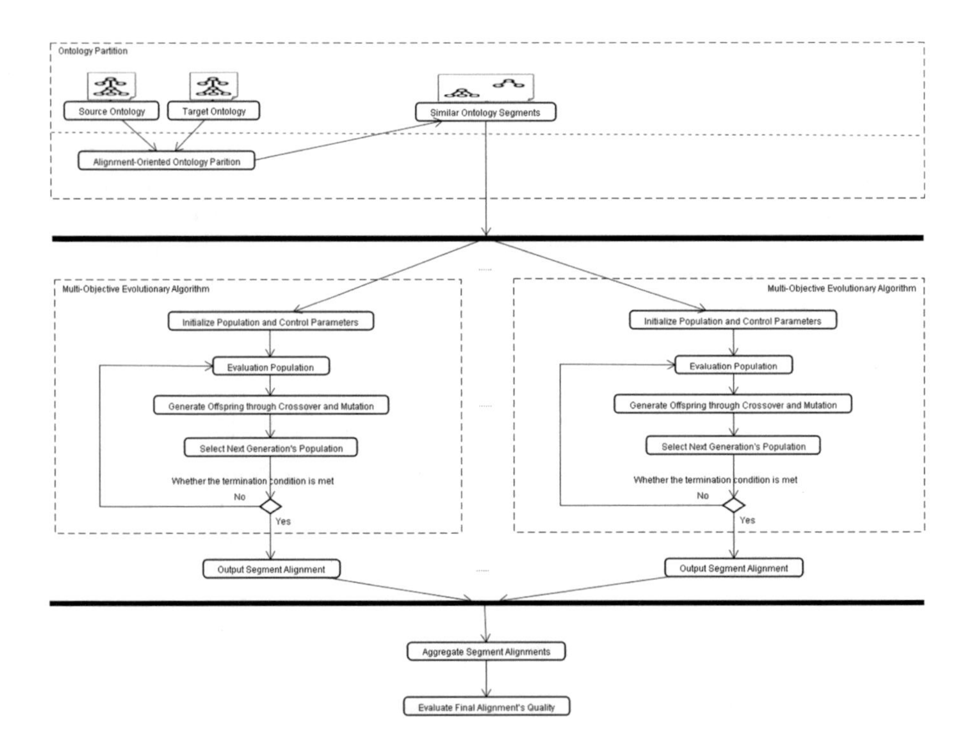

Fig. 1. Large scale multi-objective ontology matching framework

3.1 Alignment-Oriented Ontology Partition

Since it is the large scale that makes the large scale ontology matching problem very difficult to solve, it is important to reduce the scale in order to improve the efficiency and reduce unnecessary user interventions in the ontology matching process. According to [6], partitioning the ontology into various segments can effectively transform the original large scale matching problem into equivalent small segment matching problems, where the term "segment" is referred to a fragment of an ontology which stands alone on its right [8]. In this framework, an alignment-oriented ontology partition technology is recommended to pre-process the ontologies before matching. With respect to the details of the algorithm, please see also [9].

3.2 MOEA Based Ontology Matching

In this framework, NSGA-II [2] is recommended as the matching algorithm. NSGA-II is considered to be a flexible and robust technique, which is good at finding various non-dominated solutions quickly. First, the algorithm applies the standard crossover and mutation operators in the evolution of current population. Then, it uses the fast non-dominated sorting technique and a crowding distance to rank and select the next generation. Finally, the best individuals in terms of non-dominance and diversity are selected as the solutions. More details about the NSGA-II based ontology matching technology, please see also [12].

3.3 Segment Alignments Aggregation

In this framework, all the similar ontology segment pairs are matched in parallel. After obtaining each non-dominated segment alignments, the one with the best MatchFmeasure is selected as the representative solution, and aggregate all the representative segment alignments to obtain the final ontology alignment. Finally, the final ontology alignment is evaluated through f-measure for the performance evaluation.

4 Experiment

In order to study the effectiveness of our proposal, we have exploited a well-known dataset, named benchmark track, provided by the Ontology Alignment Evaluation Initiative (OAEI) [3] which are commonly used for experimentation about ontology alignment problem. In detail, each test case consists of a set of ontologies which are built around a seed ontology and many variations of it. Variations are artificially generated, and focus on the characterization of the behavior of the tools rather than having them compete on real-life problems. They are organized in three groups: Simple tests (1xx) compares the reference ontology with itself; Systematic tests (2xx) are obtained by discarding/modifying features, which include names of entities, comments, the specialization hierarchy, instances, properties and classes, from the reference ontology; Real-life ontologies (3xx) are found on the web. In this experiment, we utilize the downloadable datasets from the OAEI official website for testing purposes.

Table 1. Comparison of our proposal with OAEI participants on benchmark track

Ontology matching system	Recall	Precision	f-measure
AML	0.39	0.92	0.55
AOT	0.53	0.80	0.64
AOTL	0.53	0.85	0.65
LogMap	0.40	0.40	0.40
LogMap-C	0.40	0.42	0.41
LogMapLite	0.50	0.43	0.46
MaasMatch	0.39	0.97	0.56
OMReasoner	0.50	0.73	0.59
RSDLWB	0.50	0.99	0.66
XMap2	0.40	1.00	0.57
Our proposal	0.66	0.81	0.73

In order to compare the quality of our proposal with other approaches, we evaluate the obtained alignments with traditional recall, precision and f-measure [7], and the results in Table 1 are the mean values of all the test cases. As can be seen from Table 1 that the mean f-measure of the alignments obtained by our approach outperforms all other ontology matching systems, which shows the effectiveness of our proposal.

5 Conclusion

To improve the efficiency of MOEA based ontology matching technology, in this paper, a large scale multi-objective ontology matching framework is proposed, which works with three sequential steps: (1) partition the large scale ontologies into similar ontology segment pairs; (2) utilize MOEA to match these similar ontology segment pairs in parallel; (3) select the representative ontology segment alignments, which are further aggregated to obtain the final ontology alignment. In addition, a novel multi-objective model is also constructed for ontology matching problem and the MOEA and entity similarity measure that could be used in this framework are also recommended. The experimental result shows the effectiveness of our proposal.

Acknowledgment. This work is supported by the National Natural Science Foundation of China (No. 61503082), Natural Science Foundation of Fujian Province (No. 2016J05145), Scientific Research Startup Foundation of Fujian University of Technology (No. GY-Z15007), Fujian Province outstanding Young Scientific Researcher Training Project (No. GY-Z160149) and China Scholarship Council.

References

1. Bock, J., Hettenhausen, J.: Discrete particle swarm optimisation for ontology alignment. Inf. Sci. **192**, 152–173 (2012)
2. Deb, K., Pratap, A., Agarwal, S., Meyarivan, T.: A fast and elitist multiobjective genetic algorithm: Nsga-II. IEEE Trans. Evol. Comput. **6**(2), 182–197 (2002)
3. Dragisic, Z., Eckert, K., Euzenat, J., Faria, D., Ferrara, A., Granada, R., Ivanova, V., Jiménez-Ruiz, E., Kempf, A.O., Lambrix, P., et al.: Results of the ontology alignment evaluation initiative 2014. In: Proceedings of the 9th International Conference on Ontology Matching, vol. 1317, pp. 61–104. CEUR-WS. org (2014)
4. Maedche, A., Staab, S.: Measuring similarity between ontologies. In: Proceedings of the 14th International Conference on Knowledge Engineering and Knowledge Management, Ischia Island, Italy, pp. 251–263, July 2002
5. Martinez-Gil, J., Montes, J.F.A.: Evaluation of two heuristic approaches to solve the ontology meta-matching problem. Knowl. Inf. Syst. **26**(2), 225–247 (2011)
6. Rahm, E.: Towards large-scale schema and ontology matching. In: Bellahsene, Z., Bonifati, A., Rahm, E. (eds.) Schema Matching and Mapping, pp. 3–27. Springer, Heidelberg (2011)
7. Rijsberge, C.J.V.: Information retrieval. University of Glasgow, Butterworth, London (1975)
8. Seidenberg, J., Rector, A.: Web ontology segmentation: analysis classification and use. In: Proceedings of the 15th International Conference on World Wide Web, Edinburgh, Scotland UK, pp. 13–22, May 2006
9. Xue, X., Pan, J.: A segment-based approach for large-scale ontology matching. Knowl. Inf. Syst., 1–18 (2017)
10. Xue, X., Wang, Y.: Optimizing ontology alignments through a memetic algorithm using both matchfmeasure and unanimous improvement ratio. Artif. Intell. **223**, 65–81 (2015)
11. Xue, X., Wang, Y.: Using memetic algorithm for instance coreference resolution. IEEE Trans. Knowl. Data Eng. **28**(2), 580–591 (2016)
12. Xue, X., Wang, Y., Hao, W.: Optimizing ontology alignments by using NSGA-II. Int. Arab J. Inf. Technol. **12**(2), 175–181 (2015)

A New Evolutionary Algorithm with Deleting and Jumping Strategies for Global Optimization

Fei Wei[1(✉)], Shugang Li[2], and Le Gao[1]

[1] College of Sciences, Xi'an University of Science and Technology, Xi'an 710054, China
feiweixjf@gmail.com

[2] College of Safety Science and Engineering, Xi'an University of Science and Technology, Xi'an 710054, China

Abstract. For global optimization problems with a large number of local optimal solutions, evolutionary algorithms are efficient parallel algorithms, but they drops into local optimum easily, therefore their efficiency and effectiveness will be much reduced. In this paper, first, a new deleting strategy is proposed that can eliminate all local optimal solutions no better than this obtained local optimal solution. Second, when algorithm drops into a local optimal solution, a new jumping strategy is proposed that can jump out of the current local optimal solution and then find a better local optimal solution. Based on the above, a new algorithm called evolutionary algorithm with deleting and jumping strategies (briefly, EADJ) is proposed, and the algorithm convergence is proved theoretically. The simulations are made on 25 standard benchmark problems, and the results indicate the proposed deleting strategy and jumping strategy are effective; further, the proposed algorithm is compared with some well performed existing algorithms, and the results indicate the proposed algorithm EADJ is more effective and efficient.

Keywords: Evolutionary algorithm · Global optimization · Deleting strategy · Jumping strategy

1 Introduction

Global optimization problems with numerous local and global optima have arisen in many fields such as computer science, engineering design, and decision making. For solving global optimization problems, in recent years, many new theoretical and computational contributions have been reported. The research works on global optimization problems are mainly concerned with the efficient algorithms for the multi-modal functions. In general, the existing approaches can be classified into two categories: deterministic methods (e.g., [1–4]) and probabilistic methods (e.g., [5–9]). The typical examples of the former are filled function methods (FFM) [1, 2], trajectory methods [3], tunneling methods [4], whereas ones of the latter are clustering methods [5], evolutionary algorithms (EA) [6–8], and the simulated annealing methods [9], where evolutionary algorithms are one of the most efficient and popular algorithms.

J.-S. Pan et al. (eds.), *Advances in Intelligent Information Hiding and Multimedia Signal Processing*, Smart Innovation, Systems and Technologies 81,
DOI 10.1007/978-3-319-63856-0_32

However, for EAs, the major challenge is that EAs may be trapped in the local optima. Many researchers have tried to combine some techniques into EAs. Recently, researchers have successfully combined experimental design methods into EAs. For example, Li and Smith [11] used Latin squares to improve EAs. Wang and Dang used Latin squares and level set to enhance EAs [12].

In this paper, a new proposed deleting strategy can eliminate all such local optimal solutions no better than the best solution found so far, and can keep all better local optimal solutions than the best solution found so far unchanged. The other proposed jumping strategy can search a better local optimal solution. Thus, if we can combine EAs with deleting and jumping strategies properly, not only many local optimal solutions can be eliminated, but also the process of finding the global optimization solution will be much fast. Based on this motivation, in this paper, a new algorithm called evolutionary algorithm with deleting and jumping strategies (briefly, EADJ) is proposed. The simulations are made on 25 famous problems in CEC'2005 standard benchmark problems and the performance of the proposed algorithm is compared with that of some well performed existing algorithms. The results indicate the proposed algorithm is more efficient.

2 Basic Concepts and Some Notations

In this paper, we consider the following global optimization problem:

$$(\mathrm{P}) \quad \begin{cases} \min & f(x) \\ \text{s.t.} & x \in R^n \end{cases} \tag{1}$$

where $f(x) : R^n \to R$. Suppose $f(x)$ satisfies the condition $f(x) \to +\infty$ as $\|x\| \to +\infty$. Then there exists a closed bounded domain Ω called operating region that contains all local minima of $f(x)$. Then the global optimization problem (P) can be rewritten into an equivalent form as follows.

$$(\mathrm{P1}) \quad \begin{cases} \min & f(x) \\ \text{s.t.} & x \in \Omega = [l, u] = \{x | l \le x \le u, l, u \in R^n\}. \end{cases} \tag{2}$$

Because Ω can be estimated before problem (P) is solved, we can assume that Ω is known. We only consider problem (P1) in the following, and adopt the following symbols.

k: the iteration number;
x_k^*: the local minimizer of the objective function in the k-th iteration;
f_k^*: the function value at x_k^*;
x^*: the global minimizer of the objective function.

Assumption 2.1. The function $f(x)$ in (P1) has only a finite number of minimizers in Ω, and therefore every minimizer is isolated.

In the following, we first design a deleting technique. The details are as follows.

3 A New Deleting Strategy

The existence of multiple local minima makes global optimization become a great challenge. For the global optimization problem, the key issue is handling a large number of local optimal solutions and finding the global optimal solution as fast as possible. In order to tackle this problem, a deleting local optimum function is designed as follows:

$$S(x, x_k^*) = f(x_k^*) + g\big(f(x) - f(x_k^*)\big),$$
$$g(t) = \begin{cases} r_1, & t \geq 0, \\ r_2 \cdot t^3, & t < 0. \end{cases} \tag{3}$$

where $r_1 \geq 0$ is an adjustable real number, and $r_2 > 0$ is an adjustable positive real number. On one hand, this deleting function $S(x, x_k^*)$ will eliminate all local optimal solutions no better than the current local optimal solution x_k^*, i.e., for $\forall x \in \Omega$, if $f(x) \geq f(x_k^*)$, then $S(x, x_k^*) = f(x_k^*) + r_1$; on the other hand, $S(x, x_k^*)$ will keep any better local optimal solution than x_k^* of $f(x)$ unchanged.

The deleting function eliminates many local optimal solutions, but keep the better local optimal solutions unchanged. However, the deleting function often loses the descent direction around x_k^*. To overcome this shortcoming, we design a new jumping strategy to handle this issue. The proposed jumping strategy can be constructed as follows.

4 A New Jumping Strategy

In this section, we propose a new jumping function at a local minimizer x_k^* as follows:

$$F(x, x_k^*) = \arctan(-\|x - x_k^*\|^2) + q\big(f(x) - f(x_k^*)\big),$$
$$q(t) = \begin{cases} 0, & t \geq 0, \\ -r_3 \cdot t^2, & t < 0. \end{cases} \tag{4}$$

where r_3 is an adjustable positive real number.

Note that the proposed jumping function has the following advantages: first, it has one parameter r_3 which is a positive real number and is easy to adjust; second, $\arctan\left(-\|x - x_k^*\|^2\right) \in \left(-\frac{\pi}{2}, 0\right]$ is bounded, which ensures that the calculation of $F(x, x_k^*)$ will not overflow and is of numerical stability; third, $\arctan\left(-\|x - x_k^*\|^2\right) \in \left(-\frac{\pi}{2}, 0\right]$ makes the algorithm have the descent direction, and its value range is small so that the algorithm find the global optimum much easier.

In the following section, we will design a new algorithm: A new evolutionary algorithm with deleting and jumping strategies (EADJ).

5 A New Evolutionary Algorithm with Deleting and Jumping Strategies: EADJ

In EADJ, one most popular algorithms SaNSDE [13] is used directly in the part of evolutionary algorithm. Based on all considerations above, A new evolutionary algorithm with deleting and jumping strategies (EADJ) is proposed as follows.

Algorithm 5.1 (EADJ)

Step 1. **Initialization Step**
Let k = 0. Choose three positive real numbers r_1, r_2,r_3, and the population size is N. Generate $\lceil N/2 \rceil$ points by using a uniform design method in [14], and $N - \lceil N/2 \rceil$ points randomly, and put them into the initial population $POP(k)$.

Step 2. Evolve $POP(k)$ by SaNSDE on original function $f(x)$ for one generation and get set $OFF(k)$ of all offspring. Select the best individual among $POP(k) \cup OFF(k)$.
This best individual is denoted as x_k. Execute the local search at x_k on $f(x)$ to get a local minimizer x_k^*, then go to step 7.

Step 3. **Eliminate all local minima no better than** x_k^*: Construct a deleting function $S(x, x_k^*)$ at x_k^* by Eq. (3).

Step 4. **Escape from the minimizer** x_k^* **via jumping function:**
Construct a jumping function at x_k^* by Eq. (4)

$$F(x, x_k^*) = \arctan(-\|x - x_k^*\|^2) + q(\mathrm{S}(x, x_k^*) - f(x_k^*)), \tag{5}$$

$$q(t) = \begin{cases} 0, & t \geq 0, \\ -r_3 \cdot t^2, & t < 0. \end{cases}$$

Step 5. Evolve $POP(k)$ by SaNSDE on jumping function $F(x, x_k^*)$ for one generation and get $POP(k+1)$ and set $OFF(k)$ of all offspring, and select the best individual of the jumping function among $POP(k) \cup OFF(k)$. This best individual is denoted as y_k. Execute the local search at y_k on $F(x, x_k^*)$ by Algorithm1 to get a local minimizer y_k^* of the jumping function.

Step 6. **Update the current best solution**: Starting from y_k^*, do the local search on $f(x)$ to obtain a local minimizer x_{k+1}^* of $f(x)$. Let $k = k+1$, go to Step 7.

Step 7. If the termination condition is satisfied, $x^* = x_k^*$ is taken as a global minimizer of $f(x)$, stop; otherwise, go to Step 3.

6 Numerical Experiments

6.1 Test Problems and Parameters Setting for EADJ

- In this section, the proposed algorithm is tested on 25 benchmark functions proposed in CEC'2005, the detailed description of these benchmarks can be found in [15].
- The proposed algorithm EADJ is executed for 25 independent runs on each test problem. In experiments, EADJ was tested on an Intel(R) Core(TM) i7 CPU 870 with 2.93 GHz in Matlab R2012a.
- Population size: $NP = 100$, r_1=100, r_2=100,r_3=50.
- Parameters in Algorithm EADJ: $\varepsilon = 1.0\text{e} - 10$.

6.2 The Simulation Results

The results of EADJ are compared with those of SaNSDE, NSDE, DE and CMAES in [16–19], respectively. These results are given in Tables 1 and 2, and the best results are shown in bold. The symbols used in Tables are as follows:

No.: the number of the test problems;
f-mean: the mean function value in 25 runs;
FEs: The maximum number of fitness evaluations.

(a) **Comparison between EADJ and SaNSDE**

To demonstrate the effect of the deleting function and jumping function on evolutionary algorithm, we compare the experimental results of EADJ and SaNSDE, and the results are listed in Table 1. To be consistent with the SaNSDE experimental setting in reference [13], we only use the first 14 functions in our experiments. The number of evolution generations is set to 3000 for all functions.

Functions F1-F5 are unimodal. Both EADJ and SaNSDE can find the optimal solution on Shifted Sphere function F1. For functions F2-F5, EADJ outperformed SaNSDE.

Functions F6-F14 are multimodal. Both EADJ and SaNSDE can find the optimal solution on the function F9. EADJ outperforms SaNSDE than all the others. These experimental results show that the proposed algorithm is more efficient and effective than SaNSDE. This also indicates the deleting and jumping functions can greatly enhance evolutionary algorithms.

(b) **Comparison between EADJ and other algorithms**

To evaluate EADJ further, we compare EADJ with NSDE, DE and CMAES[16–19]. The results of 25 independent runs are summarized in Table 2.

Table 1. Comparison between different algorithms with 3000 generations on 30 dimension problems.

No.	EADJ	SaNSDE	No.	EADJ	SaNSDE
F1	**0.00e + 00**	**0.00e + 00**	F8	**2.08e + 00**	2.09e + 01
F2	**7.41e-18**	5.68e-14	F9	**0.00e + 00**	**0.00e + 00**
F3	**1.03e + 04**	5.43e + 04	F10	**2.05e + 00**	4.21e + 01
F4	**1.26e-06**	1.22e-04	F11	**8.40e + 00**	1.02e + 01
F5	**1.11e-02**	2.45e-01	F12	**2.31e + 02**	4.06e + 04
F6	**1.55e-12**	1.59e-01	F13	**1.89e + 00**	2.12e + 00
F7	**0.00e + 00**	8.57e-03	F14	**1.20e + 00**	1.27e + 01

Table 2. *f-mean* of different algorithms over 30 dimensions (averaged over 25 runs, and FEs = 3.0e + 05).

No.	EADJ	NSDE	DE	CMAES
F1	**0.00e + 00**	**0.00e + 00**	**0.00e + 00**	5.28e-09
F2	**2.06e-28**	5.62e-08	3.33e-02	6.93e-09
F3	**2.56e-15**	6.40e + 05	6.92e + 05	5.18e-09
F4	**1.26e-28**	9.02e + 00	1.52e + 01	5.18e-09
F5	1.36e-08	1.56e + 03	1.70e + 02	**8.30e-09**
F6	**7.13e-26**	2.45e + 01	2.51e + 01	6.31e-09
F7	2.22e-02	1.18e-02	2.96e-03	**6.48e-09**
F8	2.03e + 01	2.09e + 01	2.10e + 01	**2.00e + 01**
F9	**0.00e + 00**	7.96e-02	1.85e + 01	2.91e + 02
F10	**3.88e + 01**	4.29e + 01	9.69e + 01	5.63e + 02
F11	2.07e + 01	**1.41e + 01**	3.42e + 01	1.52e + 01
F12	**9.49e + 01**	6.59e + 03	2.75e + 03	1.32e + 04
F13	**4.94e-01**	1.62e + 00	3.23e + 00	2.32e + 00
F14	**1.20e + 01**	1.32e + 01	1.34e + 01	1.40e + 01
F15	4.00e + 02	3.64e + 02	3.60e + 02	**2.16e + 02**
F16	6.15e + 01	6.90e + 01	2.12e + 02	**5.84e + 01**
F17	**7.74e + 01**	1.01e + 02	2.37e + 02	1.07e + 03
F18	**9.10e + 01**	9.04e + 02	9.04e + 02	8.90e + 02
F19	**9.12e + 01**	9.04e + 02	9.04e + 02	9.03e + 02
F20	**3.12e + 02**	9.04e + 02	9.04e + 02	8.89e + 02
F21	**4.83e + 02**	5.00e + 02	5.00e + 02	4.85e + 02
F22	**8.54e + 02**	8.89e + 02	8.97e + 02	8.71e + 02
F23	**5.34e + 02**	**5.34e + 02**	**5.34e + 02**	5.35e + 02
F24	**2.00e + 02**	**2.00e + 02**	**2.00e + 02**	1.41e + 03
F25	**2.00e + 02**	**2.00e + 02**	7.30e + 02	6.91e + 02

For unimodal functions F1-F5, three algorithms (EADJ, NSDE and DE) can find the global optimal solution on Shifted Sphere function F1. EADJ performed far better than all the other algorithms on F2-F4. EADJ performed better than NSDE and DE on F5, but was slightly weaker than CMAES. In conclusion, for unimodal functions, the performance of EADJ is optimal.

For basic and expanded multimodal functions F6-F14, EADJ outperformed all others on almost all functions except F7, F8 and F11. For F7, EADJ performs similar to NSDE (the same order of magnitude), and performs a little bit worse than DE. For F8, EADJ outperforms NSDE and DE, and performs similar to CMAES (the same order of magnitude). For F11, EADJ outperforms DE, and performs similar to NSDE and CMAES (the same order of magnitude). Therefore, EADJ is more efficient than the other algorithms for basic and expanded multimodal functions.

For hybrid composition functions F15-F25, these functions are much more difficult. In Table 2, for F15, EADJ performs almost same as the compared algorithms. For F16, EADJ outperforms NSDE and DE, and performs a little bit worse than CMAES (in fact, they perform almost the same, and the analysis results are of the same order of magnitude). For F17-F25, EADJ outperforms all the compared algorithms. This also indicates that EADJ is still suitable to these composition multimodal functions, and is more efficient than the other algorithms.

7 Conclusions

In this paper, A new global optimization algorithm called evolutionary algorithm with deleting and jumping strategies (EADJ) has been proposed. The deleting strategy could filter a lot of local optimal solutions. The jumping strategy could help EAs to jump one local optimal solution and find another better one. The experiment results also indicated that EADJ was efficient for problems with a lot of local optimal solutions and more efficient than the compared algorithms.

Acknowledgement. This work was supported by the National Natural Science Foundation of China (No. U1404622) and the Cultivation Fund of Xi'an University of Science and Technology (No. 201644).

References

1. Ge, R.: A filled function method for finding a global minimizer of a function of several variables. Math. Program. **46**, 191–204 (1990)
2. Lin, H.W., Wang, Y.P., Fan, L., Gao, Y.L.: A new discrete filled function method for finding global minimizer of the integer programming. Appl. Math. Comput. **219**(9), 4371–4378 (2013)
3. Branin Jr., F.H.: Widely convergent method for finding multiple solutions of simultaneous nonlinear equations. IBM J. Res. Dev. **16**, 504–522 (1972)
4. Levy, A., Montalvo, A.: The tunneling algorithm for the global minimization of functions. SIAM J. Sci. Stat. Comput. **6**, 15–29 (1985)

5. Bai, L., Liang, J., Dang, C., Cao, F.: A cluster centers initialization method for clustering categorical data. Expert Syst. Appl. **39**, 8022–8029 (2012)
6. Lin, H.W., Gao, Y.L., Wang, Y.P.: A continuously differentiable filled function method for global optimization. Numerical Algorithms **66**(3), 511–523 (2014)
7. Dai, C., Wang, Y.P.: A new uniform evolutionary algorithm based on decomposition and CDAS for many-objective optimization. Knowl. Based Syst. **85**, 131–142 (2015)
8. Ren, A.H., Wang, Y.P.: Optimistic Stackelberg solutions to bilevel linear programming with fuzzy random variable coefficients. Knowl. Based Syst. **67**, 206–217 (2014)
9. Dang, C., Ma, W., Liang, J.: A deterministic annealing algorithm for approximating a solution of the min-bisection problem. Neural Netw. **22**, 58–66 (2009)
10. Liang, J., Qin, A., Suganthan, P.N., Baskar, S.: Comprehensive learning particle swarm optimizer for global optimization of multimodal functions. IEEE Trans. Evol. Comput. **10**, 281–295 (2006)
11. Richter, H.: Evolutionary Algorithms and Chaotic Systems (2010). Springer-Verlag Berlin and Heidelberg GmbH & Co. KG, Heidelberg (2010). ISBN 9783642107061
12. Wang, Y., Dang, C.: An evolutionary algorithm for global optimization based on level-set evolution and latin squares. IEEE Trans. Evol. Comput. **11**, 579–595 (2007)
13. Yang, Z., Tang, K., Yao, X.: Self-adaptive differential evolution with neighborhood search. In: IEEE Congress on Evolutionary Computation, pp. 1110–1116 (2008)
14. Fang, K., Wang, Y.: Number-Theoretic Methods in Statistics. Chapman & Hall, London (1994)
15. Suganthan, P.N., Hansen, N., Liang, J.J., Deb, K., Chen, Y.P., Auger, A., Tiwari, S: Problem definitions and evaluation criteria for the CEC 2005 special session on real-parameter optimization. Technical report, Nanyang Technological University, Singapore (2005)
16. Yang, Z., Yao, X., He, J.: Making a difference to differential evolution. In: Siarry, P., Michalewicz, Z. (eds.) Advances in Metaheuristics for Hard Optimization, pp. 397–414. Springer, Heidelberg (2008)
17. Ronkkonen, J., Kukkonen, S., Price, K.V.: Real-parameter optimization with differential evolution. In: IEEE Congress on Evolutionary Computation, vol. 1, pp. 506–513 (2005)
18. Auger, A., Hansen, N.: Performance evaluation of an advanced local search evolutionary algorithm. In: IEEE Congress on Evolutionary Computation, vol. 2, pp. 1777–1784 (2005)
19. Nikolaus, H.: Compilation of results on the CEC benchmark function set (2005). http://www.ntu.edu.sg/home/EPNSugan

Estimation of River Water Temperature from Air Temperature: Using Least Square Method

Heng Ouyang[1(✉)], Xingsi Xue[2], Zongxin Qiu[3], and Yongsheng Lu[3]

[1] Department of Civil Engineering, Fujian University of Technology, Fuzhou 350108, Fujian, China
heng.ouyang@fjut.edu.cn

[2] School of Information Science and Engineering, Fujian University of Technology, Fuzhou 350108, Fujian, China

[3] Fuzhou Investigation and Surveying Institute, Fuzhou 350003, Fujian, China

Abstract. The water temperature plays a major role in application of Water-Source Heat Pump (WSHP) system. The water temperature is directly related to the energy efficiency of heat pump unit and economy of WSHP system. As the air temperature is the most influential factor in the water temperature in all meteorological parameters, the effect of air temperature on the changing water temperature should be considered first when estimating the river water temperature. In this paper, we dedicate to study the changing rule of water temperature with the air temperature by analyzing over one year water temperature data. Particularly, the least square method is used for the regression analysis of 1st-order to 6th-order curve, and a fitting curve is educed, based on Cubic model of the air temperature and the average of the corresponding water temperature. The 88% of the difference values between average water temperatures and fitted water temperatures are in the range of −1.5–1.5 °C. Therefore, it can provide a simple method to determine the daily average water temperature through analyzing the relationship between the water average temperature and air temperature by using the method of Cubic fitting. The study in this paper can be used as a reference for the design and analysis of WSHP system.

Keywords: Water-source heat pump (WSHP) · Least square method · Water temperature · Air temperature · Changing relationship

1 Introduction

In recent years, the river water-source heat pump (WSHP) system has been paid more and more attention to internationally. Compared with the air-source heat pump system, WSHP system has larger heat capacity, and less temperature fluctuation. The water temperature decides the quality and stability of the river water as cold and heat sources of the water source heat pump system. But compared with the air temperature, it is difficult to forecast and access to water temperature. If we can find out the changing relationship between the water temperature and air temperature by comparing the testing water temperature and air temperature, which is easy to get from the official weather

J.-S. Pan et al. (eds.), *Advances in Intelligent Information Hiding and Multimedia Signal Processing*, Smart Innovation, Systems and Technologies 81,
DOI 10.1007/978-3-319-63856-0_33

website, then the water temperature can be calculated by air temperature, which can be applied in the water source heat pump system to achieve the better effect with less effort.

At present, the river water temperature is mainly calculated and forecast by numerical calculation and empirical formula. Based on the analysis of the main factors influencing the water temperature, the former establishes the mathematical model according to the principle of heat and mass balance, which has certain universality and high precision, but the calculation is more complicated. The latter method is based on the measured data to establish the water temperature and air temperature regression equation to find the water temperature. The method is simple, but the accuracy is slightly worse.

Since the river is characterized by the huge amount of water and fast flow, the temperature of such water source depends largely on the water temperature. The transfer process of water temperature and local heat is slow, so the second method is adopted. This study measured the water temperature in the lower Minjiang River for over a year, further analysis of the changing relationship between water temperature and air temperature. On this basis, the least square method was used to do the regression analysis of 1st-order to 6th-order curve to find out the best one. This study is to provide a simple method to determine the water temperature. The water temperature is directly related to the energy efficiency of heat pump unit and economy of WSHP system.

2 Measurement of Water Temperatures

2.1 Ocation of Measurements

The Minjiang River is a 577 kilometers-long river in Fujian Province, southeast of China. It is the largest river in Fujian, and also an important water transport channel. Fuzhou, the capital city of Fujian province, sits on the lower Minjiang River. Although currently in Fuzhou city there are few projects using raw water in Minjiang River water as cooling and heat sources, Minjiang River WSHP system has not been popularized in Fuzhou [1–3], because of the lack of survey of water temperature of Minjiang River. This test is mainly to monitor the water temperature of the lower Minjiang River in Fuzhou. With the comprehensive consideration of Fuzhou city urban planning and the promotion of the Minjiang River WSHP project, test section is through the city north port of the Minjiang River.

2.2 Description of Measurements

The HOBO data logger (model: U20-001-02) is used to test the water temperature. The temperature measurement range is from −20 °C to 50 °C. The temperature measurement accuracy is 0.37 °C. The distinguishability is 0.1 °C. The temperature reaction time is 3.5 min. The water level range is 0–30.6 m.

The air temperature is obtained from the official weather website. The air temperature observations are made in thermometer shelters about 1.5-meter above the ground where the turf is planted. Since the thermometer maintains good ventilation and avoids direct sunlight, it is well represented. The thermometer is made of mercury, alcohol or bimetallic as the thermal expansion and contraction of the sensor. The unit of temperature is in degrees Celsius (°C).

2.3 Measuring Data

Minjiang River has a continuous flow of water bodies, in which the heat exchanges, so it is not easy to form the distribution of the temperature stratification along the depth direction. The water flow in a cross section distribution is isothermal. By measuring the water temperature of the lower Minjiang River in Fuzhou in different depths and different offshore distance monitoring, the results show that different depth is less than 0.1 °C, and the maximum temperature difference of offshore distance of maximum temperature difference is less than 0.1 °C.

The water temperature of the lower Minjiang River was monitored in Fuzhou from July 2013 to March 2015. The data show that the water temperature of the lower Minjiang River in Fuzhou is from 14.2 °C to 29.3 °C. The monthly average temperature in summer is from 28.2 °C to 29.3 °C. The monthly average temperature in winter is from 14.2 °C to 17.3 °C [4].

As is shown in Fig. 1, the water temperature is more stable than air temperature. The variation of water temperature in a day is usually within 1 °C. This is especially in the winter with obvious advantages. Testing data of December 22, 2013, for example, the water temperature in 24 h is 16–16.5 °C, which is far higher than the lowest air temperature 4 °C and highest air temperature 12 °C. Testing data of August 27, 2014, for example, the water temperature in 24 h is 27.2–28.1 °C, which is much lower than the highest air temperature 34 °C on the same day but close to the lowest air temperature 27 °C.

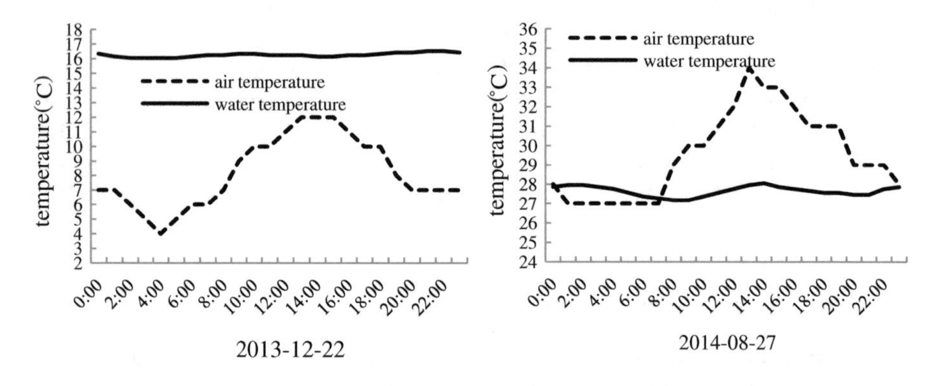

Fig. 1. The daily variation of testing water temperature and air temperature of Minjiang River

The variation of monthly average value of air temperature and water temperature of the lower Minjiang River in Fuzhou is shown in Fig. 2. From the figure, it can be find that the monthly average water temperature changes with the monthly average air temperature. From April to August, the monthly average water temperature is less than the average temperature, and the monthly average water temperature is higher than the monthly average air temperature in the rest. As can be seen from the figure, the water temperature is lower than the air temperature in summer, but higher than the air temperature in winter. This is because the water heat capacity is larger than the air, which results to the slower changes of water temperature than those of the air temperature. The above analysis shows the advantages of the water source heat pump over the air source heat pump.

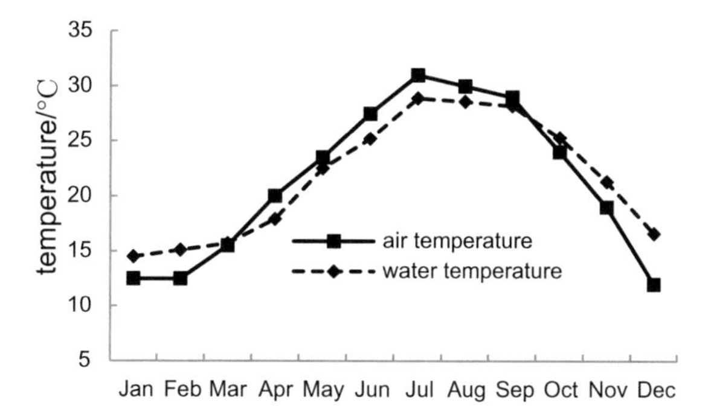

Fig. 2. The variation of monthly average value of air temperature and water temperature

3 Analyze the Relationship Between the Water Temperature and Air Temperature

The air temperature plays an important role in the water temperature. To study the changing rule of water temperature with the air temperature, the regression analysis and test are carried out between the water temperature and air temperature, which is collected from the official weather website.

3.1 The Air Temperature and the Measured Water Temperature

According to the measured data, putting aside the influence of time factor, and taking just each air temperature and its corresponding water temperature into consideration, we can see the hourly air temperature and its corresponding relationship with water temperature. The variation of the hourly water temperature with the air temperature is shown in Fig. 3. Under the same air temperature, there are multiple water temperatures, and at the same time, there are multiple air temperatures under the same water temperature. There is a nonlinear relationship between the water temperature and the air temperature, because a variety of factors will affect the water temperature, including solar radiation, relative humidity, dry bulb temperature, wind speed, air pressure and the upstream water characteristics [5, 6]. In general, in a certain temperature range the higher the air temperature, the higher the water temperature is.

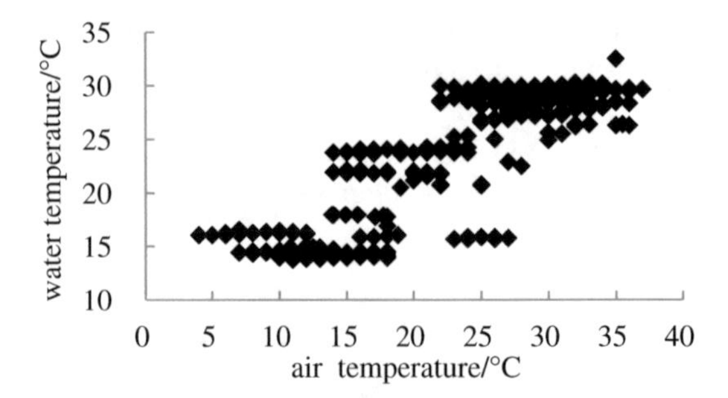

Fig. 3. The variation of the hourly water temperature with air temperature

3.2 The Average Water Temperature and the Air Temperature

When the average water temperature under the same air temperature is calculated, the relationship between air temperature and the average water temperature is more simple and clearer than the relationship between the water temperature and air temperature, as Fig. 4 shows. These two kinds of temperature present a strong coupling relationship.

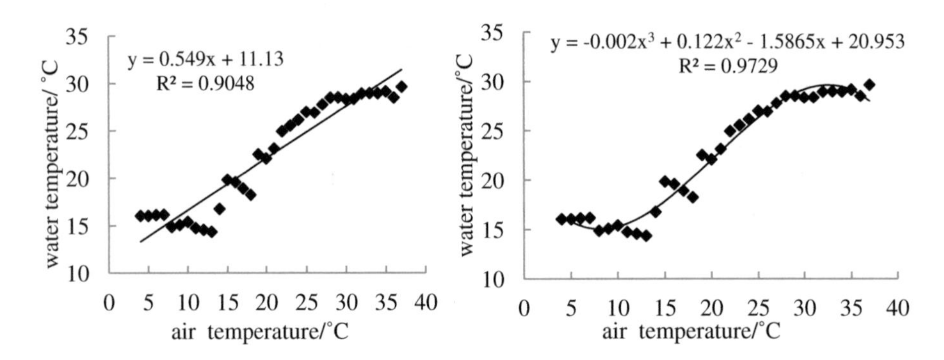

Fig. 4. The variation of the average water temperature with air temperature

Linear regression model is used to fit the air temperature and the corresponding average water temperature. The form of the linear regression model form is:

$$y = A + Bx$$

The variation of the average water temperature with the air temperature is shown in Fig. 4. The linear fitting equation is:

$$T_w = 11.13 + 0.549T_a$$

Symbols: Tw-the water temperature, Ta-air temperature, unit-°C, $R^2 = 0.904 > 0.90$.

In order to further explain the relationship between the average water temperature and the air temperature, Cubic model is used for further research. The Cubic model form is:

$$y = A + Bx + Cx^2 + Dx^3$$

The fitting analysis of the average water temperature and the air temperature in the lower Minjiang River is shown in Fig. 4. When the air temperature is lower than 20 °C, the measured average temperature around the fitted curve has large fluctuations; when the air temperature is higher than 20 °C, the measured average temperature matches the fitted curve.

The fitted equation is:

$$T_w = 20.95 - 1.586T_a + 0.122T_a^2 - 0.002T_a^3$$

Symbols: Tw-the water temperature, Ta-air temperature, unit-°C, $R^2 = 0.972 > 0.95$. This suggests that the Cubic model can explain the average measured data well.

Significance $F = 0.000000 < 0.05$. This shows that the regression equation of the regression effect is remarkable.

Then the least square method is used to do the regression analysis of 1st-order to 6th-order curve. The results are shown in Tables 1 and 2

Table 1. First to sixth order model regression equations

Model	Regression equation	Equation number
1st-order	$y = 0.549x + 11.13$	(1)
2nd-order	$y = -0.001x^2 + 0.588x + 10.82$	(2)
3rd-order	$y = -0.002x^3 + 0.122x^2 - 1.586x + 20.95$	(3)
4th-order	$y = 6E\text{-}05x^4 - 0.006x^3 + 0.258x^{2-3}.044x + 25.74$	(4)
5th-order	$y = 7E\text{-}06x^5 - 0.0006x^4 + 0.019x^3 - 0.198x^2 + 0.391x + 17$	(5)
6th-order	$y = -4E\text{-}07x^6 + 5E\text{-}05x^5 - 0.002x^4 + 0.071x^3 - 0.831x^2 + 4.0$ $32x + 9.419$	(6)

$T_w = y, T_a = x$

Table 2. First to sixth order model regression indexes

Model type	Prob (t-Statistic) maximum	R^2	Adjusted R^2	Prob (F-statistic)
1st-order(1)	0.0000 < 0.05	0.904	0.902	0.000000
2nd-order(2)	0.7961 > 0.05	0.905	0.899	0.000000
3rd-order(3)	0.0000 < 0.05	0.972 > 0.95	0.970	0.000000
4th-order(4)	0.0243 < 0.05	0.977 > 0.95	0.974	0.000000
5th-order(5)	0.7960 > 0.05	0.981 > 0.95	0.978	0.000000
6th-order(6)	0.2625 > 0.05	0.982 > 0.95	0.978	0.000000

Analysis of Tables 1 and 2:

1. T test: The maximum of Prob (t-Statistic) of type (2), type (5) and type (6) are all above 0.05. After inspection, it is found that the 2nd-order variable Prob (t-Statistic) of type (2) is above 0.05, the 2nd-order to 4th-order variable Prob (t-Statistic) of type (5) above 0.05, and the 1st-order to 6th-order variable Prob (t-Statistic) of type (6) above 0.05. Therefore, the above three types are eliminated by T test. The 1st-order to 3rd-order variable Prob (t-Statistic) of type (3) is 0.0000. The 4th-order variable Prob (t-Statistic) of type (4) is 0.0243.
2. F test: The Prob (F-statistic) of all the types are<0.05 so that the regression equation through F test.
3. R^2 of type (3) and type (4) are both>0.95, and adjusted R^2 of type (3) and type (4) are both>0.95, so that the regression effect of type (3) and type (4) is better.

3.3 Fitting Value is Compared with the Measured Values

In order to verify the applicability of the fitting curve, the comparison of the fitted and the average water temperatures of Minjiang River are shown in Fig. 5. In type (3) with the increase of air temperature, the trend of the fitted and the average water temperatures are consistent; but in type (4) the fitted temperature greatly deviates from the average water temperature. Therefore, the type (3) is the best regression equation.

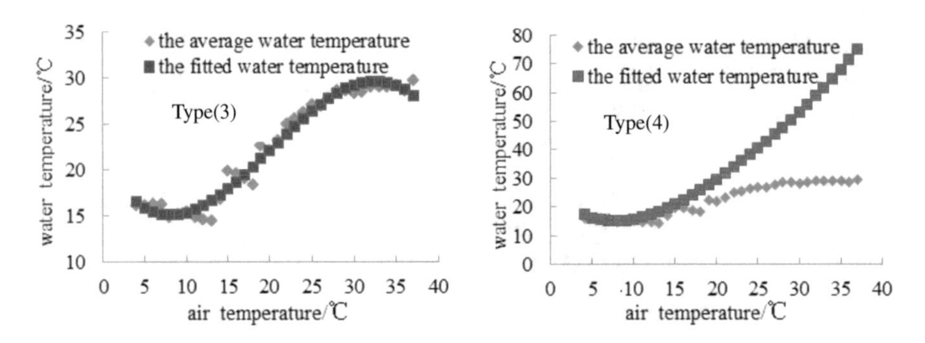

Fig. 5. Comparison of the fitted and the average water temperatures of Minjiang River

In type (3), the average water temperatures are waving around fitted water temperatures. Further studies conclusively show that the biggest difference value between the average water temperature and the fitted water temperature is about 2 °C. The difference between the average and the fitted water temperature of type (3) is shown in Fig. 6. The 88% of the difference values are in the range of −1.5~1.5 °C, among which 62% of the difference values are in the range of −1.0~1.0 °C, close to half of the difference values(in the range of −0.5~0.5 °C). This suggests that the fitting formula has relatively high precision.

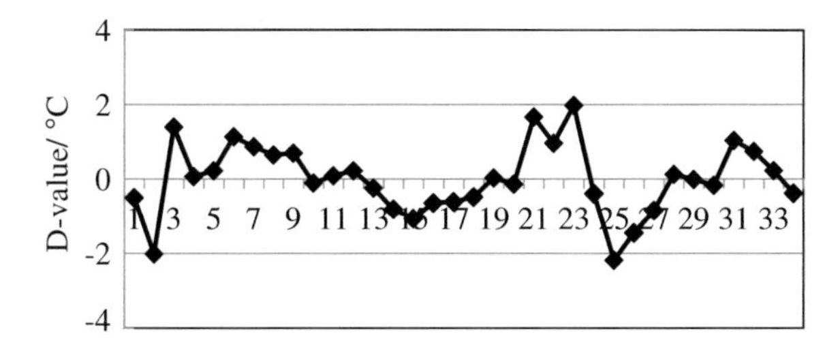

Fig. 6. The difference between the average and the fitted water temperature of type (3)

4 Conclusions

The main conclusions can be drawn as follows:

The hourly water temperature within 24 h of the design of the WSHP system can be replaced by daily average temperature because the change in daily water temperature is small. Although many factors affect water temperature, there is a remarkable relationship between the average water temperature and air temperature by data analysis, which indicates that the water average temperature can be obtained on the basis of this related coupling relationship. The fitted curve of Cubic model is obtained by fitting. The 88% of the difference values between average water temperatures and fitted water temperatures are in the range of −1.5–1.5 °C. Therefore, it can provide a simple method to determine the daily average water temperature through analyzing the relationship between the average water temperature and air temperature by using the method of Cubic fitting.

Acknowledgment. This work is supported by the Fuzhou Science and Technology Project (No. 2012-G-111).

References

1. Xiao, J.: Investigation on application of ground source heat pump technology at Fuzhou. Refrig. Air-conditioning **11**(3), 86–89 (2011)
2. Li, Y.: Analysis and discussion of HVAC energy saving design of Fuzhou strait international conference and exhibition center. Build. Sci. **26**(10), 121–125 (2010)
3. Ha, J.: Application of Fuzhou strait international conference and exhibition center by ground-source heat pump system. Shanxi Archit. **37**(22), 142–144 (2011)
4. Ouyang, H., Wang, D.: Study on the characteristics of Minjiang river as water source heat pump systems' water source. Water Wastewater Eng. **43**(2), 84–89 (2017)
5. Pengfei, S., Angui, L., Xiangyang, R., Ya, F., Zhengwu, Y., Qinglong, G.: New optimized model for water temperature calculation of river water source heat pump and its application in simulation of energy consumption. Renewable Energy **84**, 65–73 (2015)
6. Zhao, J., Liu, J., Ding, G.: Water temperature model and simulation of surface water source heat pumps. Heating Ventilating Air Conditioning **39**(10), 33–36 (2009)

Short-Term Forecasting on Technology Industry Stocks Return Indices by Swarm Intelligence and Time-Series Models

Tien-Wen Sung[1,2], Cian-Lin Tu[3], Pei-Wei Tsai[4], and Jui-Fang Chang[3(✉)]

[1] College of Information Sciences and Engineering, Fujian University of Technology, Fuzhou 350118, Fujian Province, China
tienwen.sung@gmail.com

[2] Fujian Provincial Key Laboratory of Big Data Mining and Applications, Fujian University of Technology, Fuzhou 350118, Fujian Province, China

[3] Department of International Business, National Kaohsiung University of Applied Sciences, Kaohsiung 807, Taiwan
1104346116@gm.kuas.edu.tw, rose@kuas.edu.tw

[4] Department of Computer Science and Software Engineering, Swinburne University of Technology, Victoria 3122, Australia
ptsai@swin.edu.au

Abstract. Forecasting is an important technique in many industries and business fields for reading the terrain. The category of technology industry stock, which includes 7 independent stocks, in Taiwan Stock Exchange (TWSE) is selected to be the study subject in this paper. The goal is to forecast the return index of the individual stocks base on the information observed from the trading historical da-ta of the subjects. By including the trading volume, the number of trading rec-ords, the opening price, and the closing price in the inputs to the representative models in time-series and computational intelligence: EGARCH(1,1) and the In-teractive Artificial Bee Colony (IABC), respectively, the forecasting accuracy are compared by the Mean Absolute Percentage Error (MAPE) value. The experi-mental results indicate that the IABC forecasting model with the selected input variables presents superior results than the EGARCH(1,1).

Keywords: IABC · EGARCH · Return index · Technical industry stock

1 Introduction

An accurate forecasting result is an important and crucial support for business and industry managers to make decisions. The demand for the forecasting lies in many different places such as the customers' needs, the product production capacity, the trading costs, and the expected net profit, so on so forth. The foreign exchange rate forecasting is one of them directly related to the net profit and the trading costs for the companies involving in the international trading business. In addition, the business circle is a series of fluctuated economic experiences

J.-S. Pan et al. (eds.), *Advances in Intelligent Information Hiding and Multimedia Signal Processing*, Smart Innovation, Systems and Technologies 81,
DOI 10.1007/978-3-319-63856-0_34

composed of four circulating phases: the expansion, the peak, the contraction, and the trough. Motions of the circulation provide the impact to the trend of the long-term growth in Gross Domestic Product (GDP), which is closely related to the economic cycle. During the past two decades, globe stock markets undergo two significant episodes that can use economic cycle to spell out a rough sketch about the current globe stock markets: the first event takes place in the US that the dot-com bubble took globe stock markets a nose dive in 2001, and the other is subprime mortgage crisis culminated in the financial crisis of 2007 to 2008 that lasted over seven years' bear market over the world. In accordance with the laws of the business circle, globe stock markets took a jump in 2016 and the prediction on the next tendency becomes a crucial problem to be monitored. Investors can obtain the flow of stock market by the industrial sectors' indices which commonly assemble stocks into groups and are the primary form to convey the level of prices. The indices provide more encyclopaedic comparability and the more wide-spread angle of industrial performances we can grasp.

Base on the ground truth mentioned above, a swarm intelligence algorithm called Interactive Artificial Bee Colony (IABC) is utilised to compose the forecasting model. It produces the forecasting result of the return index by receiving inputs from the trading volume, the number of trading records, the opening price, and the closing price. The same inputs are used in a time series model called EGARCH to produce the baseline for compare. The rest of the paper is composed as follows: the literature review is given in Sect. 2, the experiment design is discussed in Sect. 3, the experimental results are given in Sect. 4, and the conclusion is given in Sect. 5.

2 Literature Review

Time-series Model is a typical technique used in the financial analysis [1]. It can bring data to light and conjecture that the reactions are taking place by analysing the mass data. Autoregressive Conditional Heteroscedasticity (ARCH) is brought up by Robert F. Engle (1982) [5] and that included Mean and Variance equations. Mean equations denote that the Efficient-market, the long-term arbitrage, and the speculation are unworkable. The variance equations denote that the risks are foreseeable and handleable. In ARCH model, the conditional variance will be adjusted with the time, which provides the ability of enhancing the characteristics of accuracy of the model. In recent years, ARCH models are expanded in a bunch of linked time-series models called ARCH family under many improvements and developments made by the scholars. For example, Tim Bollerslev (1986) proposes the Generalised Auto-regressive Conditional Heteroskedastic model (GARCH) [2] based on the ARCH [3] fundamental to simplify the setting of parameters that presumes the conditional variance is affected by both the error sum of squares and the conditional variance in the previous period. The ARCH and the GARCH models are both constructed base on the symmetric in fluctuation. Different from these models, French et al. (1986) [10] consider the same subject in a different angle of view. They point out that the impacts of the markets brought by the good news and the bad news are asymmetric. The bad news always trigger more variations than the good news when the event appears. In 1991, Nelson presents the Exponential GARCH (EGARCH) [11] model to include the asymmetric fluctuation phenomenon in the consideration. It has been proofed that the EGARCH model has better explanatory power in the analysis.

Different from the time-series models, the swarm intelligence is a research field containing algorithms designed based on the observation and simulation of the collective behaviours in natural or artificial [12]. The Artificial Bee Colony Algorithm (ABC) algorithm [13], which is proposed by Karaboga (2005). A few years later, Tsai et al. (2009) embed the Newtonian law of Universal Gravitation in the ABC algorithm to reform the selection mechanism for broadening the development capability of onlooker bees, the trial probes the interactive affection of ABC algorithm with the Interactive Artificial Bee Colony (IABC). The IABC algorithms has been used in foreign exchange rate forecasting in 2014 and 2016 [8,9], renewable energy unit distribution [6] and pattern recognition [7]. And the outcomes obtained in the foreign exchange rate forecasting is satisfactory.

3 Experiment Design

In this work, the historical data of the trading volume, the number of trading records, the opening price, and the closing price in the past 30 days are collected for feeding into the EGARCH(1,1) and the IABC forecasting models, respectively. The experimental data includes those collected in January 10th in 2009 and December 31st in 2015. By the lubrication of the sliding window strategy, the forecasting models are capable to produce the outcomes for all targeting observational periods. According to the published classification in Taiwan Stock Exchange (TWSE), the category of technology industry stock contains seven independent stocks. The stocks in the technology industry category are the subject studied in this paper. The forecasting accuracy is measured via the Mean Absolute Percentage Error (MAPE) value. The MAPE tells the distance between the forecasted result and the actual return index in an objective way.

The experiment is composed of the steps listed as follows:

1. Fundamental statistical tests of the input data: The input data need to be preprocessed by the Ljung-Box Q test and the Augmented Dickey-Fuller (ADF) Unit Root Tests [4]. The details will be given in the following subsection.
2. Feeding the input data into both the EGARCH and the IABC forecasting models for training and testing.
3. Calculating the MAPE values corresponding to the forecasting results of both models.

3.1 Input Data for the Forecasting Models

The study subjects are the stocks in the technology industry category in TWSE market. This category includes seven stocks of computers and computer peripheral equipment, the optoelectronics, the communication technology and internet, the electronic parts and components, the electronic product distribution, the information service, and the other electronics.

For every stock, the collected trading volume, the number of trading records, the opening price, and the closing price are examined to make sure they are with the expected statistical characteristics and distribution. In general, the stock price should fit in the normal distribution by logarithm as a log return. The rate of the return is calculated by Eq. (1):

$$p_r = \log\left(\frac{p_t}{p_{t-1}}\right) \tag{1}$$

where p_t refers to the price on day t, and p_r denotes the rate of the return.

If the outcome of Eq. (1) fits in the normal distribution, the original input data can be further processed by the Ljung-Box Q text for analyzing the residual correlation by Eq. (2):

$$Q = n(n+2)\sum_{k=1}^{h}\frac{\hat{p}_k^2}{n-k} \tag{2}$$

where Q denotes the test result, n refers to the size of the sample, $\hat{p}_k$ indicates the autocorrelation at lag k, and h is the test number of lags.

The ADF unit root test is applied after the Ljung-Box Q test for verifying the sta-tionary of the series data by Eq. (3):

$$\triangle y_t = \alpha + \beta t + \gamma y_{t-1} + \delta_1 \Delta y_{t-1} + \cdots + \delta_{p-1}\Delta y_{t-p+1} + \varepsilon_t \tag{3}$$

where α stands for a constant, β indicates the coefficient on a time trend, and p is the lag order of the Autoregressive (AR) process.

3.2 The Time-Series Model: EGARCH

In this study, the classical EGARCH(1,1) model is used in the experiments. The EGARCH model can be depicted as follows:

$$y_t = x_t b + \varepsilon_t \tag{4}$$

where y_t refers the function of exogenous variable x_t, $x_t b$ is the conditional mean of y_t, and ε_t denotes the residuals.

$$\varepsilon_t \mid \Omega_{t-1} \sim N(0, h_t) \tag{5}$$

where Ω_{t-1} means all set of information to period $t-1$ and h_t refers conditional variance of y_t.

$$\ln(h_t) = \alpha_0 + \sum_{i=1}^{q}\alpha_i\left\{\gamma_i\frac{\varepsilon_{t-i}}{\sqrt{h_{t-1}}} + \vartheta_i\left[\frac{|\varepsilon_{t-i}|}{\sqrt{h_{t-i}}} - \sqrt{\frac{2}{\pi}}\right]\right\} + \sum_{j=1}^{p}\beta_j \ln(h_{t-j}),$$
$$\text{subject to } \alpha_0 > 0,\ \alpha_i \geq 0,\ \beta_j \geq 0,\ \forall i,j \tag{6}$$

where γ_i is the parameter of the asymmetric volatility effect, $\gamma_i \frac{\varepsilon_{t-i}}{\sqrt{h_{t-i}}}$ denotes the sign effect, ϑ_i indicates the parameter of scale of unexpected variation, and $\vartheta_i \left[\frac{|\varepsilon_{t-i}|}{\sqrt{h_{t-i}}} - \sqrt{\frac{2}{\pi}} \right]$ represents the magnitude effect.

3.3 The Interactive Artificial Bee Colony (IABC) Model

IABC is a branch of ABC algorithm, which involves the universal gravitation into the onlooker movement process in the conventional ABC algorithm. The IABC model can be depicted as follows:

1. Initialization: Randomly spread n_e percent of the population into the solution space, where n_e refers the ratio of employed bees to the total population.
2. Move the onlookers: Calculating the probability of selecting a food source by Eq. (7) and choose G of them as the reference by the roulette wheel scheme for every on-looker bees.

$$P_i = \frac{F(\Theta_i)}{\sum_{k=1}^{S} F\Theta_k} \tag{7}$$

   where P_i denotes the probability of selecting the ith employed bee, S stands for the total number of the employed bees, $F(\Theta_k)$ and $F(\Theta_i)$ represent the fitness value of employed bees k and i, respectively. Move the onlookers by Eq. (8) based on the selected G reference employed bees.

$$x_i(t+1) = \Theta_i + \sum_{k=1}^{G} \{\tilde{F}_{ik} \cdot [\Theta_i(t) - \Theta_k(t)]\} \tag{8}$$

   where x_i is the coordinate of the ith onlooker bee, t indicates the iteration number, Θ_i and Θ_j denote the coordinates of the ith employed bee and the randomly chosen employed bee, respectively, G is the number of the selected employed bees, and $\tilde{F}_{ik}$ refers to the normalized universal gravitation.
3. Move the scouts: The employed bee will become the scout and be moved again if none of the onlooker bees selected it as a reference after a predefined *Limit* iterations. Nevertheless, there are two randomly selected employee bees can be exempted from the rule. The rest of the employee bees, which matches the condition to become the scouts will be moved by Eq. (9):

$$\Theta_{ij} = \Theta_{jmin} + r(\Theta_{jmax} - \Theta_{jmin}) \tag{9}$$

   where Θ_{jmax} denotes the maximum value over all dimensions of Θ_i, Θ_{jmin} indicates the minimum value appears in all dimensions of Θ_i, and r is a random variable in the range of $[0, 1]$.
4. Update the near best solution: Keep the best fitness value and the corresponding co-ordinates of the bee.
5. Termination checking: As the termination conditions are satisfied, terminate the algorithm and output the kept near best solution; otherwise, go back to Step 2 and repeat the processes.

In the IABC forecasting model, the input data is merged from four variables into two variables by Eqs. (10) and (11):

$$TP = \frac{T_V}{V_R} \tag{10}$$

where TP denotes the trading power, T_V is the trading volume, and T_R is the number of the trading records.

$$PD = \frac{P_O}{P_C} \tag{11}$$

where PD stands for the price difference of a trading day, P_O and P_C represents the opening price and the closing price of a trading day, respectively.

In the training phase, 30-day historical data (denoted by $Data_{(t-31:t-1)}$ is feed into the IABC model for training the corresponding weighting mask (denoted by W). The desired output should be as much close to the actual return index on day $t-1$. The trained W is later used in the testing phase with the input data $Data_{(t-30:t)}$ to produce the forecasting return index for day t.

3.4 Calcuation of the MAPE Value

The MAPE is calculated by Eq. (12):

$$MAPE = \frac{1}{m}\sum_{t=1}^{m}\frac{|\hat{S}_t - S_t|}{S_t} \times 100\% \tag{12}$$

where $\hat{S}_t$ and S_t stand for the prediction and the actual value, respectively, and m is the total number of the data.

4 Experiments and Experimental Results

The experimental results include seven independent stocks in the technology industry category in 2009 to 2015. Totally, 1731 observations are included in the data. The experimental results are presented base on the stocks in Fig. 1.

According to the experimental results obtained with 7 independent stocks, we can find that the IABC forecasting model produces much smaller MAPE values than the EGARCH(1,1) model. Nevertheless, it is still suffered from trapping in the local optimum, sometimes, and produces a significant jump of the MAPE value.

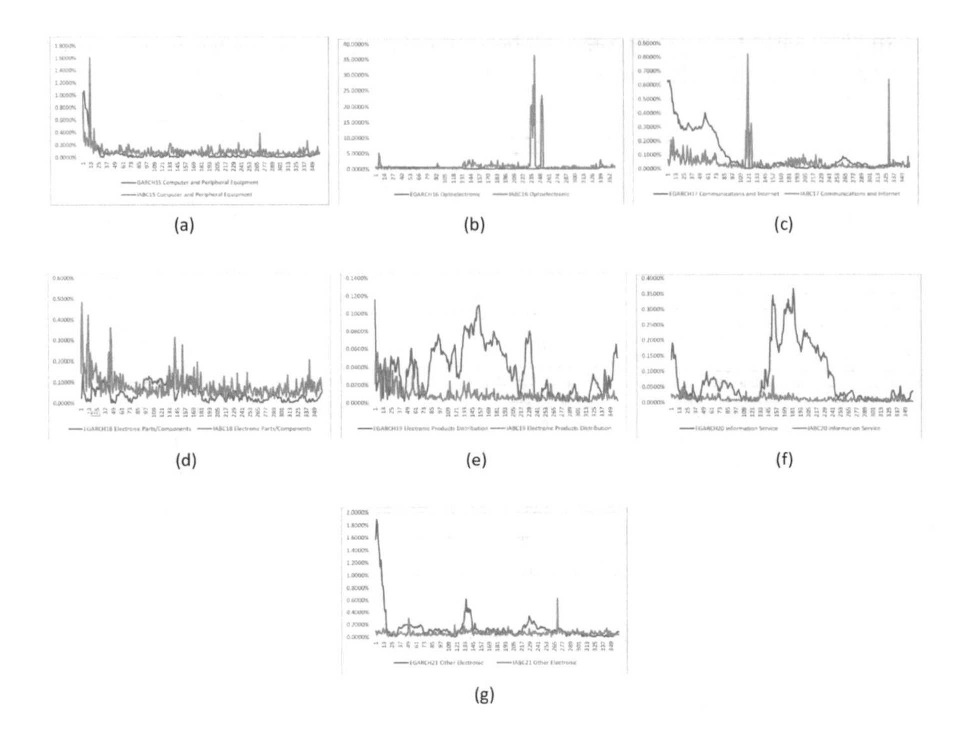

Fig. 1. MAPE values of all stocks in 2009–2015.

5 Conclusions and Future Works

The goal of this work is to forecast the return index of the individual stocks base on the information observed from the trading historical data of the subjects. The category of technology industry stock, which includes 7 independent stocks, in Taiwan Stock Exchange (TWSE) is selected to be the study subject. The EARCH(1,1) and the IABC forecasting models are utilized to generate the forecasting results. The forecasting accuracy is measured by the MAPE value. The experimental results indicate that the IABC forecasting model presents superior results than the EGARCH(1,1). In the future work, we will focus on increasing the stability of the IABC forecasting model to reduce the unsteadiness of the outcome. The forecasting models and the forecasting return index provides the shareholders additional references for operating the investments.

Acknowledgement. This work is funded by the Key Project of Fujian Provincial Education Bureau (JA15323).

References

1. Chen, S.-S.: Applied Time-series Econometrics for Macroeconomics and Finance. Tunghua, Taipei (2013)
2. Bollerslev, T.R.: Generalized autoregressive conditional heteroskedasticity. J. Econometrics **31**(3), 307–327 (1986)
3. Bollerslev, T.R.: ARCH modeling in finance: A review of the theory and empirical evidence. J. Econometrics **52**(1–2), 5–59 (1992)
4. Dickey, D.A.: Likelihood ratio statistisc for autoregression time series with a unit root. Econometrica **49**, 1057–1072 (1981)
5. Engle, R.F.: Autoregressive conditional heteroskedasticity with estimates of the variance of united kindom inflation. Econometrica **50**(4), 987–1007 (1982)
6. Abedinia, O., Barazandeh, E.S.: Interactive artificial bee colony based on distribution planning with renewable energy units. In: 2013 IEEE PES Innovative Smart Grid Technologies (ISGT), pp. 1–6. IEEE Press, Washington, DC (2013)
7. Tsai, P.-W., Pan, J.-S., Liao, B.-Y., Chu, S.-C.: Enhanced artificial bee colony optimization. J. Innov. Comput. Inf. Control **5**(12B), 5081–5092 (2009)
8. Tsai, C.-F., Hsiao, C.-T., Tsai, P.-W., Chang, J.-F.: Applying interactive artificial bee colony algorithm and the time-series methods in the foreign exchange rate forecasting. Econ. Manage. **36**(Z1), 147–150 (2014)
9. Tsai, P.-W., Liu, C.-H., Zhang, J., Chang, J.-F.: Structuring interactive artificial bee colony forecasting model in foreign exchange rate forecasting with consumer confidence index and conventional microeconomics factors. ICIC Expr. Lett. Part B: Appl. **7**(4), 895–902 (2016)
10. French, K.R., Roll, R.: Stock return variance: the arrival of information and the reaction of traders. J. Financ. Econ. **17**(1), 5–26 (1986)
11. Nelson, D.B.: Conditional heteroskedasticity in asset returns: a new approach. Economietrica **59**(2), 347–370 (1991)
12. Beni, G., Wang, J.: Swarm intelligence in cellular robotic systems. In: Dario, P., Sandini, G., Aebischer, P. (eds.) Robots and Biological Systems: Towards a New Bionics?, 1993. NATO ASI Series (Series F: Computer and Systems Sciences), vol. 102, pp. 703–712. Springer, Berlin, Heidelberg (1993)
13. Karaboga, D.: An Idea Based on Honey Bee Swarm for Numerical Optimization. In: Technical Report-TR06, 2005 (2005)

Applications of Image Encoding and Rendering

Image Segmentation for Lung Lesions Using Ant Colony Optimization Classifier in Chest CT

Chii-Jen Chen(✉)

Department of Medical Imaging and Radiological Technology, Yuanpei University of Medical Technology, Hsinchu, Taiwan
cjchen@mail.ypu.edu.tw

Abstract. The chest computed tomography (CT) is the most commonly used imaging technique for the inspection of lung lesions. In order to provide the physician more valuable preoperative opinions, a powerful computer-aided diagnostic (CAD) system is indispensable. In this paper, we aim to develop an ant colony optimization (ACO-based) classifier to extract the lung mass. We could calculate some information such as its boundary, precise size, localization of tumors, and spatial relations. Final, we reconstructed the extracted lung and tumor regions to a 3D volume module to provide physicians the more reliable vision. In order to validate the proposed system, we have tested our method in a database from 15 lung patients. We also demonstrated the accuracy of the segmentation method using some power statistical protocols. The experiments indicate our method results more satisfied performance in most cases, and can help investigators detect lung lesion for further examination.

Keywords: Lung tumor · Ant colony optimization · Segmentation · Reconstruction

1 Introduction

Pulmonary carcinoma is the first rank of cancer in Taiwan. Early detection and treatment of lung cancer can effectively prohibit its progress and decrease mortality rate. However, many diseases of lung often cannot be found and diagnosed early; the conditions of patients were become more and more serious. The chest is the most important coelom, and the heart and lungs are included in the chest. The heart controls the blood circulation, and the lungs control the gas exchange. Because the volume of chest is finite, when the diseases occurred in the heart or lungs, it may be influenced by each other. In recent years, the computed tomography (CT) imaging for chest examination become more and more important. Moreover, the quality of CT images can be guaranteed, so it usually can correctly display the location, size and shape of the organizations and tissues within the human body. Furthermore, the CT image is high DPI, and can show the very small variations between the tissues [1–4].

In this study, the procedure of lung tumor segmentation is very important, and it may affect the efficacy of system performance. Several researches have discussed automatic or

J.-S. Pan et al. (eds.), *Advances in Intelligent Information Hiding and Multimedia Signal Processing*, Smart Innovation, Systems and Technologies 81,
DOI 10.1007/978-3-319-63856-0_35

semi-automatic segmentation methods to extract known anatomic structures from the chest CT images. Some of these algorithms extract a region of interest (ROI) out and may rely on a priori shape information of features or structures [2, 5–8]. Chen, et al. proposed a CAD system based on fuzzy C-means clustering that considered the shape information for liver tumor such as area, circularity and minimum distance from liver boundary to tumor. They obtained an accuracy of 91% for classifying normal and abnormal slice [7]. In addition, their method was performed in three-dimension (3D) so as to improve its accuracy. However, it is difficult to segment liver lesion on abdominal CT using a predefined mathematical model because the shape of liver varies among humans. Conventionally, knowledge-based methods are usually used to solve this problem.

In this paper, we proposed a full automated method to detect lung lesions on chest CT scan images. The proposed method combines a segmentation approach for extracting lung mass and an ACO-based (Ant Colony Optimization) classifier for select abnormal pixels. In the addition, we used DICOM images as source images and proposed a method to compute adaptive liver window level to enhance the abdominal CT images. This study not only considers the commonly used Haralick textures, but also regards more morphological information to obtain a higher accuracy for the classification. The experiment shows that our method results in higher accuracy in most cases.

2 Materials and Methods

The proposed system flowchart is shown in Fig. 1. The original testing image slices were first processed by the pre-processing procedure to reduce the noises in images. Then an

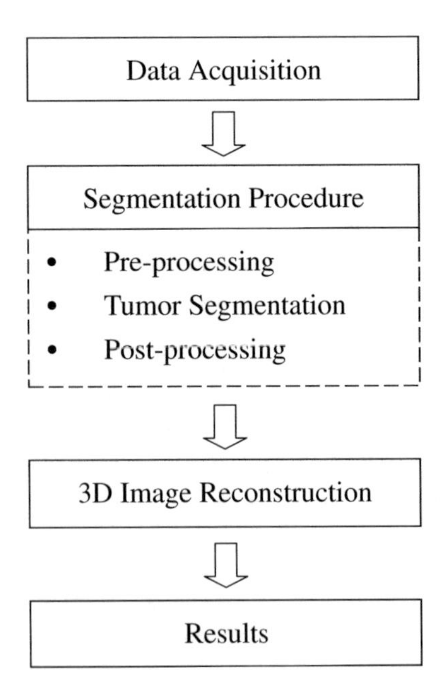

Fig. 1. The flowchart of proposed system framework.

efficient segmentation method was applied in the following image data. In order to improve the segmentation result, a post-processing procedure was used. It could reduce the redundant regions after segmentation procedure.

2.1 Image Pre-processing

The gray-value of current pixel is replaced by the median value of 8-neighbor pixels. There are several low-pass filters, such as median filter and averaging filter, can be adopted for reducing the noise in medical image. The averaging filter may efficiently reduce noise, but the boundary information and texture patterns have also been blurred, which would makes it difficult on extracting contour information. Hence, for reducing noise and preserving the object information, we apply median filter to eliminate the speckle noise.

The pixel values in binary image are only 0 (black) and 255 (white). We can set a threshold for a gray image to divide all pixel values into black and white, and the function is defined as follows

$$\begin{cases} p(x,y) = 255, & \textit{if } p(x,y) \geq th \\ p(x,y) = 0, & \textit{if } p(x,y) < th \end{cases}, \tag{1}$$

where th is threshold; then the image will become a binary image.

Through the binary threshold processing, the image will become simpler than original gray image. Then the following procedure, segmentation, will also easy to process and recognize.

For medical images, there are some fundamental requirements of the noise filtering methods should be noticed. First, do not lose the important information such as object boundaries and detailed structures. Second, reduce the noise in the homogeneous regions efficiently. Because the quality of CT images is better than other medical images, such as X-ray and ultrasound, the conventional smooth filtering and linear diffusion can be used to reduce noise before segmentation. Therefore, the pre-processing in this study was applied two methods: median filtering and binary threshold.

2.2 Tumor Segmentation by ACO-Based Classifier

The proposed ant colony optimization (ACO-based) classifier is used to track the tumor contour in the lung CT images. This algorithm is simulated as ant investigation from the initial start point and end point [9, 10], as described in Fig. 2, where R is the number of rounds and N is the number of ants that can update the pheromone in each round. The ACO-based Classifier obtained the optimize contour for N ants that can track the lobe fissures on the sagittal view of lung CT. First, the input image is a sagittal view of lung CT image. Second, we select start point, end point and pheromone table for the initial procedure. Then, the process found several candidate contours that choose the best contour to update pheromone for each round. Finally, we can use the updating rules of pheromone and find out the best contour which could be the lung tumor. The results of lung mass segmentation by ACO-based classifier is shown in Fig. 3.

```
procedure ACO ( )
   initialize_ant( );
   initialize_pheormone( );
   for R=1 to number_of_rounds
      for N=1 to number_of_ants
         while (current_point!= end_point)
            ant_following_types;
         end while
         local_updating_ rule;
         record_optimal_path;
      end for
      global_updating_ rule;
   end for
end procedure
```

Fig. 2. The algorithm of ant colony optimization (ACO-based) classifier.

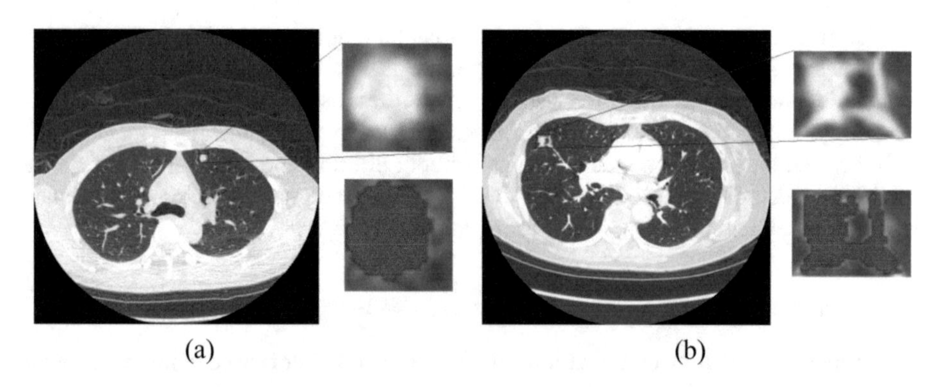

Fig. 3. The results of lung mass segmentation by ACO-based classifier.

2.3 Post-processing for Outside Contour Excision

In previous procedure, the segmentation regions are included the contours of body (outside) and lung (inside). However, we only want to obtain the contours of lung in this study, so we proposed the following method to reduce the outer contour of human body for all segmented images.

2.4 Tumor Reconstruction

There are 520 continuous images for each case, so we could combine the information of 2D continuous images into a 3D stereo picture. In this paper, we applied the 3D structured-grid method to reconstruct the 3D volume [11–14]. We depended on the contour images after the ACO-based classifier segmentation function, and found the corresponding relationships between the neighboring images to reconstruct the triangle structured-grids. Then, we used the adequate filter to modify the non-smooth surfaces, such as marching cubes algorithm, which is a high-resolution 3D surface construction algorithm. For marching cubes algorithm, there are 15 surface filters to modify the non-smooth surfaces. The flowchart of 3D structured-grid method is shown in Fig. 9. The two examples for 3D reconstruction (with tumor) by the structured-grid method are shown in Fig. 4.

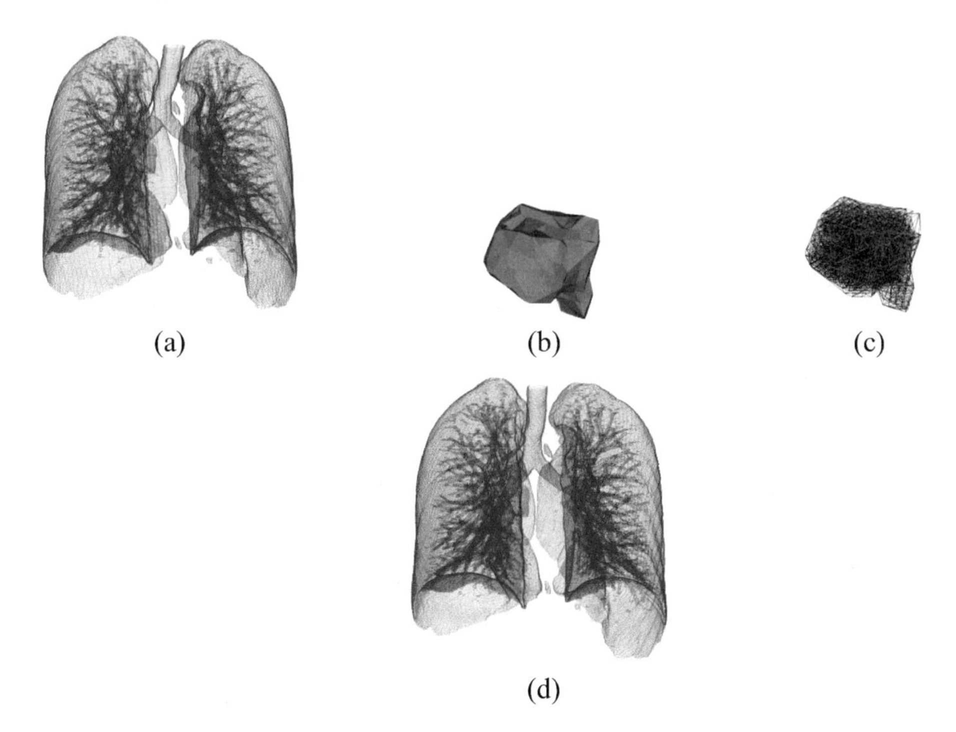

Fig. 4. The first example for 3D reconstruction (with tumor) by the structured-grid method.

3 Results and Conclusion

In the experiment, there were 15 testing cases and each case has 520 continuous testing images with axial-view. All images were processed by the proposed segmentation and reconstruction framework. The results of precision, recall and F-measure is listed in Table 1; the experiments indicate our method results more satisfied performance. Because all of features in the lung were preserved, we can integrate all the segmented images to reconstruct a three-dimensional (3D) module to process the further observation and diagnosis in the future.

Table 1. The results of precision, recall and F-measure.

Lung	Mean (%)		
	Precision	Recall	F-measure
Left	80.32 ± 4.62	81.79 ± 5.72	82.94 ± 5.18
Right	88.72 ± 2.34	85.73 ± 3.16	85.14 ± 3.37

In this study, we will integrate the computer-aided diagnosis (CAD) system to provide the physician the more reliable vision for preoperative diagnosing and analyzing. We develop a novel method for segmenting lung and lung tumors to assist the physician diagnosing and evaluating preliminary. We also reconstruct the extracted lung and tumor regions to a three-dimensional (3D) image module to provide physicians the more reliable vision. The current results show that the contours of organ and body can be detected and segmented precisely.

In the future, we will integrate all the segmented and reconstructed results to develop a preoperative 3D computer-aided diagnosis system and help the preoperative procedure more precisely.

References

1. Awai, K., et al.: Pulmonary nodules: estimation of malignancy at thin-section helical CT–effect of computer-aided diagnosis on performance of radiologists. Radiology **239**(1), 276–284 (2006)
2. Anitha, S., Sridhar, S.: Segmentation of lung lobes and nodules in CT images. Sig. Image Process. Int. J. (SIPIJ) **1**(1), 1–12 (2010)
3. Suzuki, K., Li, F., Sone, S., Doi, K.: Computer-aided diagnostic scheme for distinction between benign and malignant nodules in thoracic low-dose CT by use of massive training artificial neural network. IEEE Trans. Med. Imaging **24**(9), 1138–1150 (2005)
4. Iwano, S., Nakamura, T., Kamioka, Y., Ikeda, M., Ishigaki, T.: Computer-aided differentiation of malignant from benign solitary pulmonary nodules imaged by high-resolution CT. Comput. Med. Imaging Graph. **32**(5), 416–422 (2008)
5. Haris, K., Efstratiadis, S.N., Maglaveras, N., Katsaggelos, A.K.: Hybrid image segmentation using watersheds and fast region merging. IEEE Trans. Image Proc. **7**(12), 1684–1699 (1998)
6. Ladak, H.M., et al.: Prostate boundary segmentation from 2D ultrasound images. Med. Phys. **27**(8), 1777–1788 (2000)
7. Chen, T., Metaxas, D.: A hybrid framework for 3D medical image segmentation. Med. Image Anal. **9**(6), 547–565 (2005)
8. Horsch, K., Giger, M.L., Venta, L.A., Vyborny, C.J.: Automatic segmentation of breast lesions on ultrasound. Med. Phys. **28**(8), 1652–1659 (2001)
9. Benatcha, K., Koudil, M., Benkhelat, N., Boukir, Y.: ISA an algorithm for image segmentation using ants. In: IEEE International Symposium on Industrial Electronics (ISIE 2008), Cambridge, UK, pp. 2503–2507 (2008)

10. Chen, C.J., Wang, Y.W., Shen, W.C., Chen, C.Y., Fang, W.P.: The lobe fissure tracking by the modified ant colony optimization framework in CT images. Algorithms **7**(4), 635–649 (2014)
11. Werahera, P.N., et al.: A 3-D reconstruction algorithm for interpolation and extrapolation of planar cross sectional data. IEEE Trans. Med. Imaging **14**(4), 765–771 (1995)
12. Sha, Y., et al.: Computerized 3D-reconstructions of the ligaments of the lateral aspect of ankle and subtalar joints. Surg. Radiol. Anat. **23**(2), 111–114 (2001)
13. Qiu, M.G., et al.: Plastination and computerized 3D reconstruction of the temporal bone. Clin. Anat. **16**(4), 300–303 (2003)
14. Anderson, J.R., Wilcox, M.J., Barrett, S.F.: Image processing and 3D reconstruction of serial section micrographs from Musca Domestica's biological cells responsible for visual processing. Biomed. Sci. Instrum. **38**, 363–368 (2002)

Auto-Recovery from Photo QR Code

Shang-Kuan Chen(✉)

Department of Applied Mobile Technology,
Yuanpei University of Medical Technology, Hsinchu, Taiwan
skchen@mail.ypu.edu.tw

Abstract. A novel idea of automatic recovery from photo QR Code is proposed. The photo QR code can be generated by the proposed pyramid QR coding method. The original photo is then embedded into the generated QR code. The experimental result demonstrates that the original photo is with very high quality and the photo QR code can be used for decoding the embedded words or URLs by common QR code decoders.

Keywords: Block matching · Data hiding · Image hiding · Steganography

1 Introduction

Nowadays, users adopt the mobile phones browsing the web pages, which has brought much more conveniences to their life. However, the time consumption of typing URL (Uniform Resource Locator) on the mobile phone brings less conveniences and difficulties for linking to the web pages compared to traditional keyboard. To solve this problem, the quick response (QR) code [1] which stores textual information can be read by any smart device including most mobile phones.

The QR codes can be usually found from web pages, products, and posters. It is a two-dimensional code seen as a square shape, and it is mostly represented by binary form (black and white pixels) which looks like a random pattern. QR code can be also regarded as the visible watermark. Due to the fact that visible watermark cause the degradation of image quality, Huang et al. [2] proposed a reversible data hiding method to recover the original image from an image with QR code as a visible watermark.

For visual art, Chu et al. [3] proposed a halftone QR coding method to generate an art QR code in gray scale. Figure 1 shows the generated art QR code by their method.

QR code is a kind of error-tolerant code. Lin et al. [4] use the encoding characteristics to design a visual QR code with high quality visual content. However, none of these photo QR codes can be reused as its original view. The data hiding methods [5–8] can be adopted for hiding the original photo into the corresponding QR code. In this paper, a hybrid encoding method for generating photo QR code is proposed. It will be introduced later.

The rest of this paper is organized as follows. Section 2 describes how the photo QR code is generated. Section 2 states the process that enables the photo QR code with

J.-S. Pan et al. (eds.), *Advances in Intelligent Information Hiding and Multimedia Signal Processing*, Smart Innovation, Systems and Technologies 81,
DOI 10.1007/978-3-319-63856-0_36

ability of auto recovery. Experimental results are demonstrated in Sect. 3. Finally, the conclusions are given in Sect. 4.

Fig. 1. The halftone QR code generated by Chu et al. [3]

2 Photo QR Code Generation

For each input text, an on-line QR code generator can generate a standard QR code. A standard QR code is a matrix consisting of a series of black and white squares, for example, shown in Fig. 2.

Fig. 2. The example of QR code

The position-detection patterns and the alignment pattern, shown in Figs. 3 and 4, are located in three corners of each symbol and are to align symbol, respectively.

Fig. 3. Finder patterns and alignment pattern

Fig. 4. Finder patterns and alignment patterns of larger QR code

Given a user-selected photo, a photo QR code can be simply generated by the following process:

1. Keep finder patterns and alignment patterns of the standard QR code in the photo QR code.
2. Shortened other pixels of QR code and fill them into the photo QR code.
3. The unused region of the photo QR code are painted by the corresponding pixels of original photo.

Figure 5(a) shows the tested photo and Fig. 5(b) shows an example of the photo QR code generated by the above process.

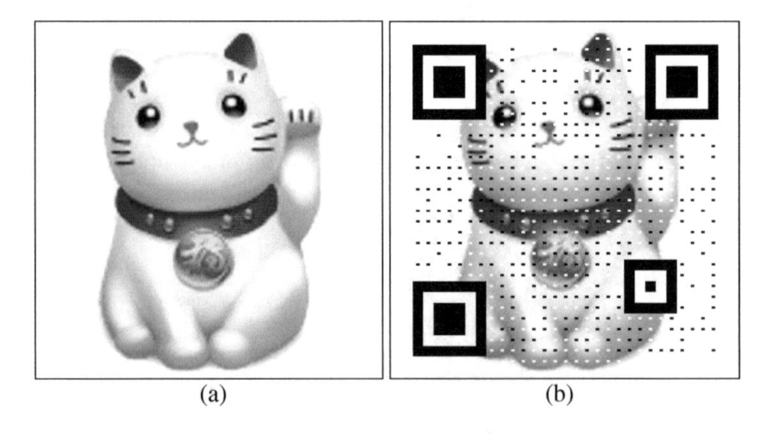

Fig. 5. (a) The original photo. (b) The photo QR code.

In Fig. 5(b), the black and the white pixels are so obvious that the look of the photo QR code is not satisfactory. To improve the situation, the following process is to lighten the black and the white pixels of the photo QR code. Figure 6 shows an example of the photo QR code generated by the below improved process.

1. Keep finder patterns and alignment patterns of the standard QR code in the photo QR code.
2. Shortened other pixels of QR code and fill them into the photo QR code.
3. The unused region of the photo QR code are painted by the corresponding pixels of original photo.

Fig. 6. The photo QR code generated by improved code

To recover the original photo from the photo QR code, the regions of finder patterns, alignment patterns, and the shortened areas of other pixels of the QR code should be restored. The data hiding method that uses optimized modulus least significant bits is adopted for hiding these areas.

3 Experimental Results

In the experiment, firstly, a string should be input for generating standard QR code. A given photo, then, is resized by the size of the standard QR code by the image resizing method, for example, Rubinstein et al. [9] and Dong et al. [10]. The string used for generating standard QR code is "http://www.google.com". The generated QR code is shown in Fig. 7(a). Figure 7(b) shows the resized original photo.

Figure 8(a) shows the result that hides the QR code regions of a photo QR code with lighter black pixels into the photo QR code itself. Figure 8(b) shows the recovering result from the photo QR code. The recovered result has very good photo quality with PSNR = 45.21.

(a) (b)

Fig. 7. **(a)** The standard QR code of string "http://www.google.com" (b) The resized original photo

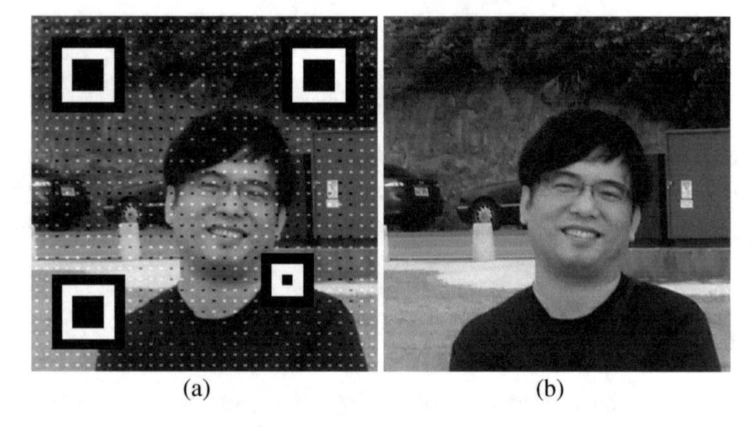

(a) (b)

Fig. 8. (a) The embedded photo QR code with lighter black pixels (b) The recovered photo with PSNR = 45.21.

4 Conclusions

This paper has introduced a kind of photo QR code generating method and furtherly hides the original photo into the photo QR code without extra storage. The experimental result demonstrates that the recovering result is with very high quality compared to the original photo.

References

1. Tan, J.S.: QR code. Synth. J. **3**, 59–78 (2008)
2. Huang, H.C., Chang, F.C., Fang, W.C.: Reversible data hiding with histogram-based difference expansion for QR code applications. IEEE Trans. Consum. Electron. **57**(2), 779–787 (2011)
3. Chu, H.K., Chang, C.S., Lee, R.R., Mitra, N.J.: Halftone QR codes. ACM Trans. Graph. **32**(6), 217.1–217.8 (2013)
4. Lin, S.S., Hu, M.C., Lee, C.H., Lee, T.Y.: Efficient QR code beautification with high quality visual content. IEEE Trans. Multimed. **17**(9), 1515–1524 (2015)
5. Chang, C.C., Lin, M.H., Hu, Y.C.: A fast and secure image hiding scheme based on LSB substitution. Int. J. Pattern Recogn. Artif. Intell. **16**(4), 399–416 (2002)
6. Chang, C.C., Hsiao, J.Y., Chan, C.S.: Finding optimal least-significant-bit substitution in image hiding by dynamic programming strategy. Pattern Recogn. **36**(7), 1583–1595 (2003)
7. Chan, C.K., Cheng, L.M.: Hiding data in images by simple LSB substitution. Pattern Recogn. **37**(3), 469–474 (2004)
8. Thien, C.C., Lin, J.C.: A simple and high-hiding capacity method for hiding digit-by-digit data in images based on modulus function. Pattern Recogn. **36**(12), 2875–2881 (2003)
9. Rubinstein, M., Shamir, A., Avidan, S.: Improved seam carving for video retargeting. ACM Trans. Graph. **27**(3), 16:1–16:10 (2008)
10. Dong, W., Zhou, N., Paul, J.C., Zhang, X.: Optimized image resizing using seam carving and scaling. ACM Trans. Graph. **28**(5), 125:1–125:10 (2009)

Using Color Converting to Hide Image Information

Wen-Pinn Fang(✉), Yu-Feng Huang, Lu-Hsuan Li, and Yan-Ru Pan

Yuan Ze University, No. 135, Yuandong Road, Zhongli District, Taoyuan City 320, Taiwan (R.O.C.)
wpfang@saturn.yzu.edu.tw

Abstract. Recently, it is very popular to convert image or video to different format. For example, copy a color printed paper to gray scale paper or show video with different devices. This paper proposes a fast removable watermark generating method. The watermark disappears after the watermarked image converts to another format. As a gray image can be generated from different color images with same converting approach, the goal of this paper has been achieved. An example shows that it is possible to create recoverable contour style watermark by control hue and saturation properly. A vision quality optimum approach also presented. The advantages of proposed method includes the recover image is identical to the converted image without watermark, people can remove the watermark without computer or the computational cost is very low. The result of this paper is useful for secret removing or demo video transmitting.

Keywords: Lossless recovery · Watermark · Fragile · Color model · Color converting

1 Introduction

Nowadays, a lot of multimedia transmits frequently. People use many kinds of media in everyday. The style of media includes text, audio, video and animator. During the data storing as the form of the images and videos, it is possible to remove some information for copyright protection or security issue. For example, it will be convenience for teachers when the correct answers disappear after the sheets have been picture copy as shown in Fig. 1. Another example is we can remove the watermark but blur a demo video for product preview as Fig. 2. This paper provides a fast, user-friendly approach for these situations.

There are many researchers study the relative approaches. For example, Yang, Hengfu and Yin [1], proposed a secure removable visible watermarking for BTC compressed images. Barni and Bartolini [2] proposed the watermarking system. Lin, Zhao and Liu [3] proposed theoretic strategies and equilibriums in multimedia fingerprinting social networks. Rial et al. [4] actual implement protocols for realistic multimedia contents. However, these discuss seldom remove the gap of digital and printed watermarking scheme.

J.-S. Pan et al. (eds.), *Advances in Intelligent Information Hiding and Multimedia Signal Processing*, Smart Innovation, Systems and Technologies 81,
DOI 10.1007/978-3-319-63856-0_37

The rest of this paper is organized as follows: the background knowledge is shown in Sect. 2; the method is proposed in Sect. 3; Experimental results are shown in Sect. 4. Finally, the discussion is represented in Sect. 5.

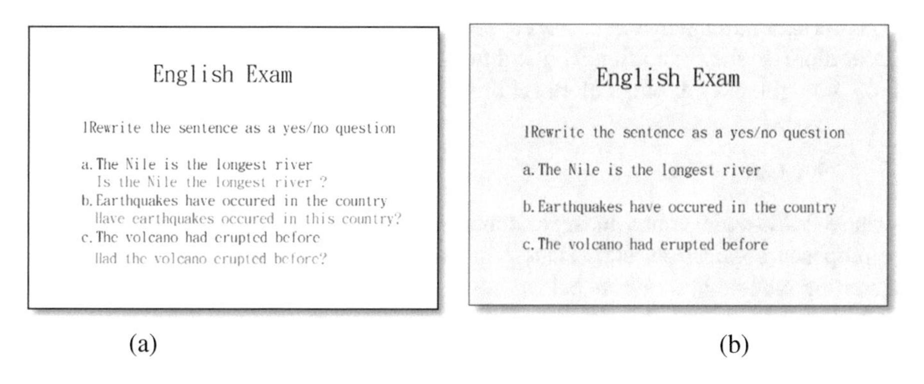

(a) (b)

Fig. 1. The application of proposed method. (a) is the original image (b) is the image after specific filter.

(a) (b) (c)

Fig. 2. The application of proposed method (a) is the watermark (b) is an original image of (c) is the video after specific filter which watermark disappear but became gray. (The copyright of snapshot is belongs to Alejandro González Iñárritu)

2 Relative Research

2.1 RGB Color Space

The RGB color space [2] is an additive color model in which red, green and blue light are added together in various ways to reproduce a broad array of colors. The name of the model comes from the initials of the three additive primary colors, red, green and blue (named r, g and b).

2.2 HSL Color Space

HSL color space [5] is similar to the way of how human's eyes "feel" colors. H represents hue, value from 0 to 360, refers to a pure color. S represents saturation, value from 0 to 100%. As saturation increases, colors appear more pure. As saturation decreases, colors appear more washed-vomiting. H is lightness (also called luminance), value from 0 to 100%. As lightness goes up, colors get brighter.

2.3 Color Converting

Because RGB model is not linear for human vision, HSL model is a better approach in the proposed application. HSL stands for hue, saturation, and luminosity. The color converting method is shown as below:

RGB to HSL conversion formula:

$$M = max\{r, g, b\} \tag{1}$$

$$m = min\{r, g, b\} \tag{2}$$

$$H = \begin{cases} 0^\circ, & \text{if } M = m \\ 60^\circ \times \frac{g - b}{M - m} + 0^\circ, & \text{if } M = r \text{ and } g \geq b \\ 60^\circ \times \frac{g - b}{M - m} + 360^\circ, & \text{if } M = r \text{ and } g < b \\ 60^\circ \times \frac{b - r}{M - m} + 120^\circ, & \text{if } M = g \\ 60^\circ \times \frac{r - g}{M - m} + 240^\circ, & \text{if } M = b \end{cases} \tag{3}$$

$$\mathrm{S} = \begin{cases} 0, & \text{if } l = 0 \text{ or } M = m \\ \frac{M - m}{M + m} = \frac{M - m}{2l}, & \text{if } 0 < l \leq \frac{1}{2} \\ \frac{M - \mathrm{m}}{2 - (M + m)} = \frac{M - \mathrm{m}}{2l}, & \text{if } l > \frac{1}{2} \end{cases} \tag{4}$$

$$L = \frac{1}{2}(M + m) \tag{5}$$

On the contrast, the formula that convert HSL to RGB is shown as below.

As $S = 0$, the result of color is achromatic or gray.

In this case R, G and B are equal to L. The H value in this situation is undefined. When $S \neq 0$ use following as below:

$$q = \begin{cases} L \times (1 + S), & if\ L < \frac{1}{2} \\ L + S - (L \times S), & if\ L \geq \frac{1}{2} \end{cases} \tag{6}$$

$$p = 2 \times l - q \tag{7}$$

$$h_k = \frac{h}{360} \tag{8}$$

$$t_R = h_k + \frac{1}{3} \tag{9}$$

$$t_G = h_k \tag{10}$$

$$t_B = h_k - \frac{1}{3} \tag{11}$$

$$\text{if}\, t_C < 0 \rightarrow t_C = t_C + 1.0 \quad \text{for each}\ C \in \{R, G, B\} \tag{12}$$

$$\text{if}\, t_C > 1 \rightarrow t_C = t_C - 1.0 \quad \text{for each}\ C \in \{R, G, B\} \tag{13}$$

$$Color_C = \begin{cases} p + ((q - p) \times 6 \times t_C), & \text{if}\, t_C < \frac{1}{6} \\ q, & \text{if}\, \frac{1}{6} \leq t_C < \frac{1}{2} \\ p + \left((q - p) \times 6 \times (\frac{2}{3} - t_C)\right), & \text{if}\, \frac{1}{6} \leq t_C < \frac{1}{2} \\ q, & \text{otherwise} \end{cases} \tag{14}$$

$$\text{for each}\ C \in \{R, G, B\}$$

A common method is to use the principles of photometry to match the luminance of the grayscale image to the luminance of the original color image. The gray value can be gotten by Eq. 6.

$$\text{Gray} = 0.2126\,R + 0.7152\,G + 0.0722\,B \tag{15}$$

2.4 Visible Watermarking

A digital watermark [6] is a kind of marker covertly embedded in a noise-tolerant signal such as an audio, video or image data. It is typically used to identify ownership of the copyright of such signal. The visible watermarking scheme used to embedding an image which has changed its opacity or tone. In order to conserve feature of the image and protect the content of digital media from unauthorized use.

A general notion of reversible visible watermark was proposed by IBM in 1997 [7] and the concept of invertible authentication was first presented by Honsinger et al. [8]. In spatial domain, a visible watermarking algorithm has been proposed, embedding a binary logotype on the host image, which is a JND [9] (Just Noticeable Distortion) multiplied by a constant. Furthermore, though the managing visibility [8], quality of the watermarked image can improve.

2.5 Picture Copy Model

Today, it is popular to use copier to reproduce a gray color image. A simple testing is shown as Fig. 3. It is obvious that it is possible generate similar printed image from different color images.

(a) (b) (c) (d)

Fig. 3. The test of lemma 1. (a) is the original image (b) is the yellow version image (c) is the gray image which converted from (a) or (b) (d) is the printed and then hard copy image

3 Proposed Method

This paper proposes a fragile watermarking method for visual watermarking. The goal of the proposed method is shown as Fig. 4. The goal of the proposed method is to design a watermarking scheme which the watermark is easy to remove by optical operation. The notation of the algorithm is shown in Table 1.

$$\mathbf{O}' = \mathbf{O} \oplus \mathbf{W} \tag{7}$$

$$\mathbf{T}' = \mathrm{F}(\mathbf{O}') \tag{8}$$

$$\mathbf{T} = \mathrm{F}(\mathbf{O}) \tag{9}$$

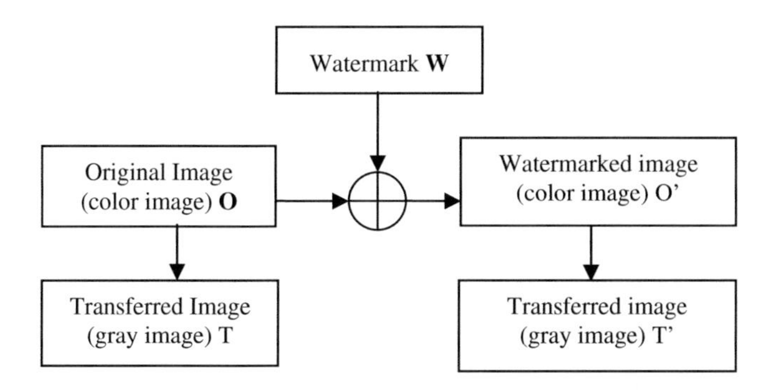

Fig. 4. The goal of proposed method.

Table 1. Notation list

Notation	Explain
O	Original image
O'	Watermarked image
W	Watermark
T	The transferred image which generated from original image
T'	The transferred image which generated from watermarked image
$\oplus$	Embed watermark
F	Watermark remove filter

Here, **T** is identical to $\mathbf{T}'$

The choice of filter depends on the applications. The common filters include gray value, saturation modification and picture copy.

If the filter is convert color image to gray value image, the watermark embedding algorithm is shown as Embedding Algorithm 1:

Embedding algorithm1

Input: Original image (color image) which size is $w \times h$, watermark (binary image) which is $m \times n$, location of watermark (x_0, y_0), predefine degree d.

Output: Watermarked image

Determine thegray valueof original image in the region P(x, y) which $x= x_0\sim x_0+m$, $y= y_0\sim y_0+n$by equation (6), named **Gray**(x, y).

Let O'=O

For x=1 to m

For y=1 to n

if **W**(x, y)=white then

O'(x_0+x, y_0+y)=**O**(x_0+x, y_0+y)

else

random select **R'**(x_0+x, y_0+y) **and G'**(x_0+x, y_0+y)

where Gray-R'-G'>0

B'$(x_0+x, y_0+y)=\frac{\mathbf{Gray}(x0+x,y0+y)-0.2126\mathbf{R}(x0+x,y0+y)+0.7152\mathbf{G}(x0+x,y0+y)}{0.0722}$

end if

end for

end for

---end of algorithm

However, it is difficult to control the quality of watermarked image. Another filter approach is proposed as below. Assume the filter is picture copy, the watermark embedding algorithm is shown as Embedding Algorithm 2:

Embedding algorithm2

Input: Original image (color image) which size is $w\times h$, watermark (binary image) which is $m\times n$, location of watermark (x_0, y_0), quality parameter degree d.

Output: Watermarked image

First, Determine the h(hue), s(saturation), and l(luminosity) values of original image in the region P(x, y) which $x= x_0\sim x_0+m$, $y= y_0\sim y_0+n$by equation (3)-(5), named **L**(x, y), **H**(x, y), **S**(x, y).

For x=1 to m

For y=1 to n

if **W**(x, y)=white then

O'(x_0+x, y_0+y)=**O**

else

L'(x_0+x, y_0+y)= **L**(x_0+x, y_0+y)

S'(x_0+x, y_0+y)= **S**(x_0+x, y_0+y)

H'(x_0+x, y_0+y)= **H**$(x_0+x, y_0+y)+ d$ Mod 360

end if

end for

end for

---end of algorithm

Watermarking remove method

The watermark can be removed by just picture copy. However, it is possible that there is no solution to generate the watermark. For example, if the original is a pure white image. Although it is seldom to generate a blank image which just has watermark in real application. The problem can be solved by shift all pixel values as a preprocessor. And then adopt the Algorithm 2.

If the filter is the saturation modifier the embedding algorithm is shown as below:

Embedding algorithm3

Input: Original image (color image) which size is $w \times h$, watermark (binary image) which is $m \times n$, location of watermark (x_0, y_0), quality parameter d.

Output: Watermarked image

First, Determine the h(hue), s(saturation), and l(luminosity) values of original image in the region P(x, y) which x= x_0~ x_0+m, y= y_0~ y_0+$n$by equation (3)-(5), named **L**(x, y), **H**(x, y), **S**(x, y).

For x=1 to m

For y=1 to n

if **W**(x, y)=white then

O'(x_0+x, y_0+y)=**O**

else

L'(x_0+x, y_0+y)= **L**(x_0+x, y_0+y)

S'(x_0+x, y_0+y)= **S**(x_0+x, y_0+y)+d

H'(x_0+x, y_0+y)= **H**(x_0+x, y_0+y)

end if

end for

end for

---end of algorithm

Watermarking remove method

The watermark can be removed by just picture copy. Or just transfer the color image to gray level image.

4 Experimental Result

There are three cases in the experiments. The first case is to generate watermarked image by changing the corresponding pixel values with Algorithm 1 as shown in Fig. 5. The printed image is the same as the gray version of original image. The second case is to changing the corresponding hue to generate watermarked image. As shown in Fig. 6. The reversed method is to change the hue. The third case is to change the saturation and hue to make the watermarked image and the reverse method is hard copy the color image by copier (Fig. 7).

Case 1: Gray values modification
Case 2: Hue modification
Case 3: Same luminosity

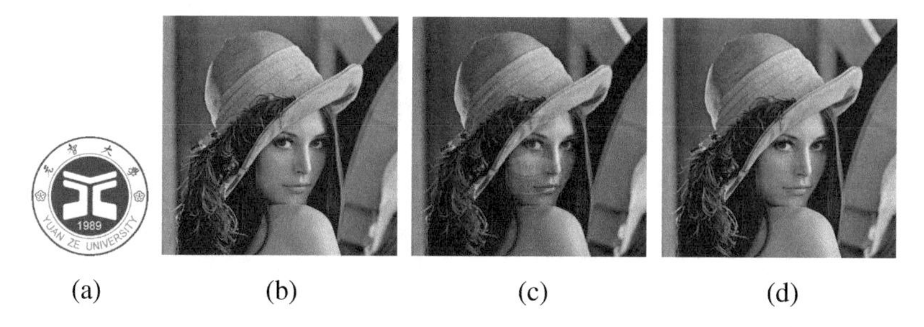

Fig. 5. The experiment of case 1. (a) is the watermark (b) is original image (c) is the watermarked image (d) the hard printed result.

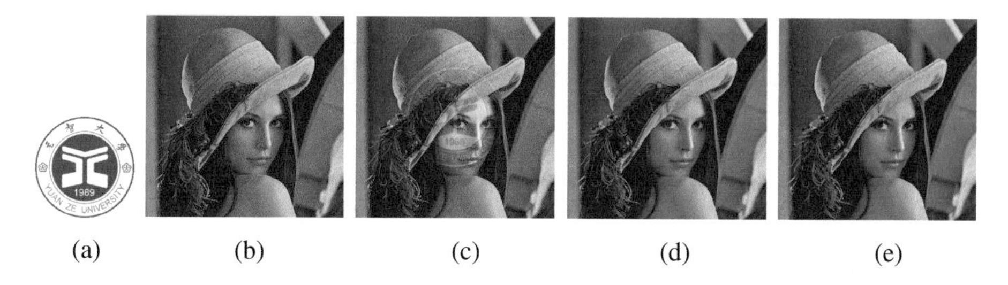

Fig. 6. The experiment of case 2. (a) is the watermark (b) is original image (c) is the watermarked image (d) is the gray image converted from original image. (e) is the image that set the saturation to zero.

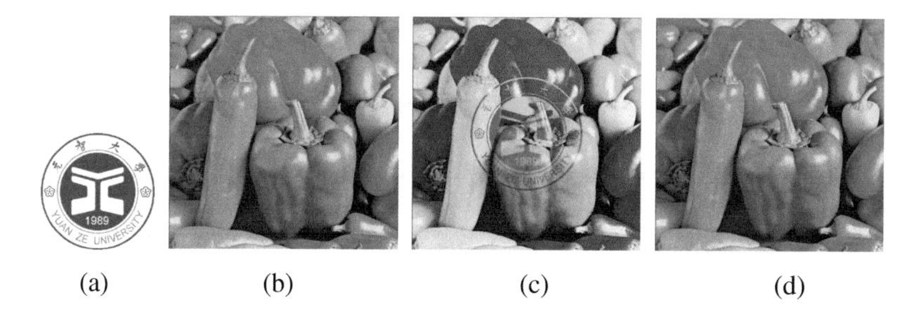

Fig. 7. The experiment of case 3. (a) is the watermark (b) is original image (c) is the watermarked image (d) the hard printed result.

5 Discussion and Conclusion

This paper proposed a fast removable watermarking embedding method. The method is suitable for digital and printed application. There are three cases to demonstrate the approach. The case 1 can be applied to pure digital image processing. The case 2 and case 3 can be applied to printed application. The result of this paper can be used in hand write examination and copyright protection of multimedia.

Furthermore, the visual quality of watermark can be controlled by selecting different hue. For example, Fig. 8(c) and (d) are generated by hue and saturation modification. This need more human vision model testing. In the future, an optimum approach will be study.

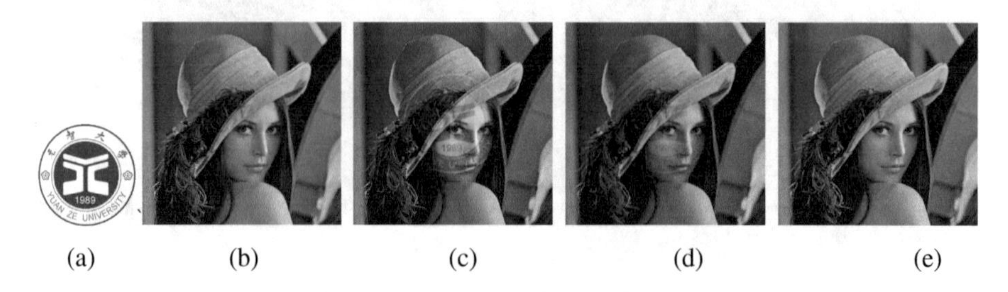

(a) (b) (c) (d) (e)

Fig. 8. The experiment of case3 (a) is the watermark (b) is original image (c) is the watermarked image with hue modification (d) is the watermarked image with saturation modification (d) the hard printed result.

Compare with exist watermark embedded methods, as shown in Table 2, the application of this proposed method can both adopt with and without computer. It is suitable for printed image or demonstration of preview video as mention in introduction.

Table 2. Comparison with exist study

Method	Medea	Feature
[7]	Digital image	Cannot remove
This paper	Color watermarked image	Remove with and without computer

References

1. Yang, H., Yin, J.: A secure removable visible watermarking for BTC compressed images. Multimed. Tools Appl. **74**(6), 1725–1739 (2015)
2. Barni, M., Bartolini, F.: Watermarking Systems Engineering: Enabling Digital Assets Security and Other Applications. Marcel Dekker, New York (2004)
3. Lin, W., Zhao, H., Liu, K.: Game-theoretic strategies and equilibriums in multimedia fingerprinting social networks. IEEE Trans. Multimedia **13**(2), 191–205 (2011)
4. Rial, A., Deng, M., Bianchi, T., Piva, A., Preneel, B.: A provably secure anonymous buyer-seller watermarking protocol. IEEE Trans. Inf. Forensics Secur. **5**(4), 920–931 (2010)
5. RGB color model. https://en.wikipedia.org/wiki/RGB_color_model
6. HSL color model. https://en.wikipedia.org/wiki/HSL_and_HSV
7. Cox, I.J.: Digital watermarking and steganography. Morgan Kaufmann, Burlington (2008)
8. Mintzer, F., et al.: Safeguarding digital library contents and users. D-lib Mag. **3**(7/8) (1997)
9. Honsinger, C.W., et al.: Lossless recovery of an original image containing embedded data. U.S. Patent No. 6,278,791, 21 August 2001
10. Agarwal, H., Sen, D., Raman, B., Kankanhalli, M.: Visible watermarking based on importance and just noticeable distortion of image regions. Multimed. Tools Appl. **75**, 7605–7629 (2015). 25 p.
11. Fragoso-Navarro, E., et al.: Visible watermarking technique in compressed domain based on JND. In: 2015 Proceedings of the 12th International Conference on Electrical Engineering, Computing Science and Automatic Control (CCE). IEEE (2015)

A Novel Visible Watermarking Scheme Based on Distance Transform

Guo-Jian Chou[1], Ran-Zan Wang[1,2(✉)], Yeuan-Keun Lee[3], and Ching Yu Yang[4]

[1] Department of Information Communication, Yuan Ze University, Taoyuan, Taiwan
zwang@saturn.yzu.edu.tw

[2] Department of Computer Science and Engineering, Yuan Ze University, Taoyuan, Taiwan

[3] Department of CSIE, Ming Chuan University, Taipei, Taiwan

[4] Department of CSIE, National Penghu University of Science and Technology, Magong, Taiwan

Abstract. Visible watermarking is a common intellectual property protection mechanism for digital images. It is fulfilled by translucently overlaying a visual logo and/or a segment of text onto an image to declare the ownership of the image and thus deterring unauthorized usage to the image. This paper presents a novel visible watermarking technique by considering HVS characteristics and applying a distance transform function to evaluate appropriate stamping strength for individual watermark pixels. It exhibits lucid watermark pattern under moderate modification to the host image. Compare to uniform-contrast watermarks generated using conventional HVS based visible watermark schemes, the proposed method achieves better balance to the visual quality between the watermark and the image content beneath it.

Keywords: Digital watermarking · Visible watermark · Distance transform · Human Visual System

1 Introduction

Visible watermark [1] is an artificial logo or text translucently overlaid on a host image, which can directly be viewed by naked eye. It is typically utilized to notify and declare the ownership of the host image. In order to display the watermark lucidly while maintaining the context of the original the image, contemporary visible watermarking schemes [2–5] usually apply characteristics of Human Visual System (HVS) to evaluate appropriate strength for stamping the watermark. These methods often exhibit the watermark with uniform visual contrast, and small alternation to the image is required to maintain the fidelity of the original content.

The visible watermarking algorithm proposed by Kankanhalli et al. [2] divides the image into non-overlapping blocks with 8 × 8 pixels, and each block is classified into one of 8 classes depending on the sensitivity of block distortion parameters include texture, edge and luminance information. The watermark bits are embedded in the discrete cosine transform (DCT) domain and the embedding strength of the watermark in a block depends on the class to which the block belongs, which obtains pleasant and

J.-S. Pan et al. (eds.), *Advances in Intelligent Information Hiding and Multimedia Signal Processing*, Smart Innovation, Systems and Technologies 81,
DOI 10.1007/978-3-319-63856-0_38

unobtrusively watermarked image irrespective of the type of image. Huang et al. [3] proposed a visible watermarking technique by analyzing the image content such as textures, edges, smooth areas, and wavelet coefficients contrast-sensitive function (CSF) of perceptual importance weight. The watermark in different regions of the image is stamped with varied strength depending on the underlying content of the image and human's sensitivity to spatial frequencies, which ensures the perceptual uniformity of the embedded watermark over different regions of the image. Zeng and Wu [5] proposed an image adaptive watermarking method by considering the HVS characteristics. In their method, the scaling factor and the embedding factor for watermarking inserting is calculated by combining image features such as luminance and texture from both the host image and the watermark image, and the watermark is embedded into the host image by adjusting adaptively the scaling factor and the embedding factor. The method obtains the watermarked image with good visual perceptual quality and exhibits the watermark in high contrast. Agarwal et al. [6] proposed a visible watermarking algorithm in which watermarking positions are automatically found using visual saliency or eye fixation density map. They presented a mathematical model in terms of importance of portion of an image, which automatically determines the optimal embedding energy during the implementation. The method stamps the watermark in such a way that important portions of image are not occluded by watermarks.

In general, an efficient visible watermark should have the following properties:

1. The watermark should be visible to naked eye and must not significantly obscure the image details beneath it.
2. The watermark should spread in a large and important area of the image in order to prevent its deletion by clipping.
3. The watermark must be hard to remove automatically by computer programs.

The remainders of this paper are organized as follows. Section 2 introduces the distance transform for digital image. The details of proposed method are described in Sect. 3. Experimental results are shown in Sect. 4, and conclusions are made finally in Sect. 5.

2 Distance Transform

The distance transform (DT) [7] is a derived representation for binary images, it maps each pixel into its smallest distance to region of interests. Typical DT represents a black/foreground pixel by the distance from it to the closest boundary, and represents a white/background pixel by value 0. The result of DT is usually shown using a gray level image with the same scale to the input image, in that the gray level intensities of points inside foreground regions depicts the distance to the closest boundary from each point.

DT can be defined in terms of arbitrary distance metrics. Three common metrics are (1) Euclidean distance, (2) City block distance (or Manhattan distance), and (3) Chess-board distance. The equation to evaluate the distance between two pixels $p = (x_1, y_2)$ and $q = (x_2, y_2)$ for the three metrics are summarized in Eqs. (1) to (3).

$$D_{Euclidean}(p,q) = \sqrt{(x_1 - x_2)^2 + (y_1 - y_2)^2}, \tag{1}$$

$$D_{CityBlock}(p,q) = |x_1 - x_2| + |y_1 - y_2|, \tag{2}$$

$$D_{Chessboard}(p,q) = \max\{|x_1 - x_2|, |y_1 - y_2|\}. \tag{3}$$

3 The Proposed Method

In general, the process of stamping watermark W on host image H to obtain the watermarked image H_w can be expressed by the equation

$$\mathbf{H} + \mathbf{W} \rightarrow \mathbf{H}^w. \tag{4}$$

In this paper we focus on the issue of stamping a binary visible watermark on a true color host image. The flowchart of the proposed watermarking stamping process is depicted in Fig. 1.

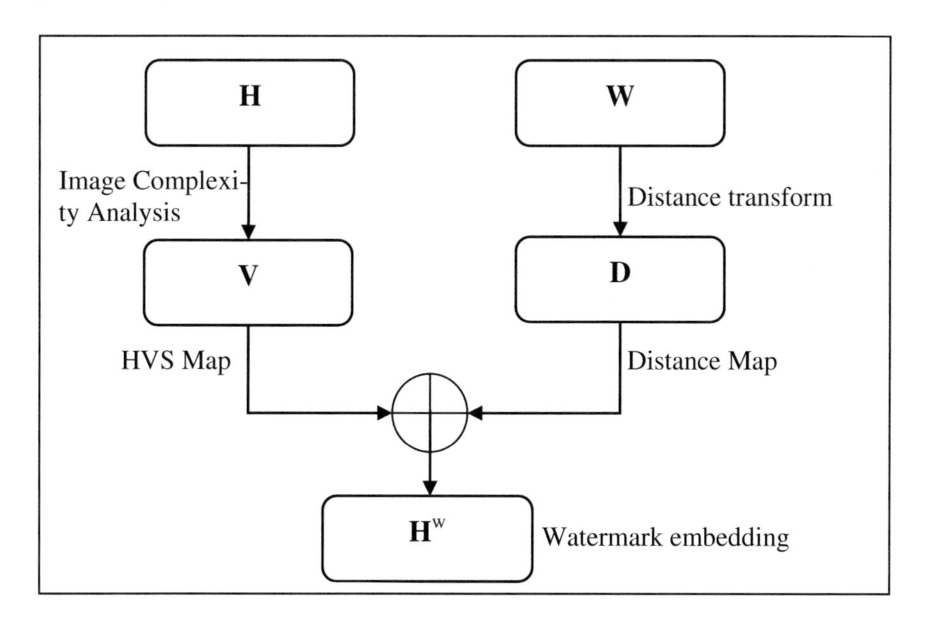

Fig. 1. Flowchart of the proposed visible watermarking scheme.

In this study it is assumed the host image **H** and the watermark **W** have the same dimension M × N, i.e. there is one to one correspondence between image pixels p_{ij} on **H** and watermark bits w_{ij} on **W**. The host image **H** is first transformed to CIE Lab color space, and all the operations are applied on the L channel. The watermark stamping process is characterized by the function about modifying the luminance component l_{ij} of pixel p_{ij} according to the watermark bit $w_{ij} = \{0, 1\}$, which is given below:

$$l_{ij}^{w} = \begin{cases} l_{ij} + t_{ij} \times \Delta l, & \text{if } w_{ij} = 1, \\ l_{ij}, & \text{if } w_{ij} = 0, \end{cases} \tag{5}$$

where l_{ij} and l_{ij}^{w} are luminance values before and after the watermark stamping process, respectively; Δl is a user-specified threshold that controls the amount of change to the luminance, and t_{ij} is the watermark stamping strength for p_{ij}.

The watermark stamping strength t_{ij} controls the degree of change to the luminance of pixel p_{ij}. It consists of two parts: the HVS term $\mathbf{V} = \{v_{ij}\}$ and the distance term $\mathbf{D} = \{d_{ij}\}$, which is combined by a weighted sum function as follows:

$$t_{ij} = \alpha \times v_{ij} + (1 - \alpha) \times (1 - d_{ij}), \tag{6}$$

where α is a real number in the range [0, 1]. The HVS term v_{ij} is applied to evaluate the visual impact to the luminance change of image pixels, and the distance term d_{ij} is applied to derive the visual stress of watermark pixels. Both of them are real numbers with value normalized to the range from 0 to 1.0.

In terms of HVS property, there were many researches [8–10] devoted to understand the HVS. The sensitivity and masking properties of HVS are usually applied in the design of digital watermark [11, 12], and most of them have chosen to work only on the luminance. In this paper we measure the masking effect of luminance change by estimating the neighborhood complexity of a pixel, and design a function to decide the modification strength. It follows the property of HVS that human eyes are less sensitivity to the luminance change in noisy region than smooth region, hence the modification to the pixel in smooth region should be smaller in order to keep the fidelity to the content of the host image. The HVS term v_{ij} is determined using the following equation

$$v_{ij} = \exp\left(\frac{\sigma_{ij}^2 - \sigma_{\min}^2}{\sigma_{\max}^2 - \sigma_{\min}^2} - 1\right), \tag{7}$$

where σ_{ij}^2 is the variance of 3 × 3 block with center point p_{ij}, and $\sigma_{\max}^2$ and $\sigma_{\min}^2$ are the maximum value and minimum value of variance.

The distance term d_{ij} is applied to derive the visual stress of the pixels on the watermark, which is calculated using the Euclidean distance transform shown in Eq. (1) to the watermark image. Let the distance obtained by applying Euclidean distance transform to **W** is denoted by $\mathbf{E} = \{e_{ij}\}$, and the maximum distance obtained is $e_{\max}$, all of the distance is divided by $e_{\max}$ to get the normalized distance in the range [0, 1]. A pixel with small distance value represents the pixel is located nearby the boundary of the watermark pattern, which should has enough contrast in order to exhibit the contour of the watermark. The normalized distance transform term d_{ij} is calculated using the following equation

$$d_{ij} = \exp\left(\frac{-2 \times e_{ij}}{e_{\max}}\right). \tag{8}$$

4 Experimental Results

The test program was implemented in C++ language, and ran on a personal computer with Intel Core i7 CPU with 2.80 GHz and 4 GB memory. Figure 2 illustrates watermark stamping examples of the proposed scheme. The parameter α in Eq. (6) is set 0.2. The images shown in Figs. 2(a) and (b) are the true color host image and the binary watermark used in this test, both of them are with size 512×512. Figure 2(c) is the stamping strength map evaluated using Eq. (6), which has higher stamping strength in the regions corresponding to the boundary region of the watermark and the noisy region of the host image.

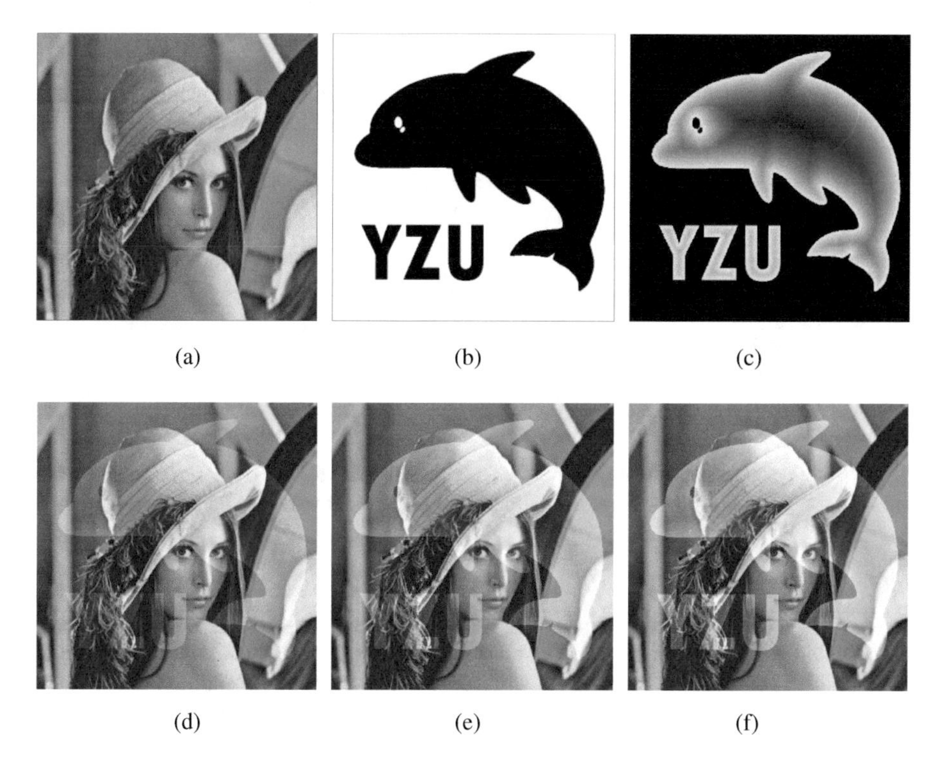

Fig. 2. Illustration of the proposed visible watermark. (a) Host image, (b) Watermark, (c) Stamping strength map, (d)–(f) are watermarking results obtained by setting Δl to 40, 60 and 80, respectively.

The watermarked images obtained by setting Δl to 20, 40 and 60, respectively, are shown in Figs. 2(d) to (f). It can be seen from these Figs. that the contrast of the watermark gets higher when the stamping strength increases, and the content of the host image beneath the watermark can be viewed lucidly.

Figures 2 and 3 compare the three watermarking schemes: (a) Uniform strength ($\Delta l/2$) watermarking, (b) HVS watermarking, (c) the proposed method. The parameter Δl set to 60. It can be seen that the proposed watermarking method by combining the HVS term and the DT term has the best visual quality, both on the visible watermark and the content of the host image beneath (Fig. 4).

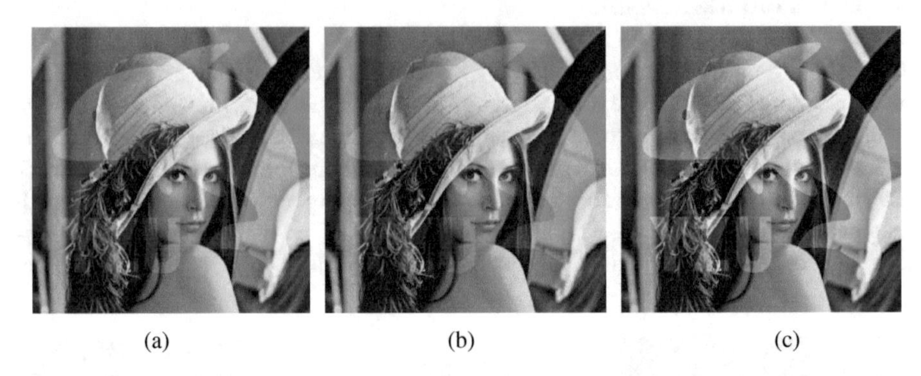

(a) (b) (c)

Fig. 3. Comparison of visible watermarks obtained using different stamping schemes: (a) Uniform strength watermarking, (b) HVS watermarking, (c) the proposed method.

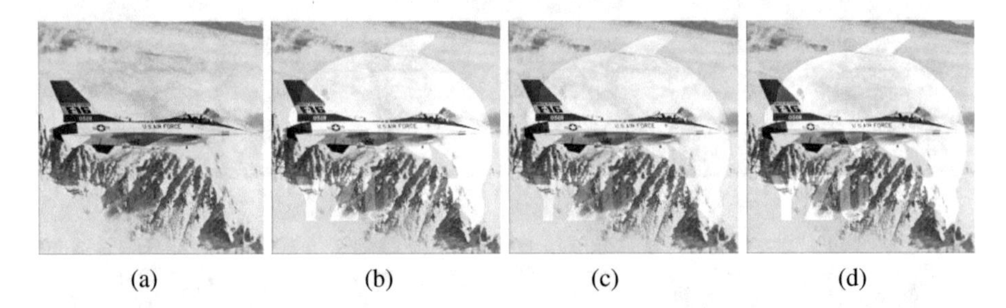

(a) (b) (c) (d)

Fig. 4. Another example of visible watermarks obtained using different watermarking schemes: (a) host image, (b) Uniform strength watermark, (c) HVS watermark, (d) the proposed method.

5 Conclusion

Most visible watermark schemes in the literature exhibit watermark in uniform contrast to make it be perceived clearly by viewers; however, the image content underneath the watermark is obscure, especially in the region containing objects in low resolution. In this paper we presents an efficient watermarking scheme by considering the visual clarity both to the watermark and to the watermarked image. The proposed method apply the DT to evaluate the stamping strength for the watermark, which yields a good compromise between the lucidness to the watermark and the visibility to image content beneath the watermark.

References

1. Yeung, M.M., Mintzer, F.C., Braudaway, G.W., Rao, A.R.: Digital watermarking for high-quality imaging. In: Proceedings of IEEE First Workshop on Multimedia Signal Processing, pp. 357–362 (1997)
2. Kankanhalli, M.S., Rajmohan, Ramakrishnan, K.R.: Adaptive visible watermarking of images. In: IEEE International Conference on Multimedia Computing and Systems, vol. 1, pp. 568–573 (1999)
3. Huang, B.B., Tang, S.X.: A contrast-sensitive visible watermarking scheme. IEEE Multimed. **13**(2), 60–66 (2006)
4. Tsai, M.J.: A visible watermarking algorithm based on the content and contrast aware (COCOA) technique. J. Vis. Commun. Image Represent. **20**(5), 323–338 (2009)
5. Zeng, W., Wu, Y.: A visible watermarking scheme in spatial domain using HVS model. Inf. Technol. J. **9**(8), 1622–1628 (2010)
6. Agarwal, H., Sen, D., Raman, B., Kankanhalli, M.: Visible watermarking based on importance and just noticeable distortion of image regions. Multimed. Tools Appl. **25**, 7605–7629 (2015)
7. Fabbri, R., et al.: 2D euclidean distance transform algorithms: a comparative survey. ACM Comput. Surv. **40**(1), Article 2, February 2008
8. Mannos, J.L., Sakrison, D.J.: The effects of a visual fidelity criterion on the encoding of images. IEEE Trans. Inf. Theory **IT-20**(4), 525–536 (1974)
9. Daly, S.: A visual model for optimizing the design of image processing algorithms. Proceedings IEEE International Conference on Image Processing, vol. II, pp. 16–20 (1994)
10. Eckert, M.P., Bradley, A.P.: Perceptual quality metrics applied to still image compression. Sig. Process. **70**(3), 177–200 (1998)
11. De Vleeschouwer, C., Delaigle, J.F., Maco, B.: Invisibility and application functionalities in perceptual watermarking - An overview. Proc. IEEE **90**(1), 64–77 (2002)
12. Singh, P., Chadha, R.S.: A survey of digital watermarking techniques, applications and attacks. Int. J. Eng. Innov. Technol. (IJEIT) **2**(9), 165–175 (2013)

Using Digital Hiding to Revitalize Traditional Chinese Proverb

Wen-Pinn Fang[1(✉)], Yan-Jiang[2], Jiu-Sheng Kuo[1], and Verna Ip[1]

[1] Yuan Ze University, No. 135, Yuan-Tung Road, Chung-Li District, Taoyuan City 32003, Taiwan
wpfang@saturn.yzu.edu.tw

[2] Xiamen University of Technology, No. 600 Ligong Road, Jimei District, Xiamen 361024, Fujian Province, China

Abstract. This paper proposes architecture to hide proverb data in the corresponding animator image. The method not only stores data but also prevents the relation between image and proverb data lose. The hiding data includes the word of proverb, pronunciation and the explanation. Base on the digital data hiding algorithm, the goal has been achieved. A prototype system is also established in this study. The result can help the traditional culture preserve in the internet.

Keywords: Lossless recovery · Watermark · Fragile · Color model · Color converting

1 Introduction

Chinese traditional proverb is the shortest literary style, it can describe a natural phenomena or life truth with fluent words that can be read smoothly, Chinese traditional proverb contain Chinese traditional sense of worth, scientific truth, it is the great intelligence of the people. Chinese traditional proverb was carried by word of mouth in the interpersonal communication, it met also be inherited by the next generation for thousand years. But in recent years, with the develop of internet technology, change of the society and the migration of the population, Chinese traditional proverb began to lose their condition for communication. There are many countries and international organizations try to use digital technology to save traditional culture in the world [1], such as American's Memories project [2]. At the same time, we have gotten so much experience about digitizing traditional culture. As the most projects about digitizing traditional culture, traditional culture were recorded as in text, picture, video and audio. The digital files should be saved with reasonable structure. American Memories set up standard document library in Encoded Archival Description (EAD) and Text Encoding Initiative (TEI), which provide efficient files for users. EAD was applied for describe documents and manuscript resource, including text, video and scanning document, it can set up a standardized coding system. TEI is a text document encoding standard which is applied for digital document exchanging. It provides rules for describe text, markup tags and structure of record. Every record of traditional culture has metadata and content data,

J.-S. Pan et al. (eds.), *Advances in Intelligent Information Hiding and Multimedia Signal Processing*, Smart Innovation, Systems and Technologies 81,
DOI 10.1007/978-3-319-63856-0_39

all the information such as the edition, author, notes can be saved in the library, so that the library can be flexible, extendable and integration. In Bulgarian national folklore digital archive library, they use image watermarking in spatial domain to keep the images from unauthorized distribution. They also used method for audio watermarking with amplitude modulation and text watermarking with encoding by row offset. Their functional specification based on the internet environment for registration, documentation, access and exploration of a practically unlimited number of Bulgarian folklore artefacts and specimens digitally included in the "Bulgarian Folklore Heritage" archive, including content annotation, preview, complex search, selection, group and management. The above two methods bring us so much experience, they convert intangible heritage into digital documents, then the information was saved as a structured library, presented on the webpage for users. Water marking was used the copyright of the digital document from unauthorized use. This paper use digital watermarking for spreading Chinese traditional proverb, there are two methods used to hide information in the pictures. (a) the information is hide in meta header, it can be reveal with common tools, but it might be modified easily. (b) hiding the information by modifying local palette, it cannot reveal directly, it is a loosely method and the capacity of hiding information is limited (Fig. 1). The rest of this paper is organized as follows: the background knowledge is shown in Sect. 2; the method is proposed in Sect. 3; Experimental results are shown in Sect. 4. Finally, the discussion is represented in Sect. 5.

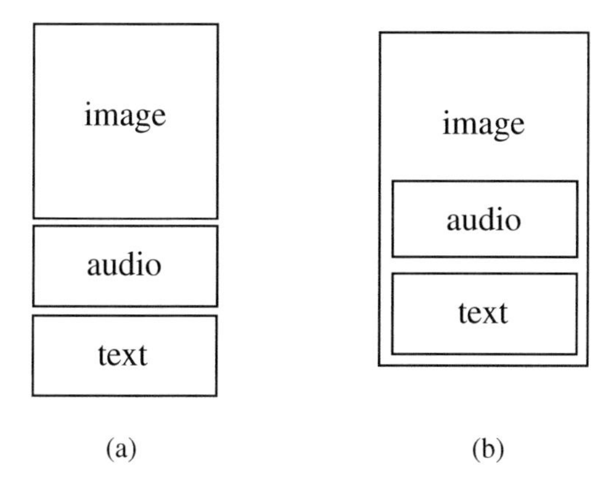

Fig. 1. The different between traditional store format and the proposed method. (a) is the traditional format that save image, audio, text in individual position. (b) is the proposed format that save all in one file.

2 Relative Research

2.1 Digital Archives

Digital preservation [3] is the method of keeping digital material alive so that they remain usable as technological advances render original hardware and software specification obsolete. In library and archival science, digital preservation is a formal endeavor to ensure that digital information of continuing value remains accessible and usable. The goal of digital preservation is the accurate rendering of authenticated content over time. The one approach of digital preservation method is digital archive. A digital archive is a repository that stores one or more collections of digital information objects with the intention of providing long-term access to the information.

2.2 Data Hiding

Digital watermarking [4] appears mainly to protect and guarantee the ownership of intellectual property rights. Digital watermarking refers to embed a message into a picture, movie or audio. [5] There are two types of digital watermarking: visible watermarking and invisible watermarking. Visible watermarking uses to display the creator or copyright owner's name or trademark. Invisible watermarking does the opposite; insert information into an image that cannot be seen, although it can be detected with the right software. There are four categories of using digital watermarking, including: Copyright protection watermarking, tampering tip watermarking, anti-counterfeiting watermarking and anonymous mark watermarking.

2.3 The Format of Graphics Interchange Format

The Graphics Interchange Format [6] is a bitmap image format that has since come into widespread usage on the World Wide Web due to its wide support and portability. The format supports up to 8 bits per pixel for each image, allowing a single image to reference its own palette of up to 256 different colors chosen from the 24-bit RGB color space. It also supports animations and allows a separate palette of up to 256 colors for each frame. It is well-suited for simpler images such as graphics or logos with solid areas of color. The format of GIF is shown as Fig. 2.

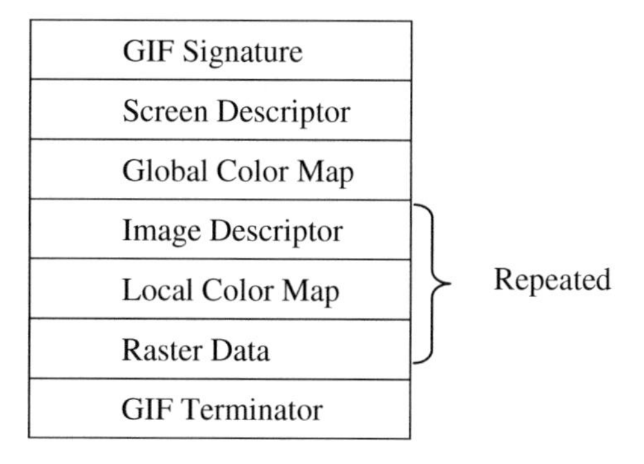

Fig. 2. The file format of GIF.

An example of local color table of GIF is shown in Fig. 3. There are 256 colors per color map. The depth of color is 24 bits. There are three channels stores interleave. The sequence is red, green and blue.

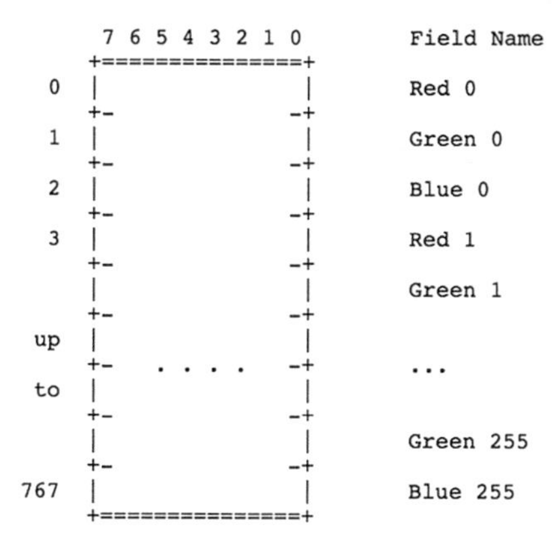

Fig. 3. An example of local color table syntax

2.4 The Characteristic of Traditional Chinese Proverb

A proverb is a simple and concrete saying, popularly known and repeated, that expresses a truth based on common sense or experience. They are often metaphorical. A proverb that describes a basic rule of conduct may also be known as a maxim. An example of traditional Chinese proverb is shown as below. The characteristic of traditional Chinese proverb is brief and more meaning. Often, the pronunciation of traditional Chinese proverb is different from the official format. It is necessary to preserve the pronunciation and meaning of it (Fig. 4).

http://history.bayvoice.net/b5/whmt/2015/10/29/139079.htm

Transliteration (pinyin): Shùdǎohúsūnsàn.

Traditional: 樹倒猢猻散

Simplified: 树倒猢狲散

Translation: When the tree falls, the monkeys scatter.

English equivalent: Rats desert a sinking ship.

"When a leader loses power, his followers become disorganized.

This proverb is often used to describe fair-weather friends.

Fig. 4. An example of traditional Chinese proverb

3 Proposed Method

This paper proposes a method to hide data in animated image such as GIF, that is suitable for hiding text data. The text data is hided in color maps of Gifs. There are two situations for data hiding: lossy and lossless hiding. The difference between lossy and lossless hiding is based on the image characteristic. If the number of colors is less than 256, it is possible to create a lossless stego-image. As shown in Fig. 5, the information is hided in many frames. The hiding data will be convert to bit stream, hiding material might be text, audio and video.

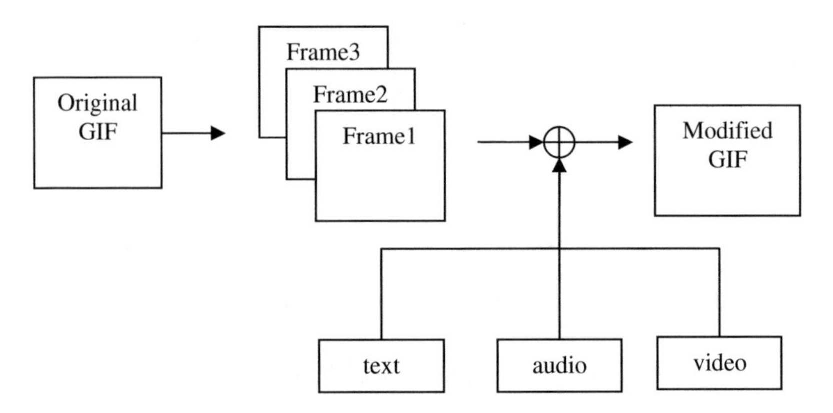

Fig. 5. The proposed method

Before hiding procedure is performed, the necessary number of color map shall be predicted. Assume that the size of hiding data (bit stream) is l, the necessary number of color map is n. The minimum value of n can be determined in Eq. 1.

$$n = \left\lceil \frac{l}{256 \times 3} \right\rceil \tag{1}$$

After predict the number of necessary color map, a least significant bit hiding method will be performed. If the original color maps are $\mathbf{M}_j(i)$ and the modified color maps are

$\mathbf{M}'_j(i)$, $j \geq n$ and $1 \leq i \leq 256 \times 3$ (if the animator is GIF, the size of every color map is 256 with 3 channels). The hiding data bit stream is B.

The hiding method is shown as Eq. 2.

$$\mathbf{M}'_j(i) = \left\lfloor \frac{M_j(i)}{2} \right\rfloor \times 2 + \mathbf{B}(\mathrm{k}) \tag{2}$$

Where k is from 0 to l.

$$j = \left\lfloor \frac{k}{256 \times 3} \right\rfloor \tag{3}$$

$$i = k mod(256 \times 3) \tag{4}$$

The detail is shown as below:

<u>Hiding data in GIF</u>

Input:original color map $\mathbf{M}_j$, $j \leq m$, bit stream data $\mathbf{B}(i)$, $i \leq l$

Output: modofied color map $\mathbf{M}'$

Convert data into bit stream B

Get n by equation 1.

If $n > l$then

  Return false

else

  $k \leftarrow 1$

  $while\ (k < l)$

    $j = \left\lfloor \frac{k}{256 \times 3} \right\rfloor$

    $i = k\ mod\ (256 \times 3)$

    $\mathbf{M}'_j(i) = \left\lfloor \frac{M_j(i)}{2} \right\rfloor \times 2 + \mathbf{B}(k)$

    $k \leftarrow$ k+1

  End while

End if

-------------- End of method

The information can be gotten by reverse the hiding method.

4 Experimental Result

This paper proposes a method to hide information in animator image, such as GIF. There are two situation: lossless and lossy recovery. If the information is limited, a lossless recovery result has been gotten by reorder the index of color map. An example of lossy experimental result is shown in Table 1. There are four types of experimental images, including grayscale image, color image, smoothing image and binary image. The difference between original image and stego-image is determined by Peak signal-to-noise ratio (PSNR). It is hard to find out the difference between the original image and stego-image by naked eye because the PSNR is bigger than 51.61 dB. It is the minimum value of the frames.

Table 1. Example hiding text message in GIF result

(a)	(b)	(c)	(d)
Shuǐ néng zài zhōu, yì néng fù zhōu. https://www.cloudyquotes.com/quote/Chinese_proverbs/11072	Tiān gāo huángdì yuǎn. https://www.cloudyquotes.com/quote/Chinese_proverbs/11074	Bù wén bù ruò wén zhī, wén zhī bù ruò jiàn zhī, jiàn zhī bù ruò zhīzhī, zhīzhī bù ruò xíng zhī; xué zhìyú xíng zhī ér zhǐ yǐ. https://www.cloudyquotes.com/quote/Chinese_proverbs/11053	Liángyào kǔkǒu. https://www.cloudyquotes.com/quote/Chinese_proverbs/11064
52.53dB	54.17dB	51.61dB	55.71dB

5 Discussion and Conclusion

This paper proposed a method to hide Traditional Chinese Proverb in an animated image(in the format of GIF) based on the LSB hiding method. Data are hided in the color maps of the Gif. The goal has been achieved. Different from the traditional LSB hiding method, this method is possible to make the original image and stego-image identical. If the number of colors are less than 256, the data can be hide in the useless fields.

Moreover, because the goal of this paper is to prevent placing wrong pair of image and its content together, the security method is not the issue in this paper. Data size and easy to manage is more important in this application. As shown in Table 2, two exist methods and the propose method characteristics are shown. Traditional method saved the information in separate position and present in the webpage, the method in this paper can save the information in a picture. To summary of this paper, this paper proposed a mistake prevent embedding method for Revitalize Traditional Chinese Proverb.

Table 2. Comparison with exist study

	Format	Purpose	Water marking
American memories	Webpage	Storage	No
Bulgarian Folklore digital library	Webpage	Storage	Protecting intellectual rights
Method in this paper	Picture	Spreading	Saving information

References

1. Pavlov, R., Bogdanova, G., Paneva-Marinova, D., Todorov, T., Rangochev, K.: Digital archive and multimedia library for bulgarian traditional culture and Folklor. Inf. Theor. Appl. **18**(3), 276–288 (2011)
2. 刘燕权 and 韩志萍, 美国记忆-美国历史资源数字图书馆.7, 66–70 (2009)
3. Prytherch, R.J.: Digital preservation. In: Harrod's Librarians' Glossary and Reference Book (10th edn.) (2005)
4. Averkiou, M.: Digital watermarking. http://s3.amazonaws.com/academia.edu.documents/6338147/10.1.1.163.5770.pdf?AWSAccessKeyId=AKIAIWOWYYGZ2Y53UL3A&Expires=1493692896&Signature=UsSCUpJ0ziKRj95fdEn6wn3VbBM%3D&response-content-disposition=inline%3B%20filename%3DDigital_Watermarking.pdf
5. Durvey, M., Satyarthi, D.: A review paper on digital watermarking. Int. J. Emerg. Trends Technol. Comput. Sci. **3**(4), 99–105 (2014)
6. Graphics Interchange Format Version 87a. W3C (1987)

Robust Unseen Visible Watermarking for Depth Map Protection in 3D Video

Zhaotian Li(✉), Yuesheng Zhu, and Guibo Luo

Communication and Information Security Lab, Shenzhen Graduate School, Institute of Big Data Technologies, Peking University, Shenzhen 518055, Guangdong, China
{lizhaotian,zhuys,luoguibo}@pku.edu.cn

Abstract. In 2D-to-3D video conversion process, 3D video can be generated from 2D video and its corresponding depth map by depth image based rendering (DIBR). The depth map is the key in the conversion process as it provides immersive experience to viewers. So the copyright protection for depth map must be considered. Traditional unseen visible watermarking (UVW) for depth map protection cannot resist filtering attacks. In this paper, a robust unseen visible watermarking (RUVW) scheme is proposed, in which the watermark regions without interference are detected for embedding, the copyright information is enhanced with Discrete Cosine Transformation (DCT) and watermark can be seen directly when the rendering conditions are changed. The experimental results show that the proposed method has good robustness against various attacks such as scaling, filtering, noises and compression.

Keywords: Depth map · Copyright protection · Robust unseen visible watermarking (RUVW) · Depth image based rendering (DIBR) · DCT

1 Introduction

The popularity of 3D videos is on the rise in recent years, therefore the demand for 3D content has increased gradually. There are mainly two major ways to produce 3D videos. One is directly capturing left view and right view by two cameras simultaneously. The other method is using 2D-to-3D conversion system. Compared with the first method, the second one can directly use 2D video and corresponding depth map to synthesis left-eye and right-eye view, which can enrich the existing 3D resources.

As we known, the quality of depth map determines the three dimensional effect of 3D video. So the depth map plays an important role in the 2D-to-3D conversion system and the protection for depth map cannot be ignored.

As for information security and content protection for 3D videos (2D video plus depth map format), digital watermarking is widely used to avoid the illegal distribution of depth map and 3D video. There are mainly two ways to protect the depth map: one is unseen visible watermarking (UVW) scheme [1], unseen visible watermarking (UVW) information is a hiding scheme which is unseen for human visual system (HVS)

J.-S. Pan et al. (eds.), *Advances in Intelligent Information Hiding and Multimedia Signal Processing*, Smart Innovation, Systems and Technologies 81,
DOI 10.1007/978-3-319-63856-0_40

in the normal rendering condition. But when the rendering condition changes, the UVW information will appear in the carrier and can be easily recognized by naked eyes.

Researches [2, 3] applied UVW watermark to protect the depth map. The embedding regions in method [3] are the farthest regions in depth map which can avoid that the disparity value is overflowed. The method [3] proposed a D-nose model to modify the suitable regions mentioned in method [2]. Both schemes embed the watermark into depth map in the spatial domain, so they have poor robustness to various attacks. In several attacks, the watermark may be destroyed.

The other one is the scheme based on Quantization Index Modulation (QIM) algorithm [4]. In this method, the copyright information is embedded in the DCT coefficients of depth map and the watermark can be extracted from depth map with IDCT. This method shows a good performance in perceptibility and robustness compared with method [2]. But both the embedding and extraction process is complex compared with UVW methods [2, 3].

To improve the robustness of the traditional UVW methods, In this paper, we propose a robust UVW module based on DCT which has good robustness compared with method [2, 3] and the embedding and extraction process is more easier than method [4].

2 Proposed Method

The proposed RUVW method is similar with the UVW scheme [2, 3] for depth map protection that utilizes the non-linear property of DIBR to embed copyright information into depth map. In this part, we mainly focus on how to detect watermark embedding regions without interference and to embed UVW appropriately into depth map. The framework of the proposed method is shown in Fig. 1.

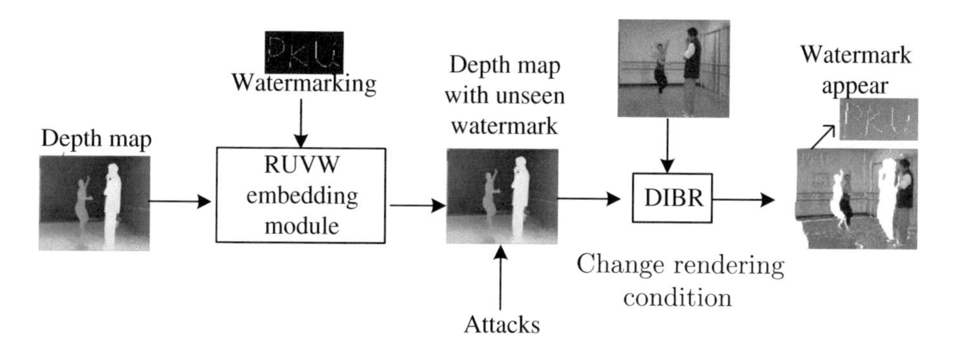

Fig. 1. The framework of proposed method

A. Theoretical analysis for DCT

In the proposed method, the DCT is exploited since embedding watermarking in frequency domain coefficients is more robust than that embedded in spatial domain coefficients. Two-dimensional 4*4 DCT and IDCT can be mathematically expressed as Eqs. (1) and (2):

$$F(u,v) = \frac{1}{2} c(u)c(v) \sum_{x=0}^{3} \sum_{y=0}^{3} f(x,y) \cos\frac{(2x+1)u\pi}{8} \cos\frac{(2y+1)v\pi}{8} \tag{1}$$

$$f(x,y) = \frac{1}{2} \sum_{u=0}^{3} \sum_{v=0}^{3} c(u)c(v)F(u,v) \cos\frac{(2x+1)u\pi}{8} \cos\frac{(2y+1)v\pi}{8} \tag{2}$$

where $f(x, y)$ is the depth value of the points in the spatial domain, $F(u, v)$ is the corresponding value of the points in the frequency domain. When $u = 0$ and $v = 0$, the $F(0, 0)$ can be calculated as the DC coefficient. According to the Eq. (2), if DC coefficient varies Δ, $f(x, y)(x = 0\ldots3; y = 0\ldots3)$ vary $\Delta/4$.

B. Embedding regions selection

As there are some holes in the virtual view after DIBR process, and the depth value may overflow after the embedding process, the watermark cannot be seen when embedded in these regions. To avoid this happens, it is necessary to find the watermark regions without interference to embed the watermark.

Assuming that the size of the depth map is $w_d{*}h_d$, and the size of the watermark is $w_w{*}h_w$. The suitable appearance region of watermark is $S = \{(x, y)|x_m \leq x < x_m + w_w, y_m \leq y < y_m + h_w\}$, $h_n(S)$ is the number of holes in area S, the region S should satisfy two limitations: (1) all the pixels' depth value in S cannot exceed 255-$\Delta/4$ to avoid the depth value overflow. (2) the hole percentage in S cannot exceed $\beta(\beta = 0.05$ in our experiment), in other word, $h_n(S) < \beta{*}w_w{*}h_w$. Then the suitable embedding region G in original depth map is inverse warped from S. Figure 2 shows the suitable starting position regions to embed UVW information.

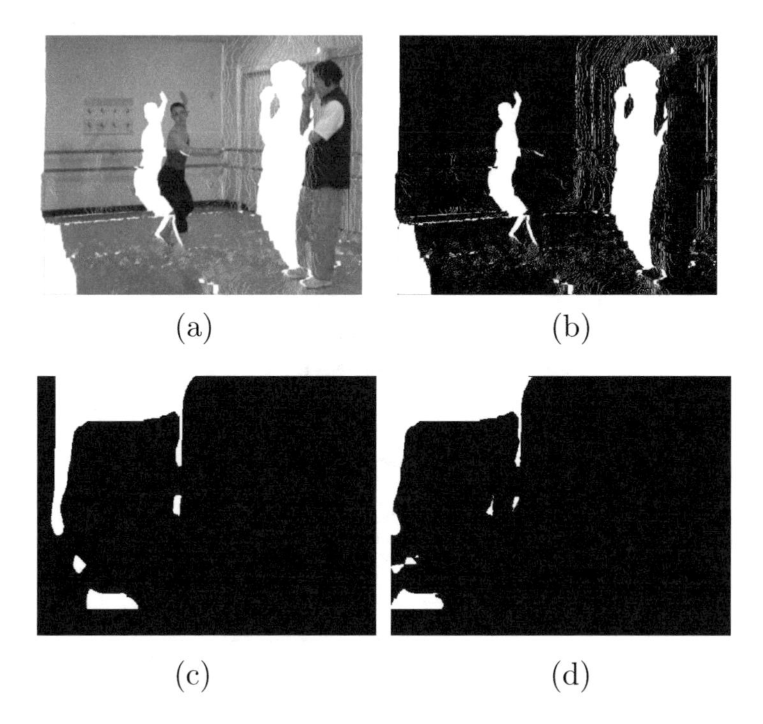

Fig. 2. The embedding region selection result of test data *Ballet*. (a) the synthesized view under extraction condition. (b) the hole regions of (a). (c) the suitable start position for embedding. (d) the suitable region in original depth map inverse warped from (c).

C. The embedding module

This part focuses on how to embed UVW watermarking to rise the robustness against the attacks. The embedding flowchart of the proposed robust UVW method is shown in Fig. 3. Firstly, calculating the right regions to embed UVW information according to the embedding regions mentioned above. Secondly, the embedding positon can be randomly selected in the suitable regions. Thirdly, depth frame is split into several non-overlapping 4*4 blocks and determine if the block contains watermark pixels. If it contains UVW pixels, the DCT conversion is conducted to the block and the DC coefficient of the block is increased by Δ. Finally, IDCT conversion is conducted and the watermarked depth map is obtained.

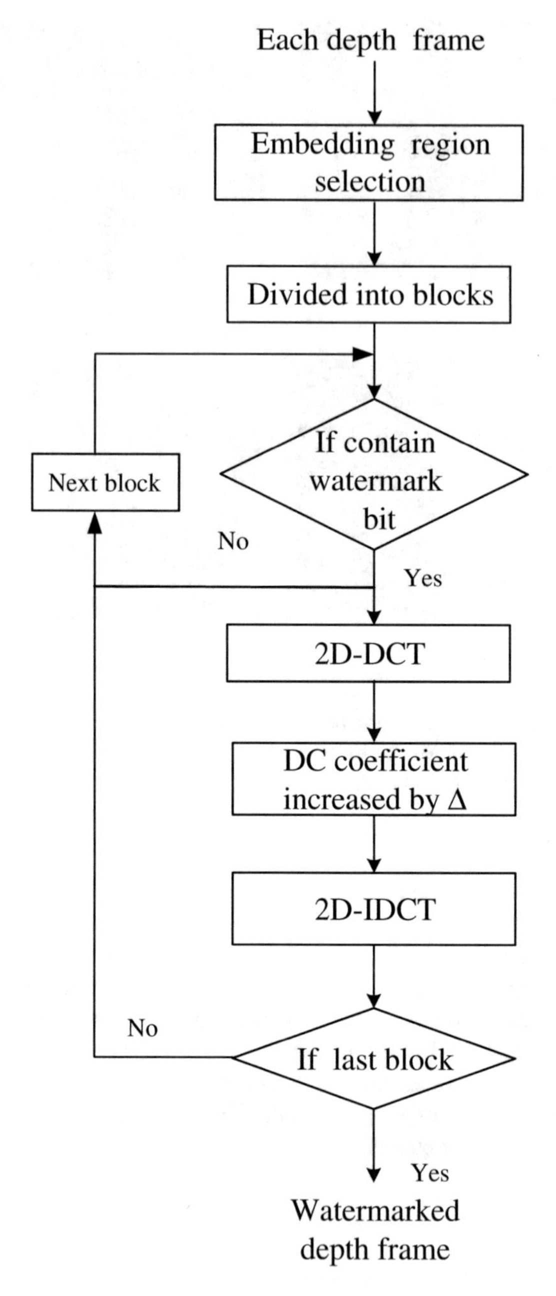

Fig. 3. The flowchart of the embedding process

3 Experiment Results and Discussion

In our experiment, the color images and corresponding depth maps are used to evaluate the performance of the proposed robust UVW method. The test datasets are *Interview* [5], *Ballet* [6], *Middlebury* [7].

In the DIBR process, we set $f = 1910$, $Z_{far} = 130$, $t_c = 0.5$, $Z_{near} = 42$ for normal rendering condition, we set $Z'_{near} = 0.1*Z_{near}$ for watermark extraction process. Figure 4 shows the watermarks we used in this paper. The extraction results of *book* and *reindeer* are shown in Fig. 5.

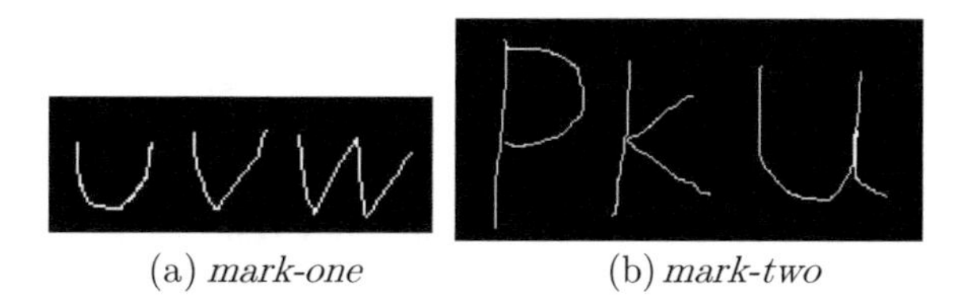

(a) *mark-one* (b) *mark-two*

Fig. 4. Test watermark *mark-one*, *mark-two*.

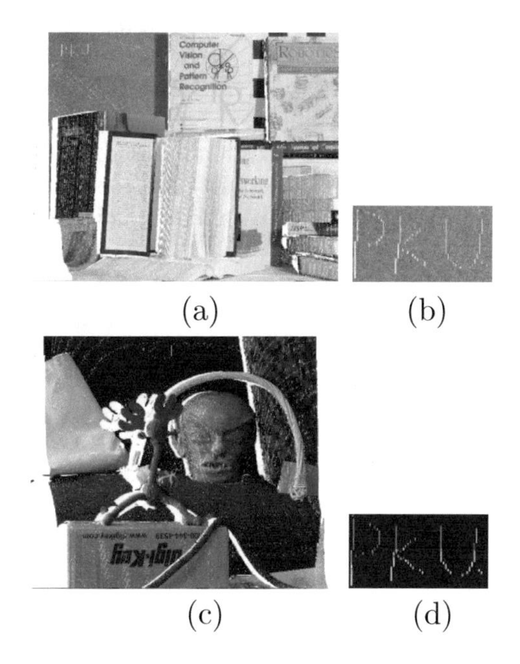

(a) (b)

(c) (d)

Fig. 5. The extraction result of test data *book*, *reindeer*.

Table 1. PSNR of original depth map and watermarked depth map.

Data	Stones	Reindeer	Books	Toys	Interview	Ballet
PSNR	70.2774	71.0809	72.1751	72.1751	67.0558	66.3305

A. Invisibility test

Figure 6 shows the original and the watermarked depth map. The change of the watermarked depth map is near to none, and the watermark cannot been seen in the synthesized view under normal rendering condition. Meanwhile, the peak of signal-to-noise ratio (PSNR) is used to measure the quality of the watermarked depth map in comparison with the original one. In our experiments, every PSNR value is larger than 65 dB as shown in Table 1 which meet the requirements of invisibility.

B. Robustness test

In this part, we make a robustness comparison between the proposed method and Lin's method [2]. Test data *Ballet* and *Interview* are adopted to test the robustness. The signal distortions are applied to the watermarked depth map to evaluate the robustness of watermark method. The signal distortions consist of scaling, filtering, salt noise, and JPEG compression. For scaling, scaling factor 0.8 and 0.5 are tested respectively. For filtering, median filter, mean filter, and Gaussian filter are used for both 3*3 and 5*5 filtering windows. For salt noise, noise density 0.04 and 0.08 are tested respectively. For JPEG compression, the compression ratio is 1/32 and 1/26. The comparison results are shown in Tables 2 and 3. (The extraction effect are enlarged). The results indicate that our method can provide strong robustness against all attacks cases referred above, while the watermarks cannot been seen under most of attacks in Lin's method.

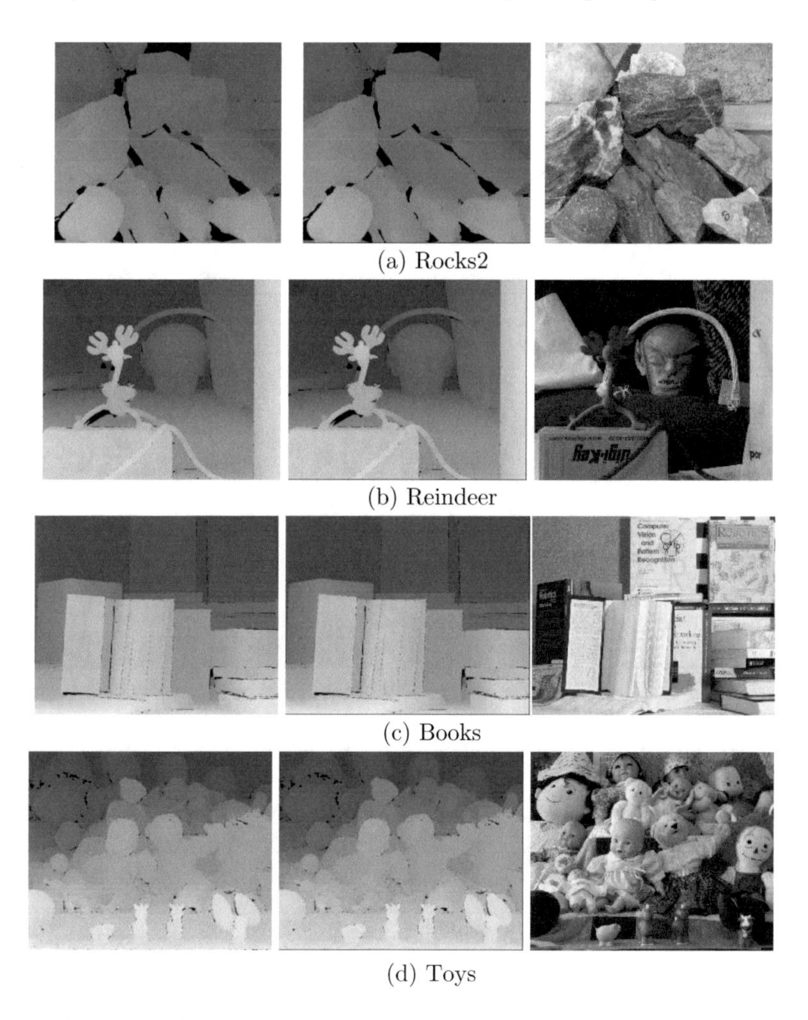

Fig. 6. The imperceptible test. The first column shows the original depth map; the second column shows the depth map with watermark *mark-two*; the third column shows the virtual views under normal rendering condition.

Table 2. Result of *Ballet*.

Scaling			
0.5		0.8	
Proposed method	Lin's method	Proposed method	Lin's method

Gaussian filter			
3*3		5*5	
Proposed method	Lin's method	Proposed method	Lin's method

Median filter			
3*3		5*5	
Proposed method	Lin's method	Proposed method	Lin's method

JPEG compression			
1/32		1/26	
Proposed method	Lin's method	Proposed method	Lin's method

salt_pepper_noise			
0.04		0.08	
Proposed method	Lin's method	Proposed method	Lin's method

Table 3. Result of *Interview*.

scaling			
0.5		0.8	
Proposed method	Lin's method	Proposed method	Lin's method
Gaussian filter			
3*3		5*5	
Proposed method	Lin's method	Proposed method	Lin's method
Median filter			
3*3		5*5	
Proposed method	Lin's method	Proposed method	Lin's method
JPEG compression			
Proposed method	Lin's method	Proposed method	Lin's method
salt_pepper_noise			
0.04		0.08	
Proposed method	Lin's method	Proposed method	Lin's method

4 Conclusion

In this paper, a robust UVW method for depth map protection is proposed, in which the copyright information is enhanced in special areas with Discrete Cosine Transformation (DCT). The experimental results show that the proposed method has good robustness against various attacks such as scaling, filtering, noises and compression.

Acknowledgments. This work is supported by the Shenzhen Municipal Development and Reform Commission (Disciplinary Development Program for Data Science and Intelligent Computing), and the Shenzhen Engineering Laboratory of Broadband Wireless Network Security.

References

1. Chuang, S., Huang, C.H., Wu, J.L.: Unseen visible watermarking. In: IEEE International Conference on Image Processing, ICIP 2007, pp. 261–264, 16–19 October 2007
2. Lin, Y.H., Wu, J.L.: Unseen visible watermarking for color plus depth map 3D images. In: IEEE International Conference on Acoustics, Speech and Signal Processing, pp. 1801–1804 (2012)
3. Pei, S.C., Wang, Y.Y.: Auxiliary metadata delivery in view synthesis using depth no synthesis error model. IEEE Trans. Multimedia **17**(1), 128–133 (2015)
4. Guan, Y., Zhu, Y., Liu, X., et al.: A digital blind watermarking scheme based on quantization index modulation in depth map for 3D video. In: International Conference on Control Automation Robotics & Vision, pp. 346–351 (2014)
5. Zitnick, C.L., Kang, S.B., Uyttendaele, M., Winder, S., Szeliski, R.: High-quality video view interpolation using a layered representation. In: ACM Transactions on Graphics (TOG), vol. 23, pp. 600–608. ACM (2004)
6. Fehn, C.: Depth-image-based rendering (DIBR), compression, and transmission for a new approach on 3D-TV. In: Stereoscopic Displays and Virtual Reality Systems XI, SPIE, vol. 5291, no. 1, pp. 93–104 (2004)
7. Scharstein, D., Szeliski, R.: Middlebury Stereo Datasets. http://vision.middlebury.edu/stereo/data/

An Improved ViBe Algorithm Based on Salient Region Detection

Yuwan Zhang and Baolong Guo(✉)

School of Aerospace Science and Technology, Xidian University, Xian 710071, China
xd_zyw@foxmail.com, blguo@xidian.edu.cn

Abstract. The ViBe algorithm is a powerful technique for the background detection and subtraction in video sequences. Compared to the state-of-the-art algorithms, ViBe algorithm is better in fast speed and less memory consumption. However, when applying ViBe algorithm to the moving objects appeared in the first frame of videos, all pixels in first frame will be used to build the background model that will result in the foreground pixels in sample set. This problem causes the ghost areas emerge. And it will remain for a long time. In this paper, a salient region detection based ViBe algorithm is proposed to eliminate the ghost areas fast. First, the foreground region is extracted from the first frame of videos using the salient region detection algorithm. According to the result of salient region detection, the background area of image is separated from the foreground area. The foreground pixels are dislodged from the sample set. Then, only background pixels are used for background model initialization. The experimental result shows that the improved algorithm can eliminate ghost in few frames quickly.

Keywords: ViBe algorithm · Ghost area · Background model · Object detection

1 Introduction

With the development of computer vision technology, target detection technology is widely used in intelligent video surveillance system. The effective structure is to build background model, and use the frame difference method in the current frame and the target detection. Because of the differences in application scenarios, many scholars propose different algorithms. The common algorithms are inner-frame difference algorithm, Gauss Mixed Model (GMM) algorithm, Optical Flow algorithm and the Code-Book algorithm. The inner-frame difference algorithm is simple and with small calculation amount [1]. It can meet real-time requirement. However if the foreground objects moving speed is low, the algorithms accuracy rate decreases. The Optical Flow algorithm has high accuracy rate. However the algorithm is complex and with high computation

This work was supported by the National Natural Science Foundation of China under Grants No. 61571346.

J.-S. Pan et al. (eds.), *Advances in Intelligent Information Hiding and Multimedia Signal Processing*, Smart Innovation, Systems and Technologies 81,
DOI 10.1007/978-3-319-63856-0_41

cost [2]. The Gauss Mixed Model algorithm also has high accuracy rate [3]. The disadvantage of GMM is the difficulty to meet the real-time requirement because of its large computational cost. The Code-Book algorithm has high robustness and high computational efficiency [4]. However, the fixed learning-threshold will cause code unlimited expansion.

The ViBe algorithm is proposed by Oliver Barnich and other scholars in 2011 [5]. It is a pixel-level video background model for foreground detection algorithm. The basic thought of ViBe uses the first frame of video sequence to build background model. The ViBe algorithm stores a sample set for each pixel. The sample value is the pre-pixel value of the pixel and the pixel value of the neighbor pixels (generally 20 pixels). Because of the mode of building background model, compared to the state-of-the-art algorithms, the ViBe algorithm is of fast speed, high accuracy rate, and less memory consumption. However, when applying the ViBe algorithm to the moving objects appeared in the first frame of videos, the ghost areas will emerge and remain for a long time. In this paper, the principle of ViBe algorithm is introduced firstly, followed by its disadvantage analysis. Then, the improved ViBe algorithm is detailed narrated. Last, the results of the different algorithms experiment show the availability of the improved algorithm. The follow is the detailed description.

2 ViBe Algorithm Analysis

2.1 The Principle of Vibe Algorithm

The ViBe algorithm mainly includes three steps: the background model definition, the model initialization method, the updating strategy of the model [6].

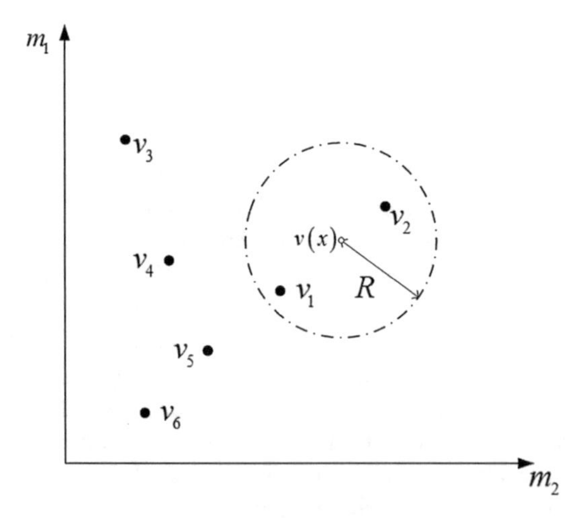

Fig. 1. ViBe background model

(a) Background model definition
The background object is a stationary or low-speed object. The foreground object correspond a moving object. So the object detection is considered as a classification problem. A background model stores a sample set for each background pixel. For each new pixel value and the sample set are compared to determine whether it be-longs to the background or foreground.

$$M(x) = \{v_1, v_2, ...v_N\} \tag{1}$$

Each pixel in the background model is composed of N background samples (N = 20 in the paper). In each frame image, the point x at the pixel value is denoted as v(x). v_i represents the background sample value of the index i.

$$m = M(x)\left[\{SR(v(x)) \cap M(x)\}\right] >= 2 \tag{2}$$

The background model M is defined as Fig. 1, let the area centered at x with a radius of R denote as SR(v(x)). The intersection of M(x) and SR(v(x)) satisfies a certain threshold m. If m >= 2, x is a background pixel, otherwise x is the foreground pixel.

(b) Background Model Initialization
The ViBe algorithm only needs the first frame of video sequence to initialize back-ground model. Due to one frame can not contain 20 values of the pixel. So with the spatial consistency principle of pixels, the ViBe algorithm selects 20 sample values from the 8 neighborhood N_G(x) of x to initialize the background model.

$$M^0(x) = \{v^0(y|y \in N_G(x))\} \tag{3}$$

(c) Updating Strategy
In order to adapt to the background change, such as illumination change, background object change and so on, the background model update itself. The ViBe algorithm adopts the strategy of random update background model. When a pixel x is classified as background pixel, it has $1/a$ probability to update the background model. At the same time, it has $1/a$ probability to update neighborhood background model from 8 neighborhood of it. After d_t time, the probability of one sample in background model is shown as mathematical Eq. 4.

$$P(t, t + d_t) = e^{-ln(N/(N-1))d_t} \tag{4}$$

2.2 The Disadvantage of ViBe Algorithm

The ViBe algorithm only uses the first frame of video sequence to initialize background model. On the one hand, the algorithm accelerates the initialization, on the other hand it will produce ghost when the first frame includes moving object, the ViBe algorithm takes the moving object as the background to build

Fig. 2. The ghost of Vibe algorithm

the model. After the moving object exits the frame, the real background pixel can not match the back-ground model. This will cause the background points are incorrectly detected as the foreground points. Then the ghost will appear as shown in Fig. 2.

The original ViBe algorithm simply treats the ghost as the foreground area. The ghost will influence the target detection, tracking and count. Many scholars have proposed solution to counter the ghost problem. The algorithm based on inner-frame difference is proposed to suppress ghost problem [7]. However, this method relative high computational cost to complete the detection of moving target. Inspired by the above article, the algorithm is proposed based on salient region detection.

3 Proposed Method

As mentioned above, when the first frame has moving object, ViBe algorithm regards the foreground pixels as background pixels to build background model. In order to eliminate this drawback, the following work is distinguished the foreground pixels from background pixels in the image. Then the improved ViBe

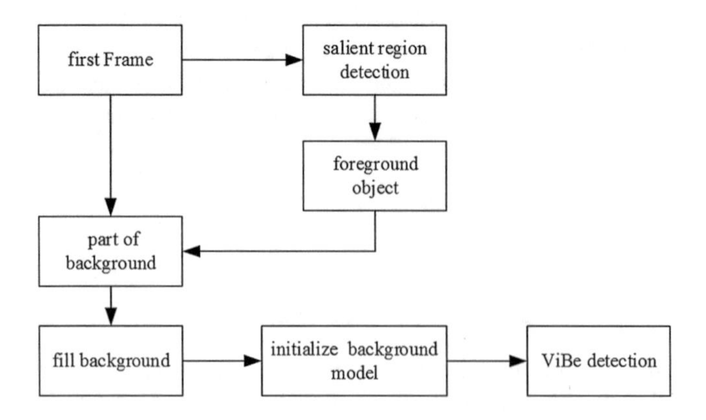

Fig. 3. The procedure of improved algorithm

algorithm only uses the background pixels to complete initialization. The integrated procedures of our algorithm are shown in Fig. 3.

Detailed steps are as follows.

(1) The algorithm gets first frame image from video, denoted as F1. The salient region detection will be used for the frame. This paper uses Doctor MingMing-Chengs salient region detection method [8]. The method is based saliency computation methods, namely histogram based contrast and spatial information-enhanced region based contrast. The algorithm has high precision and better recall rates. After the salient region detection, the foreground region gained.
(2) From the result of salient region detection, the foreground pixels are marked. The other points are background pixels. After this step, the algorithm gets part of background pixels. In order to get a full background, the algorithm needs a further procedure to fill the moving object area with background.
(3) According to the spatial consistency principle of adjacent pixels, the near pixels are approximately equal at the same time. So the algorithm uses the background pixels near the foreground region to replace of the foreground pixels. After this step, the algorithm gets the full background. Then the algorithm uses this pixels build back-ground model.
(4) After the model initialization, the algorithm uses ViBe algorithm to detect the moving object in video.

4 Experiments and Analysis

The experiments are run on PC with an i5-4590@3. 30 GHZ CPU, 8 GB memory and a Windows 7 operating system. All the algorithms are developed in c++ by using Microsoft Visual Studio 2013 and openCV 2.4.9. The video sequence is from a digital camera. The frame rate is 25 frame/s. The resolution ratio is $320 * 240$. Four algorithms are used to do this experiment. In order to analyze the real-time performance of different algorithms, the cost of time is got by different algorithms. The detection results and cost-time result of experiments are shown as Fig. 4 and Table 1.

From the experimental results, the inner-frame difference algorithm will not produce the ghost. But the detection object has cavities. And in frame 61, the moving object is almost undetectable. The GMM algorithm also produces the ghost area. And the ghost will not disappear until 61 frame. The result of GMM algorithm appears a lot of hot pixels because of the swaying leaves in the video. The proposed algorithm for the inhibitory effect of ghost is significantly better than the original ViBe algorithm. In the 61 frame of the video, the ViBe algorithm can still detect the complete ghost area, while the improved algorithm will completely eliminate ghost in 7 frames.

From the data of Table 1, the Optical Flow algorithm costs most time when it applies in object detection. GMM algorithm and inner-frame algorithm also cost more time than the improved ViBe algorithm. The conclusion is got that our algorithm can quickly suppress ghost when the moving object in the first frame. It can meet real-time requirement.

Table 1. Time-cost of different algorithms (Units: ms)

Data-sets	Inner-frame	Optical flow	GMM	ViBe	This paper
1	4736	6256	4451	2582	2569
2	11562	17023	11556	6162	6069
3	10009	14147	9579	4866	4712
4	27914	41663	27932	14055	14000
5	19033	111611	19351	10035	9981
6	36922	63746	37265	18640	18082
7	7668	11053	7664	4179	3991
8	12111	76697	12387	6278	6050
9	34544	51398	34711	17511	17211
10	64227	311670	72478	31436	31818
11	47875	257340	67230	24619	25200
12	31110	46722	31440	15656	15625

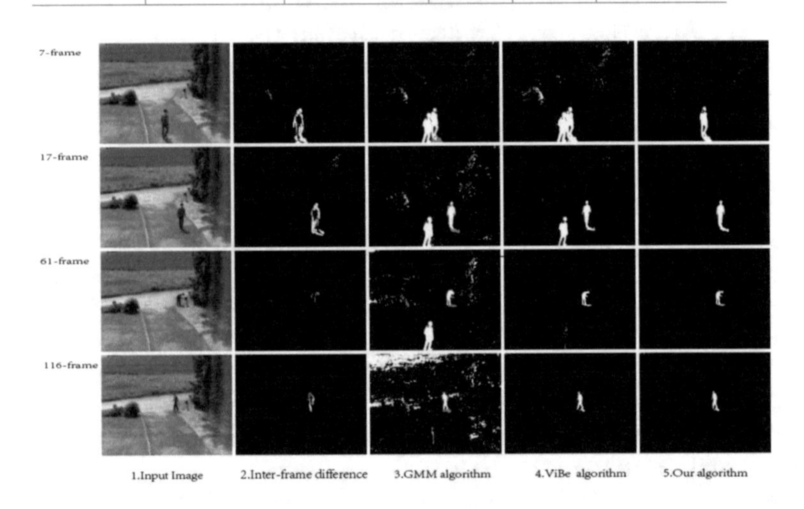

Fig. 4. Detection of different algorithms

5 Conclusion

The original ViBe algorithm has no special treatment of the ghost. When the first frame has moving object, the ghost area will appear. The elimination of ghost requires a long process. This will effect detection, tracking and classification. To solve this problem, this paper proposes an improved ViBe algorithm based on salient region detection. The algorithm can eliminate ghost in few frames. The experiment shows that this method can suppress ghost quickly. However, the algorithm still has some shortcomings, such as the shadow problem. This problem should be further study in the future work.

References

1. Wren, C.R., Azarbayejani, T., Pfinder, D.: Real-time tracking of human body. IEEE Trans. Pattern Anal. Mach. Intell. **19**, 780–785 (1997)
2. Jianbo, X., Li, J.: Moving target tracking algorithm based on scale invariant optical flow method. In: Proceedings of the 3rd International Conference on Information Science and Control Engineering, Beijing, pp. 195–197 (2016)
3. Stauffer, C., Grimson, W.E.L.: Adaptive background mixture models for real-time tracking. In: IEEE Computer Society Conference on Computer Vision and Pattern Recognition, USA, pp. 246–252 (1999)
4. Ilyas, A., Scuturici, M., Miguet, S.: Real time foreground-background segmentation using a modified codebook model. In: Sixth IEEE International Conference on Advanced Video and Signal Based Surveillance, Genvoa, pp. 454–459 (2009)
5. Barnich, O., Van Droogenbroeck, M.: ViBe: a powerful random technique to estimate the background in video sequences. In: Acoustics, Speech and Signal Processing, pp. 945–948 (2011)
6. Introduction of ViBe. http://www.telecom.ulg.ac.be/research/vibe/
7. Li, Y., Chen, W., Jiang, R.: The integration adjacent frame difference of improved ViBe for foreground object detection. In: Proceedings of the 7th International Conference on Wireless Communications, Networking and Mobile Computing, pp. 1–4 (2011)
8. Cheng, M.-M., Mitra, N.J., Huang, X.: Global contrast based salient region detection. IEEE Trans. Pattern Anal. Mach. Intell. **37**, 569–582 (2015)

Boosted HOG Features and Its Application on Object Movement Detection

Junzo Watada[1], Huiming Zhang[2(✉)], Haydee Melo[2], Diqing Sun[2], and Pandian Vasant[3]

[1] Computer and Information Sciences Department, PETRONAS University of Technology, 32610 Seri Iskandar, Perak Darul Ridzuan, Malaysia
junzow@osb.att.ne.jp

[2] Graduate School of Information, Production and Systems, Waseda University, 2-7 Hibikino, Wakamatsu, Kitakyushu 808-0135, Japan
huimingde@gmail.com, melo.haydee@asagi.waseda.jp

[3] Fundamental and Applied Sciences Department, PETRONAS University of Technology, 32610 Seri Iskandar, Perak Darul Ridzuan, Malaysia

Abstract. Nowadays, traffic accidents is universally decreasing due to many advanced safety vehicle systems. To prevent the occurrence of a traffic accident, the first function that a safety vehicle system should accomplish is the detection of the objects in traffic situation. This paper presents a popular method called boosted HOG features to detect the pedestrians and vehicles in static images. We compared the differences and similarities of detecting pedestrians and vehicles, then we use boosted HOG features to get an satisfying result. In detecting pedestrians part, Histograms of Oriented Gradients (HOG) feature is applied as the basic feature due to its good performance in various kinds of background. On that basis, we create a new feature with boosting algorithm to obtain more accurate result. In detecting vehicles part, we use the shadow underneath vehicle as the feature, so we can utilize it to detect vehicles in daytime. The shadow is the important feature for vehicles in traffic scenes. The region under vehicle is usually darker than other objects or backgrounds and could be segmented by setting a threshold.

Keywords: Pedestrian detection · Vehicle detection · Hog feature

1 Introduction

Pedestrian and Vehicle detection is an important issue for enhancing traffic safety in Intelligent Transportation Systems. Detecting pedestrians and vehicles in static image is a challenging task. Unlike other road-users like automobiles, pedestrians are highly articulated objects with various statures, shapes and postures. In recent years, a plenty of pedestrian and vehicle detection approaches have been

J.-S. Pan et al. (eds.), *Advances in Intelligent Information Hiding and Multimedia Signal Processing*, Smart Innovation, Systems and Technologies 81,
DOI 10.1007/978-3-319-63856-0_42

proposed. They could be approximately classified into two categories, holistic and part-based approaches. Holistic approaches discriminate human from background by features extracted from the whole human body. Dalal and Triggs first proposed the use of dense grids of Histogram of Oriented Gradient (HOG) feature descriptors and linear SVM classifiers for human detection [1]. A similar but much faster approach is proposed by Zhu et al. [2]. It uses a wild range of HOG features in different sizes and locations. Tuzel et al. [3] further improved the performance by utilizing the covariance matrix as the descriptor and learning on nonlinear space. Wojek and Schiele [4] used multi-features including HOG, shape-context and Haar features to reach an outperforming performance.

However, we found that in all related works, up to our knowledge, there was no attempt to use boosting with Local binary pattern as discriminative features, and no investigation on how to combine detection results at classifiers level for further improving detection rate.

Recently, a popular method is called boosted HOG features on the basis of HOG feature to detect pedestrians and vehicles in static images. Although various algorithms proposed by researchers have good performance in solving pedestrian and vehicle detection, they are not effective. After comparing the differences and similarities of detecting pedestrians and vehicles, we selected boosted HOG features finally to get an satisfying result. We proposed new methods which can be just adapted in traffic scenes to improve accuracy of the detection. The experiment shows that our proposed method has advantage in detection accuracy comparing with conditional HOG method.

The remainder of this paper is structured as follows: The background knowledge are illustrated in the Sect. 2. In the Sect. 3, the authors propose a new method to detect vehicle quickly and effectively. At the end of detection, we remove some wrong detection by using pedestrian-vehicle relationship when mixing pedestrian detection and vehicle detection. The experimental results are shown in Sect. 4, while Sect. 5 draws the conclusion.

2 Background Knowledge

2.1 Boosted-HOG Feature

Boosting is a method in machine learning that combines weak learners to form strong ones. It has been extensively studied and applied in pedestrian detection tasks. Viola et al. train simple rectangle features to build a classifier cascade [5]. More recently, Sabzmeydani and Mori propose the shapelet feature (derived from low-level gradients with adaboost) which sheds new light on the issue.

The resulted HOG feature in each cell contains important information on how to separate pedestrians from other objects, yet redundant information may also be included in the feature. Now the adaboost is applied to learn a new feature from the HOG feature at hand. For each cell, the set of weak classifiers should be firstly created. As the HOG is a histogram with bins indicating local gradient distribution, we compare the value on one bin with a threshold to determine whether the image contains pedestrian. This forms our weak classifiers in adaboost, which are decision stumps. If the histogram has ten bins, then we have ten weak classifiers corresponding to each bin. As the following formula shows, the weak classifier is defined by two parameters: the threshold q_i, and the parity p_i.

$$h_{k;j}(image) = \begin{cases} 1\ if\ \ p_i Hist_{k,i}(image) < p_i\theta_i \\ 0\ otherwise \end{cases} \tag{1}$$

In the above equation, image signifies input image. For the cell k, the feature $Hist_{k,i}(image)$ presents the value on i-th bin of the histogram in that cell. The value is regarded as a feature to detect objects in the image. The threshold qi is used to make a decision according to the $Hist_{k,i}(image)$. The parity p_i can be either -1 or $+1$, which is used to change the direction of the inequality.

The adaboost algorithm starts with assigning weights to the training samples. In each iteration, the weak classifier with the least error rate is selected, and is given a weight to determine its importance in the final classifier. Before the next iteration begins, the weights of those misclassified samples are increased so that the algorithm can focus on those hard samples. The final classifier is constructed as a weighted combination of weak classifiers selected in each iteration.

The details of the algorithm are illustrated as follows:

1. All samples are presented in the form of $(X_i; Y_i)$, where X_i indicates sample images and Y_i signifies the category of the sample:
   1 for pedestrians or vehicles,
   0 for non-pedestrians or non-vehicles.
2. For each sample, initialize a sample weight; supposing the training set includes m positive samples and n negative samples, the corresponding weight for each sample is $\frac{1}{m+n}$.
3. Enter the adaboost iteration, set the iteration times as T; Normalize the sample weight; All the weak classifiers are tested on the training sample set. For each classifier, collect all the misclassified samples, and sum up their weights to obtain classification error ϵ_j, where j indicates the index of the weak classifiers. Choose the one with the least error. Use the chosen classifier to perform classification on all samples. If the sample is correctly classified, maintain its weight.
4. After all T iterations, the final strong classifier obtained from cell k is as follows:

$$H_k(X_i) = \begin{cases} 1\ if\ \ \sum_{i=1}^{N} \alpha_t h_{k,j}(x_i) > \frac{1}{2}\sum_{t=1}^{N} \alpha_t \\ 0\ otherwise \end{cases} \tag{2}$$

   where $\alpha_t = \log \frac{1}{\beta_t}$

As the output $H_k(X_i)$ of the adaboost is a strong classifier, the weighted sum of weak classifiers $\sum_{t=1}^{N} \alpha_t h_{k,j}(X_i)$ in cell k must contain important information that differentiates pedestrians from other objects. This forms the new feature in our approach, which we name as boosted HOG feature. The feature vector is built by combining them from all cells. A linear SVM is used to train on the feature vector for the final classification.

3 Proposal Method

In this paper, we first use boosted-hog descriptor extract feature points of pedestrians and vehicles respectively and train them by linear support vector machine.

3.1 Detection of Shadow Under the Vehicle

The shadow is the main characteristic for forward vehicles. The region under vehicle is usually darker than other objects or backgrounds and could be segmented by setting a threshold. While a fixed threshold applies to particular light and weather conditions only. In order to adapt to different lighting and weather conditions, the statistical histogram was used to estimating the upper limit of the threshold. Assume that the gray value of road obey to normal distribution with the mean m and standard deviations. So, the statistics histogram could be described by a Gaussian curve approximately [6]. The probability distribution function of gray-scale value $f(x)$ could be approximated as:

$$f(x) \approx \frac{1}{k\sqrt{\pi x}} e^{-\frac{-x-m^2}{2s^2}} \tag{3}$$

where k is constant.

The gray value statistical analysis of shadow area shows that the proportion of underneath vehicle pixel is less than 10 percent under different light conditions.

The shadow regions could be obtained after the above processing, the initial candidate of vehicle was generated by combining horizontal and vertical edge feature of shadow.

First, the row by row search was done in shadow image. The number of pixels was counted, which grayscale value was greater than zero. If the number of pixels less than 20 and greater than 200, the row i_{k_1} would be considered as the likely bottom location of vehicle. Then, in order to find the left vertical edge feature of shadow, the col by col search was done in rectangle region which located at the left starting point of i_{k1} with a range of 10 pixels. The number of pixels was counted, which grayscale value was greater than zero. If the number of pixels was not less than 5, the row i_{k_1} would be considered as the left border of vehicle. Next, in order to find the right vertical edge feature of shadow, the col by col search was done in rectangle region which located at the right end point of i_{k_1} with a range of 10 pixels. The number of pixels was counted, which grayscale value was greater than zero. If the number of pixels was not less than 5, the col j_{k_2} would be considered as the right border of vehicle. At last, the top location of vehicle was determined by the left border j_{k_1}, the right border j_{k_2} and the bottom location ik1 with an aspect of 1:1. The initial candidate of vehicle was represented as $(j_{k1}, j_{k2}, i_{k1}, i_{k2})$.

For the aspect of HOG, First, the gradient of an image obtained by applying two 1- D filters, which are $(-1, 0, 1)$ for the horizontal direction and $(1; 0; 1)^T$ for the vertical direction. The signed gradient was used whose values form $-\pi$ to π. Then, the signed gradient was divided into nine bins with the same range, and the range of each bin was 20. Next, the histogram of orientation was computed in each bin. The size of cell was 6 by 6, each block was consisted of four cells, and values were imposed on the histogram of each block. The number of blocks was 3×3 in a 24 by 24 pixels image. HOG feature vectors were extracted by concatenating the four new cells of histogram of oriented gradient in one block, so, the dimension of HOG feature vector was $4 \times 9 \times 9 = 324$ in a training sample. At last, the L_2- norm was used to normalize HOG feature vector.

Fig. 1. Shadow segmentation in the traffic scene

As the Fig. 1 shows, we can detect the shadows easily due to the shadow is brighter than other background in grey-scale image. The images of vehicles that are used the experiment are compiled partly from the internet and partly from the images in Caltech Pedestrian Dataset. A total of 525 images of the vehicles are collected and 125 of them are negative samples. These cropped sample images are resized to 45×45 in the training. We also apply the 100 frames from the testing set on Caltech Pedestrian Dataset as the testing images.

3.2 Pedestrian-Vehicle Context in Traffic Scenes

A lot of pedestrian detections are located around vehicles in traffic scenes. Since it is difficult to capture the complex relationship by handcraft rules, we build a context model and learn it automatically from data.

We split the spatial relationship between pedestrians and vehicles into five types, including: We denote the feature of pedestrian vehicle context as $g(p, v)$. If a pedestrian detection p and a vehicle detection v have one of the first four relationships, the context features at the corresponding dimensions are defined as $(\sigma(s), \nabla c_x, \nabla c_y, \nabla h, 1)$, and other dimensions retain to be 0. If the pedestrian detection and vehicle detection are too far or no vehicle, all the dimensions of its pedestrian-vehicle feature is 0. Here $\nabla c_x = |c_{\nu_x} - c_{p_x}|$, $\nabla c_y = c_{\nu_y} - c_{p_y}$, and $\nabla h = \frac{h_\nu}{h_p}$, where $(c_{\nu_x}; c_{\nu_y})$, $(c_{p_x}; c_{p_y})$ are the center coordinates of vehicle detection v and pedestrian detection p, respectively. $\sigma(s) = 1\frac{1}{1+exp(2s)}$ is used to normalize the detection score to $[0,1]$. For the left-right symmetry, the absolute operation is conducted for ∇c_x. Moreover, there also has a relationship between the coordinate and the scale of pedestrians under the assumption that the cameras is aligned with ground plane. We further define this geometry context feature for pedestrian detection p as $g(p) = (s(s); c_y; h; c_y^2; h^2)$, where s, c_y, h are the detection score, y-center and height of the detection respectively, and c_y and h are normalized by the height of the image (Fig. 2).

Fig. 2. The image shows that the detection above a vehicle, and detection at the wheel position of a vehicle can be safely removed.

To fully encode the context, we defined the model on the whole image. The context score is the summation of context scores of all pedestrian detections, and context score of a pedestrian is further divided to its geometry and pedestrian-vehicle scores. Suppose there are n pedestrian detections $P = p_1; p_2; \cdots; p_n$ and m vehicle detections $V = (v_1; v_2; \cdots; v_m)$ in an image, the context score of the image is defined as:

$$S(P;V) = \sum_{i=1}^{n}(\omega_p^T g(p_i) + \sum_{j=1}^{m} \omega_v^T g(p_i; v_j)) \tag{4}$$

where w_p and w_v are the parameters of geometry context and pedestrian-vehicle context, which ensure the truth detection (P, V) has larger context score than any other detection hypotheses.

Given the original pedestrians and vehicles detection P and V, whether each detection is a false positive or true positive is decided by maximizing the context score:

$$argmax \sum_{i=1}^{n}(t_{pi} w_p^T g(p_i) + t_{pi} \sum_{j=1}^{m} t_{vj} w_j^T g(p_i; v_j)) \tag{5}$$

where t_{p_i} and t_{v_j} are the binary value, 0 means the false positive and 1 means the true positive.

By training the samples, we can obtain an threshold of the pedestrian-vehicle context score and recognize whether the detection is right or not. It improve the accuracy of our detection observably.

4 Experimental Results

Testing Datasets. There are a lot of training sample libraries in the pedestrian detection field. We choose Caltech Pedestrian Detection Benchmark [7,8] as the database. The Caltech Pedestrian Dataset consists of approximately 10 h of 640×480 30 Hz video taken from a vehicle driving through regular traffic in an urban environment. About 250,000 frames (in 137 approximately minute long segments) with a total of 350,000 bounding boxes and 2300 unique pedestrians were annotated. The annotation includes temporal correspondence between bounding boxes and detailed occlusion labels. At last, we choose 100 images in which pedestrians and vehicles both exist as testing materials.

Fig. 3. The sample of Caltech Pedestrian Detection Benchmark

As the Fig. 3 shows, Calthch Pedestrian Detection Benchmark includes many photos related to traffic scenes. Therefore, we decided to use this dateset to help us accomplish our research.

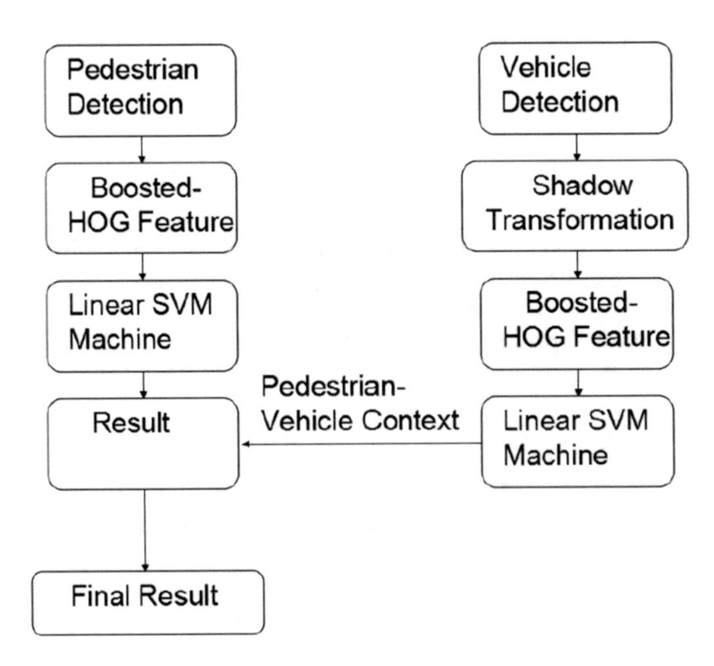

Fig. 4. The procedures of our experiment

As Fig. 4 shows, our method is based on HOG feature descriptor, adaboost method and linear SVM.

Pedestrian Detection. We use Caltech Pedestrian Dataset as the training date. The training data (set00-set05) consists of six training sets (1 GB each), each with 6–13 one-minute long seq files, along with all annotation information. We choose 500 positive image and 300 negative image from the video and all their size is 75 × 75 pixels. After training, we select 100 frame from testing data(set06-set10). Besides, we test these images using conditional HOG feature, the result is as Table 1.

Table 1. Performance of the pedestrian detection in traffic scenes

Feature type	Testing number	Positive result	Negative result	Detected rate
Conditional HOG	100	75	25	75%
Boosted HOG	100	86	14	86%
Proposed method	100	93	7	93%

Experimental results shows that proposed method has better performance than other conditional methods in traffic scenes.

4.1 Vehicle Detection

In the detection of vehicles, The shadow is the main characteristic for the detection of vehicles. The region under vehicle is usually darker than other objects or backgrounds and could be segmented by setting a threshold. Our work is to detect these shadow so we can get the contour of vehicles. Some of the successful detection is shown in Fig. 5.

Fig. 5. Successful detection in Caltech Pedestrian Detection Benchmark

5 Conclusions

In this paper we propose a combined method for pedestrian detection and vehicle detection in static images. The original HOG feature is adopted first and we train it with adaboost algorithm to obtain the boosted HOG feature. A linear SVM is trained to accomplish the detection. Shadow feature underneath vehicle is the important feature of vehicle detection in road image and can be used to generate hypotheses quickly and reliably. HOG feature and SVM has good generalization ability and can remove the non-vehicle regions such as buildings, clouds, flowers, fence presuming effectively. Experiments on Caltech Pedestrian Benchmark show that proposed method can get better performance than conditional HOG method. Shadow Detection made us detect vehicles more conveniently and effectively. The accuracy of our pedestrian detection increased significantly due to Pedestrian-Vehicle Context which removed plenty of wrong detections after all the detection.

Acknowledgment. This work was supported in part by Grants-in-Aid for Scientific Research, MEXT (No.23500289), and parially by Peronas Corpolation, Petroleum Research Fund (PRF) No.0153AB-A33.

References

1. Dalal, N., Triggs, B.: Histograms of oriented gradients for human detection. In: Computer Vision and Pattern Recognition, vol. 1, pp. 886–893 (2005)
2. Zhu, Q., Yeh, M.-C., Cheng, K.-T., Avidan, S.: Fast human detection using a cascade of histograms of oriented gradients. In: Computer Vision and Pattern Recognition, vol. 2, pp. 1491–1498 (2006)
3. Tuzel, O., Porikli, F., Meer, P.: Human detection via classification on riemannian manifolds. In: Computer Vision and Pattern Recognition, pp. 1–8 (2007)
4. Wojek, C., Schiele, B.: A performance evaluation of single and multi-feature people detection. In: Proceedings of the 30th DAGM Symposium on Pattern Recognition, pp. 82-91 (2008)
5. Viola, P., Jones, M.J., Snow, D.: Detecting pedestrians using patterns of motion and appearance. In: The 9th ICCV, Nice, France, vol. 1, pp. 734–741 (2003)
6. Li, X., Guo, X.-S., Guo, J.-B.: A Multi-feature fusion method for forward vehicle detection with single camera. In: The International Conference on Mechatronics and Industrial Informatics, pp. 998–1004 (2003)
7. Dollar, P., Wojek, C., Schiele, B., Perona, P.: Pedestrian detection an evaluation of the state of the art. Pattern Anal. Mach. Intell. **34**, 743–761 (2012)
8. Dollar, P., Wojek, C., Schiele, B., Perona, P.: Pedestrian detection: a benchmark. In: Computer Vision and Pattern Recognition, pp. 304–311 (2009)

SURF Algorithm-Based Panoramic Image Mosaic Application

Junzo Watada[1], Huiming Zhang[2](✉), Haydee Melo[2], Jiaxi Wang[2], and Pandian Vasant[3]

[1] Computer and Information Sciences Department, PETRONAS University of Technology, 32610 Seri Iskandar, Perak Darul Ridzuan, Malaysia
junzow@osb.att.ne.jp

[2] Graduate School of Information, Production and Systems, Waseda University, 2-7 Hibikino, Wakamatsu, Kitakyushu 808-0135, Japan
huimingde@gmail.com, melo.haydee@asagi.waseda.jp

[3] Fundamental and Applied Sciences Department, PETRONAS University of Technology, 32610 Seri Iskandar, Perak Darul Ridzuan, Malaysia

Abstract. Panoramic image mosaic is a technology to match a series of images which are overlapped with each other. Panoramic image mosaics can be used for different applications. Image mosaic has important values in various applications such as computer vision, remote sensing image processing, medical image analysis and computer graphics. Image mosaics also can be used in moving object detection with a dynamic camera. After getting the panoramic background of the video for detection, we can compare every frame in the video with the panoramic background, and finally detect the moving object. To build the image mosaic, SURF (Speeded Up Robust Feature) algorithm is used in feature detection and OpenCV is used in the programming. Because of special optimization in image fusion, the result becomes stable and smooth.

Keywords: Panorama · SURF · Stitching · Feature point · Video frame

1 Introduction

Nowadays, in our daily life computers and cameras become cheaper and more widely used, and we are now using digital image almost everywhere. Panoramic image mosaic plays a pivotal role in this large use of digital equipment. A panorama is a single wide-angle image of the environment around the camera. However, as the size of a single photo and a single frame in a video is limited, image mosaic may be necessary in getting a panoramic background. With a panoramic background of a video or a series of images, we can get a complete image of the scene or get some help in object tracking with a dynamic camera. Image mosaic has important value in applications such as computer vision, remote sensing image processing, medical image analysis, and computer graphics. Image mosaics also can be applicable to moving object detection with a

J.-S. Pan et al. (eds.), *Advances in Intelligent Information Hiding and Multimedia Signal Processing*, Smart Innovation, Systems and Technologies 81,
DOI 10.1007/978-3-319-63856-0_43

dynamic camera. After getting the panoramic background of the video for detection, we can compare every frame in the video with the panoramic background, and finally detect the moving object.

Any expression of wide angle or physical space can be summarized as Panorama. In the past, when digital information had not been developed, panoramic drawing is the most common manifestation to show the landscape and historical events. In recent years, Lopez-Molina, B. De Baets and H. Bustince studied generating fuzzy edge images from gradient magnitudes in 2011 [1]. Alireza Kasaiezadeh and Amir Khajepour studied multi-agent stochastic level set, and used the method in image segmentation [2] in 2013. In 2014, Gianluigi Ciocca, Claudio Cusano, Simone Santini and Raimondo Schettini used supervised features for unsupervised image categorization and Minsik Lee, Chong-Ho Choi tried to do the real-time facial shape recovery for unknown lighting by rank relaxation from a single image under general [3,4].

The remainder of this paper is structured as follows: The background knowledge are illustrated in the Sect. 2. In the Sect. 3, the authors propose SURF algorithm-based Panoramic Image Mosaic method to optimization in image fusion, the real cases test and numerical results analysis are shown in Sect. 4, while Sect. 5 draws the conclusion.

2 Background Knowledge

2.1 SIFT Feature Detector

SIFT (Scale-invariant feature transform) algorithm is a Local feature descriptor proposed by David G. Lowe in 1999, and improved in 2004 [5]. Sift feature detector and SIFT feature matching algorithm can deal with the matching of translation, rotation, affine and other transformation between two images. Overall, SIFT operator has the following features:

1. SIFT feature is the local feature of a image. It has good invariance of the translation, rotation, scaling, brightness change, occlusion and noise. To a certain extent, it also has stability against visual changes, affine transformations.
2. Good specificity, informative, suitable for fast and accurate matching in large-scale property database.
3. Even a small amount of objects may also produce a large number of SIFT eigenvectors.
4. Relatively fast, optimized SIFT matching algorithm can even achieve real-time requirements.
5. Scalability is strong, and can easily be combined with other forms of vectors.

After extracting key points, SIFT adds local feature to the key points and SIFT uses the feature vectors. In order to ensure that the extracted feature vectors have “rotation invariant”, the first thing to do is to find a “direction” in each feature point, and this direction should be same with the local image properties (Fig. 1).

Below is the equation to calculate the angle and size.

$$m(x;y) = \sqrt{(L(x+1;y) - L(x-1;y))^2 + (L(x;y+1) - L(x;y-1))^2}$$
$$\Theta(x;y) = \tan^{-1}\left(\frac{(L(x;y+1) - L(x;y-1))}{(L(x+1;y)L(x_1;y))}\right)$$

In order to get the vector, we "quantize" the angle(for example, we divide 360 degrees into 8 equal parts), and then add them up according to the position. Figure 2 is a 2×2 eigenmatrix after local sampling calculation of an 8×8 eigenmatrix. Actually, we generally use a 4×4 eigenmatrix 16 Chap. 2 Research Design and Methodology

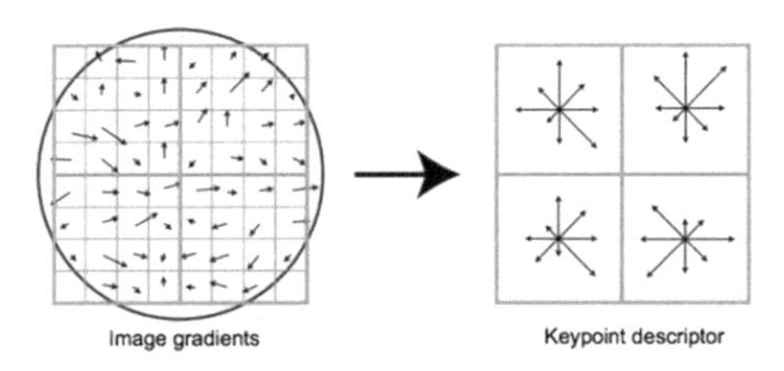

Fig. 1. Gradient magnitude and gradient direction in SIFT&SURF feature detector

After calculating the feature vector, the next step is to match the key points. The most basic way is "Nearest Neighbor". In fact, the searching method is to find a feature vector which has the nearest straight line distance in the 128-dimensional space. The way to calculate straight line distance has no difference with that in 2-dimensional space and the nearest eigenvectors are considers as corresponding points.

2.2 SURF Algorithm

SURF, Speeded Up Robust Features in other words, is the speeded up version of scale invariant feature transform algorithm (SIFT algorithm). SURF algorithm can deal with most real-time object matching under moderate condition.

Construct Hessian matrix Hessian matrix is the core of SURF algorithm. First we suppose the function f(z, y), and the partial derivative is:

$$H(f(x;y)) = \begin{bmatrix} \frac{\partial^2 f}{\partial x^2} & \frac{\partial^2 f}{\partial x \partial y} \\ \frac{\partial^2 f}{\partial x \partial y} & \frac{\partial^2 f}{\partial y^2} \end{bmatrix}$$

The matrix discriminant is:

$$det(H) = \frac{\partial^2 f}{\partial x^2}\frac{\partial^2 f}{\partial y^2} - \left(\frac{\partial^2 f}{\partial x \partial y}\right)^2$$

Value of discriminant is the eigenvalue of Matrix H and we can classify all the points according to plus or minus of the discriminant and also determine whether a point is an extreme point or not. In SURF algorithm, we use the pixel of image I(x,y) instead of function f(x,y) and use second order Gauss function as a filter. After calculating second-order partial derivative by specific kernel convolution, we can calculate the parameters of the matrix and get matrix **H**.

$$H(x;\sigma) = \begin{bmatrix} L_{xx}(x;\sigma) & L_{xy}(x;\sigma) \\ L_{xy}(x;\sigma) & L_{yy}(x;\sigma) \end{bmatrix} \tag{1}$$

L(x,t) is an image in different resolutions. In this way, we can calculate the value of H determinant of every pixel in the image and use the value to determine feature points. Again we look back to Fig. 2. We make a 8×8 window taking the key point as the central point. The center of the left figure is the position of the key point; each small cell means a pixel in the scale space of the neighborhood of the key point.

2.3 Corresponding Methods

If we need to get the transformation matrix to stitch two frames, we need information of corresponding points to do the calculation. Here X' is one of the feature points of the target frame in the search area. We use the following equation to compare the two points:

$$\rho(x;x') = \frac{\sum_{-n}^{n}\sum_{-m}^{m}[I(u+i;v+j)-E(x)][I'(u'+i;v'+j)-E(x')]}{\sqrt{\sum_{-n}^{n}\sum_{-m}^{m}[I(u+i;v+j)-E(x)]^2[I'(u'+i;v'+j)-E(x')]^2}} \tag{2}$$

$$E(x) = \frac{\sum_{-n}^{n}\sum_{-m}^{m}[I(u+i;v+j)]}{(2n+1)(2m+1)}$$

where $I(u,v)$ is a gray level of (u,v) in the frame $E(x)$ is the average value of gray histogram at point $X(u,v)$ and $\rho(x;x')$ is the coefficient correlation of x and x'. The value range of coefficient correlation is usually $[-1,1]$. If the coefficient correlation is larger than the given threshold, the two feature points would be detected as corresponding points.

Choose some data to estimate the model parameters; after this, it compares the data error and take the one which has smallest error as the best one to separate inliers and outliers. RANSAC can be used to find the parameter of the line. The points on or around the line are inliers and the others are outliers.

After detecting the corresponding points, we can use their coordinates to get the transformation matrix and do the image fusion.

2.4 Image Fusion

Image fusion is one kind of data fusion. It means to automatically analyze and synthesize the image data in certain rules using computer image processing technology. We should try to avoid losing original image information due to image fusion as far as possible.

1. Directly average fusion method Directly average fusion method directly add the grey level of pixels in overlapping area and calculate the average value. This method is simple but the trace of stitching may be obvious. f_1 and f_2 means the two original images and f is the result image.

$$f(x;y) = \begin{cases} f_1(x;y) & (x;y) \in f_1 \\ \dfrac{(f_1(x;y) + f_2(x;y))}{2} & (x;y) \in f_1 \cap f_2 \\ f_2(x;y) & (x;y) \in f_2 \end{cases} \tag{3}$$

2. Weighted average fusion method Weighted average fusion method does not simply add the grey level of the pixels together. It weights the data and then do the average calculation. As the equation below says, w1 and w2 are the weight of corresponding points in original images and $w1 + w2 = 1, 0 < w1 < 1, 0 < w2 < 1$.

$$f(x;y) = \begin{cases} f_1(x;y) & (x;y) \in f_1 \\ w_1 f_1(x;y) + w_2 f_2(x;y) & (x;y) \in f_1 \cap f_2 \\ f_2(x;y) & (x;y) \in f_2 \end{cases} \tag{4}$$

To choose the weight value, we can use the following two methods:

$$w_i(x;y) = \left(1 - \left|\frac{x}{width_i} - \frac{1}{2}\right|\right)\left(1 - \left|\frac{y}{height_i} - \frac{1}{2}\right|\right)$$

In this method, pixels near the central area have higher weight value and the ones near the edge have lower weight value.

$$f(x;y) = \begin{cases} f_1(x;y) & (x;y) \in f_1 \\ d_1 f_1(x;y) + d_2 f_2(x;y) & (x;y) \in f_1 \cap f_2 \\ f_2(x;y) & (x;y) \in f_2 \end{cases} \tag{5}$$

where $d1 + d2 = 1, 0 < d1 < 1, 0 < d2 < 1$. In this method, the more the pixel is near to the central area of f1, d1 is larger and d2 is the opposite. In the overlapping area, d1 grades from 1 into 0 and d2 grades from 0 into.

3 Real Scenario Cases and Numerical

The computer we used to do the experiment used Intel Core 15-3470S CPU 2.90 GHz, 4.00 GB memory and 64-bit Window7 operating system.

3.1 SURF Feature Detection Matching

Steps: 1. Use SURF feature detection to get the feature points. 2. Use interface DescriptorExtractor to find the feature vector of key points. 3. Use SurfDescriptorExtractor and its function compute to accomplish the calculation. 4. Use BruteForceMatcher to match the feature vector. 5. Use the drawMatches function to draw the corresponding points detected.

3.2 Image Fusion and Stitching

$$\begin{bmatrix} x' \\ y' \\ z' \end{bmatrix} R \times K^T \times \begin{bmatrix} x \\ y \\ z \end{bmatrix} \tag{6}$$

$$U = scalea \tan\left(\frac{x'}{y'}\right) \tag{7}$$

$$W = \frac{y'}{\sqrt{x'^2 + y'^2 + z'^2}} \tag{8}$$

$$V = scale \times (PI - a\cos(W)) \tag{9}$$

where (x, y, z) is the original point, R is the rotation matrix, K is the camera parameter matrix, (x', y', z') is the point after transformation and (u,v,w) is the coordinate of the point mapping to spherical coordinates.

Fig. 2. Original frames of Example-2

3.3 Multi-band Fusion and Illumination Compensation

$$I(i;j) = \frac{\sum\limits_{N(i.j)} \sqrt{\sqrt{R} + \sqrt{G} + \sqrt{B}}}{N(i;j)} \tag{10}$$

where $N(i.j)$ means the number of points that coincide with each other and $I(i,j)$ is the average brightness of i and j in the coincide area.

Figure 2 shows the frames used in modeling and result is Fig. 3. It takes 21 s. Example-3 a video which has 250 frames is used. Use 4 frames(size: 856×480) to build the panoramic image. Figure 4 shows the frames used in modeling and result is Fig. 5. It takes 22 s.

3.4 A Video Which Has 260 Frames Is Used

Use 9 frames(size: 432×240) to build the panoramic image.

Fig. 3. Example-2 of panoramic modeling

Fig. 4. Original frames of Example-3

Fig. 5. Example-3 of panoramic modeling

Fig. 6. Original frames of Example-4

Fig. 7. Example-4 of panoramic modeling

Figure 6 shows the frames used in modeling and result is Fig. 7. It takes 38 s.

4 Discussion

This program can be applied to colorful images or a colorful video and get a colorful panoramic image. Because the program uses several smooth method while doing the fusion such as waveform correction, multi-band fusion and illumination compensation, the result is more smooth and stable. Figures 8 and 9 are examples of former work that did less in image fusion such as fusion method and wave form correction and Fig. 10 is one example of result of our program.

The speed of the program is greatly affected by the quality or the size and number of pixels of original images. Of course the more images waiting to be transformed, the longer time it costs. The detailed time of some examples are in Table 1.

Table 1. Detailed time of examples

Number of images	Size	Time
6 frames	400×240	16 s
4 frames	480×856	21 s
4 frames	856×480	22 s
9 frames	432×240	38 s

Fig. 8. Example-1 of former image fusion

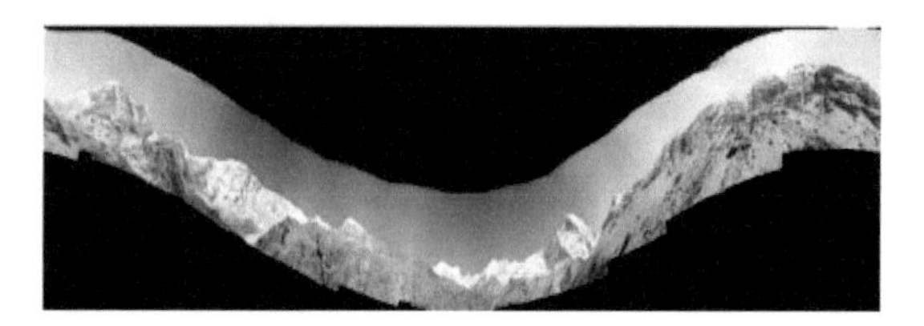

Fig. 9. Example-2 of former image fusion

Fig. 10. Example of image fusion in this program

5 Conclusions

This thesis introduces some methods of feature point detection and panoramic image mosaic using OpenCV. To do the panoramic image mosaic, after image acquisition, preprocessing is necessary so that the following registration and fusion will be easier and more accurate. As the camera is moving, transformation becomes necessary in stitching and to get the transformation matrix of two images, corner points are detected first, and then do the comparing so that some pairs of corresponding points can be found and finally we use their coordinates to get the transformation matrix. After getting the transformation matrix, we choose a fusion method to do the image fusion and use methods such as waveform correction, multi-band fusion and illumination compensation to make the result clearer and more smooth.

Acknowledgment. This work was supported in part by Grants-in-Aid for Scientific Research, MEXT (No.23500289), and parially by Peronas Corpolation, Petroleum Research Fund (PRF) No.0153AB-A33.

References

1. Gao, J.: Self-occlusion immune video tracking of objects in cluttered environments. In: Proceedings of the IEEE Conference on Advanced Video and Signal Based Surveillance, p. 79 (2003)
2. Kitahara, R., Nakamura, T., Katayama, A., Yasuno, T.: Real-time rectangle tracking method for geometric correction on mobile terminals. Technical report of IEICE. ISEC, vol. 106, no. 351, pp. 1–6 (2006)

3. Ying, S., Yang, Y., Ying, S.: Study on vehicle navigation system with real-time traffic information. In: International Conference on Computer Science and Software Engineering, pp. 1079–1082 (2008)
4. Isard, M., Blake, A.: Contour tracking by stochastic propagation of conditional density. In: Buxton, B., Cipolla, R. (eds.) ECCV 1996. LNCS, vol. 1064, pp. 343–356. Springer, Heidelberg (1996). doi:10.1007/BFb0015549
5. Cheng, M., Pham, B., Tjondronegoro, D.: Tracking and video surveillance activity analysis. In: Proceedings of the 4th International Conference on Computer Graphics and Interactive Techniques, pp. 367–373 (2006)

Information Hiding and Its Criteria

Simulation of Long-Distance Aerial Transmissions for Robust Audio Data Hiding

Akira Nishimura(✉)

Department of Informatics, Faculty of Informatics, Tokyo University of Information Sciences, 4–1 Onaridai, Wakaba, Chiba 265-8501, Japan
akira@rsch.tuis.ac.jp

Abstract. This paper proposes an evaluation framework for an audio data hiding system that can be used for long-distance aerial transmissions where the stego speech signal is emitted by outdoor loudspeakers of voice evacuation and mass notification systems. Typical disturbances of aerial transmissions are modeled and implemented by signal processing units in series. The results of computer simulations for long-distance (70–800 m) aerial transmissions show that a bilateral time-spread echo hiding combined with a novel frame synchronization technique exhibits better detection performance than the conventional time-spread echo hiding under extremely low signal-to-noise ratios of 0 and −5 dB. Moreover, high-frequency attenuation and frequency shifts at the receiving side degrade the detection performance of the time-spread echo hidings.

Keywords: Time-spread echo hiding · Bilateral time-spread echo · Frame synchronization · Bit error rate · Frequency response

1 Introduction

This paper proposes an evaluation framework for audio data hiding systems that can be used for long-distance aerial transmissions where the stego speech signal is emitted from outdoor loudspeakers of voice evacuation and mass notification systems. The evaluation process is based on a computer simulation that models the typical disturbances caused by the long-distance (from 50 to 1,000 m) outdoor aerial transmissions of sounds and audio equipment for reproduction and receiving.

A number of audio data hiding technologies that embed some information into aerial sounds and utilize the decoded information at the smartphones of users have been proposed. Some of the studies intended to use audio data hiding technologies for an evacuation and mass notification system [7,8]; however, their evaluation of the proposed system was not based on outdoor long-distance transmissions. These studies did not consider background noises and their relative intensities, the frequency response of outdoor horn loudspeakers, and the absorption of sound by the atmosphere. Other studies on audio data hiding for aerial transmission [1,4,5]

J.-S. Pan et al. (eds.), *Advances in Intelligent Information Hiding and Multimedia Signal Processing*, Smart Innovation, Systems and Technologies 81,
DOI 10.1007/978-3-319-63856-0_44

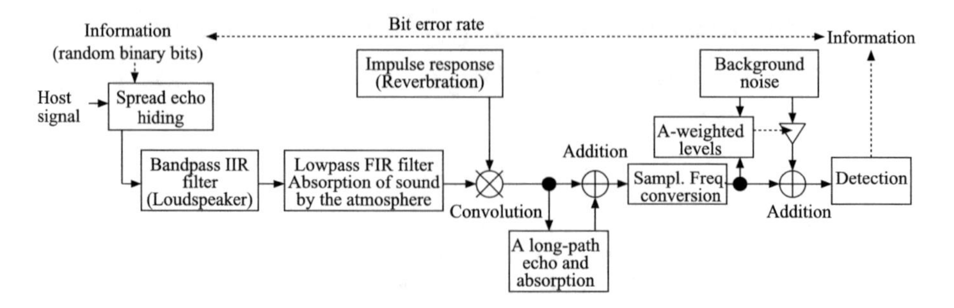

Fig. 1. A block diagram of the simulated environment.

conducted experiments in actual space. However, these studies tested only indoor environments, where the transmission distance was less than 10 m, and the background noise power was not strictly controlled [1,4].

The following section proposes a realistic simulated environment that takes the typical disturbances in the long-distance aerial transmission of sounds emitted from outdoor loudspeakers of voice evacuation and mass notification systems into account. The next section describes the time-spread hiding technology and its improved method combined with a novel synchronization technique. Additionally, the results of computer simulations of audio data hiding for long-distance aerial transmissions are presented.

2 Simulated Environments

Figure 1 presents a block diagram of the simulated environment. The models and implementations of signal processing for each disturbance are described in this section. All signal processing units are connected in series.

2.1 Frequency Response of a Distant Horn-Array Loudspeaker System

To estimate the frequency response of the horn-array loudspeaker system used in actual space, a testing speech signal of a female was reproduced from the loudspeaker system produced by TOA Corp. located on the roof of an elementary school in Itami City, Japan. The signal was recorded on a road located approximately 70 m away from the loudspeaker system in its frontal direction. Therefore, the absorption of the sound by the atmosphere at a distance of 70 m was included in the recorded sound. The speech waveform and recorded speech waveform were synchronized, and logarithmic power spectra of both waveforms were calculated after applying a synchronized 16-s Hanning window. Subsequently, the logarithmic power spectrum of the original signal was subtracted from that of the recorded signal to obtain an estimated detailed frequency response. The spectral envelope of the frequency response was estimated by taking the inverse

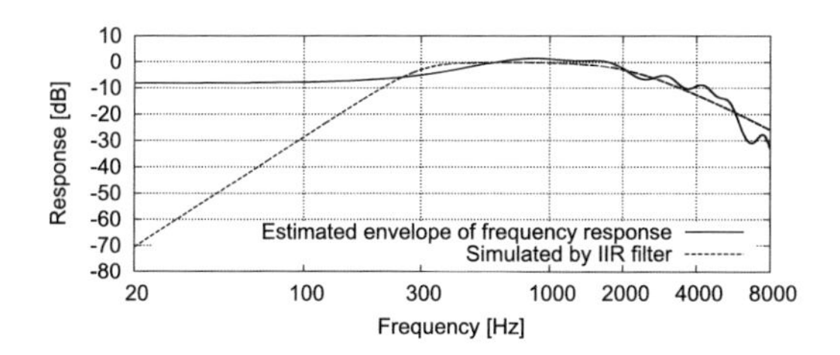

Fig. 2. Estimated spectral envelope of the loudspeaker system at a distance of 70 m, including the frequency characteristics of absorption by the atmosphere and the frequency response of the IIR filter that simulates the loudspeaker system.

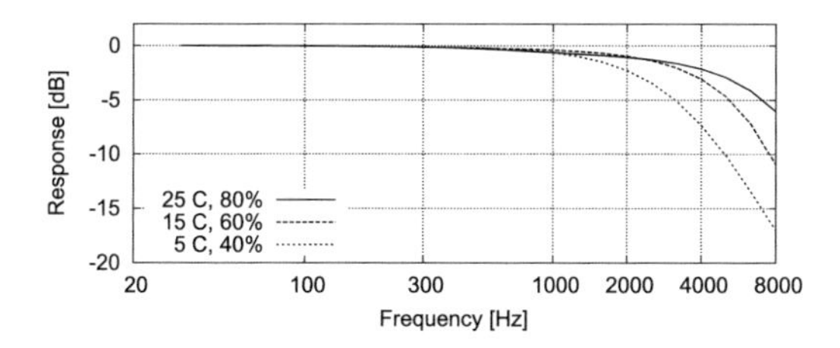

Fig. 3. Frequency attenuation characteristics per 100 m calculated using the ISO 9613-1 method under typical atmospheric conditions.

Fourier transform of the cepstrum of the detailed frequency response, liftered above the 40-th coefficient.

The high-frequency envelope of the loudspeaker system is approximated by designing a second-order low-pass Butterworth IIR filter with a cutoff frequency of 2 kHz. The low-frequency characteristics are designed using a third-order high-pass Butterworth IIR filter with a cutoff frequency of 300 Hz, simulating a typical low-frequency response of the horn-array loudspeaker system (−18 dB at 150 Hz). Figure 2 shows the estimated spectral envelope and the frequency response of the IIR filter that simulates the loudspeaker system.

2.2 Absorption of Sound by the Atmosphere

It is important to consider the absorption of sound by the atmosphere in long-distance sound transmission. ISO 9613-1 defines the standard attenuation characteristics by absorption as a function of frequency. Figure 3 presents the frequency attenuation characteristics per 100 m calculated using the ISO 9613-1 method under typical conditions, that is, temperature, relative humidity, and air pressure of the atmosphere. The attenuation characteristics at 15 degrees Celsius, 60 % relative humidity, and 1013 hPa pressure are applied for the following simulation experiment. It is simulated using a 33-tap FIR filter.

2.3 Reverberation and Long-Path Echo

Sounds emitted from a loudspeaker system are received as a directly transmitted sound and its reverberation, which include a large number of reflected sounds

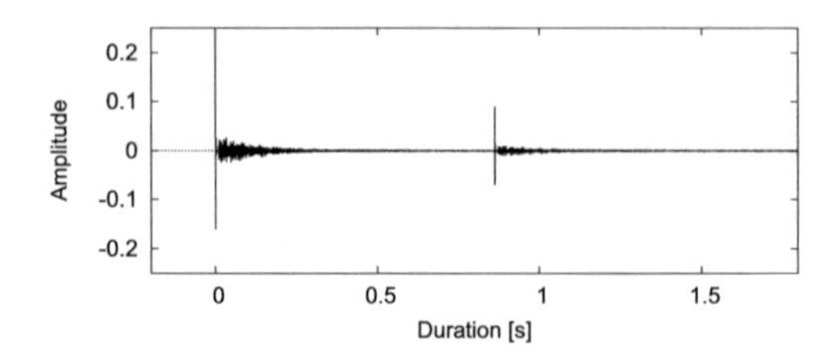

Fig. 4. An example of an impulse response consisting of a direct sound succeeding its reverberation and a long-path echo succeeding its reverberation.

from the ground, surrounding buildings, and trees. The reverberation is simply modeled by the impulse response, which consists of an exponentially attenuated Gaussian random noise. Long-path echoes from buildings of flat structures or another loudspeaker in the same mass notification network are mixed with the received sounds. The number of echoes and their relative amplitudes depend on the outdoor environment and the distribution of the located loudspeaker system.

In the following simulation, the reverberation time is set to 1.0 s, and the direct to reverberant ratio is set to 4 dB, which realizes an STI (speech transmission index) [6] of 0.6. Additionally, a single long-path echo with a random delay time from 0.1 to 1.0 s is added to the direct sound. Figure 4 presents an example of an impulse response consisting of a direct sound succeeding its reverberation and a long-path echo succeeding its reverberation.

2.4 Frequency Shift

The frequency of the received sound is slightly fluctuated or shifted compared with the emitted sound due to changes in speed and wind direction and a Doppler shift caused by movement of the receiver. The following simulation introduces a +0.1 % constant frequency shift before mixing of the background noise.

2.5 Background Noise

The background noise is 'SPSQUARE/01.wav', which was recorded at a public town square with many tourists and is selected from DEMAND (Diverse Environments Multichannel Acoustic Noise Database) [9]. A randomly selected segment is mixed with the stego speech signal at an SNR (signal-to-noise ratio) that is calculated as the difference between A-weighted equivalent levels of the signal and the noise. In the following experiment, 0 and −5 dB SNRs are simulated.

3 Audio Data Hiding Technology for Evaluation

The payload data embedded in the speech signal are information for refuges from disasters that is coded into a small number of bits. Therefore, the bit rate of the payload can be smaller than 10 bps. Combined with an error correction code, the actual bit rate is considered to be from 1 to 2 bps.

In this section, the conventional time-spread echo hiding and its improvement of bilateral time-spread echo hiding are described. Then, a novel frame synchronization technique that can be applied to both hiding methods is proposed.

3.1 Conventional Time-Spread Echo Hiding

The embedding process segments the host signal $s(n)$ into $F + T_r$ samples with an overlap of T_r samples. The stego signal $r(n)$ is obtained by the convolution of the framed host signal $s(n)$ with the impulse response $k(n)$, consisting of Dirac delta function $\delta(n)$ and an echo kernel $P(n)$ of length L and a delay time of $d0$.

$$k(n) = \delta(n) + \alpha P(n - d0), \tag{1}$$

$$r(n) = s(n) * k(n), \tag{2}$$

where $*$ means convolution and α is the echo gain, which is maximized to $\alpha = 1/\sqrt{L}$. Subsequently, the framed stego signals are concatenated with an overlap of T_r samples. The payload data are embedded in every segmented frame signal.

The original echo hiding [3] represents a binary bit by switching the delay time between $d0$ and $d1$. The current hiding scheme circularly shifts $P(n)$ to obtain $P'(n)$, depending on the integer value m encoded by payload bits, as follows:

$$P'(n) = \begin{cases} P(n + L - m) & (1 \leq n \leq m), \\ P(n - m) & (m + 1 \leq n \leq L). \end{cases} \tag{3}$$

m is quantized by m' steps, in other words, $m \in \{0, m', 2m', ..., \lfloor L/m' \rfloor m'\}$, to be robust against frequency shift to the stego signal. Consequently, the amount of payload is $\log_2(\lfloor L/m' \rfloor + 1)$ per segmented frame.

In the detection process, the cepstrum transform ˜ of the stego signal $r(n)$ is calculated using DFT (discrete Fourier transform). $\tilde{r}(n) = \mathrm{IDFT}(\log(\mathrm{DFT}(r(n))))$, where IDFT represents the inverse DFT. Equation (4) is the cepstrum transform of Eq. (2). The cross-correlation (xcorr) of $\tilde{r}(n)$ and $P(n)$ exhibits a peak at the amount of circular shifting m.

$$\tilde{r}(n) = \tilde{s}(n) + \tilde{k}(n), \tag{4}$$

$$x(n) = \mathrm{xcorr}(\tilde{r}(n), P(n)). \tag{5}$$

3.2 Bilateral Symmetric Time-Spread Echo Kernel

The bottom panel of Fig. 5 presents an example of the impulse response of the bilateral symmetric time-spread echo kernel [2]. The detection gain, which represents a gain of the peak of the cross-correlation (Eq. 5), exhibits a maximum value when the echo gain $\alpha = 1/(2\sqrt{L})$. It achieves approximately 1.7 times greater detection gain than that of the conventional unilateral condition, resulting in better performance.

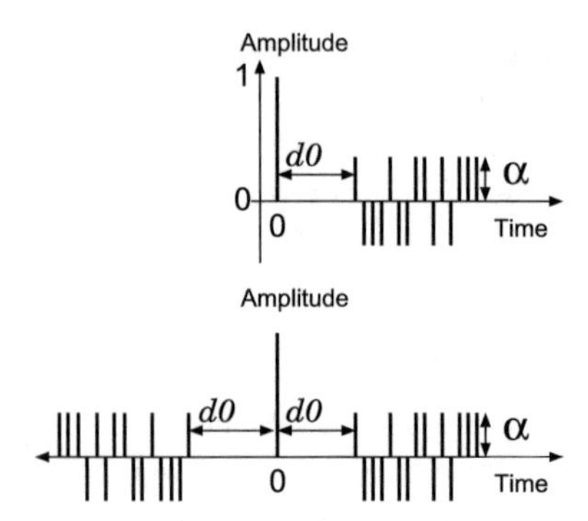

Fig. 5. The top panel shows an example of the impulse response of time-spread echo hiding, and the bottom panel shows that of bilateral echo kernel.

3.3 Frame Synchronization

Previous works on echo hiding technologies have not addressed the frame synchronization because a detection frame that is shorter than the size of the embedded frame F can extract the payload data. To maximize the detection performance, the following method is proposed and used for frame synchronization.

During the detection process, payload detection is conducted for shifting the frame of the stego signal by every $F/8$ samples. The probability of correct detection is maximized when the stego frame is synchronized to the embedded frame, and the same payload can be extracted from before and after the synchronized frame. Therefore, differentiation of the extract payload exhibits one or more zeros in every eight frames. The period of eight frames can be detected by Fourier analysis of the differentiated payload data. The correct payload can be detected from the zero-phase frame of the period.

4 Simulation and Evaluation

4.1 Conditions of Simulation

A total of 753 speech files spoken by 10 male speakers, of which two speech files are concatenated, and 903 speech files spoken by 12 female speakers serve as cover data. These files are recorded in the Continuous Speech Database for Research (Vol. 1) published by the Acoustical Society of Japan. All speech files are sampled at 16 kHz and 16-bit quantization.

The conventional and bilateral time-spread echo hidings combined with the frame synchronization technique, which are described in Sect. 3, are evaluated by the computer simulation described in Sect. 2. The baseline condition provides all factors of disturbance. The other conditions modify one of the disturbances. The simulated conditions of ‘no loudspeaker response’, ‘no echo’ and ‘no reverberation’ represent the conditions excluding each factor from the baseline condition, and 400 and 800 m distance conditions and a 0.5 % frequency shift condition are also individually simulated. The parameter values for the spread echo hiding are shown in Table 1. Watermarks of a random bit are embedded in.

4.2 Results

The bit error rates obtained from the male speakers are generally better than those from the female speakers from 1% to 7% because of the lower fundamental frequencies of the male speakers. The following results are the results obtained from thc female speakers.

Figures 6 and 7 show the bit error rates obtained from the simulation of SNRs of 0 dB and −5 dB, respectively. The bilateral spread echo hiding outperforms in all conditions. Under an extremely low SNR of −5 dB baseline condition, the bilateral method can transmit 87.5% of the payload at 3.9 bps for half of the female speech files. A long-path echo and reverberation slightly degrade the detection performance.

The frequency responses of aerial transmission are a crucial factor for the time-spread echo hiding because removing the loudspeaker response significantly improves the detection performance. Moreover, the attenuation of the high-frequency signals induced by absorption by the atmosphere at a distance of greater than 400 m deteriorates the detection performance even under constant SNR conditions. The effect of absorption by the atmosphere at a distance of 70 m, which is included in the baseline condition, is not serious, as estimated from the amount of absorption shown in Fig. 3.

A frequency shift also degrades the detection performance because a slight temporal misalignment between $\tilde{r}(n)$ and $P(n)$ in Eq. (5) caused by a frequency shift causes the peak of the cross-correlation function to be less strong.

5 Discussion

The benefit of the time-spread echo hiding is that the timbre formed by the echo kernel is similar to that formed by random reflections observed at indoor and outdoor environments; thus, the quality of the stego speech is not severely degraded. The current study did not confirm the subjective quality of the stego speech signal. This topic is one of the research interests for further work.

Table 1. Parameter values of the spread echo hidings.

Parameter	Values	
Delay time ($d0$)	80 samples (5 ms)	
Frame overlap ($T_r/2$)	50 samples	
Frame length (F)	16,384	32,768
Bit rate [bps]	7.8	3.9
Length of PN series (L)	1,023	
Payload bits per frame	8	

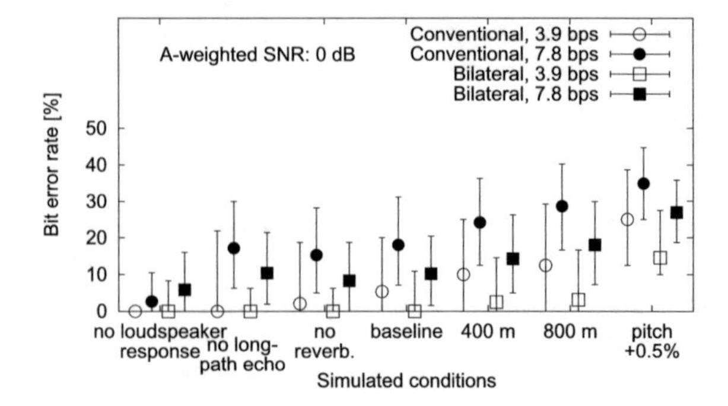

Fig. 6. Median, 10th, and 90th percentiles of bit error rates for the simulated conditions of SNR 0 dB.

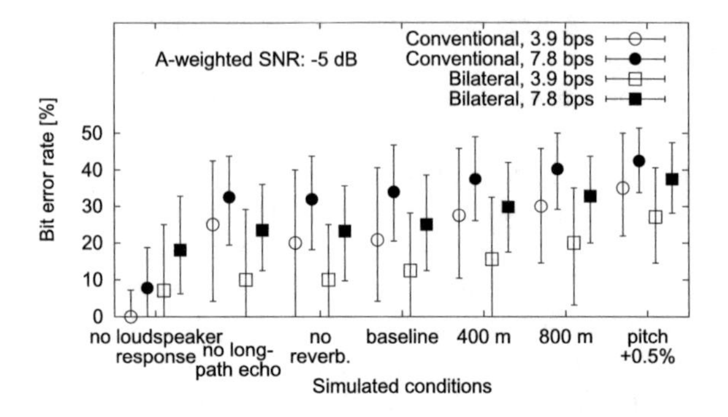

Fig. 7. Bit error rates for the simulated conditions of SNR −5 dB.

The present results of the simulated long-distance aerial transmission suggest that the development of technologies for robust aerial audio data hiding requires evaluation using computer simulations of concatenated disturbances in an actual environment.

6 Summary

This paper proposed an evaluation framework for audio data hiding systems that can be used for long-distance aerial transmissions where the stego speech signal is emitted by outdoor loudspeakers of voice evacuation and mass notification systems. The results of computer simulations for long-distance (70—800 m) aerial transmissions showed that a bilateral time-spread echo hiding combined with a novel frame synchronization technique realized better detection performance than the conventional spread echo hiding under extremely low SNR of 0 and −5 dB. Additionally, high-frequency attenuation and frequency shifts at the receiving side degrade the detection performance of the time-spread echo hidings.

Acknowledgment. This study was partially supported by TOA Corporation.

References

1. Cho, K., Choi, J., Jin, Y.G., Kim, N.S.: Quality enhancement of audio watermarking for data transmission in aerial space based on segmental SNR adjustment. In: Proceedings of IIHMSP 2012, pp. 122–125 (2012)
2. Chou, S.A., Hsieh, S.F.: An echo-hiding watermarking technique based on bilateral symmetric time spread kernel. In: Proceedings of ICASSP 2006 III, pp. 1100–1103 (2006)
3. Gruhl, D., Lu, A., Bender, W.: Echo hiding. In: Anderson, R. (ed.) IH 1996. LNCS, vol. 1174, pp. 295–315. Springer, Heidelberg (1996). doi:10.1007/3-540-61996-8_48
4. Matsuoka, H., Nakashima, Y., Yoshimura, T.: Acoustic OFDM system and performance analysis. IEICE Trans. Fundam. E91-A(7), 1652–1658 (2008)
5. Nishimura, A.: Audio data hiding that is robust with respect to aerial transmission and speech codecs. Int. J. Innov. Comput. Inf. Control 6(3(B)), 1389–1400 (2010)
6. Schroeder, M.R.: Modulation transfer functions: Definition and measurement. Acustica **49**, 179–182 (1981)
7. Tetsuya, K., Akihiro, O., Udaya, P.: Properties of an emergency broadcasting system based on audio data hiding. In: Proceedings of the IIHMSP2015, pp. 142–145 (2015)
8. Tetsuya, K., Kan, K., Udaya, P.: A disaster prevention broadcasting based on audio data hiding technology. In: Proceedings of Joint 8th International Conference on Soft Computing and Intelligent Systems and 17th International Symposium on Advanced Intelligent Systems, pp. 373–376 (2016)
9. Thiemann, J., Ito, N., Vincent, E.: DEMAND: Diverse environments multichannel acoustic noise database (2013). http://parole.loria.fr/DEMAND/

Digital Watermarking Scheme Based on Machine Learning for the IHC Evaluation Criteria

Ryo Sakuma[1(✉)], Hyunho Kang[2(✉)], Keiichi Iwamura[1(✉)], and Isao Echizen[3(✉)]

[1] Department of Electrical Engineering, Tokyo University of Science, 6-3-1 Niijuku Katsushika-ku, Tokyo 125-8585, Japan
{sakuma,iwamura}@sec.ee.kagu.tus.ac.jp

[2] National Institute of Technology, Tokyo College, 1220-2 Kunugida-machi Hachiouji-shi, Tokyo 193-0997, Japan
kang@tokyo-ct.ac.jp

[3] National Institute of Informatics, 2-1-2 Hitotsubashi Chiyoda-ku, Tokyo 101-8430, Japan
iechizen@nii.ac.jp

Abstract. Digital watermarking is a technique used for embedding information in digital content and protecting its copyright. The important issues to be considered are robustness, quality and capacity. Our goal is to satisfy these requirements according to the Information Hiding and its Criteria for evaluation (IHC) criteria. In this study, we evaluate our watermarking scheme along the IHC criteria Ver. 3 as the primary step. Although image watermarking techniques based on machine learning already exist, their robustness against desynchronization attacks such as cropping, rotation, and scaling is still one of the most challenging issues. We propose a watermarking scheme based on machine learning which also has cropping tolerance. First, the luminance space of the image is decomposed by one level through wavelet transform. Then, a bit of the watermark and the marker for synchronization are embedded or extracted by adjusting or comparing the relation between the embedded coefficients value of the LL space and the output coefficients value of the trained machine learning model. This model can well memorize the relationship between its selected coefficients and the neighboring coefficients. The marker for synchronization is embedded in a latticed format in the LL space. Binarization processing is performed on the watermarked image to find the lattice-shaped marker and synchronize it against cropping. Our experimental results showed that there were no errors in 10HDTV-size areas after the second decompression.

Keywords: Digital watermarking · Machine learning · Binarization · Lattice-shaped marker · Cropping

1 Introduction

Today, pictures and sounds are digitalized and such digital content is easily available from various sources. It is easy to copy such digital content without degradation and distribute it on the Internet. Therefore, the problem of unauthorized use of digital content

J.-S. Pan et al. (eds.), *Advances in Intelligent Information Hiding and Multimedia Signal Processing*, Smart Innovation, Systems and Technologies 81,
DOI 10.1007/978-3-319-63856-0_45

is very serious. Digital watermarking is a technique that is used for protecting the copyright of digital content such as images, audio, and videos. This technique changes the structure of digital content in such a manner that it cannot be perceived by humans. The important issues of digital watermarking to be considered are robustness, quality, and capacity. However, evaluation for watermarking schemes was not standardized until recently. The Information Hiding and its Criteria for evaluation (IHC) committee was established for the purpose of unification of the evaluation standards and their revision. Our goal is to satisfy these requirements according to the IHC evaluation criteria [1]. In this study, we evaluate our watermarking scheme along the IHC criteria Ver.3, as the primary step [2–4].

To achieve the IHC evaluation criteria, we focus on machine learning. Machine learning is a technology that replicates the process of human learning on the computer [5, 6]. The invariable features are found from the original image, the watermarked image, and the attacked image, and a well-trained model is obtained by learning these features. We propose a new, robust digital watermarking scheme using this model.

An image watermarking scheme based on machine learning has been proposed by very few researchers [7–10] till date. One of the first watermarking schemes based on machine learning was proposed by Chun-hua Li et al. [7]. In this method, a watermark bit is embedded or extracted by adjusting or comparing the relation between the embedded pixel value and the output pixel value of the trained machine learning model. This model memorizes the relationship between the selected pixel and its neighboring pixel because of its good learning and generalization capability.

However, in this area, robustness against desynchronization attacks such as cropping, rotation, and scaling is still one of the most challenging issues.

In this study, we propose a watermarking scheme having cropping tolerance and based on machine learning. The luminance space of the image is decomposed one level through wavelet transform [11, 12], and the LL space of this luminance space is divided into contiguous 3×3 blocks. Then, a watermark bit and the synchronization marker are embedded or extracted by adjusting or comparing the relation between the embedded coefficient values and the output coefficient values of the trained machine learning model. We use BCH code [13, 14], a type of error-correcting code for the watermark. The synchronization marker is embedded in a lattice-shape in the LL space. Binarization processing is performed on the watermarked image to find the lattice-shaped marker and synchronize against cropping.

The remainder of this paper is organized as follows: Sect. 2 describes the synchronization to cropping attacks. Section 3 describes the proposed watermarking scheme, including a concise explanation of the BCH code, wavelet transform, machine learning, embedding and extraction processes. The experimental results and conclusion are presented in Sects. 4 and 5, respectively.

2 The Synchronization to Cropping Attacks

In this section, we explain the synchronization to cropping attacks in detail.

First, we uniformly embed the same information as the bit '1' or '0' in a latticed position in the LL space of the luminance of the image, which is referred to as the marker. Figure 1(a) shows the watermarked image. The watermarked image is binarized by comparing the coefficients of the watermarked image to the coefficients obtained from the machine learning model. Figure 1(b) shows the binarized image. A lattice-shaped marker is observed in this binarized image. Even when the watermarked image is cropped, this lattice-shaped marker is still retained. Figure 2 shows the cropped image from the watermarked image. Our watermark scheme gains cropping tolerance by embedding the information as a watermark inside the lattice. Furthermore, we believe that this marker can henceforth be used for synchronization against scaling and rotation attacks.

(a) (b)

Fig. 1. Watermarked image (a) and Binarized image (b) with 4608 × 3456 pixels

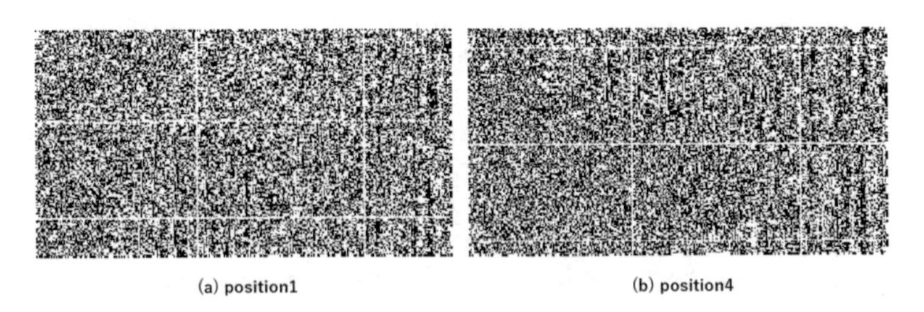

(a) position1 (b) position4

Fig. 2. Cropped image with 1920 × 1080 pixels

3 Proposed Watermarking Scheme

In our scheme, a watermark bit and a synchronization marker are embedded or extracted by adjusting or comparing the relation between the embedded coefficients in the LL space of the luminance of the image and the output coefficients of the trained machine learning model. Due to its good learning and generalization capability, this model can well memorize the relationship between its selected coefficients and the neighboring coefficients.

3.1 BCH Code

The BCH code [13, 14] is an error-correcting code named after Bose, Ray-Chaudhuri, and Hocquenghem. They published research in 1959 and 1960 that describes a method for designing codes over GF(2) with a specified design distance.

The BCH encoder generates a BCH code with message length K and codeword length N. For a given codeword length N, only specific message lengths K are valid for a BCH code. The error-correcting capability is decided by N and K.

3.2 Wavelet Transform

The wavelet transform is a strong tool for image analysis [11, 12]. Wavelet transform decomposes the image into four sub-bands denoted by LL, LH, HL, and HH at one level transform in the wavelet domain. The LL space contains the low frequency wavelet coefficients while the LH, HL, and HH spaces contain the detailed wavelet coefficients. The LL space has stronger signals than any other space. Low frequency coefficients of the LL space are less sensitive to image processing operations compared to detailed coefficients of the other spaces.

3.3 Machine Learning

Machine learning is a technology that replicates the process of human learning on a computer [5, 6].

It is assumed that the output data is determined by the input data. The machine learning model developed through training data predicts the output for any new data that is input to the model.

The invariable features are found from original image, the watermarked image and the attacked image, and we obtain a well-trained model by learning its features. We thought a new robust digital watermarking scheme could be proposed using this model.

3.4 Watermark Embedding

We describe the watermark embedding method below.

Let C be a host color image of size $m \times n$. Suppose that W is a watermark bit that has L bits.

1. Encode the watermark

To improve the robustness of the watermarking system, the watermark W is encoded by BCH code.

The BCH encoder generates a BCH code with message length K and codeword length N. For a given codeword of length N, only specific message lengths K are valid for a BCH code. The error-correcting capability T is decided by N and K. This error-correcting capability T of the valid $[N, K]$ pair used in this study can be represented as follows: $[N, K, T]$.

The encoded watermark has $L1$ bits, denoted by

$$W1 = \{W1_t | 1 \leq t \leq L1\}. \tag{1}$$

2. Obtain the space used for embedding and learning

The host image C is converted from the RGB color space to the YCbCr color space. Accordingly, the luminance component Y is obtained, such that

$$Y = \{Y_{i,j} \mid 1 \leq i \leq m, 1 \leq j \leq n\}. \tag{2}$$

The luminance component Y is decomposed one level through wavelet transform, and the LL space is obtained:

$$LL = \{LL_{i,j} \mid 1 \leq i \leq m/2, 1 \leq j \leq n/2\}. \tag{3}$$

The LL space is divided into contiguous 3×3 blocks.

3. Train the machine learning model

To model the relationship between the selected coefficients and their neighboring coefficients in the LL space, select some blocks as the dataset of training patterns, Ω. Select the center positions of some blocks as

$$\rho 1_t = (i_t, j_t)(1 \leq i_t \leq m, 1 \leq j_t \leq n). \tag{4}$$

The dataset Ω can be represented by:

$$D_t = \left\{ \begin{array}{c} LL_{i_t-1 j_t-1}, LL_{i_t-1 j_t}, LL_{i_t-1 j_t+1}, LL_{i_t j_t-1}, LL_{i_t j_t+1}, \\ LL_{i_t+1 j_t-1}, LL_{i_t+1 j_t}, LL_{i_t+1 j_t+1} \end{array} \right\}_{t=1,\ldots,k} \tag{5}$$

$$r_t = \{LL_{i_t j_t}\}_{t=1,\ldots,k} \tag{6}$$

$$\Omega = \{D_t, r_t\}_{t=1,\ldots,k} \tag{7}$$

where k stands for the number of training patterns, r_t is the desired output of the machine learning model, D_t is the input vector of the machine learning model. According to our previous research, we know that the number of training patterns do not have a big influence on the watermark scheme, so it is possible to reduce the number of training patterns to a minimum and increase the number of embedded positions.

In this scheme, we use support vector regression (SVR), which is a type of machine learning. A proper kernel function and model parameters are selected and SVR model F is trained to learn the dataset Ω. The model F can predict the output corresponding to the input vector.

$$F(x) = \sum_{t=1}^{k} \left(\beta_t^* - \beta_t\right) Ker\left(D_t, x\right) + b1 \tag{8}$$

where β stands for the coefficients of Lagrangian, Ker is kernel function, $b1$ is the value of bias.

4. Embed the marker and the watermark

To protect against cropping attacks, the markers are embedded as shown in Fig. 3. If the watermarked image cropped with the size $m'' \times n''$, the size of the watermarked area in the LL space is $a \times b$. The limits of a and b are $a \le m'' / 4$ and $b \le n'' / 4$ respectively. The markers are embedded in the blocks at the upper end and the left end in the watermarked area. All embedded markers contain the same bit, '1' or '0'.

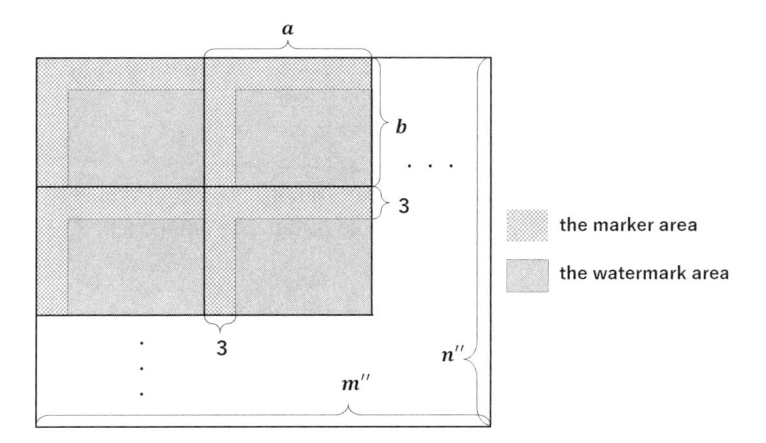

Fig. 3. Representation of the areas of the marker and the watermark

Randomly select $L1$ blocks for embedding the encoded watermark. The center position of the selected blocks is

$$\rho 2_t = \left(i_t, j_t\right) \left(1 \le i_t \le a, 1 \le j_t \le b\right). \tag{9}$$

As the input to the trained model F, construct the dataset Φ of input vectors. The dataset Φ can be represented by:

$$\Phi = \left\{D_t\right\}_{t=1,\ldots,L1} \tag{10}$$

Calculate the output $V_{\rho 2_t}$ using the dataset Φ as input for the model F. A watermark bit and the synchronization marker are embedded by adjusting the relation between the coefficients of the center of the embedded block in the LL space and the output coefficients of the trained model. The embedding operation is performed by:

$$LL_{\rho 2_t} = \begin{cases} max\left(LL_{\rho 2_t}, V_{\rho 2_t} + \alpha\right) & if\ W1_t = 1 \\ min\left(LL_{\rho 2_t}, V_{\rho 2_t} - \alpha\right) & otherwise \end{cases} \quad (t = 1, \ldots, L1) \tag{11}$$

where α is the positive constant that determines the watermark strength. Note that a larger value of α can offer better robustness; however, it degrades the visual quality of the watermarked image. α is determined by balancing the robustness and the quality of the watermarked image. In the same manner, the marker is embedded.

5. Reconstruct the watermarked image

We obtain the watermarked luminance component by inverse wavelet transform and convert colors from YCbCr to RGB. Finally, we construct the watermarked image C'.

3.5 Watermark Extraction

We describe the watermark extraction method below.

The cropped image from the watermarked image C' is represented by C''. The size of the cropped image C'' is $m'' \times n''$.

1. Binarization

Load the trained model F, which is used for embedding. The cropped image C'' converts colors from RGB to YCbCr. Accordingly, the luminance component Y'' is obtained, where

$$Y'' = \left\{ Y''_{i,j} \mid 1 \leq i \leq m'', 1 \leq j \leq n'' \right\}. \tag{12}$$

The luminance component Y'' is decomposed through one level of wavelet transform to obtain the LL'' space, where

$$LL'' = \left\{ LL''_{i,j} \mid 1 \leq i \leq m''/2, 1 \leq j \leq n''/2 \right\}. \tag{13}$$

The obtained LL'' space is divided into contiguous 3×3 blocks. The coordinates of the centers of all blocks are

$$\rho 3_t = \left(i_t, j_t\right) \left(1 \leq i_t \leq m''/2, 1 \leq j_t \leq n''/2\right). \tag{14}$$

For input to the trained model F, construct a dataset Ψ of input vectors. The dataset Ψ can be represented by:

$$\Psi = \left\{ D''_t \right\}_{t=1,\ldots,A} \tag{15}$$

where A stands for the number of all blocks. The output $V''_{\rho 3_t}$ is calculated using the input dataset Ψ and the model F.

The binarized image B'' is obtained by comparing the coefficients of the center of the blocks in the LL'' space with the output coefficients $V''_{\rho3_t}$. The binarization operation is performed by:

$$B''_{\rho4_t} = \begin{cases} 1 & if\ LL''_{\rho3_t} > V''_{\rho3_t} \\ 0 & otherwise \end{cases} \quad (t = 1, \ldots, A) \tag{16}$$

where B'' is the binary image, and

$$\rho4_t = (i_t, j_t)\ (1 \le i_t \le m''/6, 1 \le j_t \le n''/6). \tag{17}$$

2. Synchronization and Extract watermark

To synchronize against cropping attacks, we find the vertex of the lattice-shaped marker from the binary image B''. At least 4 vertices should be found to obtain the watermarked area. In this method, we find these vertices geometrically. The obtained watermarked area is represented by B'''. Load the embed position $\rho2_t$. The watermark extraction is performed by:

$$W1'_t = B'''_{\rho2_t} \quad (t = 1, \ldots, L1') \tag{18}$$

Finally, the watermark $W1'$ is decoded by the BCH code to obtain the decoded watermark W'.

4 Experimental Results

We used six IHC [1] standard images to carry out the watermark scheme evaluation experiment using MATLAB 2016a [6, 12, 14]. The embedding parameter was strictly adjusted to satisfy both the image quality and the robustness criteria. Our goal is to satisfy these requirements according to the IHC evaluation criteria. In this study, we evaluate our watermarking scheme against the IHC criteria Ver.3. These criteria require a minimum coding tolerance and cropping tolerance. The specified image sources were six color images with more than 10M pixels each. Ten types of information (200 bit) were embedded into the whole image as a watermark, and the compressing-decompressing cycle was performed twice. The file size should less than $1/15^{th}$ the original size after the first compression, and the decompressed images should again be compressed, which is referred to as the second compression. After the second compression, the file size should be less than $1/25^{th}$ the original size. The original unwatermarked images were also be compressed using the same coding rate. The peak signal to noise ratio (PSNR) and the mean structural similarity (MSSIM) were calculated for each pair. The PSNR of each pair, which is calculated with the luminance signal, should be higher than 30 dB. The file was decompressed after the second compression, HDTV-size (1920×1080) images were cropped from each decompressed 4608×3456 image. Ten types of vertices are present in these cropped images. The watermark embedded in each cropped image should be detectable with no reference information, and without using the original image for its detection.

According to the IHC evaluation criteria Ver.3, there are two competition categories: highest tolerance and highest image quality. We focus on the competition for the highest image quality (Table 1).

Table 1. Cropping positions used in the IHC evaluation criteria

Position	(x_1, y_1)	(x_2, y_2)	(x_3, y_3)	(x_4, y_4)
1	(16, 16)	(1935, 16)	(1935, 1095)	(16, 1095)
2	(1500, 16)	(3419, 16)	(3419, 1095)	(1500, 1095)
3	(2617, 16)	(4536, 16)	(4536, 1095)	(2617, 1095)
4	(16, 770)	(1935, 770)	(1935, 1849)	(16, 1849)
5	(1500, 770)	(3419, 770)	(3419, 1849)	(1500, 1849)
6	(2617, 770)	(4536, 770)	(4536, 1849)	(2617, 1849)
7	(1344, 768)	(3263, 768)	(3263, 1847)	(1344, 1847)
8	(16, 1520)	(1935, 1520)	(1935, 2599)	(16, 2599)
9	(1500, 1520)	(3419, 1520)	(3419, 2599)	(1500, 2599)
10	(2617, 1520)	(4536, 1520)	(4536, 2599)	(2617, 2599)

Table 2 shows the results for the highest image quality, including the average compression ratio, the PSNR value and the MSSIM value. The average for the first compression ratio is 6.54%, under the given compression condition of 6.67% (1/15). The average for the second compression ratio is 3.82%, which is under the given compression condition of 4.00% (1/25). The image qualities in PSNR are 41.2579 dB for the first compression and 40.6678 dB for the second compression. They exceed the minimum condition of 30 dB. As shown in Table 2, our scheme satisfied the image quality criteria. The image quality in MSSIM is 0.9816 for the first compression and 0.9787 for the second compression.

Table 2. Average compression ratio, PSNR value, and MSSIM value for the highest image quality

	Compression ratio		PSNR		MSSIM	
	1st coding	2nd coding	1st coding	2nd coding	1st coding	2nd coding
Image1	0.0655	0.0387	42.2507	41.1787	0.9909	0.9889
Image2	0.0649	0.0381	41.9447	42.0595	0.9801	0.9796
Image3	0.0635	0.0376	38.4955	38.6804	0.9689	0.9683
Image4	0.0665	0.0367	42.0696	40.7871	0.9869	0.9820
Image5	0.0663	0.0400	41.0439	40.0965	0.9769	0.9683
Image6	0.0659	0.0385	41.7427	41.2044	0.9859	0.9851
Average	0.0654	0.0382	41.2579	40.6678	0.9816	0.9787

Table 3 shows the average error rates for 10 HDTV-size areas after the second decompression. No errors were detected in any case. As shown in Table 3, our scheme satisfied the robustness criteria.

Table 3. Average error rates for 10 HDTV-sized areas after the second compression of the highest image quality [%]

	Position									
	1	2	3	4	5	6	7	8	9	10
Image1	0	0	0	0	0	0	0	0	0	0
Image2	0	0	0	0	0	0	0	0	0	0
Image3	0	0	0	0	0	0	0	0	0	0
Image4	0	0	0	0	0	0	0	0	0	0
Image5	0	0	0	0	0	0	0	0	0	0
Image6	0	0	0	0	0	0	0	0	0	0
Average	0	0	0	0	0	0	0	0	0	0

5 Conclusion

In this study, we have proposed a digital watermarking scheme with cropping tolerance based on machine learning. The synchronization marker is embedded in a lattice-shape in the whole LL space. Binarization processing is performed on the watermarked image to find the lattice-shaped marker and synchronize against cropping. As a result, our scheme achieved sufficient image quality to satisfy the IHC evaluation criteria Ver.3.

References

1. Information hiding and its criteria for evaluation, IEICE. http://www.ieice.org/iss/emm/ihc/en/
2. Hirata, N., Kawamura, M.: Digital watermarking method using LDPC code for clipped image. In: The First International Workshop on Information Hiding and its Criteria for Evaluation, pp. 25–30. ACM Press (2014)
3. Kang, H., Iwamura, K.: Watermarking based on the difference of discrete cosine transform coefficients and error-correcting code. In: The First International Workshop on Information Hiding and its Criteria for Evaluation, pp. 9–17. ACM Press (2014)
4. Kang, H., Iwamura, K.: Information hiding method using best DCT and wavelet coefficients and its watermark competition. Entropy **17**(3), 1218–1235 (2015)
5. Scholkopf, B., Smola, A.J.: Learning with Kernels: Support Vector Machines, Regularization, Optimization, and Beyond. MIT Press, Cambridge (2001)
6. Math Works, Statistics and Machine Learning Toolbox. https://jp.mathworks.com/products/statistics.html
7. Li, C., Lu, Z., Zhou, K.: An image watermarking technique based on support vector regression. In: IEEE International Symposium on Communications and Information Technology, Proceedings of ISICT 2005, pp. 183–186 (2005)
8. Sanping, L., Yusen, Z., Hui, Z.: A wavelet-domain watermarking technique based on support vector regression. In: 2007 International Conference on Grey Systems and Intelligent Services, Nanjing, pp. 1112–1116 (2007)
9. Lv, X., Bian, H., Yu, B., Quan, X.: Color image watermarking scheme based on support vector regression. In: 2009 Fifth International Conference on Intelligent Information Hiding and Multimedia Signal Processing, Kyoto, pp. 144–147 (2009)

10. Mehta, R., Rajpal, N., Vishwakarma, V.P.: A robust and efficient image watermarking scheme based on Lagrangian SVR and lifting wavelet transform. Int. J. Mach. Learn. Cybern. **8**(2), 379–395 (2017)
11. Antonini, M., Barlaud, M., Mathieu, P., Daubechies, I.: Image coding using wavelet transform. IEEE Trans. Image Process. **1**(2), 205–220 (1992)
12. Math Works, Wavelet Toolbox. https://jp.mathworks.com/products/wavelet.html
13. MacWilliams, F.J., Sloane, N.J.A.: The Theory of Error-Correcting Codes, pp. 257–291. Elsevier (1977)
14. Math Works, Block Coding. https://jp.mathworks.com/help/comm/block-coding.html

SIFT Feature-Based Watermarking Method Aimed at Achieving IHC Ver.5

Masaki Kawamura(✉) and Kouta Uchida

Graduate School of Sciences and Technology for Innovation,
Yamaguchi University, Yamaguchi-shi 753-8512, Japan
m.kawamura@m.ieice.org

Abstract. We propose a watermarking method using scale-invariant feature transform (SIFT) features that have both scale and rotation invariance, and evaluate our method in accordance with the information hiding criteria (IHC) ver. 5. It is defined as evaluation criteria against several possible attacks; these attacks are JPEG compression and geometric attacks, e.g., scaling, rotation, and clipping. In our method, we use local feature regions located around the SIFT features that are robust against scaling and rotation. The regions are normalized in size and selected as marked regions. Watermarks are embedded in the marked regions. We also introduce two error-correction techniques: weighted majority voting (WMV) and low-density parity-check (LDPC) code. When a stego-image is attacked by scaling or rotation, the image is spatially distorted. WMV and LDPC code can correct errors of extracted watermarks in the distorted stego-image. On the other hand, it is not easy to detect rotated marked regions. Therefore, the correct orientation is searched for by brute force. We evaluated the proposed method in accordance with IHC ver. 5. Our method can achieve robustness against scaling and rotation attacks in the highest tolerance category.

Keywords: Feature-based watermarking · SIFT · IHC · Geometric attack

1 Introduction

Usage of digital watermarking techniques include protecting digital content (e.g., image, video, or audio data) from being modified either legally or illegally. Even if the digital content were attacked, the watermark could be extracted from the attacked content. Therefore, a robust watermarking method should be developed. By focusing on watermarking for still images, we want to extract watermarks from a processed image. Image processing, e.g., lossy compression, clipping, scaling, or rotation, is usually applied to an image. Therefore, the image processing is regarded as an *attack* on the image.

Evaluation standards are required to evaluate robustness and image quality for the watermarking techniques. Therefore, the Institute of Electronics, Information and Communication Engineers (IEICE) proposed information hiding criteria (IHC) [1]. In the IHC ver. 5, for example, the stego-images are attacked

J.-S. Pan et al. (eds.), *Advances in Intelligent Information Hiding and Multimedia Signal Processing*, Smart Innovation, Systems and Technologies 81,
DOI 10.1007/978-3-319-63856-0_46

by JPEG compression, clipping, scaling, rotation, and combinations of these attacks. The watermarks should be extracted from attacked images with almost no errors. The extraction should be performed blind, that is, the decoder cannot use any information about both the original images and the attacks. IHC ver. 5 [1] is summarized as follows. There are three roles in the model: watermarker, attacker, and detector. It is supposed that the six original IHC standard images are provided with a 4608×3456 pixel size. Ten messages are generated by using an M-sequence. The message length is 200 bits. The watermarker encodes the messages for the purpose of correcting errors. An encoded message is embedded into the original image. Since the JPEG format is popular, the image is compressed with JPEG to be distributed (1st compression). The file size should be less than 1/15 of the original size. The compressed image is called a stego-image in this model. The image quality of the stego-image is measured by peak signal-to-noise ratio (PSNR) and mean structural similarity (MSSIM) [2]. The PSNR of the stego-image should be more than 30 dB.

The attacker performs a geometrical attack, such as scaling, rotation, and combinations of these two attacks on the stego-image. After the geometrical attack, the image is clipped to an HDTV-size area (1920×1080) at four specified coordinates. The clipped images are compressed with JPEG again to be saved in the JPEG format (2nd compression). The obtained image is called an attacked image. The detector extractes the message (200 bits) from a given attacked image without referring to the original image, attack parameters, and any related information. The watermarks should be decoded from the attacked image with an almost zero bit error rate (BER). The accuracy of the decoded message is measured by the BER [1].

Once an image has been geometrically attacked (i.e. rotation, scaling, and clipping have been performed on the image), the positions of the pixels in the image would be moved, and then the marked regions would also be moved. To extract watermarks from the marked regions, the original marked positions need to be detected. This is called *synchronization* and is used to find the marked positions. Watermarking methods using a marker or synchronization code have been proposed [3,4]. In these methods, a watermark consists of an encoded message and a marker. The marker is a specific bit sequence, and is effective for a clipping attack to find marked positions. When the image is subjected to geometric attacks, the synchronization is performed by searching for markers in the attacked image. However, it is hard to find the marked positions due to distortion of the image. Since the attack parameters of the scaling ratio and rotation angles are unknown, the marked positions can only be searched for by brute force. Therefore, we focus on a synchronization technique using rotation- and scale-invariant features.

We propose a watermarking method using scale-invariant feature transform (SIFT) features for synchronization in accordance with IHC ver. 5. A message is encoded by using low-density parity-check (LDPC) code [5] as a watermark. Marked regions are selected around the SIFT feature points. The size of the regions is normalized for robustness against scaling. The watermark is embedded into the discrete cosine transform (DCT) domain of each region for robustness against JPEG compression. In the detecting process, the extracted regions are rotated due to synchronization. The correct orientation is searched for by brute force.

2 Related Works

Scale-invariant feature transform (SIFT) [6] is a promising feature detector. Even if rotation and scaling attacks are applied, the same feature points can be extracted by the SIFT detector, and the scale parameters, which are proportional to the scaling ratio, can also be extracted. Therefore, the SIFT features are represented by circular patches or disks whose centers are feature points, and their radii are equal to the scale parameters. Since the circular patches are rotation- and scale-invariant, we can obtain the same marked regions. There are two major embedding domains for invariant feature-based methods: pixel-value domain and frequency domain. For pixel-value domain watermarking, the circular patches are extracted by SIFT or other invariant feature detectors. The bit sequence of watermarks are converted to polar coordinates, and then they are embedded into the patches directly [7–11]. The image quality of stego-images created by these methods is not good due to directly changing of the pixel values. Also, the watermarks in the stego-image are vulnerable to compression. Moreover, the conversion to polar coordinates involves loss of the watermarks.

For frequency domain watermarking, watermarks are embedded into some frequency domains in marked regions around the feature points. Since the discrete Fourier transform (DFT) has invariance for scaling, the methods involving embedding in the DFT domain are robust against scaling attacks [12–15]. However, they are sensitive to compression. Embedding in the DCT domain is effective for robustness against compression [16,17], but existing methods are non-blind ones. The method of Pham *et al.* [16] needs a database to store the SIFT features. Wei and Yamaguchi's method [17] needs the original watermark to find the watermarks.

Since a rotation attack changes the marked regions, robustness against rotation should be considered. The Fourier-Mellin transform domain is effective for rotation, scaling, and translation (RST) [18]. The Fourier-Mellin transform is equivalent to computing the Fourier transform of a log-polar mapping. Tone and Hamada's method [12] uses a Harris-Affine detector and a log-polar mapping as the invariant feature detector. It can extract scale- and rotation-invariant features. However, the log-polar mapping distorts the watermarks, resulting in the number of errors of the watermarks becoming high. Therefore, these methods could not achieve the IHC.

Methods without the log-polar mapping seem to be better for loss-less embedding. The orientation of the marked regions is worthy of consideration. In one study, the dominant orientation, which is the peak of the histogram of the gradient direction of the pixels, was used [19]. In another study, the dominant orientation was obtained by SIFT detector to align the direction of the features [14].

In two studies, the characteristic orientation based on the moments of a local region was used [10,20]. These methods detect the significant orientation and rotate the local regions. Therefore, the region has robustness against rotation. Since the process rotating the region involves distortion of the image, watermarks are also damaged in the embedding process. Moreover, since the orientation is low precision, the probability of failing to detect the orientation is higher in the detecting process.

3 Proposed Feature-Based Watermarking Method

3.1 Watermarking Process

In the proposed method, a message, $\boldsymbol{m}$, of length $N_m = 200$ is encoded to a codeword, $\boldsymbol{c}$ of length $N_c = 300$ by using LDPC code [5,21]. Since difficult attacks will be performed on the stego-images in accordance with IHC ver. 5, a lot of errors will appear in the extracted codewords. To measure the error rate, a check bit, $\boldsymbol{s}$, of length $N_s = 87$ is introduced, where it is given by $\boldsymbol{s} = (1, 1, 1, \cdots, 1)$. Therefore, the watermark, $\boldsymbol{w}$, consists of the codeword $\boldsymbol{c}$ and check bits $\boldsymbol{s}$, and is given by

$$w_i = \begin{cases} c_i, 1 \leq i \leq N_c \\ 1, \ N_c < i \leq N_c + N_s \end{cases} . \tag{1}$$

The marked regions are selected around the SIFT feature points. The feature points (x_i, y_i) and the scale parameters σ_i are obtained by using the SIFT algorithm [6], and then the circular patches are constructed. However, circular forms should be avoided due to distortion of the watermarks. Therefore, the bounding squares of the circular patches are selected as marked regions in our method. The bounding squares are squares of side $2d\sigma_i$ pixels, where d is the radius magnification. In this paper, we set $d = 7$ in view of the length of the watermark. The scale parameters in the range of $\sigma_L < \sigma_i < \sigma_U$ are selected, since large bounding squares will overlap each other, and small ones will vanish as a result of shrinking [10]. We set $\sigma_L = 4$ and $\sigma_U = 10$. Even if the range of the scale parameters are limited, some of the squares may be overlaped each other. In this case, the feature point which has the largest value of difference of Gaussian (DoG) filter is selected. Each marked region is normalized to a square of side $h = 96$ pixels in preparation for the scaling attack.

The watermarks $\boldsymbol{w}$ are embedded in the DCT domain of the normalized regions. In the method of Hirata and Kawamura [3,4], the normalized region is divided into 8×8 pixel blocks. One bit of the watermark is embedded in one block. In our method, the region is divided into 32×32 pixel blocks, since the 8×8 pixel region was too small to embed a watermark in it. Each block is transformed by using the 2D DCT. A more than one bit ($N_B > 1$) watermark is embedded in the DCT coefficients of the block by using quantization index modulation (QIM) [22], where N_B is given by

$$N_B = \left\lceil \frac{32 \times 32}{h \times h}(N_c + N_s) \right\rceil , \tag{2}$$

where $\lceil x \rceil$ stands for the ceiling function. In the case of $h = 96, N_c = 300$, and $N_s = 87$, the number of the bits is $N_B = 43$. Since a stego-image will be clipped, the same watermarks are repeatedly embedded throughout an image. After embedding the watermarks, the stego-image is compressed by JPEG compression to be less than 1/15 of the original size.

3.2 Extraction and Decoding Process

The stego-image is attacked by JPEG compression, clipping, scaling, rotation, and combinations of these attacks in accordance with IHC ver. 5. Watermarks must be extracted from the attacked image. The synchronization is performed by SIFT. The process to obtain the marked regions is the same as the watermarking process except for the range of the scale parameters. In extraction, the scale parameters in the range of $0.8\sigma_L < \sigma_i < 1.2\sigma_U$ are selected, since the stego-image is magnified or shrunk by the scaling attack. We assume P marked regions are extracted from the attacked image and P candidates for watermarks are obtained by QIM [22].

Since the marked region may be rotated by the rotation attack, the region is rotated to find the correct marked position in the extraction process. Let a candidate for the p-th watermark rotated by a θ-degree angle be $\hat{\boldsymbol{w}}^p(\theta) = (\hat{\boldsymbol{c}}^p(\theta), \hat{\boldsymbol{s}}^p(\theta))$. Note that the first N_c bits correspond to the codeword bits, and the residual N_s bits correspond to the check bits. Now, the matching ratio for the check bits $\hat{\boldsymbol{s}}^p(\theta)$ is defined by

$$R^p(\theta) = \frac{1}{N_s} \sum_{i=1}^{N_s} \hat{s}_i^p(\theta). \tag{3}$$

Since all of the check bits are 1, the estimated degree of the angle, $\hat{\theta}^p$, can be calculated by

$$\hat{\theta}^p = \arg \max_{0 \le \theta \le 90} \{R^p(\theta)\}. \tag{4}$$

By using the SIFT features and searching for the rotation angle, the synchronization for clipping, rotation, and scaling can be performed.

After the synchronization, a message will be estimated from the candidates $\hat{\boldsymbol{c}}^p(\hat{\theta}^p)$. The candidates include bit sequences extracted from incorrect feature points, in which the watermarks are not embedded. Even if candidates are the bit sequences in which the watermarks are embedded, they may be distorted by the attacks. Therefore, the candidates contain a lot of errors. To remove incorrect candidates, we introduce a weighted majority voting (WMV) algorithm [3,4]. The estimated codeword $\hat{\boldsymbol{c}} = (\hat{c}_1, \hat{c}_2, \cdots, \hat{c}_{N_c})$ can be calculated from the weighted candidates by the WMV, that is,

$$\hat{c}_i = \Theta\left(\sum_{p=1}^{\hat{P}} \alpha\left(R^p(\hat{\theta}^p)\right)\left\{\hat{w}_i^p(\hat{\theta}^p) - 0.5\right\}\right), 1 \le i \le N_c, \tag{5}$$

where $\alpha(x)$ is the weight function of the ratio $R^p(\hat{\theta}^p)$ and is defined by

$$\alpha(x) = \begin{cases} \tanh\left(\beta\left(x - T\right)\right), T \leq x \\ 0, \qquad\qquad\qquad x < T \end{cases}, \tag{6}$$

where T is the threshold and β is the weight coefficient. The function $\Theta\left(x\right)$ is the step function defined by

$$\Theta(x) = \begin{cases} 1\,, x \geq 0 \\ 0\,, x < 0 \end{cases}. \tag{7}$$

After obtaining the estimated codeword $\hat{\boldsymbol{c}}$, the estimated message $\hat{\boldsymbol{m}} = (\hat{m}_1, \hat{m}_2, \cdots, \hat{m}_{N_m})$ can be calculated by the sum-product algorithm of the LDPC code [23]. The BER is defined by

$$\text{BER} = \frac{1}{N_m} \sum_{i=1}^{N_m} m_i \oplus \hat{m}_i, \tag{8}$$

where $\oplus$ stands for exclusive OR (XOR).

4 Evaluation by Computer Simulations

We evaluate our method by computer simulation on the basis of the IHC ver. 5. The attack procedure is shown in Fig. 1. Before attacking the stego-image, the Q-value of the JPEG compression is computed in advance to be less than 1/25 of the original size. The stego-image is attacked by scaling, rotation, and combinations of these two attacks. The scale parameter is $s \in \{80, 90, 110, 120\%\}$, and the rotation angle is $\theta \in \{3, 5, 7, 10°\}$. The combinations of the scaling and the rotation attack are $(s, \theta) \in \{(80, 9), (90, 7), (110, 5), (120, 3)\}$. These parameters are defined in the IHC ver. 5 [1]. The term 'No attack' in Fig. 1 means that the stego-image is not attacked by any geometric attacks. The next stage is clipping; the image is clipped by an HDTV-size area at four specified coordinates. The center points of the four clipped areas are $(x_c \pm 700, y_c \pm 500)$, where the point (x_c, y_c) is the center point of the stego-image. Each clipped area is saved as a new image by using the Q-value, which is computed in advance. The compressed image is called the attacked image.

The step size of the QIM is $\Delta = 72$. The threshold T and the weight coefficient β in the weight function are $T = 0.7$ and $\beta = 0.7$, respectively. We use a $(3, 4)$-regular LDPC code generated by the progressive edge-growth (PEG) algorithm [4,21]. The channel error rate in the sum-product algorithm [23] is 0.05. Ten different messages are generated by the M-sequence, they are embedded in six IHC standard images, and then four areas are clipped. Therefore, 240 attacked images are generated.

There are two categories of evaluation criteria for comparing methods: "highest image quality" and "highest tolerance." Table 1 shows the average compression ratio, PSNR, and MSSIM. Since ten different messages were embedded, the

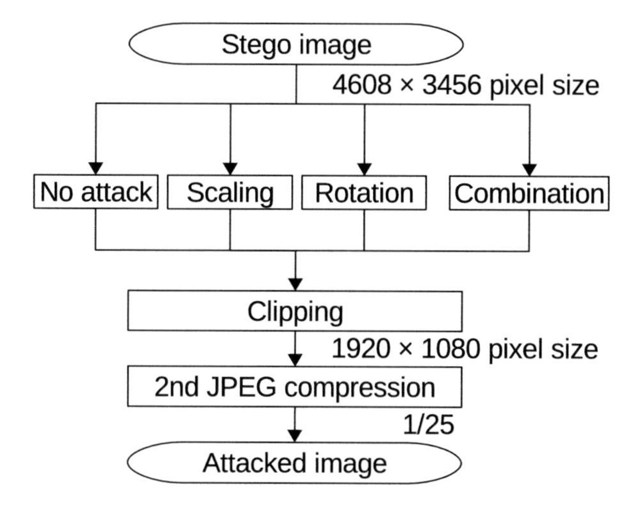

Fig. 1. Attack procedure

Table 1. Average compression ratio, PSNR, and MSSIM

	Compression ratio [%]	PSNR [dB]	MSSIM
Image 1	6.628	33.313	0.917
Image 2	6.588	33.800	0.917
Image 3	6.609	35.462	0.937
Image 4	6.621	36.851	0.949
Image 5	6.629	35.108	0.926
Image 6	6.656	33.718	0.919
Average	6.622	34.709	0.927

values are the average values for ten trials. The compression ratio was under $1/15 = 6.67\%$ for the first compression. The PSNRs and MSSIMs were calculated for the luminance signal of the generated stego-images before these images were attacked. All PSNRs were over 30 dB.

Table 2 shows the average error rate for the attacked images with additional attacks. In accordance with the highest tolerance category of the IHC [1], there are four clipped areas in each stego-image, and the BERs for three of the four areas must be zero. In other words, one area can be discounted. Therefore, the best three of the four BERs were used for the evaluation. The compression ratio in Table 2 was less than $1/25$ of the original size for the second compression. As a result, the proposed method could achieve a BER of 0.0 for scaling or rotation attacks. However, the BERs for combined attacks were over 0.0 for many images. For the highest image-quality category of the IHC, the BER must be no more than 1% on average. In the worst case, the BER for the three clipped images must be less than 2%. The proposed method could not achieve this criterion.

Table 2. Average error rate for attacked images with additional attacks (%)

Image no.	No attack	Scaling (%)				Rotation (°)				Combination (s, θ)			
		80	90	110	120	3	5	7	10	(80, 9)	(90, 7)	(110, 5)	(120, 3)
1	0.000	0.000	0.000	0.000	0.000	0.000	0.000	0.000	0.000	0.002	0.000	0.000	0.000
2	0.000	0.000	0.000	0.000	0.000	0.000	0.000	0.000	0.000	0.000	0.000	0.000	0.000
3	0.000	0.000	0.000	0.000	0.000	0.000	0.000	0.000	0.000	0.005	0.003	0.004	0.001
4	0.000	0.000	0.000	0.000	0.000	0.000	0.000	0.000	0.000	0.002	0.004	0.003	0.001
5	0.000	0.000	0.000	0.000	0.000	0.000	0.000	0.000	0.000	0.003	0.001	0.000	0.000
6	0.000	0.000	0.000	0.000	0.000	0.000	0.000	0.000	0.000	0.002	0.000	0.002	0.002

However, the BERs for almost 80% of the attacked images were zero. In a few cases, the BERs were over 2%.

5 Conclusion

We proposed a SIFT feature-based watermarking method aimed at achieving the IHC ver. 5. To accomplish the criteria, the method has to be robust against compression, clipping, scaling, rotation, and combinations of these attacks. Since the IHC ver. 5 is the newest criteria, we demonstrated the ability of our method to operate in current conditions. We introduced SIFT against scaling and rotation attacks and WMV and the LDPC code against compression and clipping. As a result, our method does not have enough robustness against combined attacks of a scaling attack and a rotation attack but does have robustness against an individual attack.

Acknowledgments. This work was supported by JSPS KAKENHI Grant Number JP16K00156, and was performed in the Cooperative Research Project of the RIEC, Tohoku University. The computer simulations were carried out on PC clusters at Yamaguchi University.

References

1. Information hiding and its criteria for evaluation, IEICE. http://www.ieice.org/iss/emm/ihc/. Accessed 10 Mar 2017
2. Wang, Z., Bovik, A.C., Sheikh, H.R., Simoncelli, E.P.: Image quality assessment: from error visibility to structural similarity. IEEE Trans. Image Process. **13**(4), 600–612 (2004)
3. Hirata, N., Kawamura, M.: Watermarking method using concatenated code for scaling and rotation attacks. In: 14th International Workshop on Digital-Forensics and Watermarking (IWDW 2015) (2015)
4. Hirata, N., Nozaki, T., Kawamura, M.: Image watermarking method satisfying IHC by using PEG LDPC code. IEICE Trans. Inf. Syst. **E100–D**(1), 13–23 (2017)
5. Gallager, R.G.: Low-density parity-check codes. IRE Trans. Inf. Theory **IT–8**(1), 21–28 (1962)
6. Lowe, D., David, G.: Distinctive image features from scale-invariant keypoints. Inter. J. Comput. Vis. **60**(2), 91–110 (2004)

7. Lee, H.Y., Kim, H., Lee, H.K.: Robust image watermarking using local invariant features. Opt. Eng. **45**(3), 037002 (2006)
8. Verstrepen, L., Meesters, T., Dams, T., Dooms, A., Bardyn, D.: Circular spatial improved watermark embedding using a new global SIFT synchronization scheme. In: 16th International Conference on Digital Signal Processing, pp. 1–8 (2009)
9. Li, L.D., Guo, B.L.: Localized image watermarking in spatial domain resistant to geometric attacks. Int. J. Electron. Commun. **63**(2), 123–131 (2009)
10. Yu, Y., Ling, H., Zou, F., Lu, Z., Wang, L.: Robust localized image watermarking based on invariant regions. Digit. Signal Proc. **22**(1), 170–180 (2012)
11. Tsai, J.S., Huang, W.B., Kuo, Y.H., Horng, M.F.: Joint robustness and security enhancement for feature-based image watermarking using invariant feature regions. Sig. Process. **92**(6), 1431–1445 (2012)
12. Tone, M., Hamada, N.: Scale and rotation invariant digital image watermarking method. IEICE Trans. Inf. Syst. (Japanese Edition) **J88–D1**(12), 1750–1759 (2005)
13. Wang, X.Y., Hou, L.M., Wu, J.: A feature-based robust digital image watermarking against geometric attacks. Image Vis. Comput. **27**(7), 980–989 (2008)
14. Gao, X., Deng, C., Li, X., Tao, D.: Local feature based geometric-resistant image information hiding. Cogn. Comput. **2**(2), 68–77 (2010)
15. Yang, H., Xia, Z., Sun, X., Luo, H.: A robust image watermarking based on image restoration using SIFT. Radioengineering **20**(2), 525–532 (2011)
16. Pham, V.Q., Miyaki, T., Yamasaki, T., Aizawa, K.: Geometrically invariant object-based watermarking using SIFT feature. In: IEEE International Conference on Image Processing (ICIP), pp. 473–476 (2007)
17. Wei, N., Yamaguchi, K.: Image watermarking resistant to geometric attacks based on SIFT feature. IEICE Tech. Rep. EMM **111**(496), 43–48 (2012)
18. O'Ruanaidh, J., Pun, T.: Rotation, translation and scale invariant digital image watermarking. In: International Conference on Image Processing, vol. 1, pp. 536–536. IEEE Computer Society (1997)
19. Deng, C., Gao, X., Li, X., Tao, D.: A local Tchebichef moments-based robust image watermarking. Sig. Process. **89**(8), 1531–1539 (2009)
20. Nasir, I., Khelifi, F., Jiang, J., Ipson, S.: Robust image watermarking via geometrically invariant feature points and image normalisation. IET Image Proc. **6**(4), 354–363 (2012)
21. Hu, X.Y., Eleftheriou, E., Arnold, D.M.: Regular and irregular progressive edge-growth tanner graphs. IEEE Trans. Inform. Theory **51**(1), 386–398 (2005)
22. Chen, B., Wornell, G.W.: Quantization index modulation: a class of provably good methods for digital watermarking and information embedding. IEEE Trans. Inform. Theory **47**(4), 1423–1443 (2001)
23. Wadayama, T.: A coded modulation scheme based on low density parity check codes. IEICE Trans. Fundam. **E84–A**(10), 2523–2527 (2001)

Data Hiding for Text Document in PDF File

Minoru Kuribayashi(✉), Takuya Fukushima, and Nobuo Funabiki

Graduate School of Natural Science and Technology,
3-1-1, Tsushima-naka, Kita-ku, Okayama 700-8530, Japan
{kminoru,funabiki}@okayama-u.ac.jp, fukushima.takuya@s.okayama-u.ac.jp

Abstract. Among data hiding methods for documents, the spaces between words and paragraphs are popular features for embedding information. However, in order to avoid the visible distortions, the embedding capacity is limited to be small in conventional methods. In this paper, we regard a collection of space lengths as one dimensional feature vector and apply signal processing approaches for the vector to embed more information with less distortions. We also focus on the Portable Document Format (PDF) files and propose a new data hiding method for PDF files with large capacity. Considering the secrecy of the embedded information, a random permutation and dither modulation are introduced in the operation.

Keywords: Portable Document Format · QIM · Dither modulation · Permutation

1 Introduction

Data hiding technique enables us to embed information into digital content without causing perceptual degradation. There are many media for digital content such as image, video, audio, and text document. Depending on the purpose of its application, the data hiding technique is classified into two classes; steganography and watermark. The former conceals secret communication using digital content, while the latter protects the digital content by embedding information. In both classes, it is desirable to embed more information with less distortions. For convenience, the embedding information is called "watermark" in this paper.

Portable Document Format (PDF) is developed by the Adobe Systems Society as a page description language, and it is an evolution of PostScript format which can preserve the formatting of a file. Because of its popularity, some data hiding methods for PDF files have been studied. Zhong et al. [8] changes the preset distance between each word in PDF files. In [7], the spaces between words and paragraphs are used for embedding, which is extended from the basic embedding algorithm in [1]. In the PDF file, the TJ operator displays text strings with position and space lengths between characters. So, it is easy to apply the method to the PDF file. The main drawback in this method is one-to-one correspondence between a space length and embedding bit. A malicious party will be able

J.-S. Pan et al. (eds.), *Advances in Intelligent Information Hiding and Multimedia Signal Processing*, Smart Innovation, Systems and Technologies 81,
DOI 10.1007/978-3-319-63856-0_47

to edit the embedding watermark by modifying specific space lengths. As the hexadecimal text encoding of "20" and "A0" are displayed as blank characters in PDF files, such specific characters are used for embedding watermark in [5]. Lin et al. [6] proposed an interesting method based on PDF files of ISO-8859 encoding. It combines a data hiding method for PDF files and an encryption method based on quadratic residue. The method is, however, suffered from the removal of inserted watermark. In [2], the spread transform dither modulation is used for embedding information into PDF files. Each bit of watermark is embedded into some x-coordinate (horizontal position) values. Because of its spreading effects, the method retains transparency and robustness. However, the capacity is small. In [4], the object in each line of PDF files are divided into some groups according to watermark. The advantage of the method is that no visual distortion is appeared on the display of PDF viewer. However, the capacity is not large, and a malicious party will easily find the irregularities in the watermarked PDF file if the descriptions of the document are observed.

In this paper, we propose a new data hiding method for PDF files with large embedding capacity and low distortions, by embedding watermark into spaces among characters in each lines in consideration of inside structure of PDF file. Different from the conventional methods, the proposed method regards a collection of the space lengths as a host vector, and modifies the vector using signal processing techniques. To control the changes of space lengths produced by the embedding, the watermark is inserted into the frequency components converted from the vector by DCT(Discrete Cosine Transform). The embedded watermark signal is spread over the vector after IDCT. In order to keep the sum of the space lengths in each line, the watermark is embedded only into AC components avoiding its DC component. Considering the secrecy of the watermark, the Dither Modulation Quantization Index Modulation(DM-QIM) [3] is employed, which introduces randomness on the embedding process based on a secret key. In addition, before performing the DCT, the host vector is permuted using a secret key in order to enhance the secrecy of the watermark and its existence in a PDF file.

2 Preliminaries

2.1 PDF File Structure

PDF is based on a structured binary file format composed of 4 components; Header, Body, Cross-Reference Table, and Trailer. The Header is the first line of PDF file and indicates the version of the PDF specification. In the Body, there are objects including text streams, images, other multimedia elements, etc. The Body is used to hold all the data of documents to be shown on a PDF viewer. The Cross-Reference Table contains the references to all the objects in the document. It allows random access to objects in the file. Each object is represented by one entry in the table. The Trailer is used to find the Cross-Reference Table, which is like a dictionary indicating a link to each object.

```
? 0 obj
<</Length ??>
stream

BT
/F1 24 Tf
10 500 Td
[(Hello)]TJ
0 -100 Td
[(He)-50(llo)]TJ
ET

endstream
endobj
```

Fig. 1. Example of text syntax in an object.

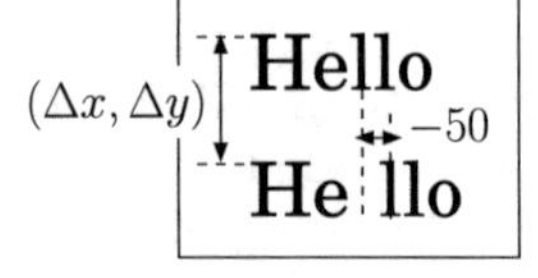

Fig. 2. Displayed example of the object in Fig. 1.

An example of PDF syntax in object is shown in Fig. 1. Each object has its unique number, and it starts from "obj" to "endobj". The script and data for displaying text, figure, images are appeared in between "stream" and "endstream". "BT" and "ET" represent Begin Text and End Text, respectively. There are some operators to represent text document. "Tf" operator specifies the text style and font size, and "Td" operator specifies the offset of the beginning of the current line. When a current coordinate is (x, y), the Td operator shift the coordinate into $(x+\Delta x, y+\Delta y)$. "TJ" operator shows the text characters and spaces between characters. In this example, the syntax "[(He)−50(llo)]TJ" represents the characters "He" and "llo", and the space length "−50" between them. The displayed example of this object is shown in Fig. 2. Generally, the spaces between characters are observed at English sentences in PDF files.

2.2 DM-QIM

Let $w \in \{0, 1\}$ be a watermark bit, and δ be a quantization step size to control the distortion level. Assume that d is an element of selected host signal and it is a real number. In the QIM method [3], d is rounded into a nearest odd/even quantized value $d^{\star}$ according to the watermark bit w using the step size δ. It is noticed that $d^{\star}$ becomes a multiple of δ. If a malicious party knows the embedding algorithm, the DCT coefficients selected for embedding can be found by simply observing the values. In order to avoid such an attack, a random dither modulation should be introduced into the above algorithm. Such a method is called RDM-QIM in this paper.

Let r be a real number randomly selected from a range $[-\delta/2, \delta/2]$. The embedding operation is performed as follows.

$$d' = \begin{cases} \delta D & \text{if } D \bmod 2 = w \\ \delta(D+1) & \text{otherwise} \end{cases} \tag{1}$$

where

$$D = \left\lfloor \frac{d+r}{\delta} \right\rfloor, \tag{2}$$

The above operation quantizes $d+r$ into nearest even/odd value according to the watermark bit w. If $w = 0$, $\lfloor d'/\delta \rfloor$ becomes an even value; otherwise, it becomes an odd value. Finally, the random number r is subtracted from d' to obtain the output $d^\star$.

$$d^\star = d' - r \tag{3}$$

It is remarkable that $d^\star$ is not the multiple of δ. As the watermark bit w cannot be extracted without r, it is regarded as a secret key in this method.

At the detection, the watermark w is extracted as follows.

$$w = \bar{D} \bmod 2, \tag{4}$$

where

$$\bar{D} = \left\lfloor \frac{d^\star + r + \frac{\delta}{2}}{\delta} \right\rfloor. \tag{5}$$

3 Proposed Data Hiding Method for PDF File

In conventional studies, the spaces among characters and words are regarded as scalars and each watermark bit is embedded into each selected space length. The proposed method regards a collection of space lengths in each line as one dimensional signal, and introduces signal processing techniques to embed multi-bit watermark information into the signal.

Assume that the number of spaces among characters in t-th line, denoted by ℓ_t, is more than 1. The collection of space lengths in t-th line is denoted by a vector $\boldsymbol{s_t} = (s_{t,0}, s_{t,1}, s_{t,2} \ldots, s_{t,\ell_t-1})$. We can embed at most $(\ell_t - 1)$-bit watermark information into $\boldsymbol{s_t}$.

Let $\boldsymbol{w_t} = (w_{t,1}, w_{t,2}, \ldots, w_{t,\ell_t-1})$ be $(\ell_t - 1)$-bit watermark information for t-th line, where $w_{t,j} \in \{0, 1\}$, $(1 \leq j \leq \ell_t - 1)$. The following operation is performed to embed the watermark.

1. Extract the collection of space lengths $\boldsymbol{s_t}$ from t-th line.
2. Permute randomly the elements of $\boldsymbol{s_t}$ according to a secret key.
3. Perform one dimensional DCT to $\boldsymbol{s_t}$. The DCT coefficients are denoted by $\boldsymbol{d_t} = (d_{t,0}, d_{t,1}, d_{t,2}, \ldots, d_{t,\ell_t-1})$.
4. Embed j-th watermark bit $w_{t,j}$ into the corresponding DCT coefficient $d_{t,j}$ for $1 \leq j \leq \ell_t - 1$ using the RDM-QIM method with a secret key. The watermarked DCT coefficients are denoted by $\boldsymbol{d_t^\star} = (d_{t,0}, d_{t,1}^\star, d_{t,2}^\star, \ldots, d_{t,\ell_t-1}^\star)$.
5. Perform IDCT to $\boldsymbol{d_t^\star}$, and permute the result in the inverse order to obtain the watermarked space lengths $\boldsymbol{s_t^\star}$.
6. Replace the original space lengths in t-th line with $\boldsymbol{s_t^\star}$.

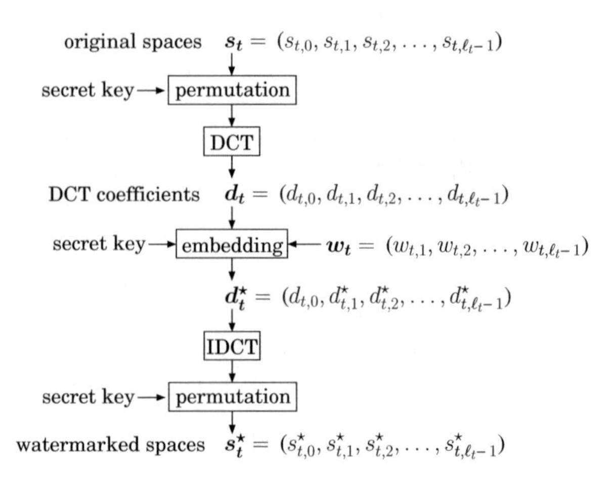

Fig. 3. Procedure of the proposed embedding operation.

Figure 3 illustrates the above operation. It is remarkable that the DC component $d_{t,0}$ is avoided for embedding in the above operation. It assures that the total length of the spaces are unchanged after the embedding, namely, $\sum s_{t,j} = \sum s^\star_{t,j}$. If the total length is changed, irregularities are clearly appeared at the end of lines, which will become noticeable distortions. By excluding the DC component, we avoid such distortions. The capacity C of the above embedding method is

$$C = \sum_{t=1}^{n} (\ell_t - 1), \tag{6}$$

where n is the number of lines in a PDF file.

Because of the permutation of $\boldsymbol{s_t}$ and the RDM-QIM method, it is difficult for malicious party to modify the embedded information without the secret key. Even if the number of space lengths is small in a certain line, the RDM-QIM method prevents a malicious party from getting useful information about the permutation and watermark. It is possible to be robust against addition of noise by reducing the bit-length of the watermark. Without the secret key used for the permutation, the energy of additive noise will be spread over all DCT coefficients because the direct modification of DCT coefficients is difficult. If the bit-length of the watermark is $(\ell_t - 1)/2$, only half of the noise energy is expected to affect the embedded signal, which reduces the effects of additive noise. By properly selecting the quantization step size δ and the payload (equal to or less than C), it is possible to control the trade-off between the perceptual distortions and the robustness.

4 Experiment

We create a PDF file from a document file produced by the Microsoft Word 2013 ver.15.0.4893.1000 using a Century font with 10.5pt. The document is an English version of Japanese old story "Peach Boy"[1], which created PDF file has 36 lines with 3,097 characters. Since the document of PDF file is compressed, we use the "pdftk" toolkit[2] to decompress the file before embedding watermark. The size of original PDF file is 55,164 Bytes, while the size becomes 104,369 Bytes after the decompression. The embedding capacity in the PDF file is $C = 1,448$ bits. We use a random sequence as watermark, and embed into the PDF file using the quantization step size $\delta = \{3, 6, 12\}$. When $\delta < 3$, the watermark cannot be extracted correctly even if no attack is performed. It is because of the round-off noise at the embedding operation. With the increase of δ, the round-off noise does not cause the detection error.

Figure 4 shows the first 3 lines of the original and watermarked files. No visible distortion is appeared in space lengths among characters. Because the DC component is avoided for embedding, we can see that the total length of

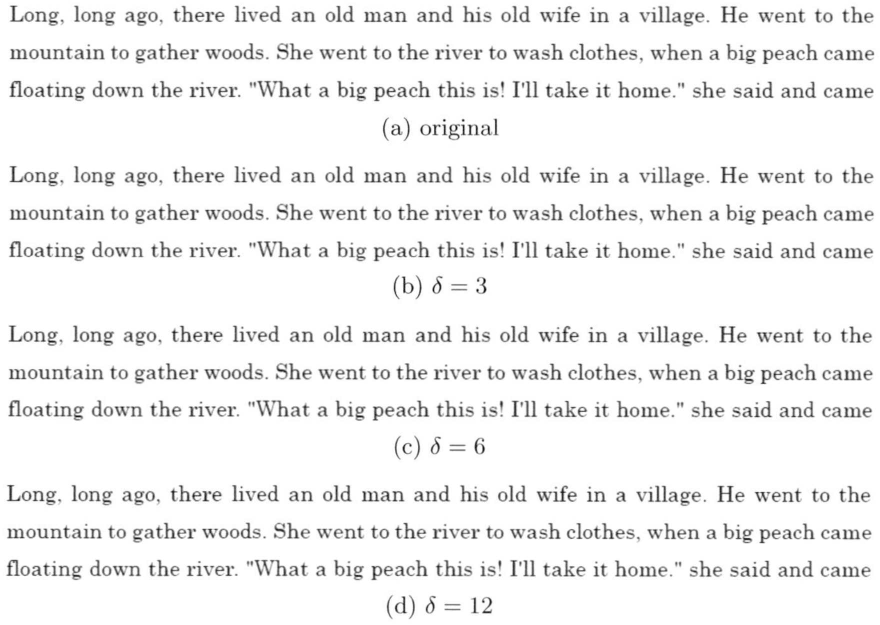

Long, long ago, there lived an old man and his old wife in a village. He went to the mountain to gather woods. She went to the river to wash clothes, when a big peach came floating down the river. "What a big peach this is! I'll take it home." she said and came

(a) original

Long, long ago, there lived an old man and his old wife in a village. He went to the mountain to gather woods. She went to the river to wash clothes, when a big peach came floating down the river. "What a big peach this is! I'll take it home." she said and came

(b) $\delta = 3$

Long, long ago, there lived an old man and his old wife in a village. He went to the mountain to gather woods. She went to the river to wash clothes, when a big peach came floating down the river. "What a big peach this is! I'll take it home." she said and came

(c) $\delta = 6$

Long, long ago, there lived an old man and his old wife in a village. He went to the mountain to gather woods. She went to the river to wash clothes, when a big peach came floating down the river. "What a big peach this is! I'll take it home." she said and came

(d) $\delta = 12$

Fig. 4. Visual differences in the first 3 lines.

[1] http://www.geocities.co.jp/HeartLand-Gaien/7211/momo.html.
[2] https://www.pdflabs.com/tools/pdftk-the-pdf-toolkit/.

each line is not changed. Figure 5 shows the descriptions of spaces at the first line of the PDF files. In case of $\delta = 3$, the changes in the spatial domain are very small. Although the changes becomes large with the increase of δ, no irregularity is observed from these descriptions. Table 1 enumerates the file size of the PDF files. As expected, the file size is increased with the step size δ because the distortions caused by embedding is increased accordingly.

```
[(L)-2(ong,)6( )-130(l)-2(o)11(ng )-130(ago,)7( )-130(ther)12(e
)-130(l)-2(i)8(ved )-133(an)-2( )-130(o)11(l)-2(d)-5( )-130(m)14(an)-2(
)-130(a)11(nd)-7( )-130(h)9(i)-2(s)8( )-130(ol)-2(d)-5(
)-130(w)5(i)-2(f)3(e )-130(i)8(n )-133(a )-131(vil)-2(l)8(age.)7(
)-130(H)3(e )-130(w)5(ent )-130(to )-127(th)11(e)11( )] TJ
```

(a) original

```
[(L)-2(ong,)5( )-130(l)-1(o)12(ng )-130(ago,)8( )-130(ther)9(e
)-132(l)-2(i)8(ved )-134(an)-1( )-131(o)11(l)-2(d)-8( )-131(m)17(an)-3(
)-132(a)9(nd)-10( )-128(h)10(i)-1(s)7( )-129(ol)1(d)-2(
)-131(w)7(i)-3(f)2(e )-131(i)8(n )-131(a )-132(vil)-1(l)5(age.)8(
)-132(H)4(e )-127(w)1(ent )-131(to )-125(th)12(e)15( )] TJ
```

(b) $\delta = 3$

```
[(L)-1(ong,)14( )-130(l)3(o)11(ng )-132(ago,)8( )-121(ther)15(e
)-134(l)-2(i)5(ved )-134(an)-3( )-126(o)7(l)-3(d)-7( )-131(m)9(an)-6(
)-134(a)11(nd)-6( )-126(h)9(i)-8(s)5( )-135(ol)2(d)-6(
)-127(w)6(i)0(f)2(e )-128(i)8(n )-131(a )-129(vil)2(l)6(age.)8(
)-134(H)1(e )-132(w)9(ent )-130(to )-125(th)11(e)7( )] TJ
```

(c) $\delta = 6$

```
[(L)-12(ong,)7( )-128(l)2(o)12(ng )-125(ago,)9( )-135(ther)4(e
)-140(l)-9(i)10(ved )-134(an)1( )-132(o)14(l)2(d)-11( )-134(m)23(an)-3(
)-108(a)9(nd)-19( )-131(h)22(i)6(s)20( )-129(ol)2(d)-8(
)-136(w)10(i)-7(f)-11(e )-135(i)3(n )-126(a )-130(vil)6(l)10(age.)10(
)-138(H)11(e )-135(w)1(ent )-139(to )-124(th)13(e)9( )] TJ
```

(d) $\delta = 12$

Fig. 5. Changes of space lengths in the first line.

Due to the permutation and RDM-QIM method using a secret key, a malicious party will not be able to find any useful information from the observation of frequency components without the secret key. Hence, the proposed method could be used for the application of steganography if the step size δ is small.

Table 1. Comparison of the average file size [Bytes] when 1,448 bits watermark is embedded.

	uncompressed	compressed
original	104,369	55,164
$\delta = 3$	104,348	55,964
$\delta = 6$	104,346	56,156
$\delta = 12$	104,440	56,319
$\delta = 24$	104,676	56,512
$\delta = 48$	104,951	56,726

5 Concluding Remarks

In this paper, we proposed a novel data hiding method for PDF files. The main idea is to regard a collection of space lengths in each line as one dimensional signal, and to use the signal processing operations in order to embed watermark with less distortions. By avoiding a DC component, the total length of each line is kept unchanged so as to control the distortions caused by the embedding. We also consider the secrecy of the watermark by introducing the permutation and the dither modulation according to a secret key. If the quantization step size is small, we will be able to apply for steganography. If the amount of embedding information is reduced, it will be used for robust watermarking method.

It is possible for malicious party to insert random noise to the watermarked space lengths. If the noise energy is smaller, the watermark can be extracted correctly, and vice versa. For the further investigation on the robustness, the evaluation criteria should be considered for watermarking methods for PDF documents. Similar to multimedia watermarking schemes, there is a trade-off among transparency, robustness, and capacity. The numerical measurement of transparency in PDF documents is not studied though it will be measured by the energy of signal-to-noise (watermark) ratio in the proposed method. The robustness should be also measured under possible attacks on PDF documents.

Acknowledgements. This research was partially supported by JSPS KAKENHI Grant Number JP16K00185.

References

1. Bender, W., Gruhl, D., Morimoto, N., Lu, A.: Techniques for data hiding. IBM Syst. J. **35**(3–4), 313–336 (1996)
2. Bitar, A.W., Darazi, R., Couchot, J.F., Couturier, R.: Blind digital watermarking in PDF documents using spread transform dither modulation. Multimed. Tools Appl. **76**(1), 143–161 (2017)

3. Chen, B., Wornell, G.Q.: Quantization index modulation: a class of provably good methods for digital watermarking and information embedding. IEEE Trans. Inf. Theory **47**(4), 1423–1443 (2001)
4. Iwamoto, T., Kawamura, M.: Proposal for invisible digital watermarking method for PDF files by text segmentation (in Japanese). In: IEICE Technical Report, EMM2016-59, vol. 116, pp. 31–35 (2016)
5. Lee, I.S., Tsai, W.H.: A new approach to covert communication via PDF files. Signal Process. **90**(2), 557–565 (2010)
6. Lin, H.F., Lu, L.W., Gun, C.Y., Chen, C.Y.: A copyright protection scheme based on PDF. Int. J. Innov. Comput. Inf. Control **9**(1), 1–6 (2013)
7. Por, L.Y., Delina, B.: Information hiding: a new approach in text steganography. In: Proceedings of the ACACOS 2008, pp. 689–695 (2008)
8. Zhong, S., Cheng, X., Chen, T.: Data hiding in a kind of PDF texts for secret communication. Int. J. Netw. Secur. **4**, 17–26 (2007)

Tally Based Digital Audio Watermarking

Kotaro Sonoda(✉) and Shu Noguchi

Graduate School of Engineering, Nagasaki University, 14-1, Bunkyo-machi, Nagasaki-shi, Nagasaki 852-8521, Japan
sonoda-iihmsp17@cis.nagasaki-u.ac.jp

Abstract. In this paper, we propose a novel digital audio watermarking system inspired by the tally trade. In ordinary digital audio watermarking, a single stego signal is produced through the embedding process and the hidden message is extracted from the stego signal. In tally-based system we propose, multiple stego signals are tallied to produce and the hidden message which is extracted from the temporally mixed signal composed of the required number of stego tallies. When the extractor can't mix the required number of stego tallies, it misses the hidden message. The system's performances are evaluated in terms of the number of tallies (shares).

Keywords: Tally · Audio secret sharing · Time-spread echo method · Digital audio watermarking

1 Introduction

Due to the widespread presence of video sharing sites, anyone can easily share videos and music. Such illegal uploading, ignoring the copyright of contents, illegal copies without permission from the producer, illegal buying and selling correspond to infringement of copyright, etc. are regarded as problems. Copyright holders have used similar music search technology and embedding of digital watermarks of contents in order to detect such illegal usage and protect legitimate earnings from the distribution of the material.

Recently many algorithms for embedding watermarks in acoustic digital watermarks have been studied. However, even if the algorithm is complicated, once the algorithm becomes known, confidential information may be extracted. For that purpose, we have been taking countermeasures using secret parameters such as random numbers, but the random numbers are easily generated from predefined seed value. Once the seed value is leaked, the system becomes defenseless. In this paper, we aim to create a new audio tally which was inspired by tally-trade. A kind of tally is shown in Fig. 1, which was used for trade in medieval Japan. In tally trade, a watermark letter is written on a piece of wood, or paper, etc. and it is split/torn in two. The both parties held one piece each and then combined it to confirm it's validity. Here, a watermark letter is a kind of secret, but both of the corresponded connect pair and appropriate connect way are still required for evidence even if the watermark letter is known.

J.-S. Pan et al. (eds.), *Advances in Intelligent Information Hiding and Multimedia Signal Processing*, Smart Innovation, Systems and Technologies 81,
DOI 10.1007/978-3-319-63856-0_48

Fig. 1. Tally trade in medieval Japan

2 Tally-Based Watermarking System

In this section, we propose a tally-based watermarking system. The tally-based system aims to make the hidden message by decomposing it and embeddeding it to N pieces of stego-tally signals. The K pieces of stego-tally signals are mixed temporally and the hidden message is extracted by decoding the mixed stego signal.

Similar strategies are studied in a secret sharing system. The secret sharing is a distributed management method of hidden data and it was independently proposed by Blakley [1] and Shamir [2] in 1979. In (K, N) – threshold secret sharing, secret datum s is divided into N shares of data. You can restore secret datum s when K shares are combined but cannot restore from $K - 1$ shares or less.

With regard to an acoustic secret sharing method studied in order to make use of the characteristics of sound, Desmedt et al. [3] and Lin et al. [4] proposed an acoustic secret sharing of binary bit strings. Desmedt's method is $(2, 2)$-threshold secret sharing and they construct two shared signals such that any two shares have counter phases if the secret data is '0' or the same phases if the secret data is '1'. Lin extends Desmedt's method to $(2, N)$-threshold by representing a sharing bit with the main counter/same phase in consequent multiple sub-blocks. In Lin's study, this was implemented in Rock music.

These acoustic secret sharing methods are methods of acoustic information hiding. However these approaches do not consider the quality of the audio signals that are shared and results in quality degradation. Therefore, we introduce this concept to an audio watermarking system in this report.

In some audio watermarking methods, the hidden message is spread by a pseudo-noise (PN) sequence (key), and the spread data is added or convolved with a host signal to produces a stego signal. The embedded hidden message is extracted by auto-correlation of time-sequences or the cepstrum of the stego data and the predefined pseudo-noise.

The PN sequence $p(n) \in \{-1, +1\}$ is used as a key here can be composed by the summation of N pieces of $p_k(n) \in \{-1, +1\}$ as $p(n) = \sum_{k=1}^{N} p_k(n)$. A signal group spread using $p_k(n)$ instead of $p(n)$ is the same as the signal diffused using $p(n)$ by addition synthesis. Therefore, by mixing and synthesizing all the stego signal groups $s_k(t)$ generated by using $p_k(n)$, the stego signal $s(t)$ generated by using $p(n)$ can be reproduced. Due to the noise immunity of the base watermarking technique, it is possible to detect the watermarks even when K or more pieces of $s_k(n)$ are used without using all N pieces. However, it should be noted that $\hat{s(t)}_K$ obtained by mixing and synthesizing K items has an embedding strength of $1/K$ as compared with the base watermarking method.

2.1 Decomposion of Pseudo-noise Sequence

We decompose each element of the PN sequence $p(n)$ so that it is generated by addition of each element of N pieces of $p_k(n)$. Here, N is an odd number. This means that $(N + 1)/2$ pieces in $p_k(n)$ $(k = 1 \ldots N)$ are equal to $p(n)$ and the remaining $(N - 1)/2$ pieces are $-p(n)$. Therefore $(N - 1)/2$ pairs equaling $+p(n)$ and $-p(n)$ are canceled in summation and only one piece equaling $p(n)$ remains. You can choose any tally index k of $(N + 1)/2$ randomly for each element n, but you do not need to know how to choose this on the detection side.

For example, for an original PN sequence $p(n), n = 1 \ldots 5$ is decomposed to $N = 3$ sequences $p_k(n), k = 1 \ldots 3$ as following:

$$
\begin{array}{c}
p(n) \\
\downarrow \\
\begin{bmatrix} +1 \\ +1 \\ -1 \\ +1 \\ -1 \end{bmatrix}
\end{array}
=
\begin{array}{c}
p_1(n)\ p_2(n)\ p_3(n) \\
\downarrow \quad \downarrow \quad \downarrow \\
\begin{bmatrix} +1 & +1 & -1 \\ +1 & +1 & -1 \\ -1 & -1 & +1 \\ -1 & +1 & +1 \\ +1 & -1 & -1 \end{bmatrix}
\end{array}
\begin{bmatrix} 1 \\ 1 \\ 1 \end{bmatrix}
$$

3 Message Detection Performance in Tally-System

Because the proposed tally-based system has not changed the base embedding scheme, the sound quality of stego-tally signals conform to the performance of the embedding scheme. On the other hand, the message detection performance is expected to depend on the proportion k/N. k is the number of mixing stego-tally signals, and N is the whole number of stego-tally signals.

In this section, we evaluate the message detection performance with several k values.

Remark 1. In this report, we use the Time-spread Echo method [5] as a base watermarking method. In the time-spread echo method, a PN sequence is used

Table 1. Parameters of time-spread echo method

Echo-gain β	0.08
Time-delay of watermark bit "0" τ_0	4.5 ms
Time-delay of watermark bit "1" τ_1	6.8 ms
Length of PN L_{PN}	1023 samples (23.2 ms)
Frame length L_{Fr}	8192 samples (185.7 ms)
embedding bit rates R_{emb}	5.38 bps

Table 2. SQAM tracks and genres

SQAM track #	Genre
35	Glockenspiel
69	ABBA

as a component of the echo generating kernel with predefined time-delays representing the hidden bits. The component "+1" yields same phase echo and the component "−1" yields anti-phase echo. Therefore mixing the echo of a tally with the component "+1" and one of another tally with component "−1" cancels the echo. The parameters used in this method are listed in Table 1.

Remark 2. Payload data embedded in stego signal is 0/1 binary random bit sequence and it is published on the IHC website [6].

Remark 3. Tested host signals were monaural and selected from SQAM (Sound Quality Assessment Material) [7] published by EBU (European Broadcasting Union). Table 2 shows the track numbers and genres.

3.1 K Out of N Tally System

We carried out the BER evaluation test with the proposed tally-based system with $N = 11$ tallies.

Figure 2 shows the result of BER in the case of k out of 11 tallies for SQAM Track #35 (a solo performance). The result shows that BER exponentially decreases and mixing whole tallies ($K = 11$) achieved under 10% and $K = 6$ ($K/N = 0.6$) or more tallies are required for a BER of under 20%.

While in the case of using another type of music (club music), similar results are achieved as shown in Fig. 3. $K = 7$ ($K/N = 0.6$) or more tallies are required for under a BER of less than 20% for this case.

Regarding these results, mixing whole tallies ($K = N$ condition) resulted in missing a few hidden bits and BER is not zero. It is assumed that the embedding strength is insufficient to detect successfully after mixing tallies. The echo gain $\beta = 0.08$ set in the time-spread echo method was degraded to $\hat{\beta} = \beta/K$ in

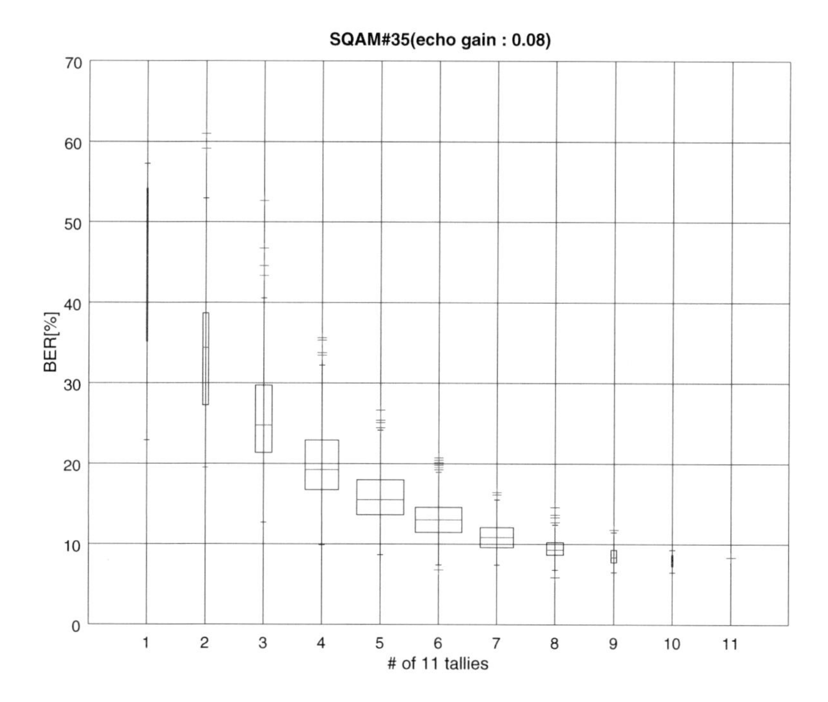

Fig. 2. BER in k out of 11 tallies; SQAM Track: 35, echo gain: 0.08

mixing K tallies. In condition of $K = N = 11$, that $\hat{\beta}$ is degraded to 0.08/11. Larger echo-gain, longer PN sequences, or smaller sharing size are required for a lower BER.

We will expect the required relationships between the embedding strength β, threshold K and size of shares N from the experimental results in Sect. 3.1.

The no-bit-error least strength of our base time-spread echo method is represented by $\beta^2 L \geq S_0$ depending on a length of PN sequence (L) and a echo gain (β). In our N tally-based system, the echo can detect from N-mixed signal. N-mixed signal has N times bigger signal component and same level of echo component compared with normal stego signal. Thus, the tally signal must be embedded by the strength of greater than $\beta^2 L \geq N \cdot S_0$.

Here, echo gain β is usually bounded by required imperceptibility. Therefore, our system required to embed the echoes of N times longer PN sequence than base time-spread echo method.

Reason why the result of BER is not zero in $K = N = 11$ in figures of Sect. 3.1 might because the experiment condition: $N = 11, \beta = 0.08, L = 1023$, has not suffered the relationship.

From the result of Sect. 3.1, our tally-based system can detect many of bits from K-mixed signal ($K \leq N$). K-mixed signal has statistically correct $L_K = L \cdot \frac{K+1}{N+1}$ components of embedded PN sequence. Therefore the embedding strength is $\beta^2 L_K$ in K-mixed. Supposing all-bit-error threshold strength S_{100}, (K, N) tally system requires $\beta^2 L_{K-1} \leq (K-1) \cdot S_{100}$ and $\beta^2 L_K \geq K \cdot S_0$.

The practical setting of these parameters are feature works.

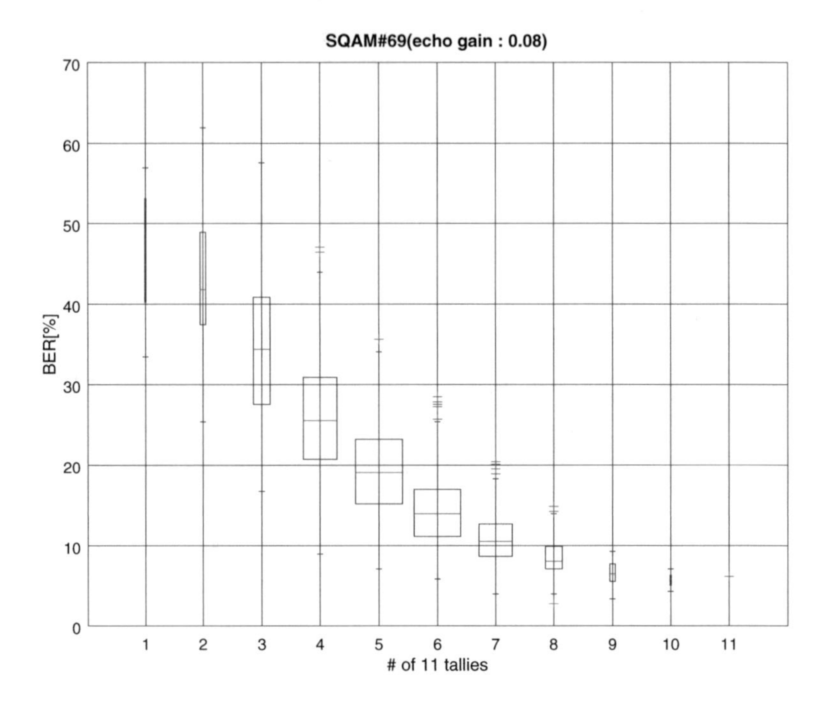

Fig. 3. BER in k out of 11 tallies; SQAM Track: 69, echo gain: 0.08

4 Discussion

Proposed our tally-based watermarking system is managing a hidden message by N tally audio signals in while conventional watermarking system managing by just one stego signal. Moreover, in the proposed method, larger echo-gain and longer PN sequence, those deteriorate the audio quality, are required for attaining the same correct detection rates with the conventional time-spread echo watermarking. However, the risk of leakage is expected to ramp down compared with the conventional one, because the proposed one requires the cooperative other tally signal physically for message detection.

5 Conclusion

In this report, we propose a novel tally-based audio watermarking system. Our system decomposes a PN sequence, which is used as a spread kernel for payload bits in a hybrid audio watermarking method, for N tally sequences. A stego signal watermarked using the tally sequence restores the stego signal to the original watermarking method by mixing. The results of BER for some music sources show that the proposed system supports 60% whole tallies to restore the hidden message.

References

1. Blakley, G.-R.: Safeguarding cryptographic keys. In: International Workshop on Managing Requirements Knowledge, p. 313. IEEE Computer Society (1899)
2. Shamir, A.: How to share a secret. Commun. ACM **22**(11), 612–613 (1979)
3. Desmedt, Y., Hou, S., Quisquater, J.-J.: Audio and optical cryptography. In: Advances in Cryptography, ASIACRYPT 1998, pp. 392–404. Springer (1998)
4. Lin, C.-C., Laih, C.-S., Yang, C.-N.: New audio secret sharing schemes with time division technique. J. Inf. Sci. Eng. **19**(4), 605–614 (2003)
5. Ko, B.S., Nishimura, R., Suzuki, Y.: Time-spread echo method for digital audio watermarking using PN sequences. In: IEEE International Conference on Acoustics, Speech and Signal Processing, ICASSP 2002, pp. II-2001–II-2004 (2002)
6. IHC Evaluation Criteria and Competition (2017). http://www.ieice.org/iss/emm/ihc/IHC_criteriaVer5.pdf
7. European Broadcasting Union (EBU): SQAM (Sound Quality Assessment Material) (2008). http://tech.ebu.ch/publications/sqamcd

Variable-Length Key Implementation Based on Complex Network WSN Clustering

Hongbin Ma, Wei Zhuang, Yingli Wang(✉), Danyang Qin, and Xiaojie Xu

School of Electronic Engineering,
Heilongjiang University, Harbin 150000, Heilongjiang, China
wangyingli@hlju.edu.cn

Abstract. We use the adaptive clustering method that is the division of network by following the complex network theory model to manage nodes in case of WSN node scale changes and node data traffic surges, which complicates the node relationship at the same time and puts forward higher requirements for security management. In this paper, it is proposed to generate a variable-length random key to manage the relationship between the nodes to ensure that the complex relationship between the node changes and the security of the entire network, and provide an easy-to-implement variable security management mechanism.

Keywords: WSN · Security · Chaos · Clustering · Variable-length

1 Introduction

With the increasingly extensive application of wireless sensor networks, the layout of the network is also growing, and the early plane network structure has been difficult to meet the various requirements of wireless sensor networks, and clustered network structure has occupied a dominant position. Therefore, the key management scheme that corresponds to the requirements based on clustering has been proposed in order to meet the security requirements of wireless sensor networks [1].

Key technologies of node scale, security control and energy management in WSN are interrelated, inter-conditioned and interacted, when the node scale changes, flexible control of WSN security and energy distribution, rational use of tripartite constraints, to ensure the relevance and robustness of the tripartite, which is the key technology in this paper. When the node scale and data flow are changing constantly, the topology and management mechanism of the nodes is constantly updated. In order to deal with the changing node relations, a variable-length random key generation algorithm based on complex network clustering is proposed, and the relationship between nodes is managed by variable-length random keys to ensure the security of the whole network.

The traditional encryption method that does not distinguish the security level uses the fixed key length to encrypt, which lacks application elasticity and affects communication efficiency, meanwhile wastes the sensor energy, making

J.-S. Pan et al. (eds.), *Advances in Intelligent Information Hiding and Multimedia Signal Processing*, Smart Innovation, Systems and Technologies 81,
DOI 10.1007/978-3-319-63856-0_49

the transmission delay increases. The variable-length random key proposed can change the key length according to the change of the external environment, and reduce the transmission delay while improving the communication efficiency.

2 Clustering Based on Complex Network

Complex networks can be widely reflected in a large number of actual networks and in real life. The structure of large-scale complex networks often changes with time, and it has limitations in obtaining network global information for assessing the importance of nodes [2]. At present, the research of scale-free evolution model of wireless sensor networks often regard the network as homogeneous network, and do not fully consider the evolving characteristics of the network under the real situation, which leads to the obvious difference between the network topology and the actual network [3]. Therefore, based on the theory of the complex networks, the clustering evolution model of wireless sensor networks is proposed considering the typical clustering structure of wireless sensor networks, the sensitivity of energy consumption and the dynamic behavior of node and link exit which are prevalent in real networks [4].

Barabasi and Albert proposed the formation mechanism of complex network with power law degree distribution from a dynamic and growing view, called scale-free network and established B-A scale-free network model. The scale-free network node degree distribution $P(k)$ is consistent with the power law distribution, which determines that a small number of nodes in the network occupies most of the connections, and the vast majority of nodes in the network are low. Thus, scale-free network has better fault-tolerant performance. Scale-free network and its related theory provide a new thinking and methods for establishing a new fault-tolerant WSN topology structure [3]. According to the reasons of the above network congestion, it can be found that network congestion is more in the case of WSN traffic imbalance, when part of the node traffic is too large, the data transmission in a small number of nodes, i.e., nodes with higher node degrees are more prone to cause congestion. Because the data of B-A are easy to converge to higher node degree, the probability of congestion in the network structure formed by B-A model will be greatly enhanced. Therefore, based on the scale-free network topology model, combined with the D.Shi model and WSN itself Characteristics, taking into account WSN energy consumption control, we use an energy-aware survival of the fittest evolution model [5]. The specific algorithm based on this model is as follows.

1. In the initialization process, assume that WSN initially has M_0 nodes, e_0 edges.

2. The preferred mechanism is enabled, that is, a new node will join the wireless sensor network within each timestamp, and the new node will be given energy value according to specific probability model, next it is connected to m($m \leq m_0x$) connectable old nodes. By using the fittest mechanism of the D.Shi model, when the new nodes in the cluster are connected with the cluster head to establish the data link, the selection probability is selected according to the merit probability. The cluster head selection is performed according to selected probability $\prod(k_i)$. And $\prod(k_i)$ is still determined by the energy, the degree of the selected node, and the communication distance, so that the probability of the newly added node and the connectable node establishing the link is subject to the following distribution,

$$\prod(k_i) = \frac{M}{m_0 + t} \cdot \frac{E_i k_i}{\sum_j E_i k_i} \cdot (1 - \frac{k_i}{k_{max}}) \cdot P(d_{ij} \leq d_{jmax}) \quad (1)$$

where $M(m \leq M \leq m_0 + tm)$ is the number of nodes available in the local world Ω network when evolving to time t, k_i is the degree of the node, k_{max} is the saturation of the cluster head, that is the remaining energy of the node. It can be seen that if the current cluster head degree k_i satisfies the saturation constraint k_{max}, then $1 - k_i/K_{max} = 0$, depending on the model used, the probability of this node is selected for connection is $\prod(k_i) = 0$.

3. Reverse selection mechanism, we can assume that there are m links within each timestamp, and these links are disconnected at probability p, there will be mp links disconnected and at least one link is disconnected by the probability $\prod^*(k_i)$, that is, the smaller the residual energy of the node, the greater the probability of the link being selected to be disconnected.

$$\prod^*(k_i) = \frac{(E_i k_i)^{-1} \rho_i^{-1}}{\sum_j E_i k_i^{-1}} \quad (2)$$

where, ρ_i^{-1} is the reciprocal of the node data flow, the larger the ρ_i^{-1}, the smaller the node data flow. Node j is deleted in the neighborhood $O(i)$ of the node i with the anti-preferential probability $K_i^{-1} \prod^*(k_i)$, where $K_i = \sum_{j \in O_i} \prod^*(k_i)$, and then the links that contain nodes i and j are disconnected, repeat this step mp times. It can be seen that when a cluster head node has more residual energy value, it is likely to be selected to establish a new connection, but this possibility will decrease with the increase of k_i/k_{max}, where k_i/k_{max} is the ratio of the node degree to the maximum degree, when its degree of connectivity reaches its maximum value, the probability that the node is marked is 0, that is, it will not be connected again. Therefore, the evolutionary mechanism can maintain the dynamic balance of the node energy in the whole network, which can effectively avoid some nodes died soon due to premature depletion of energy.

3 Variable - Length Chaotic Key Design

3.1 Complexity of Clustering WSN Network

It is necessary to add or delete some nodes in wireless sensor network for the requirements of different situations, which leads to the change of the cluster head and the distribution of the cluster key. Therefore, the network structure is intricate. In order to ensure the security and reliability of communication, it is necessary to classify the key and use key hierarchical management for key updates. In addition, due to the roles of different nodes are different, and the status is different, so we can design a variable-length key algorithm to manage the key between the nodes. Actually, common node is in the bottom position for small traffic, so we can use a shorter key to transmit information, both to protect the security, but also to protect the transmission efficiency and reduce unnecessary energy consumption. Cluster head is in the middle and is responsible for forwarding the node's information, requiring higher security, so it can be transmitted using a medium-length key. Base station is in the peak, occupy the most important position of the whole network, security requirements are very high, so there is need to use long keys to encrypt the transmission of data to ensure that users are not intercepted by illegal users. According to this situation, this paper designs a variable-length key generation algorithm using chaotic system.

3.2 Design of Variable - Length Chaotic Key Model

A variable-length chaotic key generation state is shown in Fig. 1. The system first randomly generates the initial value of the chaotic key x_0, and the key length is determined according to the length field of the initial value. The chaotic key is obtained by the iterations in [6], and the process is repeated. Then we can obtain the required number of variable-length keys.

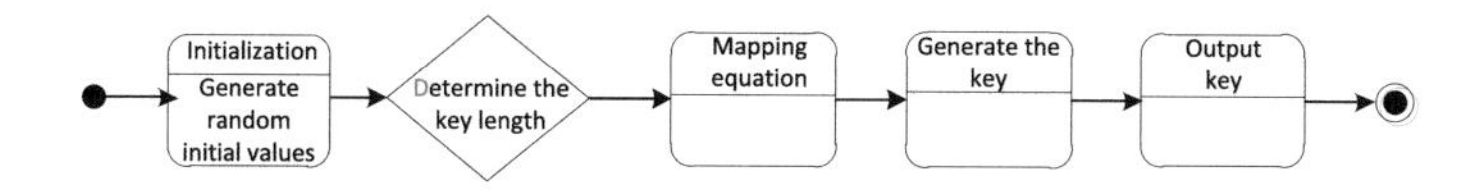

Fig. 1. The generation state based on variable-length key

3.3 Generation of Real - Valued Chaotic Sequences

By selecting the appropriate nonlinear equation, the periodicity of the sequence can be greatly improved, and the branch coefficient is adjusted so that the current state is in a chaotic state. The logistic one-dimensional mapping method is a very simple mathematical model used to explain about the growth population. It is given by the equation,

$$x_{k+1} = \mu x_k(1 - x_k) \quad x \in (0, 1) \tag{3}$$

where μ is bifurcation parameter. When $3.57 < \mu \leq 4$, the mapping is in a chaotic state.

3.4 Digital Chaotic Sequences Are Obtained from Real - Valued Sequences

Each element in the chaotic real sequences can be transformed into binary sequences via the intermediate multi-bit quantization method. The binary representation is as follows,

$$|x_n| = 0.c_0c_1c_2..., c_i \in 0, 1 \tag{4}$$

Since the resulting system is mapped over the interval (0, 1), it is necessary to convert the fractional part into a binary representation,

$$x = \sum_{v=0}^{\infty} a_v 2^{-(v+1)} = (a_0a_1a_2...), a_v = 0 \; or \; 1 \tag{5}$$

The quantization method uses the former L-bit approximate the decimal, and truncates the number of digits behind, so the quantitative sequence is

$$\widetilde{x} = \sum_{v=0}^{L-1} a_v 2^{-(v+1)} = (a_0a_1a_2...a_{L-1}) = \\ (2^{-L} \sum_{v=0}^{L-1} a_v 2^{(L-1)-v} = 2^{-L}X, \; a_v = 0 \; or \; 1) \tag{6}$$

Where, $X = \sum_{v=0}^{L-1} a_v 2^{(L-1)-v}$. X is an L-bit binary number that corresponds to the fraction. Among them, the choice of L has a strict limit, if the value of L is small, will make the binary sequence lose chaotic character due to the large cycle. Therefore, this paper adopts the quantification method of [6], which avoids the problem above by improved the intermediate multi-bit quantization method. The quantification is as follows,

1. Assume that the length of the desired sequence is M, after the iteration $M - 1$ times, the absolute value of M real chaos, including the initial value, are converted to binary,

$$\begin{bmatrix} x_1 \\ x_2 \\ ... \\ x_m \end{bmatrix} = \begin{bmatrix} c_0(x_1) & c_1(x_1) & c_2(x_1) & ... \\ c_0(x_2) & c_1(x_2) & c_2(x_2) & ... \\ ... & ... & ... & ... \\ c_0(x_m) & c_1(x_m) & c_2(x_m) & ... \end{bmatrix} \tag{7}$$

   Where c_i belongs to $\{0, 1\}$.
2. If the chaotic real value is less than 0, then take the anti-code of the binary sequence that is converted by its absolute value, and then starts from the first N bit value, where N is a random value, and truncate the L-bit, forming a matrix of $M \times N$.
3. Each column in the matrix as a chaotic sequence of length M, and we can get N sequences.

4 Simulation Analysis

4.1 Sequence Correlation Analysis

Correlation is the most significant property of verifying security of sequence. Excellent correlation is one of the important guarantees of communication security. When the correlation of the sequence is high, it will lead to a significant increase of security risks in data communication, and data interceptors can decrypt data based on data correlation, which results in data degradation.

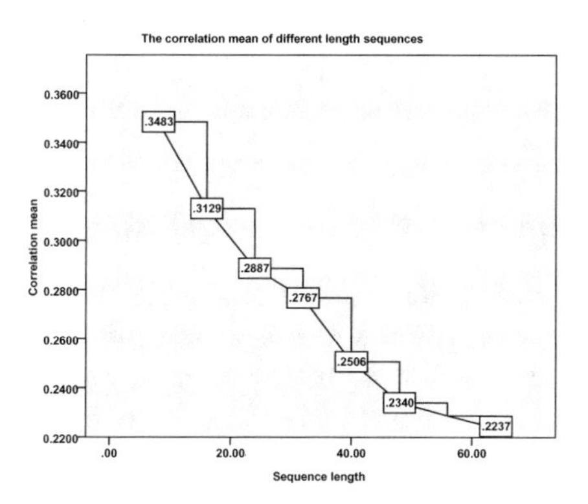

Fig. 2. Mean of correlation of variable length

In this paper, we analysed the correlation of variable length sequences by numerical methods and obtain the mean through 500 random initial values, see Fig. 2. It can be seen from the Fig. 2 that the difference between the correlations of variable length sequences is different. In order to save the hardware and software resources, the obtained data can be classified according to comparison of means.

4.2 Statistics of Sequence Security Level

The classification of sequence length can be defined according to comparison of means of correlation between neighboring length sequences.

By calculating the mean of the difference between correlation values, we can obtain the mean is 0.0178, as shown in Fig. 3. When the difference is greater than 0.0178, it can be concluded that the correlation between the two sequences is large and can be divided into two security levels. Provide that the difference is less than the average, it is necessary to accumulate the current difference and the next difference. If the accumulated value is greater than the mean, the intermediate part can be divided into the same level. As is shown in Fig. 3, We can divide security levels of sequence of different lengths into 8 bits, 16 bits, 24 bits, 40 bits, and 56 bits.

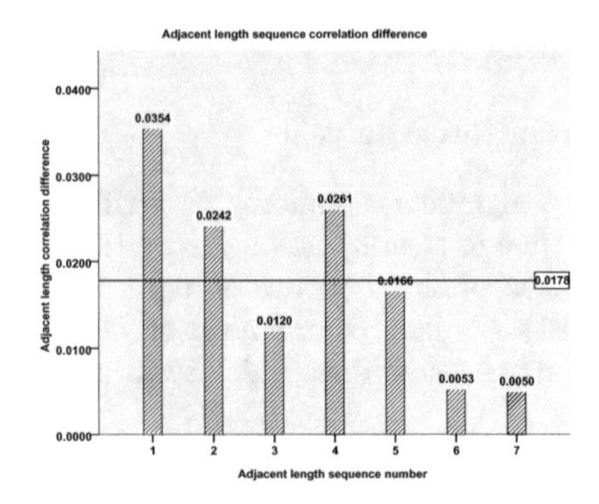

Fig. 3. Adjacent sequence correlation difference

4.3 Encryption Time of Different Length Key

In the process of data encryption simulations using the keys in different lengths, the comparison of the encryption times that can be obtained with the keys in different lengths is shown in Fig. 4. In the normal operation of wireless sensor nodes, computing time is the main factor that restricts the real time of system encryption. Surely, encryption operation for a long time, the number of iterations can also cause a large number of energy consumption. In addition, a large number of ordinary nodes send information more frequently, the use of short keys not only can guarantee the safety and reliability of communication but also can reduce the energy consumption and prolong the lifetime of nodes. Besides, the medium and long keys are used for data aggregation of cluster heads and base stations to improve security.

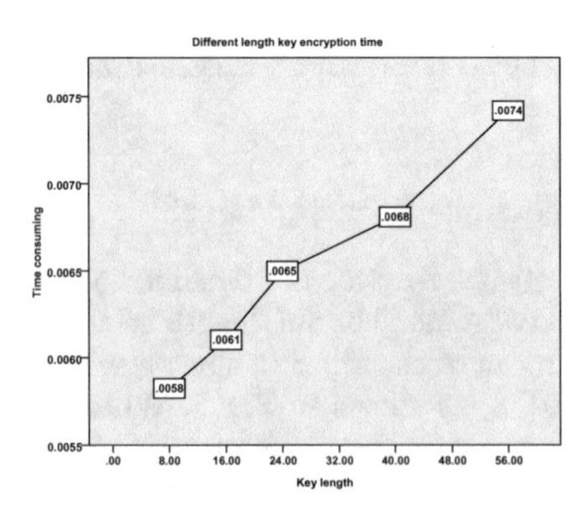

Fig. 4. Different length key encryption time consumption

5 Conclusion

In this paper, an adaptive network clustering algorithm is used to manage the nodes and to divide the network by establishing the complex network theory model under the circumstance that WSN node scale is changing constantly and the node data flow is complicated and changing unceasingly. As clustering management of the network, making the complexity of the node relationship, and the higher requirements are put forward for security management. Therefore, a variable-length key management algorithm is proposed to adapt to complex changes in node relationships, and to provide a variable security management mechanism that is easy to implement. The simulation results show that the proposed scheme has higher security, lower update cost, and support analysis of large-scale network.

Acknowledgments. Acknowledgment Project supported by the National Natural Science Foundation of China (Grant Nos. 61302074).

References

1. Qin, D.Y., Jia, S., Yang, S.X., Wang, E.F., Ding, Q.: Research on secure aggregation scheme based on stateful public key cryptology in wireless sensor networks. J. Inf. Hiding Multimedia **7**(5), 938–948 (2016)
2. Ruan, Y.R., Lao, S.Y., Wang, J.D., et al.: An algorithm for evaluating the importance of complex network node based on domain similarity. ACTA Phys. Sin. **66**(3), 371–379 (2017)
3. Fu, X.W., Li, W.F.: Clustering evolution model of wireless sensor networks based on local domain. J. Commun. **36**, 204–214 (2015)
4. Duan, H.X., Zhou, Y., Liu, M.M.: Fault tolerant scheduling algorithm in distributed sensor networks. J. Inf. Hiding Multimedia Sig. Process. **8**(1), 127–137 (2017)
5. Ma, H.B., Li, X.: Topology control based on double cluster head ellipse model in WSN. In: International Conference on Mechanical Materials and Manufacturing Engineering (2016)
6. Zhang, Y.P., Lu, R.M.: An improved chaotic sequence quantization algorithm. Commun. Technol. **42**(3), 278–281 (2016)

Virtual Test Technology and Virtual Environment Modeling

An Infrared Small Target Detection Method Based on Block Compressed Sensing

Jingli Yang(✉), Zheng Cui, and Shouda Jiang

Department of Automatic Test and Control, Harbin Institute of Technology, Xidazhistr. 92, Harbin 150080, China
jinglidg@hit.edu.cn

Abstract. Aiming at improving the real-time performance of infrared weapon systems, a method based on block compressed sensing is proposed to detect infrared small target, which is also easy to be implemented with hardware. The proposed method can detect and locate small infrared targets by classifying compressed results of blocks. In addition, in order to solve the low detection accuracy caused by using uniform block, the proposed method uses overlapping blocks to reduce the maximum distance between the center of the test sample block and that of the target. Experiments show that the proposed method can effectively improve the detection accuracy of infrared small targets.

Keywords: Compressed sensing · Infrared target detection · Overlapping blocks

1 Introduction

With advantages of good concealment, strong penetration and all-weather capabilities, infrared imaging technology has been adopted in the fields of national defense and scientific research, and has become an important component of modern photonics technology part [1]. In particular, infrared small target detection plays a critical role in large amounts of practical projects such as infrared warning and defense alertness. In the past decades, the development of infrared focal plane arrays improves the infrared image quality greatly, which helps promote the target detection accuracy of infrared weapons [2]. However, it also confers the problem of channel blocking for infrared image transmission [3].

The traditional way to solve above problem is to compress the original image data after imaging and to transfer the compressed data in the wireless channel. However, data compression of infrared image data is very inefficient because it gets involved in many complex calculations. Moreover, the data decompression procedure is also time-consuming because target detection process can only be performed on the reconstructed image data. Recently, the emerging theory of compressed sensing [4,5], which can sample and compress data with hardware-based system simultaneously, provides a new way to address this issue.

The first hardware-based system for compressed sensing theory was a single-pixel camera implemented by Richard Baraniuk in 2008 [6,7]. After that, some

J.-S. Pan et al. (eds.), *Advances in Intelligent Information Hiding and Multimedia Signal Processing*, Smart Innovation, Systems and Technologies 81,
DOI 10.1007/978-3-319-63856-0_50

researchers introduced compressed sensing theory into the field of infrared imaging and made some encouraging progress [8]. The parallel sampling technology was developed to reduce the sampling time by the combination of compressed sensing theory and infrared focal plane arrays [9,10].

With the increasing requirements for the real-time performance, the efficiency of data reconstruction in compressed sensing theory has become the primary bound to infrared weapon systems. Many algorithms have been proposed to address this issue. Li *et al.* [11] proposed a method based on the low rank recovery, which can combine image restoration with target detection. However, it still suffers from the performance degradation because of the complex data reconstruction process. Zhang *et al.* [12] developed a novel target detection method, which can implement target recognition in compressed domain. Based on that, Zhang *et al.* [13] added the local search mechanism to avoid data reconstruction process. However, most methods are based on objective image segmentation in non-even background, which means they cannot be achieved in hardware-based systems.

In an attempt to address the shortcomings of existing methods, this paper proposes an effective infrared small target detection method based on block compressed sensing theory. This method not only can be easily implemented by the hardware, but also should be able to solve the low detection accuracy caused by using uniform block in traditional block compressed sensing.

2 Target Detection Based on Block Compressed Sensing

2.1 Compressed Sensing and Target Detection

Compressed Sensing (CS) theory is a disruptive approach for signal sample device. This method, proposed by Donoho *et al.* [4], is completely different from the traditional Shannon-Nyquist sampling and coding techniques. CS can make full use of the sparseness of the signal and process high-speed signal with a low-speed sampling process.

The key technologies of CS theory, based on sparse signal processing [4], include sparse representation of the signal, the measurement matrix and reconstruction algorithm [5]. CS observes an N-dimensional signal and gets M observations ($M << N$), which can be represented as Eq. (1). The restructuring procedure uses $\Phi_{M\times N}$ to decompress Y. Equation (1) can be represented more intuitively as Fig. 1.

$$Y_{N\times 1} = \Phi_{M\times N} X_{N\times 1} \tag{1}$$

As shown in Fig. 1, the method of restructuring the original signal X from the measuring result Y is achieved by solving an under-determined equation system. CS transforms the signal to another domain, in which most of the transformed coefficients are zero [6].

The traditional method of target detection based on compressed sensing is implemented on compressed results of image sub-regions, which extracted by image segmentation or randomly sampling. The aim of the method is to identify

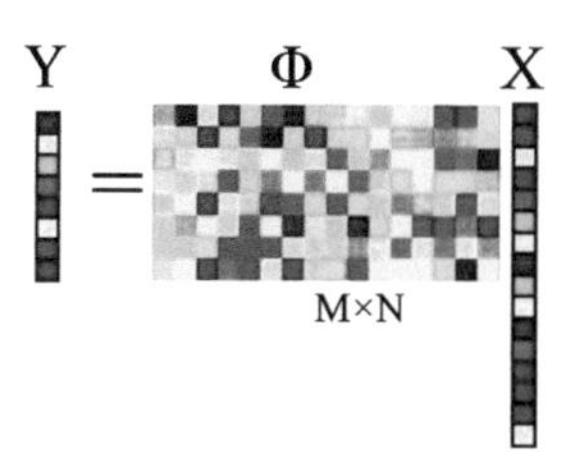

Fig. 1. Schematic diagram of compression observation matrix

the sub-regions which contain the targets. However, above method cannot be implemented in the hardware system, because there is only optical equipment in the CS hardware device, thus, it cannot perform image preprocessing and non-uniform block operation.

2.2 Block Compressed Sensing for Target Detection

Block compressed sensing (BCS) is a kind of parallel compression technology, which can fulfil the image acquisition in the form of block [9]. BCS divides the original image into uniform blocks (see Eq. (2)) and observes these sub-blocks independently (see Eq. (3)). Since the image sub-blocks have the same size, it is unnecessary to change the observation matrix.

$$X = \begin{bmatrix} g_{11} & g_{12} & \cdots & g_{1l} \\ g_{21} & \ddots & & \vdots \\ \vdots & & \ddots & g_{(r-1)l} \\ g_{r1} & \cdots & g_{r(l-1)} & g_{rl} \end{bmatrix} = \begin{bmatrix} X_1 & \cdots & X_{\frac{l}{l_b}} \\ \vdots & \ddots & \vdots \\ X_{n-\frac{l}{l_b}+1} & \cdots & X_n \end{bmatrix} \tag{2}$$

$$Y = \begin{bmatrix} Y_1 \\ \vdots \\ Y_n \end{bmatrix} = \begin{bmatrix} \Phi & & \\ & \ddots & \\ & & \Phi \end{bmatrix} \begin{bmatrix} X_1 \\ \vdots \\ X_n \end{bmatrix} \tag{3}$$

Although the whole image processing method achieves better results, BCS can bring more convenience. Large amount of information in the image makes the complexity of measurement matrix unacceptable. Therefore, the way of processing the whole image as several independent small pieces has become an effective solution.

The uniform blocking strategy of block compressed sensing makes it possible to detect and locate targets in the field of view, and the framework of the system is shown as Fig. 2. The system samples the original image by the optical equipment and uses the compressed sampling results to identify these sub-blocks. Then, all sub-blocks are classified into target blocks or background blocks. By identifying the location of the target blocks, the location of the target can also be achieved. Based on manifold learning theory, the compressed domain recognition method is employed to identify sub-blocks.

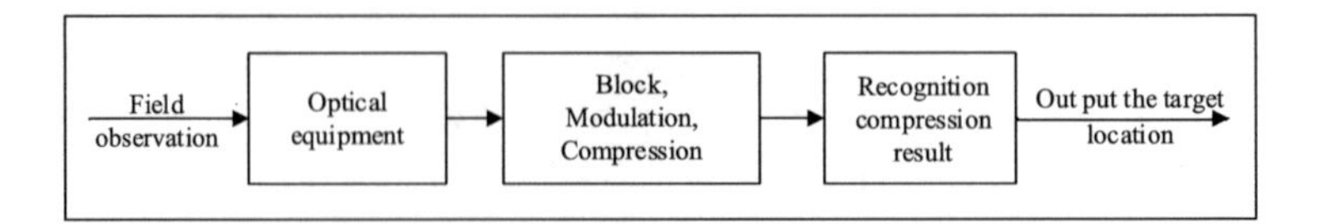

Fig. 2. The framework of target recognition based on block compressed sensing

3 Improvement of Blocking Strategy

3.1 Existing Problems

Through the theoretical analysis, the target detection strategy based on block compressed sensing is feasible in hardware-based sampling system, but still has the problem of low classification accuracy. Figures 3, 4, 5 show the results of using the method proposed in Sect. 2.2 to detect three groups of IR images sequence. When each block is identified, the classification result is distinguished by the box with different colors. The red box is correctly identified and the blue box is misidentified. It is obviously that the proposed target detection method based on block compressed sensing is inefficient on locating and classifying targets.

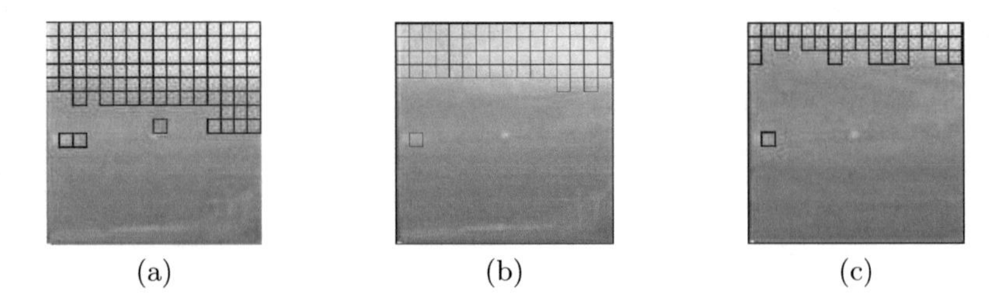

Fig. 3. Blocking recognition of image I

Because of the test sample center is not consistent with the target center, the test samples obtained using the uniform block deviate from training samples in the manifold significantly. Assuming that each image sub-block has a size of 8×8 pixels, the distance between the test sample center and the target center may reach the maximum $\sqrt{4^2 + 4^2}$ pixels. However, the distance between the training sample center and the target center is always smaller. Although the distance between the test sample center and the target center can be reduced by cutting down the size of the block, it also brings the problem of incomplete expression of the sample information.

3.2 Solution

The most straightforward way to solve this problem is to increase the number of training samples, which is difficult to be achieved in practice. Because the

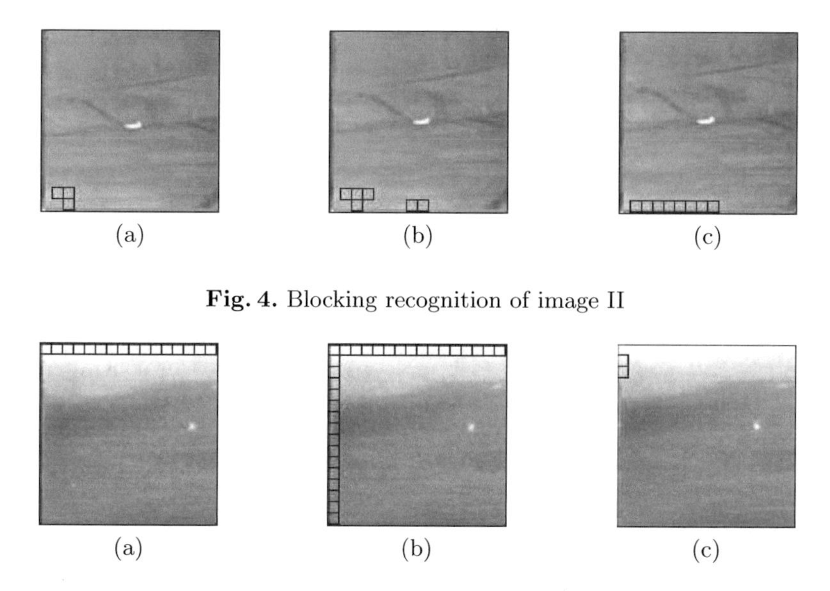

Fig. 4. Blocking recognition of image II

Fig. 5. Blocking recognition of image III

training samples is difficult to collect, and too many training samples may result in large computation complexity. Therefore, we hope to improve classification accuracy without expanding the existing training sample set.

In this paper, we use overlapping blocks to solve this problem. All blocks in the BCS do not intersect with each other. In the case that the sub-block has a size of 8×8 pixels, a 128×128 pixels image can be split into 16×16 blocks. We overlap these sub-blocks with their adjacent blocks by using a smaller step size. For instance, in the case of splitting a 128×128 pixels image by 8×8 block, we may change the block step size from 8 pixels to 4 pixels. Thus, the maximum distance between the center of the target to the sub-block center is changed from $\sqrt{4^2 + 4^2}$ to $\sqrt{2^2 + 2^2}$. The overlapping block compressed sensing is shown in Fig. 6.

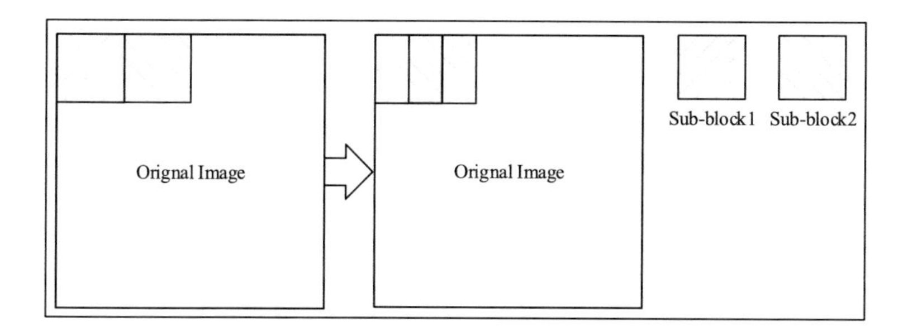

Fig. 6. Overlapping block compressed sensing

4 Performance Evaluation

Experiments on the real IR image sequence have been carried out. The experimental data includes several group of images with a resolution of 128 × 128, obtained from United States Army Aviation and Missile Command (AMCOM). All experiments were implemented by MATLAB software on a PC with 4-GB memory and 3.2-GHz Intel i5 dual processor.

According to the principle mentioned in Sect. 3.2, three groups of image sequence are selected to validate the conclusion of Sect. 3.1. As shown in Figs. 7(a), 8(a) and 9(a), the original block compression sensing method may miss target. By contrast, the overlapping block compression sensing method gets positions of the targets accurately, which can be found in Figs. 7(b), 8(b) and 9(b). Although false detections exist in both methods, missed target may lead more serious consequences in weapon systems. Therefore, the qualitative analysis shows that the classification result of the overlapped BCS is better than the original BCS.

Then, quantitative analysis is performed to show the performance of our method more intuitively. The evaluation indexes are detection rate (DR) and false alarm rate (FA). Equation (4) is the definition of detection rate. Here, AO is the number of all targets present in the image sequence, DO is the number of targets detected in the image sequence.

$$DR = DO/AO \times 100\% \tag{4}$$

The false alarm rate is defined as Eq. (5). Here, AB is the number of all blocks in the image sequence, WB is the number of background blocks detected as target blocks.

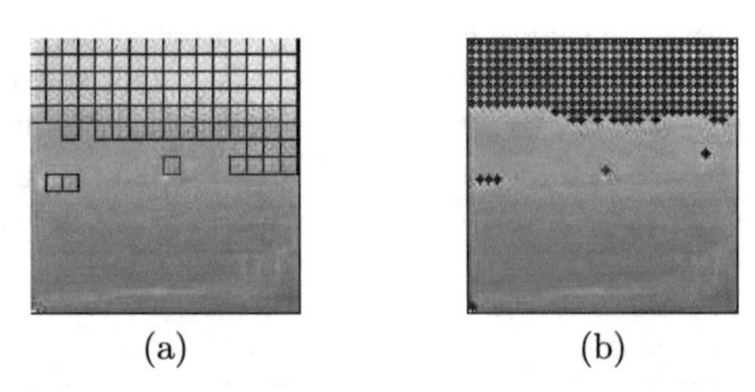

(a) (b)

Fig. 7. Overlapping block compression recognition for image I

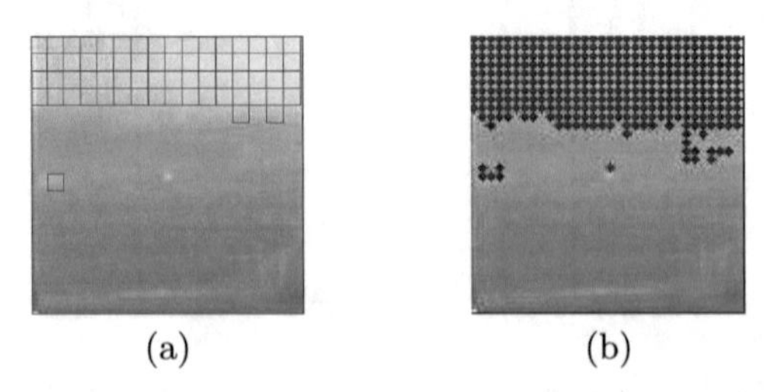

(a) (b)

Fig. 8. Overlapping block compression recognition for image II

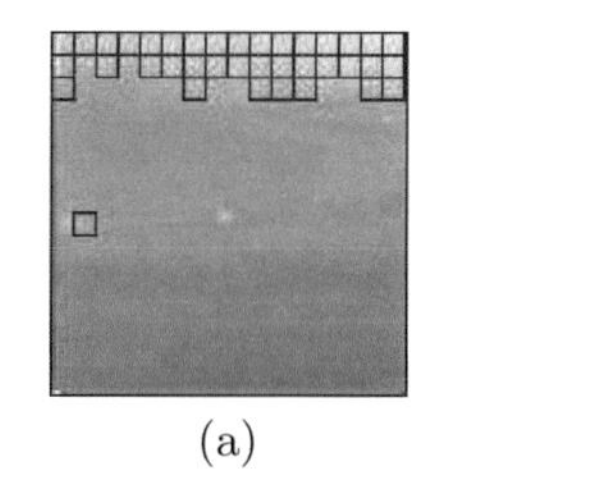

(a)

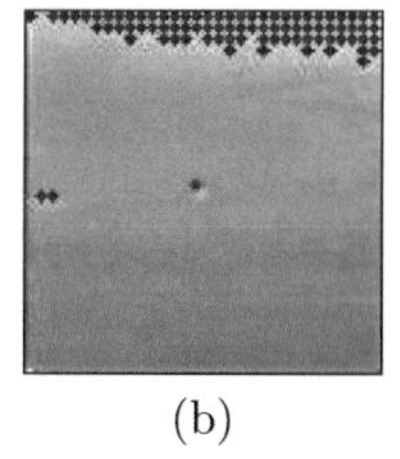

(b)

Fig. 9. Overlapping block compression recognition for image I

$$FA = WB/AB \times 100\% \tag{5}$$

To further demonstrate the advantages of the proposed method, both original and overlapping methods are evaluated with the same conditions. The block has as a size of 8×8 pixels and the sampling frequency is set to 8. The SVM classifier is employed. The penalty factor is set to 1 and the parameter of Gaussian kernel is set to 10. Results of three sequences are shown in Table 1.

Table 1. Comparison results of three sets of sequences

Image sequence	Processing method	DR	FA
L1415S	Original BCS	85.1	11.1
L1415S	Overlapping BCS	99.0	8.7
L1608S	Original BCS	100	31.5
L1608S	Overlapping BCS	100	29.9
L1720S	Original BCS	94.2	14.3
L1720S	Overlapping BCS	100	9.7

Table 1 shows that the overlapping BCS method gets a high value in the detection rate and a lower value in false alarm rate. This is due to the overlapping blocks may locate the target center easier, and the test samples of the target are more similar to the training samples in manifold.

5 Conclusion

In this paper, an infrared small target detection method based on block compressed sensing is developed. This method can detect and locate infrared small targets without reconstructing the original IR image, which ensures its real-time performance. Furthermore, it is convenient for hardware implementation which guarantee it can be employed in actual weapon systems. The overlapping mechanism makes the proposed method achieve a high detection rate and low false alarm rate without expanding the training sample set.

References

1. Cui, Z., Yang, J., Jiang, S., Li, J.: An infrared small target detection algorithm based on high-speed local contrast method. Infrared Phys. Technol. **76**, 474–481 (2016)
2. Ma, W., Wen, Y., Yu, X., Feng, Y., Zhao, Y.: Performance enhancement of uncooled infrared focal plane array by integrating metamaterial absorber. Appl. Phys. Lett. **106**, 111108 (2015)
3. Li, Q., Wang, G., Wang, Z., Zhang, S., Gao, M., Gao, F.: Missile state data compression method based on SDLST algorithm. Comput. Meas. Control **20**, 2995–2998 (2012)
4. Donoho, D.L.: Compressed sensing. IEEE Trans. Inf. Theor. **52**, 1289–1306 (2006)
5. Zhang, Y., Mei, S., Chen, Q., Chen, Z.: A novel image/video coding method based on Compressed Sensing theory. In: IEEE International Conference on Acoustics, Speech and Signal Processing, pp. 1361–1364. IEEE (2008)
6. Duarte, M.F., Davenport, M.A., Takbar, D., Laska, J.N., Sun, T., Kelly, K.F., Baraniuk, R.G.: Single-pixel imaging via compressive sampling: building simpler, smaller, and less-expensive digital cameras. IEEE Signal Process. Mag. **25**, 83–91 (2008)
7. Chan, W.L., Charan, K., Takhar, D., Kelly, K., Baraniuk, R.G., Mittleman, D.M.: A single-pixel terahertz imaging system based on compressed sensing. Appl. Phys. Lett. **93**, 121105 (2008)
8. Zhao, Y., Sui, X., Chen, Q., Wu, S.: Learning-based compressed sensing for infrared image super resolution. Infrared Phys. Technol. **76**, 139–147 (2016)
9. Dumas, J.P., Lodhi, M.A., Bajwa, W.U., Pierce, M.C.: Computational imaging with a highly parallel image-plane-coded architecture: challenges and solutions. Opt. Express **24**, 6145–6155 (2016)
10. Shepard, R.H., Cull, C.F., Raskar, R., Shi, B., Barsi, C., Zhao, H.: Optical design and characterization of an advanced computational imaging system. In: The 8th Conference of Optics and Photonics for Information Processing, vol. 9216, p. 92160A (2014)
11. Li, L., Li, T., Gao, F.: Infrared small target detection in compressive domain. Electron. Lett. **50**, 510–512 (2014)
12. Zhang, B., Li, Z., Liu, J.: A compressed sensing ensemble classifier with application to human detection. Neurocomputing **170**, 221–227 (2015)
13. Zhang, K., Zhang, L., Yang, M.H.: Fast compressive tracking. IEEE Trans. Pattern Anal. Mach. Intell. **36**, 2002–2015 (2014)

Development of Packet Codec Software Based on User Interface Protocol

Xiangyu Tian[1], ZhanQiang Ji[2], and Chang'an Wei[1](✉)

[1] Department of Automatic Test and Control, Harbin Institute of Technology, Xidazhistr. 92, Harbin 150080, China
weichangan@hit.edu.cn
[2] Shandong Institute of Aerospace Electronic Technology, No. 513 Aerospace Road, GaoXin District, Yantai 264670, Shandong, China

Abstract. At present, most of the complex electronic systems use communication protocols for data transmission. In the process of system joint test, the system usually needs to monitor the communication protocol data, that is, protocol data decoding,or replace the equipment in the system in a semi-physical way to debug the system. This requires a semi-physical device to dynamically generate protocol data frame into a complex system, that is, the encoding of the protocol data.

Keywords: Communication protocol · Protocol data decoding · Protocol data encoding

1 Introduction

At present, most of the complex electronic systems use developing special debugging or test system, through the programming way to achieve the protocol codec [1]. This method is less versatile and heavy workload. This paper designs a free programming mode of the protocol and the automatic codec of the protocol data. Through the analysis of the protocol format of the communication between the equipment of the complex electronic systems [2], the protocol data structure designed in this paper can support the description of protocols with this feature such as multi-nesting, multi-branch, designed by bit, containing dynamic elements and so on. At the same time codec software also supports the automatic codec of the above complex protocol.

2 Description of the Protocol Format

Based on the idea of protocol hierarchical management, the protocol is represented as follows: protocol item, protocol header/protocol tail, element item, element bit [3]. The format of the protocol description is shown in Fig. 1.

(1) Protocol item: The protocol item is a complete description of a protocol, including the protocol header, protocol tail, protocol element.

J.-S. Pan et al. (eds.), *Advances in Intelligent Information Hiding and Multimedia Signal Processing*, Smart Innovation, Systems and Technologies 81,
DOI 10.1007/978-3-319-63856-0_51

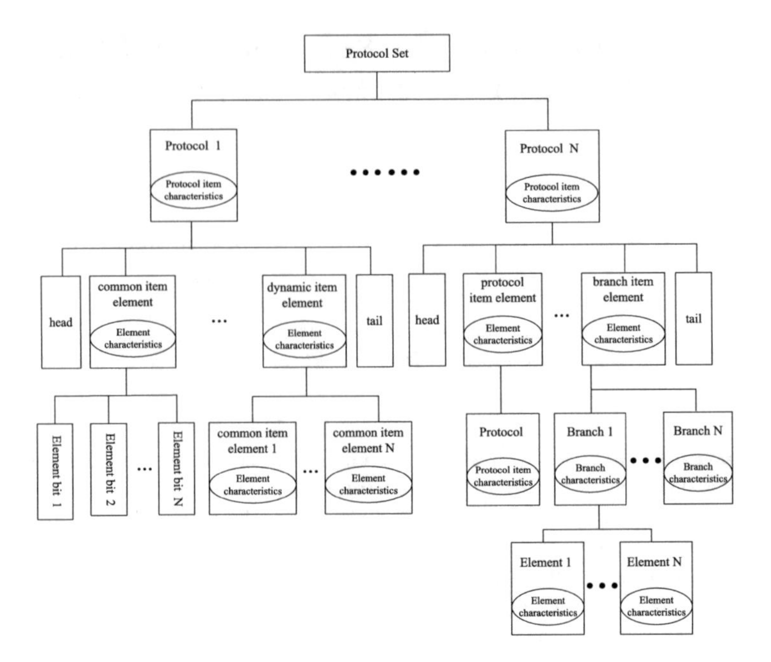

Fig. 1. Description of the protocol format

(2) Header and tail: The protocol header is used for protocol item match recognition, The protocol tail is used for protocol item validation.
(3) Element: The content specified by the protocol element is the physical data unit during the actual communication transmission. A protocol element can also contain a sub-protocol item, that is, a nested protocol item in an element, that element is called "protocol item element". The protocol element can also have a number of different branch paths depending on the element values, There can be multi-branch paths in a protocol element, Each branch can contain multiple element information. The element that contains branches is called "branch item element". There may also be repeated elements in the protocol, called dynamic elements. The element containing the dynamic element is called "dynamic item element". If the protocol element does not nest a subprotocol item and does not have branches or dynamic elements, it is called the "common item element".
(4) Element bit: It represents the meaning of some specific bit combinations in the element.
(5) Branch: It represents a branch of a "branch item element" that contains the branch item number, the jump value, and the elements in the branch.

3 Design and Development of Protocol Codec Software

General protocol codec software mainly provides users with protocol encoding and protocol decoding [4]. Protocol encoding means that this software provides users with a visual protocol data editing interface, the software also supports object model to protocol data conversion, The software can package user-edited protocol data or object model converted protocol data into protocol packets, Protocol packets can be sent through the network interface, can also be stored, loaded. Protocol decoding means that this software can receive protocol packets and parse it according to the protocol template at the same time. The software can display the parsed protocol format data and supports two types of protocol data distribution: data flow or object model.

General protocol codec software support the new radar 97, the original 97, the target track, measuring radar intelligence information frame, AIS system information transmission specification, XX control center internal information interface specification, XX model missile car communication specification and other communication protocol codec.

General protocol codec software functional requirements:

(1) The software has an Ethernet interface that supports data transfer;
(2) The software supports codecs in the following protocol formats:
  (a) The codec of the communication protocol designed by bit;
  (b) The codec of multi-branch path elements;
  (c) The codec of nested protocol;
  (d) The codec of repeated elements;
  (e) The codec of string type data elements;
  (f) The codec of array type data elements;
(3) The software supports protocol packet checksum calculations;
(4) The software supports the conversion of object model attribute values and protocol data values;
(5) The software supports two types of protocol data distribution: data flow or object model.

4 Use Case Analysis

General protocol codec software is provided in HIT-TIDE component, When the TIDE platform is in edit mode, the user needs to configure this software's network transmission parameters, At the same time the software can load the protocol description file into memory and display the loaded protocol information in a tree-like manner to provide users with a visual protocol data editing interface [5]. The software also provides the interaction between the object model and the protocol data. The user needs to configure the association between the object model and the protocol data and edit the type of protocol data distribution when editing the test scheme.

Participants in the software include the user and the HIT-TIDE platform. General protocol codec software mainly provides users with functions such as configuring network transmission parameters, viewing protocol templates, editing protocol data, configuring protocol and object model association, parsing protocol, observing running data, sending protocol data and so on. The software provides the interface with the HIT-TIDE platform, supports drawing components, saves the component parameters, stores the order information, and controls the running status of the components and other functions.

Figure 2 shows use case diagram of the software.

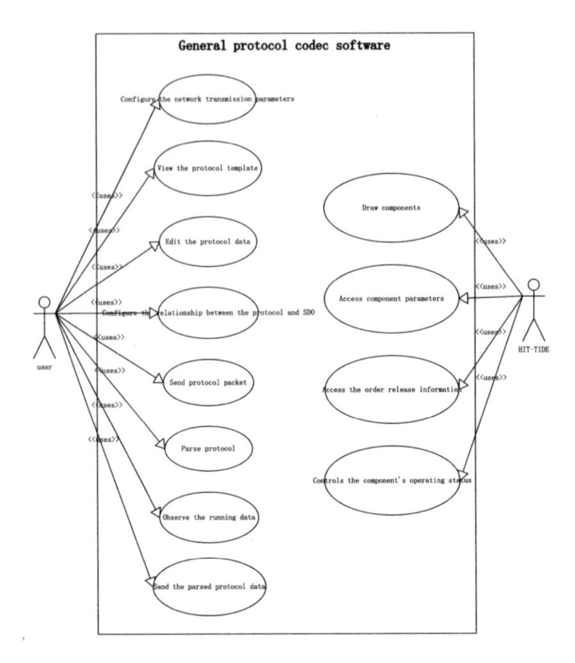

Fig. 2. Use case diagram of codec software

4.1 Static Model

(1) Component base class: Components that conform to the HIT-TIDE component interface specification are required to be derived from this class for use by the HIT-TIDE platform.
(2) Protocol codec class: This class derives from the component base class, providing protocol encoding and protocol decoding.
(3) Middleware class: This class represents middleware that provides data interaction services.
(4) Hardware communication processor class: This class provides data transfer function.
(5) Display interface class: This class provides display of protocol information and object model information.
(6) Protocol class: This class represents the protocol structure, including protocol item class, protocol header and tail class, protocol element class, protocol element bit class, branch class.

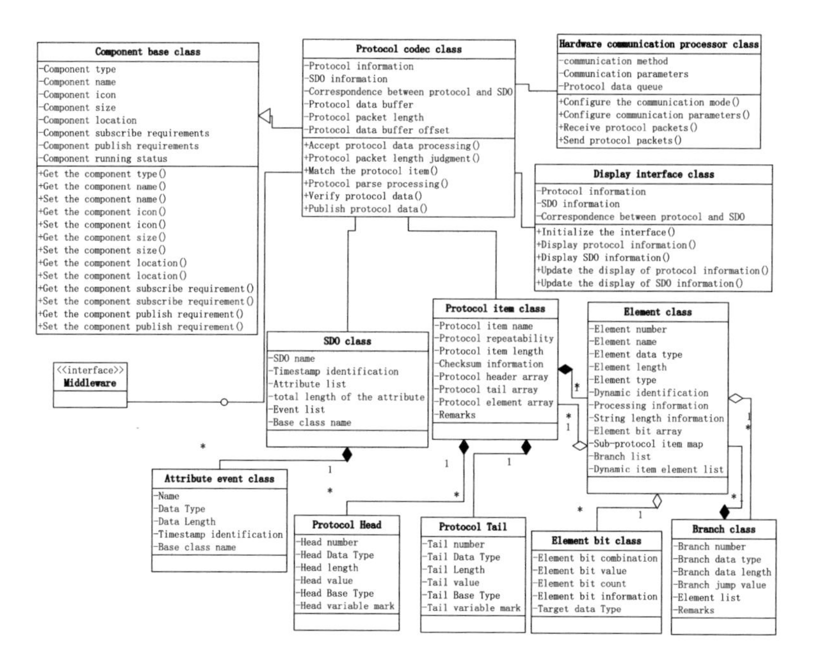

Fig. 3. Class diagram of codec software

(7) Object Model Class: This class represents the object model structure in the HIT-TENA architecture.

(8) Attribute/Event Class: This class represents the base element in the object model in the HIT-TENA architecture (Fig. 3).

4.2 Dynamic Model

When the platform is in edit mode, the sequence diagram of the user editing the component parameter process is shown in Fig. 4. When the platform is in running mode, the software receives protocol packets and parse it shown in Fig. 5.

During the protocol analysis process, the software will automatically handle the following:

(1) For bitwise design of the communication protocol, the software can parse the data by bit, and convert the data to the target data type according to the user's settings.

(2) For the branch item element, it can analyze the branch data value according to the received data, jump to the corresponding branch path according to the protocol, and analyze the subsequent protocol data;

(3) For the sub-protocol items in the protocol, the data is parsed in units of sub-protocol items, the software can automatically resolve multiple sub-protocol items that are repeated;

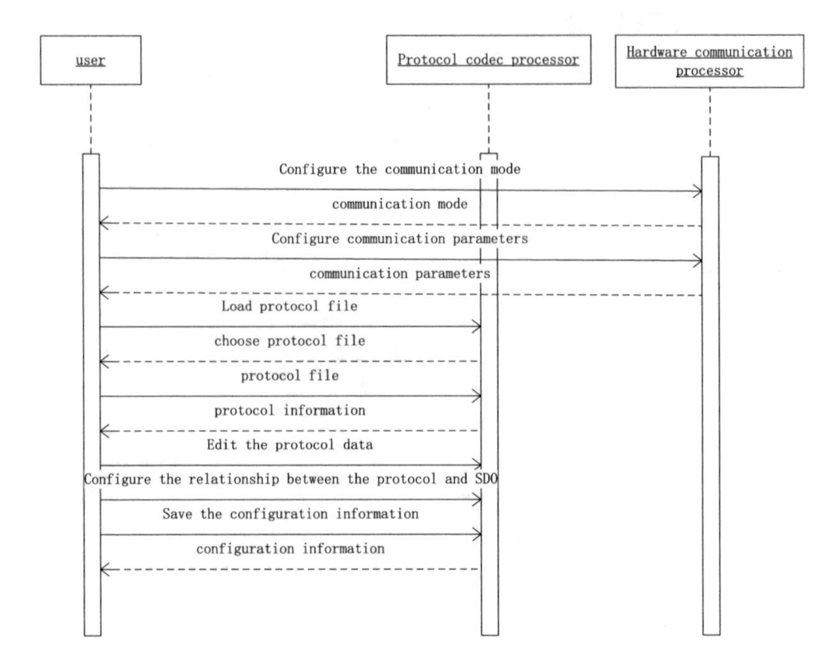

Fig. 4. Sequence diagram of configuring component parameters

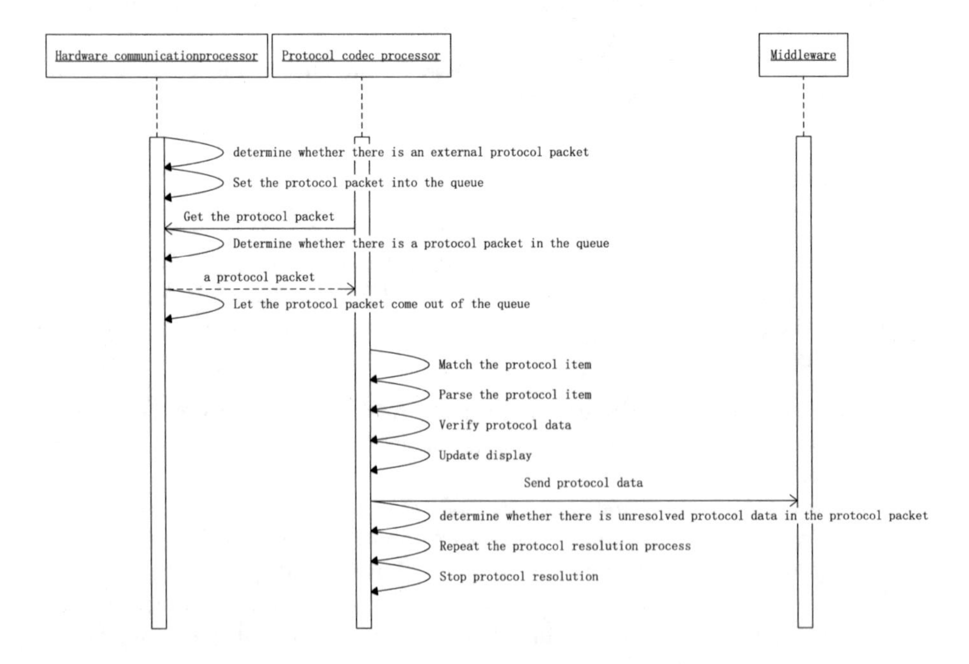

Fig. 5. Sequence diagram of protocol resolution process

(4) For the dynamic item element, the length of the element can be calculated according to the meaning of the element, and then all the dynamic elements are parsed according to the length information;
(5) For the string type data element with the string length information, the character string can be parsed according to the length information; For string type data elements without string length information, the string terminator '0' can be automatically recognized and the character string can be parsed according to it;
(6) For the array type element, the elements in the array can be resolved in turn based on the element data type of the array;
(7) The elements that need to be processed can be converted according to the processing function specified by the protocol to obtain the target data elements;
(8) For the protocol items that need to be verified, the checksum calculation can be carried out according to the calibration method set by the user.

5 System Testing and Verification

Distributed debugging verification system shown in Fig. 6. In the sending node protocol codec software constructs protocol data and sent protocol packets to the external device or HIT-TIDE other nodes through the network interface. In the receiving node protocol codec software through the network interface to receive

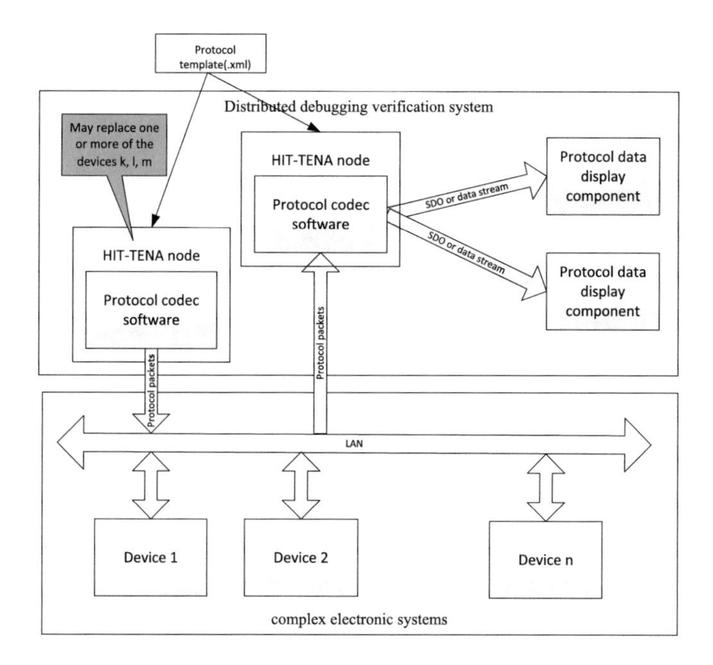

Fig. 6. Structure diagram of verification system

protocol packets and parse them. When the packet is parsed, the parsed protocol data is sent to other nodes of HIT-TIDE through SDO or data stream, the other nodes use the protocol data display software to display the protocol data.

6 Conclusion

Through the research of this paper, according to the communication protocol of the system under test and the HIT-TIDE platform, the user can quickly build the debugging and testing system, thereby greatly improving the efficiency of debugging tests.

References

1. Rajagopal, M., Miller, R.E.: Synthesizing a protocol converter from executable protocol traces. IEEE Trans. Comput. **40**, 487–499 (1991)
2. Kruglinski, D.J.: Programming Visual C++, pp. 1–1115. Microsoft Press (1999)
3. Proakis, J.: Object landscapes and Lifetimes, vol. 1, pp. 1–10. Electronic Industry Press (2003)
4. Blaha, M., Rumbaugh, J.: Object-Oritented Modeling and Desgin with UML, 2nd edn. Prentice Hall, Upper Saddle River (2005)
5. Poch, K.: The Test and Training Enabling Architecture (TENA) Interoperability at DoD Ranges [EB/OL] (2003). http://www.tena-sda.org

Hyperspectral Image Segmentation Method Based on Kernel Method

Lianlei Lin(✉) and Jingwen Du

Department of Automatic Test and Control, Harbin Institute of Technology, Xidazhistr. 92, Harbin 150080, China
linlianlei@hit.edu.cn

Abstract. Hyperspectral image Contains rich information, so improving the classification ability of hyperspectral image has become a hot spot of research recently. Image segmentation is one of the most basic and important problems of the image processing and low level computer vision and the precondition of the image processing, some hyperspectral image classification method based on image segmentation. In this study, an image segmentation method based on kernel methods was proposed. First reduce the dimension of hyperspectral images by KPCA, then the k-means method are used to cluster, finally finish the image segmentation in the high-dimensional space by gaussian kernel mapping. In simulation experiment, changing the parameters of the experiment and comparing with standard hyperspectral image segmentation, the results show that the method is good enough in image segmentation, which can be used for visual analysis and pattern recognition, and realize hyperspectral image segmentation.

Keywords: Image segmentation · Hyperspectral image · Kernel · Clustering method · Graph cut

1 Introduction

Hyperspectral images combine imaging technology and spectroscopy technology, the improvement of spectral resolution makes it is possible of the detection of complex ground material. So the hyperspectral images are widely used in civil and military fields in recent years. High dimension and low spatial resolution of hyperspectral image bring some difficulties in data processing. So it's necessary to Process the data before the hyperspectral data is applied to every field. Classification is an important link in processing, the result of which can not only be directly applied to the actual demand, but also provide useful data for the subsequent application information and technical support as the premise of a variety of applications.

In the study of hyperspectral image segmentation method, Bilgin proposed a similarity segmentation based on subtraction clustering [1] and use one class support vector machines (OC-SVM) as a new kind of clustering validation method. Erturk proposed an unsupervised hyperspectral image segmentation method based on the calibration of phase correlation [2]. In the literature [3], the author used a fuzzy C - average clustering

J.-S. Pan et al. (eds.), *Advances in Intelligent Information Hiding and Multimedia Signal Processing*, Smart Innovation, Systems and Technologies 81,
DOI 10.1007/978-3-319-63856-0_52

(FCM) algorithm to segment hyperspectral images. Literature [4] used a neural fuzzy clustering method based on weighted increment neural network to realize the unsupervised segmentation of hyperspectral images. Some scholars proposed a hyperspectral image automatic clustering method via histogram threshold [5] and a threshold segmentation method based on histogram principal component [6]. Besides hyperspectral image segmentation method based on the model has got widely attention and research, Mercier and Derrode put forward hyperspectral image segmentation using multivariate markov chain model [7]. Acito and Corsini proposed hyperspectral image segmentation methods based on gaussian mixture model [8]. Farrell put forward hyperspectral image segmentation method using the gaussian mixture model [9]. Literature [10] put forward a hybrid model based on independent component analysis (ICAMM) algorithm. Hyperspectral image segmentation based on active contour of space - spectral constraint region [11], make full use of the data space information and spectral information. In the literature [12] watershed segmentation technology is used in the classification of hyperspectral image, the segmentation tags diagram is used to further improve the classification performance. Hierarchical segmentation method provides a flexible way of image interpretation through the tree structure, which was used in the literature [13, 14]. Literature [15] is a minimum spanning tree (MSF) bases method. However the application of these methods have limitations because of high dimension and nonlinear characteristics of hyperspectral images. The purpose of this study is to investigate kernel mapping to bring the unsupervised graph cut formulation to bear on multiregion segmentation of images more general than others. The image data is mapped implicitly via a kernel function into data of a higher dimension so that the piecewise constant model, and the unsupervised graph cut formulation thereof, becomes applicable.

2 Hyperspectral Image Segmentation Using Kernel Method

Hyperspectral image segmentation method based on kernel method include four parts: pretreatment, clustering, kernel mapping, graph cut. The framework of the algorithm is shown in Fig. 1:

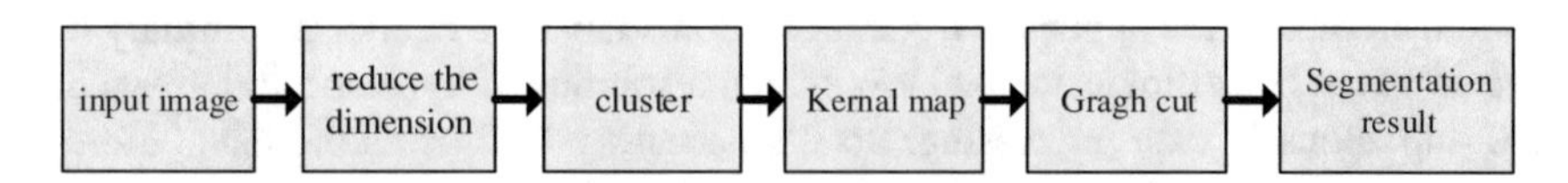

Fig. 1. Schematic diagram of hyperspectral image segmentation based on kernel method

The Specific process is as follows: Preprocess the hyperspectral image data, find the clustering centers. Map data and clustering centers to high-dimensional space, then apply mapped data to the graph cut function, image segmentation results are obtained.

2.1 Dimension Reduction

Complexity is connected with the band number and quantitative accuracy in hyperspectral data analysis. The more image band and the higher quantitative accuracy, the greater the complexity of the data. If the dimension of the data is too high or quantitative accuracy is too high will lead to the decrease of the classification accuracy. It is the famous Haghes phenomenon, and dimension reduction is one of the important method is to eliminate Haghes phenomenon. There is little data in high-dimensional space of hyperspectral image, data is usually concentrated in low dimensional space. So it's possible to reduce the dimension without loss of meaningful information and separability. The existing dimension reduction methods can be divided into two categories: Feature extraction method and wavelength selection method. The former mainly transform the original data to another space through certain transform. In another space, most of the original data information is concentrated in the low dimension, reduction can be realized by using the low dimensional data replace the original dimension data. The latter mainly choose the representative band subsets from the original band and use these subsets instead of the original image so that dimension reduction can be realized. The feature extraction method is adopted here in KPCA method for dimension reduction processing.

2.2 Clustering

Clustering is a process to divide physical or abstract objects into multiple classes, and each class are composed of similar objects. A set of data objects generate a clustering of the generated clusters. The objects in the same cluster are similar to each other, and objects in different clusters are different. We use k-means clustering method for the data processing.

The basic idea of k-means clustering method is choosing K points from the data sample as the initial value clustering center and successive move the clustering center in the iteration until the clustering center of the adjacent two iterations do not change and clustering criterion function is the optimal.

The basic steps are as follows:

(1) Randomly assign k cluster center $(m_1, m_2, \ldots, m_k)$;
(2) Find the nearest clustering center for each data and assign it to the class;
(3) Recalculate the clustering center:

$$m_1 = m_i - \frac{1}{N}\sum\nolimits_{i=1}^{N_i} X_{ij}, i = 1, 2, 3 \tag{1}$$

(4) Calculation the deviation

$$J = \sum\nolimits_{j=1}^{k} \sum\nolimits_{j=1}^{n_i} \left\| X_{ij} - m_i \right\|^2 \tag{2}$$

(5) If J is convergent, return $(m_1, m_2, \ldots, m_k)$ as the final clustering center and the algorithm stop; otherwise back to step 2.

2.3 Kernel Mapping

In low dimensional space, it is difficult to classify a set of linear inseparable vector. The image data is mapped implicitly via a kernel function into data of a higher dimension so that the piecewise constant model becomes applicable. The mapping is implicit because the dot product, the Euclidean norm thereof, in the higher dimensional space of the transformed data can be expressed via the kernel function without explicit evaluation of the transform.

$$K(y,z) = \phi(y)^T \cdot \phi(z) \tag{3}$$

where "·" is the dot product in the feature space.

Gaussian radial basis function is used as kernel function, the form is

$$k(\|x - x_c\|) = \mathrm{e}^{-\frac{\|x - x_c\|^2}{(2\sigma)^2}} \tag{4}$$

Where x_c is Kernel function center, σ is the width parameters of the function, which control the function of the radial range. Substitution of the kernel functions gives

$$J_k(I_p, \mu) = \left\|\phi(I_p) - \phi(\mu)\right\|^2 = \mathrm{K}(I_p, I_p) + \mathrm{K}(\mu, \mu) - 2\mathrm{K}(I_p, \mu) \tag{5}$$

Which is a nonEuclidean distance measure in the original data space corresponding to the squared norm in the feature space.

2.4 Graph Cut

The common image is divided into two sides and two points by graph cut algorithm: the first kind of common vertices corresponding to each pixel in the image, there are two other terminal vertex. Every two ordinary vertex connection is a kind of edge, each connection between ordinary vertex and terminal is the second edge. Cut refers to a collection of the edges, breaking of all the edges lead to "S" and "T" figure separated,

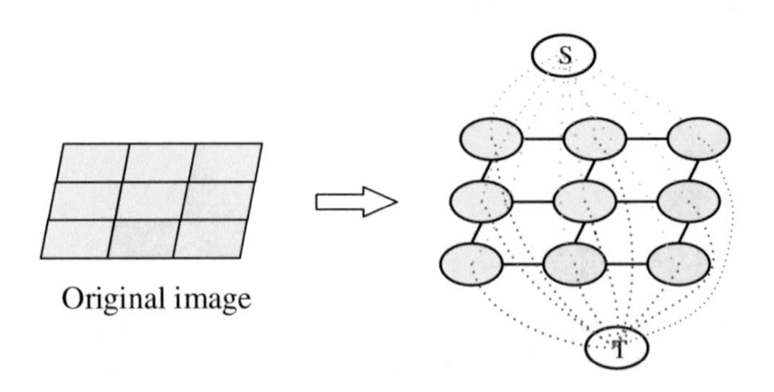

Fig. 2. Graph cut

so it is called "cut". If the edge sum of a cut is minimum, then this cut is called a minimum cut, also graph cut is finished (Fig. 2).

To write down the segmentation functional, let be an indexing function. Assigns each point of the image to a region where is the finite set of region indices whose cardinality is less or equal. The segmentation functional can then be written as

$$F(\lambda) = D(\lambda) + \alpha R(\lambda) \tag{6}$$

where $D(\lambda)$ is the data term and $R(\lambda)$ is the prior. α is a positive factor.

The MAP formulation using a given parametric model defines the data term as follows:

$$D(\lambda) = \sum\nolimits_{p \in \Omega} D_P(\lambda(p)) = \sum \sum \left(\mu_l - I_p\right)^2 \tag{7}$$

The prior is expressed as follows:

$$R(\lambda) = \sum r(\lambda(p), \lambda(q)) \tag{8}$$

$$r(\lambda(p), \lambda(q)) = \min\left(const^2, \left|\mu_{\lambda(p)} - \mu_{\lambda(q)}\right|^2\right) \tag{9}$$

$r(\lambda(p), \lambda(q))$ is a smoothness regularization function given by the truncated squared absolute difference, where const is a constant, μ_l is a region parameter. Now, the simplifications in lead to the following kernel-induced segmentation functional

$$\mathrm{F}(\{\mu_l\}, \lambda) = \sum \sum J_k\left(I_p, \mu_l\right) + \alpha \sum r(\lambda(p), \lambda(q)) \tag{10}$$

3 Experiments Results and Analysis

Experimental data was gathered by AVIRIS sensor over the Indian Pines test site in North-western Indiana and consists of 145 times 145 pixels and 224 spectral reflectance

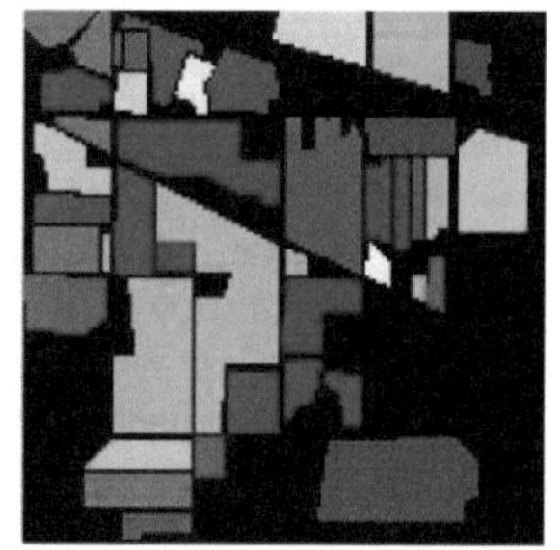

a) pseudo color image b) standard segmentation results

Fig. 3. Hyperspectral image

bands in the wavelength range 0.4–2.5 10^{-6} m. The initial dimension is 220, reduce to the dimension of 20. Set the kernel function parameter σ to 0.3, 0.4, 0.5, 0.6 respectively. Set the cluster number k to 12, the smoothing coefficient of 0.1. Figure 3 is the pseudo color image of experimental hyperspectral image and standard segmentation results. Figure 4 are the experiment results.

It can be seen from the pictures that we can separate hyperspectral image through the algorithm presented in this paper. When $\sigma = 0.5$ or 0.6, the segmentation results is close to standard segmentation results.

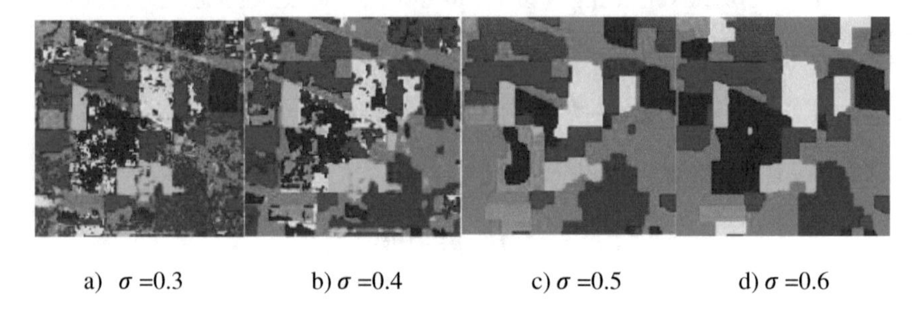

a) σ =0.3 b) σ =0.4 c) σ =0.5 d) σ =0.6

Fig. 4. Segmentation results of different parameters

4 Conclusion

This study implements the hyperspectral image segmentation based on kernel method, the algorithm uses KPCA dimension reduction method to reduce the complexity of the data at the same time keep the adequacy and effectiveness of the data. A nonlinear problem can be solved by a linear classifier using the method of kernel function. We can get a good segmentation result by adjusting the parameters of the kernel.

References

1. Bilgin, G., Erturk, S., Yildirim, T.: Segmentation of hyperspectral images via subtractive clustering and cluster validation using one-class support vector machines. IEEE Trans. Geosci. Remote Sens. **49**(8), 2936–2944 (2011)
2. Erturk, A., Erturk, S.: Unsupervised segmentation of hyperspectral images using modified phase correlation. IEEE Geosci. Remote Sens. **3**(4), 527–531 (2006)
3. Alajlan, N., Bazi, Y., Melgani, F., Yager, R.R.: Fusion of supervised and unsupervised learning for improved classification of hyperspectral images. Inf. Sci. **217**, 39–55 (2012)
4. Muhammed, H.H.: Unsupervised hyperspectral image segmentation using a new class of neuro–fuzzy systems based on weighted incremental neural networks. In: IEEE 31st Applied Image Pattern Recognition Workshop, Washington, DC, pp. 171–177, October 2002
5. Silverman, J., Caefer, C.E., Mooney, J.M., Weeks, M.M.: An automated clustering/ segmentation of hyperspectral images based on histogram thresholding. Proc. SPIE **4480**, 65–75 (2002)
6. Silverman, J., Rotman, S.R., Caefer, C.E.: Segmentation of hyperspectral images based on histograms of principal components. Proc. SPIE **4816**, 270–277 (2002)

7. Mercier, G., Derrode, S., Lennon, M.: Hyperspectral image segmentation with Markov chain model. In: IEEE Geoscience Remote Sensing Symposium, Toulouse, France, vol. 6, pp. 3766–3768, July 2003
8. Acito, N., Corsini, G., Diani, M.: An unsupervised algorithm for hyperspectral image segmentation based on the Gaussian mixture model. In: IEEE IGARSS, Toulouse, France, vol. 6, pp. 3745–3747, July 2003
9. Farrell, M.D., Mersereau, R.: Robust automatic clustering of hyperspectral imagery using non-Gaussian mixtures. Proc. SPIE **5573**, 161–172 (2004)
10. Shah, C.A., Watanachaturaporn, P., Arora, M.K., Varshney, P.K.: Some recent results on hyperspectral image classification. In: IEEE Advances Techniques Analysis Remotely Sensed Data, Greenbelt, MD, vol. 19, pp. 346–353 (2003)
11. Zhang, J., Chen, J., Zhang, Y., Zou, B.: Hyperspectral image segmentation method based on spatial-spectral constrained region active contour. In: IEEE Geoscience and Remote Sensing Symposium (IGARSS), pp. 2214–2217, July 2010
12. Tarabalka, Y., Chanussot, J., Benediktsson, J.A.: Segmentation and classification of hyperspectral images using watershed transformation. Pattern Recogn. **43**(7), 2367–2379 (2010)
13. Tarabalka, Y., Tilton, J.C., Benediktsson, J.A., Chanussot, J.: A marker-based approach for the automated selection of a single segmentation from a hierarchical set of image segmentations. IEEE Geosci. Remote Sens. **5**(1), 262–272 (2012)
14. Tarabalka, Y., Chanussot, J., Benediktsson, J.A.: Segmentation and classification of hyperspectral images using minimum spanning forest grown from automatically selected markers. IEEE Syst. Man Cybern. **40**(5), 1267–1279 (2010)
15. Bernard, K., Tarabalka, Y., Angulo, J., Chanussot, J., Benediktsson, J.A.: Spectral–spatial classification of hyperspectral data based on a stochastic minimum spanning forest approach. IEEE Image Process. **21**(4), 2008–2021 (2012)

An Encryption Algorithm for ROI Images

Chao Sun(✉), Li Li, and Yuqi Liu

Department of Automatic Test and Control, Harbin Institute of Technology, Xidazhi Street 92, Harbin 150080, China
hitsc@163.com

Abstract. According to JPEG2000 image encryption algorithm based on confusion of the wavelet coefficients cannot support region-of-interest(ROI) image coding, an encryption method for ROI images is proposed. The proposed method Generates the confusion table using Chaotic sequences, confuses the Wavelet coefficients inside each code blocks, and encrypts the sign bits using Chaotic sequences. The experimental results show that the proposed method has lower affection to the compression ratio, and has the ability to provide adequate key space to resist the brute attack. The proposed method is important to support ROI image coding.

Keywords: ROI · Image encryption · JPEG2000 · Chaotic sequences

1 Introduction

Digital image encryption technology is the important means for solving the question of digital image information security. JPEG2000 as a static image compression standard is widely and successfully applied in the field of digital images, and the research about JPEG2000 image encryption technology has been a research focus. Region-of-interest (ROI) image coding is one of the most important characteristics of JPEG2000, which has better the ROI image quality of the reconstructed image in high compression ratio. ROI coding is an important means for effectively solving the contradiction between image quality and compression ratio, and has the important value in the fields of telemedicine, image retrieval and wireless communication [1,2]. During confusing, the wavelet coefficients value is not changed and the wavelet coefficients distribution is changed only, so the impact of energy distribution and correlation is less, the impact of the compression efficiency is reduced to minimize. Currently, the research about JPEG2000 image encryption techniques is mainly based on confusion of wavelet coefficient [3–5], which is divided into two [6]: Confusion of wavelet coefficients on the whole image (CWW) and Confusion of wavelet coefficients on the sub-bands in frequency domain (CWW). For CWW and CWF are not able to support the ROI image encryption, an encryption method for ROI images is proposed in this paper, the proposed method has lower affection to the compression ratio, and has the ability to provide adequate key space to resist the brute attack. The proposed method is important to support ROI image coding.

J.-S. Pan et al. (eds.), *Advances in Intelligent Information Hiding and Multimedia Signal Processing*, Smart Innovation, Systems and Technologies 81,
DOI 10.1007/978-3-319-63856-0_53

2 CWW and CWF

CWW Confuses the wavelet coefficients in the whole image, as shown in Fig. 1. CWW is characterized by safety, but many zero and the same value of the wavelet coefficients in the high frequency sub-band, the actual image key space will be much less. For the characteristic that the energy is focused on low frequency coefficients during Wavelet transform, JPEG2000 encoder assigns more quantization bits to the low frequency coefficients, and allocates fewer quantization bits to the high frequency coefficients in the subsequent quantization process, thus better image quality is retained in higher compression ratio. While CWW confuses the wavelet coefficients in the whole image, therefore the migration between high and low frequency coefficients exists. The high frequency coefficients migration to the low frequency band will seriously affect the coding efficiency. The low frequency coefficient migration to high frequency band also inevitably causes quantization error and seriously affects the quality of the decoded image, even causes bit overflow. In short, CWW affects compression capacity greatly, and cannot be used to encrypt of lossless compressed image. More important, CWW does not support ROI image coding.

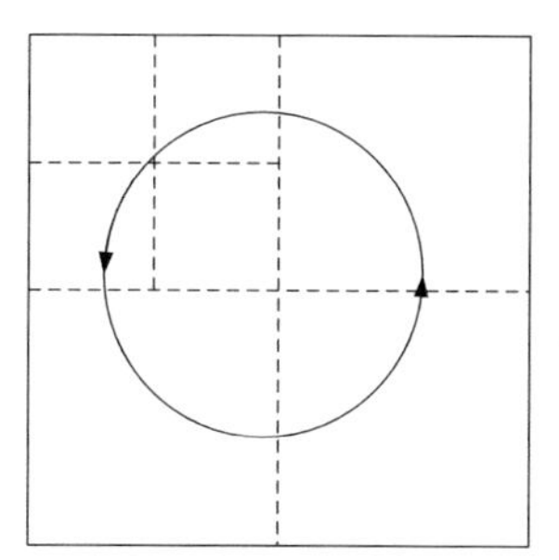

Fig. 1. CWW schematic

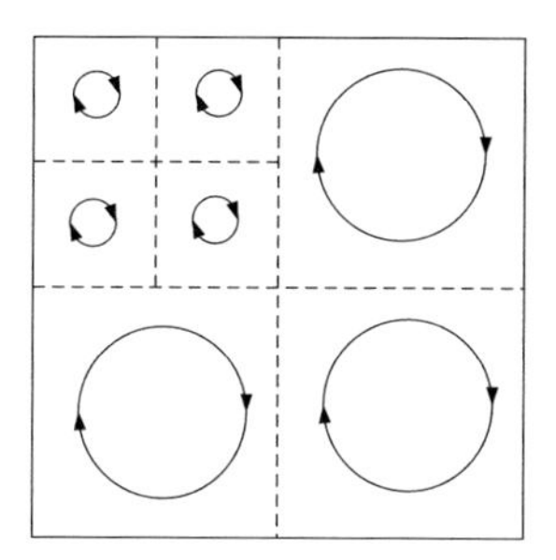

Fig. 2. CWF schematic

CWF Confuses the wavelet coefficients in each sub band, as shown in Fig. 2. The advantage of this method is to overcome the problem of CWW, ensures that all wavelet coefficients are Confused in its own band, so the migration between high and low frequency coefficients does not occur. Theoretically CWF may be used in the lossless compression of image encryption at a certain compression ratio range. Although CWF without the migration between high and low frequency coefficients, but due to Confuse the wavelet coefficients of sub-band in the algorithm, the calculation and the key overhead is very large, So CWW can not satisfy the requirements of real-time coding and transmission. CWF algorithm also does not support ROI image coding.

3 Principle of the Method

Because CWW and CWF algorithm do not support ROI image coding. In order to overcome the defects of the two algorithms, this paper proposes a ROI image encryption method, called CWB. The method Confuses the wavelet coefficients in the same block after the wavelet coefficients quantification, and meanwhile encrypts the sign bits of the wavelet coefficients. The confusion table and encryption key are both generated by chaotic system randomly. The method diagram is shown in Fig. 3.

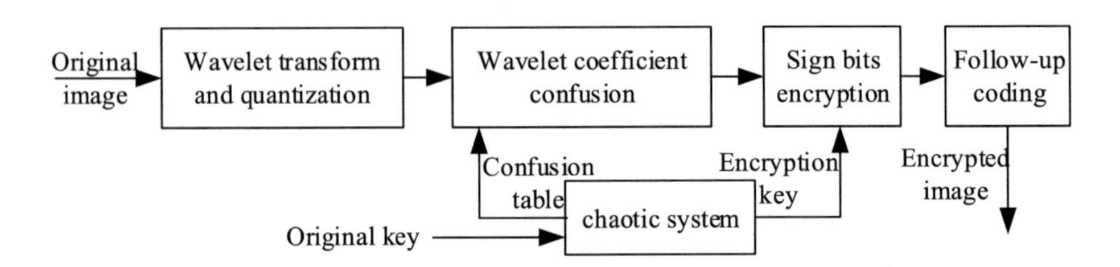

Fig. 3. The schematic of CWB

3.1 Chaotic System

Chaos is a kind of the complex dynamical behavior having special properties, it has the extreme sensitivity to initial conditions and system parameters, track irregularities, internal randomness, boundedness and ergodicity etc. [7]. A one-dimensional discrete time nonlinear dynamical system is defined as:

$$x_{k+1} = \tau(x_k) \tag{1}$$

where τ maps current state x_k to next state x_{k+1}, and the $x_k \in V(k = 0, 1, 2, ...)$ is called the state.

From a initial vale x_0, a sequence x_k can be get by repeating $\tau, k = 0, 1, 2, ...$ This sequence is called a trajectory of the discrete time dynamic system. A very simple but are used extensively in the study of dynamical systems is logistic mapping, which is defined as:

$$x_{k+1} = \mu x_k (1 - x_k) \tag{2}$$

where $x_k \in (0,1)$ and $\mu \in [0,4]$ is the bifurcation parameter. The research of chaotic dynamical system points out that logistic mapping works in chaotic state, if $3.5699456... < \mu \leq 4$. The sequence $\{x_k\}$ generating by the initial vale in the logistic mapping, $k = 0, 1, ...,$ is not periodic, convergence and very sensitive to the initial value. If $\mu = 4$, the probability distribution function of the mapping is defined as:

$$\rho(x) = \begin{cases} \dfrac{1}{\pi\sqrt{x(1-x)}}, x \in (0,1) \\ 0, \ else \end{cases} \tag{3}$$

Through the chaotic dynamic system is deterministic, but its ergodic statistical properties equivalents to white noise, which has the characteristics of simple form and sensitivity to initial conditions.

3.2 Generating Confusion Table Based on Chaotic Sequences

For the method confuses the wavelet coefficients in the same block, so the length of confusion table should be same to coding block's. In the 16×16 code block size for example. The confusion table $\{y_i\}$ is a one-dimensional array with 256 length, $i = 0, 1, ..., 255$. Using logistic mapping, data calculation method in confusion table is defined as:

$$y_i = round(x_k \times 256) \tag{4}$$

where the $round()$ means getting the integer.

The confusion table generating method is as follows:

(1) Calculating according formula 4, from in order to avoid the initial value effect;
(2) Calculating according formula 4, judging whether is equal to the before data in the table. If equal, giving up the data, k++, repeating the step; else i++, keeping calculation until generating the whole confusion table;

Improving formula 4:

$$y_i = round(x_k \times 512) mod\ 256 \tag{5}$$

After improving, equivalent of getting the $2 \sim 9$ bit after decimal point of x_i as y_i, which improves the method efficiency and has no influence to the confusion table compared with formula 4.

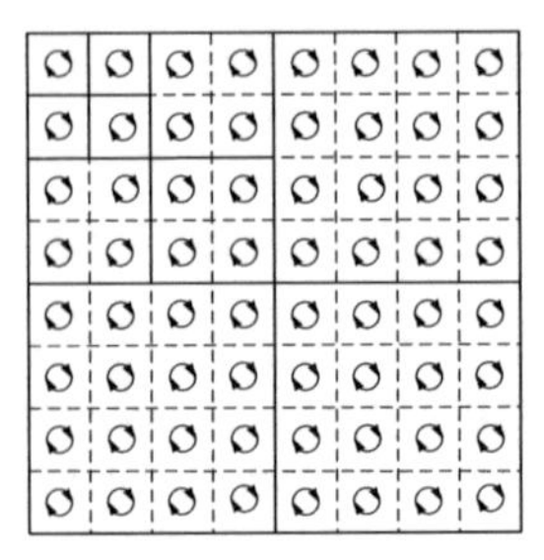

Fig. 4. The scrambling method inside code

3.3 Confusing the Wavelet Coefficients in the Code Block

Different to other confusion algorithm, CWB confuses the wavelet coefficients in each coding block, according to the confusion table generated above. Thereby achieving the purpose of image encryption, as shown in Fig. 4.

The advantages of CWB:

(1) It does not destroy the energy distribution of wavelet coefficients, and will not affect the coefficients in relation to sub-band energy, and also has no influence to the encoded weights;
(2) It can support ROI coding algorithm. JPEG2000 provides two ROI coding algorithm, a maximum bit plane method and the implicit ROI method. For the maximum bit plane method, it does not affect the ROI coding process because confusing the wavelet coefficients after the bit plane movement; for the implicit ROI algorithm, it does not affect weight division of the ROI blocks and the background blocks;
(3) The value of the wavelet coefficients does not change during confusing. The process can be embedded in JPEG2000 encoding without any effect of compression;
(4) Confusion table length is same to the code block's. The method has higher efficiency and lower key overhead.

3.4 The Sign Bits Encryption

Although having many advantages of the confusion in each coding block, but the number of the wavelet coefficients in each code block is relatively fewer, so the method are less secure relative to CWW and CWF.

For improving the secure, the method uses the sign bit encryption further on the basis of confusion, drawing lessons from the method proposed in [8]. The encryption key is also produced by the chaotic system. Because the sign bit of the wavelet coefficients have great influence to visual effects of the image, lower calculation and the key overhead can achieve to change the visual effects largely, using of the symbol encryption methods. At the same time, the sign bit of the wavelet coefficients is distributed by 0, 1 Equal probability distribution, which has high entropy low predictability characteristic, so the encrypted symbols on the compression efficiency is also less affected.

According to the JPEG2000 coding properties, the wavelet coefficients of some code blocks may be discarded in the encoding process at low bit rate. If encrypting the sign bit of these coefficients, the image can not be restored after image decoding. Therefore, in low bit rate conditions, the symbols of the higher priority code blocks should be choose and encrypted, such as the lowest sub-band coding block.

3.5 Method Steps

Image encryption algorithm complete steps are as follows:

(1) Generating the confusion table and a sign bit encryption key based on the chaotic sequences;
(2) The image preprocessing, wavelet transform and quantization;
(3) Confusing the wavelet coefficients in each coding block using the confusion table;
(4) Doing the XOR operation between the sign bits of the wavelet coefficients and the sign bit encryption key corresponding to the code blocks;
(5) Completing the following process in JPEG2000.

The decryption process is the reverse process of encryption.

4 The Experimental Results and Analysis

The experiment selects the truck image with 512×512, 8 bpp (bit/pixel) as the original image. Code block size is 16×16, 5 grade DWT by using W9X7 wavelet transform. The method is achieved by using the improved Kakadu software. For the description of the method encryption effect in different rate, the experiment respectively sets rate for 0.1bpp and 0.5bpp. In 0.5bpp, the rate is higher, so all sign bits in the code blocks are encrypted. In 0.1bpp, only the sign bits in the lowest sub-band coding block are encrypted.

4.1 The Experimental Results

Figures 5 and 6 are the experiment results respectively in the rate of 0.5bpp and 0.1bpp without the ROI coding. It can be seen that the confusion effect is good, the outline and characteristics in the encrypted image are not obvious, the image is close to the noise signal.

Figure 7 is the test results in the rate of 0.1 bpp, using the implicit ROI algorithm, and selecting the best weights coefficient of 4096. Region of interest image is the center in the 10% area of the image.

(a) the original image. (b) the encrypted image. (c) the wrong decrypted image. (d) the correct decrypted image.

Fig. 5. The results for 0.5bpp without ROI

(a) the original image. (b) the encrypted image. (c) the wrong decrypted image. (d) the correct decrypted image.

Fig. 6. The results for 0.1bpp without ROI

(a) the original image. (b) the encrypted image. (c) the wrong decrypted image. (d) the correct decrypted image.

Fig. 7. The results for 0.5bpp for the implicit ROI

4.2 Analysis About the Security

(1) Key space
In this paper, the encryption key is 64 bit, so the key space is about $2^{64} \approx 1.8447 \times 10^{19}$. Using the exhaustive attack, and assuming that a high speed computer can enumerate 109 keys per second, it takes about 585 years for all keys, so the method can effectively resist the exhaustive attack.

(2) Double security of the coefficient confusion and the sign bit encryption
The wavelet coefficients confusion and the sign bit encryption are the two independent links, which guaranteeing double safety of the method. Figure 8 gives the decoded image of the confusion table and the sign bit encryption being decrypted. Figure 8(a), (b), (c), (d) are the results at the rate of 0.5 bpp and 0.1 bpp.

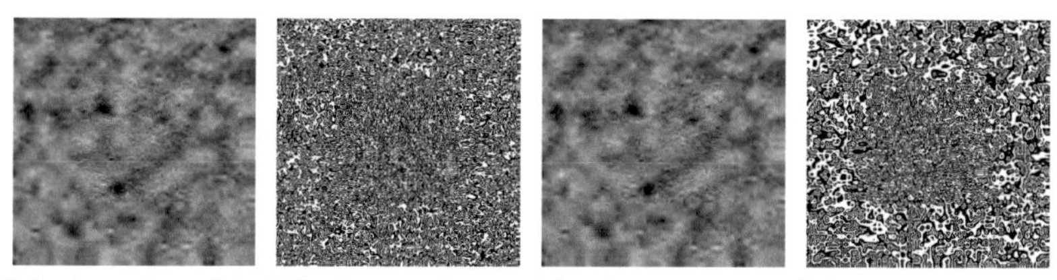

(a) the sign bit being decrypted. (b) the confusion table being decrypted. (c) the sign bit being decrypted . (d) the confusion table being decrypted.

Fig. 8. The image of the sign bit key and the confusion table decrypted

(3) The key sensitivity
With the truck image as an example, the method random changes the key in one bit at the different rate, and tests the changing percentage of the encrypted image pixels. The results as shown in Table 1.

The table shows that the encrypted image pixels change more than 60% when the key changes in 1 bit.

Table 1. The changing percent for changing one bit pixel

rate(bpp)	0.1	0.2	0.3	0.4	0.5
Pixels changing(%)	61.07	61.9	62.0	62.18	62.39

5 Conclusion

An encryption method for ROI images is proposed. The method consists of 3 steps:

(1) generating confusion table based on the chaotic sequence;
(2) confusing the wavelet coefficients in the blocks;
(3) encrypting the sign bit by the chaotic sequence.

The test results show that the proposed method can provide adequate key space to resist the brute attack and supports ROI image coding It can be predicted, for this method does not depend on the JPEG2000 encoding process followed, so the method has good adaptability.

References

1. Ke, L., Yang, D., Wang, X.: An improved target extraction algorithm based on region growing for lung CT image. In: 2007 IEEE/ICME International Conference on Complex Medical Engineering CME 2007, pp. 595–599. IEEE (2007)
2. Alim, O.A., Hamdy, N., El-Din, W.G.: Determination of the region of interest in the compression of biomedical images. In: 2007 National Radio Science Conference: NRSC 2007, pp. 1–6. IEEE (2007)
3. Wang, Y., Zhu, W., Zhan, X.: Study on scrambling capability based on image encryption. Comput. Eng. Des. **27**(24), 4730–4738 (2006)
4. Norcen, R., Uhl, A.: Performance analysis of block-based permutations in securing JPEG2000 and SPIHT compression. In: Visual Communications and Image Processing 2005. International Society for Optics and Photonics, 59602S–59602S-9 (2005)
5. Liu, J.L.: Efficient selective encryption for JPEG 2000 images using private initial table. Pattern Recogn. **39**(8), 1509–1517 (2006)
6. Ping, L., Sun, J., Zhou, J.: An algorithm for image encryption based on JPEG2000. Video Eng. **7**, 033 (2006)
7. Jia-sheng, L.: Study on chaos-based image encryption technology. Anhui Univ. **3**(2), 33–65 (2010)
8. Neto, L.G., Sheng, Y.: Optical implementation of image encryption using random phase mask and speckle-free phase fresnel hologram. In: International Society for Optics and Photonics, AeroSense 1997, pp. 389–396 (1997)

Author Index

J. Pan et al. (eds.), *Advances in Intelligent Information Hiding and Multimedia Signal Processing*, Smart Innovation, Systems and Technologies 81,
DOI 10.1007/978-3-319-63856-0